Advances in Science, Technology & Innovation

IEREK Interdisciplinary Series for Sustainable Development

Advances in Science, Technology & Innovation (ASTI) is a series of peer-reviewed books based on important emerging research that redefines the current disciplinary boundaries in science, technology and innovation (STI) in order to develop integrated concepts for sustainable development. It not only discusses the progress made towards securing more resources, allocating smarter solutions, and rebalancing the relationship between nature and people, but also provides in-depth insights from comprehensive research that addresses the **17 sustainable development goals (SDGs)** as set out by the UN for 2030.

The series draws on the best research papers from various IEREK and other international conferences to promote the creation and development of viable solutions for a **sustainable future and a positive societal** transformation with the help of integrated and innovative science-based approaches. Including interdisciplinary contributions, it presents innovative approaches and highlights how they can best support both economic and sustainable development, through better use of data, more effective institutions, and global, local and individual action, for the welfare of all societies.

The series particularly features conceptual and empirical contributions from various interrelated fields of science, technology and innovation, with an emphasis on digital transformation, that focus on providing practical solutions to **ensure food, water and energy security to achieve the SDGs.** It also presents new case studies offering concrete examples of how to resolve sustainable urbanization and environmental issues in different regions of the world.

The series is intended for professionals in research and teaching, consultancies and industry, and government and international organizations. Published in collaboration with IEREK, the Springer ASTI series will acquaint readers with essential new studies in STI for sustainable development.

ASTI series has now been accepted for Scopus (September 2020). All content published in this series will start appearing on the Scopus site in early 2021.

Jorge Chica-Olmo · Miroslav Vujičić ·
Rui Alexandre Castanho · Uglješa Stankov ·
Eliana Martinelli
Editors

Sustainable Tourism, Culture and Heritage Promotion

Development, Management
and Connectivity

A culmination of selected research papers from the
International Conference Cultural Sustainable Tourism
(CST-4th)—University of Maia 2022

 Springer

Editors
Jorge Chica-Olmo
Department of Quantitative Methods
for Economics and Business
University of Granada
Granada, Spain

Rui Alexandre Castanho ⓘ
WSB University
Dabrowa Gornicza, Poland

Eliana Martinelli
Department of Civil and Environmental
Engineering
University of Perugia
Perugia, Italy

Miroslav Vujičić
Department of Geography
Tourism and Hotel Management
Faculty of Sciences
University of Novi Sad
Novi Sad, Serbia

Uglješa Stankov ⓘ
Department of Geography
Tourism and Hotel Management
Faculty of Sciences
University of Novi Sad
Novi Sad, Serbia

ISSN 2522-8714   ISSN 2522-8722   (electronic)
Advances in Science, Technology & Innovation
IEREK Interdisciplinary Series for Sustainable Development
ISBN 978-3-031-49535-9   ISBN 978-3-031-49536-6   (eBook)
https://doi.org/10.1007/978-3-031-49536-6

This Springer imprint is published by the registered company Springer Nature Switzerland AG
The registered company address is: Gewerbestrasse 11, 6330 Cham, Switzerland

Paper in this product is recyclable.

Scientific Committee

The Editors warmly thank all the Reviewers who have contributed their authority to the double-blind review process, to ensure the quality of this publication.

Introduction

Part One: Aspects of Sustainable Tourism

Rui Alexandre Castanho
Faculty of Applied Sciences, WSB University,
Dabrowa Górnicza, Poland

CITUR—Madeira—Centre for Tourism Research, Development and Innovation,
Funchal-Madeira, Portugal
e-mail: acastanho@æsb.edu.pl

Sustainable tourism encompasses a range of critical aspects aimed at mitigating the adverse impacts of tourism while maximizing benefits for the environment, local communities, and all stakeholders involved (Rasoolimanesh & Jaafar, 2017; Castanho, Couto & Pimentel, 2020).

Some examples are: (a) Environmental Conservation: Sustainable tourism minimizes tourism's ecological footprint by conserving natural resources, ecosystems, and biodiversity (Velázquez et al., 2023); (b) Community Engagement and Empowerment: Local communities are actively involved in decision-making processes and economic benefits. In fact, sustainable tourism supports community initiatives, respects cultural traditions, and encourages local businesses' participation, fostering economic empowerment (Rasoolimanesh, & Jaafar, 2017; Couto et al., 2021a); (c) Economic Viability: Economic benefits are generated for host communities through job creation, local business support, and overall economic development (Castanho, Couto & Santos, 2023); (d) Cultural Preservation: Cultural heritage and authenticity are upheld by promoting respect for local customs, traditions, and historical sites. Here, sustainable tourism facilitates cultural exchange and interaction between visitors and locals (Couto, Castanho & Santos, 2023); (e) Visitor Education and Awareness: Sustainable tourism educates visitors about the destination's environment, culture, and responsible travel behaviors (Hall, Williams & Lew, 2014); (f) Infrastructure and Resource Management: Responsible planning and management of tourism infrastructure and resources are integral (Pimentel et al., 2022); (g) Conservation of Natural Resources: Responsible use of resources like water, energy, and land is advocated. Thus, sustainable tourism promotes practices that minimize resource consumption and waste (Bildirici et al., 2022); (h) Stakeholder Collaboration: Collaboration among government bodies, local communities, businesses, NGOs, and tourists ensures that tourism aligns with local needs and sustainability goals (Castanho et al., 2022); (i) Long-Term Perspective: Sustainable tourism takes a future-oriented approach, considering both short-term economic gains and long-term environmental and social well-being (Batista et al., 2022); (j) Responsible Marketing and Promotion: Honest and responsible marketing strategies are employed to accurately represent the destination's attributes and values, avoiding exploitative practices and emphasizing responsible travel (Couto et al., 2021b).

Therefore, sustainable tourism establishes a harmonious relationship between tourists, host communities, and the environment—mainly by equitably distributing benefits and conserving resources; besides, it aims to ensure the destination's well-being for present and future generations.

Contextually, this part is organized into eight chapters ("A Readable Wukang Building: A Case Study on Cultural Sustainable Tourism (CST)"–"CittaSlow: Hospitality and Sustainable Urban Tourism Development. The Case of Vizela (Northern Portugal)").

Chapter "A Readable Wukang Building: A Case Study on Cultural Sustainable Tourism (CST)" takes us on a thrilling expedition through the uncharted territory of the Readable Wukang Building. This building serves as a compelling case study on culturally sustainable tourism, illuminating the vast potential inherent in such investment frameworks.

In Chapter "Sustainable Tourism Development in Less Touristy Destinations; The Case of Epirus, Greece", an examination is conducted into the progression of sustainable tourism in destinations less frequented by tourists. The research delves into the specific case of Epirus, Greece, highlighting its unique attributes.

Chapter "Actual Conditions of Tourist Guides in the Covid-19 Pandemic: Ecuador Case Study" aims to formulate a set of recommended practices addressing tourist guides' challenges during the COVID-19 Pandemic. To achieve this, a case study from Ecuador illustrates these practices in action.

Chapter "Accessible Tourism Businesses as a Means of Promoting Sustainable Cultural Tourism in Thessaloniki, Greece" tackles the issues surrounding accessible tourism businesses, which serve as a vehicle for advancing sustainable cultural tourism. This comprehensive case study employs the example of Thessaloniki, Greece, to exemplify these challenges and their potential solutions.

Chapter "A Bottom-Up Approach for Sustainable Cultural Tourism in Ladakh: An Initiative Taken by Women and Homestays" presents a grassroots strategy for achieving sustainable cultural tourism in Ladakh. The emphasis is on an initiative spearheaded by women and homestays, showcasing a bottom-up approach to this endeavor.

Chapter "Promotion of World Heritage Sites in Kyoto, Japan" discusses the strategies for promoting World Heritage Sites located in Kyoto, Japan.

Chapter "CittaSlow: Hospitality and Sustainable Urban Tourism Development. The Case of Vizela (Northern Portugal)" delves into the subject of sustainable urban tourism development with a focus on hospitality. This exploration centers around a Portuguese case study in the municipality of Vizela.

References

Batista, M. G., Castanho, R. A., Sousa, Á., Couto, G., & Pimentel, P. (2023). Assessing rural tourism experiences: What can we learn from the Azores region? *Heritage 6*, 4817–4833 (2023). https://doi.org/10.3390/heritage6060256.

Bildirici, M., Castanho, R. A., Genç, S. Y., & Kayıkçı, F. (2022). ICT, energy intensity, and CO_2 emission nexus. *Energies 15*, 4567 (2022). https://doi.org/10.3390/en15134567.

Castanho, R. A., Couto, G., & Pimentel. (2020). *Principles of sustainable tourism and cultural management in rural and ultra-peripheral territories: Extracting guidelines for its application in the Azores archipelago.* Cultural Management: Science and Education (CMSE) ISSN: 2512–6962. https://doi.org/10.30819/cmse.4-1.01.

Castanho R. A., Couto, G., Pimentel, P., Carvalho, C. B., Sousa, Á., & da Graça Batista, M. (2022). A preliminary approach about the perceptions of decision makers versus entrepreneurs about rural tourism development in the Azores Autonomous region. In Á. Rocha, C. Ferrás, A. M. Porras, J. E. Delgado (Eds.) *Information technology and systems. ICITS 2022. Lecture Notes in Networks and Systems* (Vol. 414, pp. 277–285). Springer. https://doi.org/10.1007/978-3-030-96293-7_25.

Castanho, R. A., Couto, G., & Santos, C. (2023). Tourism promoting sustainable regional development: Focusing on rural and creative tourism in low-density and remote regions. *Revista de Estudios Andaluces 45*, 189–205. https://dx.doi.org/10.12795/rea.2023.i45.10.

Couto, G., Castanho, R. A., Pimentel, P., Carvalho C. B., Sousa Á. (2021a). The potential of adventure tourism in the Azores: Focusing on the regional strategic planning. In A. Abreu, D. Liberato, E. A. González, J. C. G. Ojeda (Eds.) *Advances in tourism, technology and systems. ICOTTS 2020. Smart innovation, systems and technologies* (Vol. 209). Springer. https://doi.org/10.1007/978-981-33-4260-6_2.

Couto, G., Pimentel, P., Oliveira, A., Crispim, J. P., Santos, C., Estimaand, D., & Castanho, R. A. (2021b). SmartDest: Converting the Azores into a smart tourist destination. Chapter #25 in the *IGI GLOBAL Handbook of Research: Handbook of Research on Sustainable Development goals, climate change, and digitalization* (pp. 409–432). ISBN13: 9781799884828. https://doi.org/10.4018/978-1-7998-8482-8.ch025.

Couto, G., Castanho, R. A., & Santos, C. (2023). Creative tourism, public policies and land use changes: A multi-method approach towards regional sustainable development in Azores Islands. *Sustainability 15*, 5152 (2023). https://doi.org/10.3390/su15065152.

Hall, C. M., Williams, A. M., & Lew, A. A. (2014). Tourism: Conceptualisations, disciplinarity, institutions and issues'. In A. Lew, C. M. Hall, & A. M. Williams (Eds.) *The Wiley Blackwell companion to tourism.* Wiley.

Pimentel, P., Vulevic, A., Couto, G., Behradfar, A., Gómez, J. M. N., & Castanho, R. A. (2022). Maritime transportation dynamics in the Azores Region: Analyzing the period 1998–2019. *Infrastructures 7*, 21 (2022). https://doi.org/10.3390/infrastructures7020021.

Rasoolimanesh, S. M., & Jaafar, M. (2017). Sustainable tourism development and residents' perceptions in World Heritage Site destinations. *Asia Pacific Journal of Tourism Research 22*(1), 34–48.

Velázquez, J., Infante, J., Gómez, G., Hernando, A., Gülçin, D., Herráez, F., Rincón, V., & Castanho, R. A. (2023). Walkability under climate pressure: application to three UNESCO world heritage cities in central Spain. *Land 12*, 944 (2023). https://doi.org/10.3390/land12050944.

Part Two: Future Prospects and Trends of Tourism

Uglješa Stankov, Miroslav Vujičić
Department of Geography, Tourism and Hotel Management, Faculty of Sciences,
University of Novi Sad, Novi Sad, Serbia
e-mail: ugljesa.stankov@dgt.uns.ac.rs

Miroslav Vujičić
e-mail: miroslav.vujicic@dgt.uns.ac.rs

The future of cultural sustainable tourism holds a promising tapestry woven with innovative trends and transformative possibilities (Stankov & Gretzel, 2020; Zheng et al., 2023). Central to this evolution is the seamless integration of digitalization, which is poised to reshape every facet of the tourism experience. Digital interpretation platforms will offer visitors an immersive journey into the heart of a destination's cultural heritage (Liu, 2020). Social media, a formidable force in shaping travel decisions, will continue to play a pivotal role, amplifying cultural sustainability efforts and connecting travelers with authentic local experiences (Jovanović et al., 2019). As the lines between physical and digital realms blur, virtual reality will emerge as a tour de force, enabling travelers to transcend geographical boundaries and explore cultural landmarks in unprecedented ways (Filimonau et al., 2022; Vujičić et al., 2022). These trends collectively underscore the potential for cultural sustainable tourism to not only conserve heritage but also invigorate local economies and foster cross-cultural understanding in a world where technology harmonizes with tradition.

This part of the book, titled "**Future Prospects and Trends of Tourism**", is structured into four chapters. Within the chapter titled "Contemporary Digital Age Pilgrimage in Chichibu in Japan", Nakabasami delves into the elements that render modern pilgrimage sustainable. The chapter indicates that gamified cultural communication between local stakeholders and tourists, as well as emotional mapping along cultural routes, could serve as pivotal success factors for the case study. In the chapter "Digital Interpretation as a Visitor Management Strategy: The Case of Côa Valley Archeological Park and Museum", Dinis and colleagues reaffirm that the augmented development and utilization of information technologies and multimedia have transformed the manner in which archaeological heritage is comprehended. The instance of the Côa Valley Archaeological Park emerges as an exemplary case study, illustrating both successful practices and areas open to suggestions for enhancement. The chapter titled "The Role of Social Media in the Conservation and Safeguard of Gastronomy as Intangible Cultural Heritage" authored by Ramazanova and her colleagues delves into the realm of international platforms and networks associated with gastronomy as a facet of cultural heritage, examining their presence across social media. The study reveals that numerous countries from disparate corners of the world have displayed a recent inclination towards safeguarding their distinctive cuisines as elements of cultural heritage. The concluding chapter, titled "Sustainable Cultural Tourism and Virtual Reality the Contribution of the New Technologies Applications in the Fields of Preservation and Sustainable Tourism Management of the Cultural Heritage. The Case of Greece" authored by Chatzopoulou, emphasizes the potential of information technologies to play a pivotal role in safeguarding and showcasing both natural and cultural landscapes. Furthermore, the chapter underscores how the utilization of virtual and augmented reality applications, in particular, can foster the evolution of thematic tourism and drive economic growth.

References

Filimonau, V., Ashton, M., & Stankov, U. (2022). Virtual spaces as the future of consumption in tourism, hospitality and events. *Journal of Tourism Futures*. Scopus. https://doi.org/10.1108/JTF-07-2022-0174.

Jovanović, T., Božić, S., Bodroža, B., & Stankov, U. (2019). Influence of users' psychosocial traits on Facebook travel-related behavior patterns. *Journal of Vacation Marketing*, 25(2), 252–263. https://doi.org/10.1177/1356766718771420.

Liu, Y. (2020). Evaluating visitor experience of digital interpretation and presentation technologies at cultural heritage sites: A case study of the old town, Zuoying. *Built Heritage*, 4(1), 14. https://doi.org/10.1186/s43238-020-00016-4.

Stankov, U., & Gretzel, U. (2020). Tourism 4.0 technologies and tourist experiences: A human-centered design perspective. *Information Technology and Tourism*. https://doi.org/10.1007/s40558-020-00186.

Vujičić, M. D., Kennell, J., Stankov, U., Gretzel, U., Vasiljević, Đ. A., & Morrison, A. M. (2022). Keeping up with the drones! Techno-social dimensions of tourist drone videography. *Technology in Society*, 68, 101838. https://doi.org/10.1016/j.techsoc.2021.101838.

Zheng, D., Huang, C., & Oraltay, B. (2023). Digital cultural tourism: Progress and a proposed framework for future research. *Asia Pacific Journal of Tourism Research*, 28(3), 234–253. https://doi.org/10.1080/10941665.2023.2217958.

Part Three: Cultural Heritage and Sustainable Environment

Eliana Martinelli
Department of Civil and Environmental Engineering,
University of Perugia, Perugia, Italy
e-mail: eliana.martinelli@unipg.it

When discussing cultural sustainable tourism, we are implicitly referring to two key concepts: the notion of "cultural heritage" as defined by the UNESCO World Heritage Centre in 1972, and the concept of "sustainable environment". The latter takes into consideration particularly the built environment, which encompasses all the human-made environment that provides the settings for human activities and is mostly affected by sustainability problems. The integration of these two concepts into the framework of tourism assumes a broader significance. The expression "heritage tourism", as defined by Timothy and Boyd (2002), refers to a specific segment of cultural tourism that focuses on various forms of heritage. This includes both tangible and intangible aspects, encompassing natural and cultural elements that have not yet been safeguarded or placed under protection regimes. The common thread among these heritage forms is their ability to establish a connection with a significant historical past. Within the context of promoting sustainable development in the field of tourism, it is essential to adopt an approach that prioritizes the preservation of cultural assets while simultaneously facilitating the necessary transformation of the built environment. In order to achieve this outcome, it is important to incorporate both transformative and conservative measures into appropriate design tools: on the one hand, to actively preserve heritage; on the other hand, to make it an operational instrument for developing the future city and environment. Hence, it is important to address the following questions: What investigation techniques and design actions prove beneficial in strengthening the relationships between heritage and context? What are the most effective strategies for enhancing the value of heritage sites while simultaneously establishing sustainable approaches for their utilization and implementing measures for their preservation? (Marzo et al., 2022) How can design processes contribute to the valorization of marginal touristic areas while simultaneously considering the sense of place and visitor needs and prioritizing the region's objectives and its inhabitants? (Bertini & Rocha, 2020)

This part of the book, titled "**Cultural Heritage and Sustainable Environment**", is organized into seven chapters, focusing on different geographical areas and approaches to tourism aimed at valorizing the specificity of places. Within the chapter "Ecotourism and Rural Sustainable Development, Albania Case, Blezënckë Village", Veleshnja examines the relationship between rural development, tourism, and ecotourism, challenging the necessity of sustainable development and the preservation of the local Albanian traditions. This strategy hastens the economic transition from agriculture to manufacturing and services while also enhancing farmer welfare and encouraging environmental preservation in rural areas. The chapter "Heritage in Socio Economic Sustainable Development: The Salzedas and São João de Tarouca Case", authored by Mendonça, discussed the idea of cultural tourism as it applied to a Portuguise case near the Douro Valley, an area typified as economically and socially depressed, in which heritage and culture activities can be a way to achieve sustainability. Another Portuguese case is explored in the chapter "Tourists' Perceptions of the Image of the Peneda-Gerês National Park" by Martins and Pinheiro. The purpose of this study is to comprehend the psychological and functional characteristics of a destination's image that attract the most tourists. The study method applied was quantitative methodology, and data collection was carried out through the use of a questionnaire survey. The territory under consideration was Peneda-Gerês, the only Portuguese national park, which has seen substantial growth in tourists as a result of its projection and notoriety. Paulino and Cruz, in the chapter "Cultural Heritage, Tourism and Sustainable Development. The Model of the Cultural Heritage Digital Media Lab", described the activity of CHDML, an ongoing project focused on Portuguese-speaking countries and aimed to contribute to the participatory governance process, to the rescue and enhancement of cultural heritage and memory, and to the creation of opportunities for economic development and appreciation of local products, bearing in mind the market potential. The authors of the chapter "The World Heritage Classification in Urban Tourism Destinations: Perspectives for the City of Porto, Portugal" (Pinheiro and Martins) concentrated on the idea of urban tourism and the significance of the World Heritage classification for the success of the tourist destination. The chapter titled "The Role of Community-Led Initiatives in the Circularity-Based Heritage Revitalization" by Tira and Türkoğlu presented key points on the community-led revitalization policy under the aegis of the circular city concept, applied to the Medina of Tunis UNESCO World Heritage Center, using an empirical methodology of research. Finally, in the last chapter "Addressing Connectivity Issues Between the Historical and Natural Touristic Heritage Sites of Egypt", Elkaftangui and Amer attempted to address the connectivity issues faced by travelers in Egypt to improve the tourist experience and increase the tourism business. The study makes the case that Egypt can strengthen its position as a top destination for both natural and historical heritage tourism by enhancing connectivity between the various tourist sites.

References

Bertini, V., & Rocha, J. (Eds.) (2020). *Architecture tourism and marginal areas: Research and design proposals*. LetteraVentidue.

Marzo, M., Ferrario, V., & Bertini, V. (Eds.) (2022). *Between sense of time and sense of place: Designing, heritage, tourism*. LetteraVentidue.

Timothy, D. J., & Boyd, S. W. (2002). *Heritage tourism*. Prentice Hall.

UNESCO World Heritage Centre. (1972). *Convention concerning the protection of the world cultural and natural heritage*. UNESCO World Heritage Centre. https://whc.unesco.org/en/conventiontext/.

Part Four: Social and Economic Impacts

Miroslav Vujičić, Uglješa Stankov
Department of Geography, Tourism and Hotel Management, Faculty of Sciences
University of Novi Sad, Novi Sad, Serbia
e-mail: miroslav.vujicic@dgt.uns.ac.rs

Uglješa Stankov
e-mail: ugljesa.stankov@dgt.uns.ac.rs

Cultural sustainable tourism wields a profound influence, resonating across both social and economic dimensions of a destination (Vujičić et al., 2023). On a social plane, it serves as a powerful bridge, fostering cross-cultural understanding and dialogue between visitors and local communities (Kovačić et al., 2020). As tourists engage with authentic traditions, heritage, and ways of life, a harmonious exchange of ideas could occur, cultivating mutual respect and appreciation. This intercultural interaction could bolsters social cohesion, breaking down stereotypes, and nurturing a sense of global interconnectedness (Sharma et al., 2018). Simultaneously, the economic impact of cultural sustainable tourism could be considerable. It can breathe life into local economies by creating employment opportunities, supporting small-scale enterprises, and encouraging artisanal craftsmanship. The preservation and revitalization of cultural assets could bolster cultural pride and attract visitors, generating revenue that can be reinvested in community development and heritage conservation efforts (Besculides et al., 2002). However, a delicate balance must be struck to ensure that the economic benefits are inclusive, that cultural authenticity remains intact, and that the environment is not compromised (Vasiljević et al., 2023). In this nexus of social and economic dynamics, cultural sustainable tourism unfolds as a transformative force with the potential to uplift communities and safeguard traditions while contributing to shared global prosperity.

This part of the book, titled "**Social and Economic Impacts**", is organized in eight chapters. The opening chapter, titled "Elbe—A Tourist Line Cycling Through Europe" penned by Štěbetáková and colleagues, delves into the intricate realm of resident perspectives regarding the evolution of tourism within the Lower Elbe region. It examines whether a connection exists between the burgeoning facets of tourism, developmental aspirations, and the imperative to preserve the area's pristine allure. The next chapter "Residents' Perceptions of the Socio-economic Benefits of Restaurants in the Township" by Bavuma underscores the pivotal role that restaurants play within townships, elucidating their significance in propelling tourism development within these vibrant communities in the context of South Africa. In the chapter titled "Architecture of Historical Mosques: A Typological Study on the Archaeological Site of Barobazar", Alam and colleagues center their attention on the architectural arrangement and typological attributes of ancient mosques situated within the archaeological remnants of Barobazar, examining the historical, socio-cultural, and architectural influences. The subsequent intriguing chapter, authored by Quijano Herrera and Pérez-Tapia and entitled "Tourism as a Driver of Soft Power: The Case of South Korea", delves into the way tourism has evolved into a catalyst for the nation's soft power. This analysis considers the transformation in the country's rebranding strategy, which has embraced the cultural and entertainment phenomenon "Hallyu," alongside the nation's adeptness in crafting public policies aimed at fostering appealing and accessible tourism experiences for global tourists. The chapter titled "Tourists' Perceptions of Service Quality: Using Text for Tourism Hospitality Industry Insights" authored by Melo and colleagues analyzes thousands of online tourist reviews of Portuguese rural accommodation establishments. The study indicates that the perceptions of overall service quality, as portrayed in these online reviews, are contingent upon how tourists perceive both the accommodations and the surrounding environment. The chapter titled "Religious Tourism in Covid-19 Period: The Event of the Festival of Crosses, Barcelos (Portugal)" authored by Martins identifies the impacts that COVID-19 has had on the Festival of Crosses in Barcelos. This pertains to aspects such as the occupancy rate within the city and the alterations that its programming has undergone due to the pandemic. The subsequent chapter also presents a case study from Portugal, in which Teles and colleagues in their chapter titled "Event Tourism: Analysis of Residents' Perceptions of the São Bento Festivities, Santo Tirso (Portugal)" focus on the perceptions held by residents in relation to one of the region's most iconic celebrations, the São Bento Festivities. The concluding chapter, entitled "Pilgrimage Tourism for Enhancement of Heritage Conservation Management: Study of Potential, Possibilities in Kurnool, Anantapur Districts of Andhra Pradesh, India", authored by Turaga, investigates the potential to expand the benefits of religious (Hindu pilgrimage) tourism to overlooked historical monuments and non-religious sites in the vicinity.

References

Besculides, A., Lee, M. E., & McCormick, P. J. (2002). Residents' perceptions of the cultural benefits of tourism. *Annals of Tourism Research 29*(2), 303–319.

Kovačić, S., Vujičić, M., Čikić, J., Šagovnović, I., Stankov, U., & Vasiljević, T. Z. (2020). Impact of the European Capital of Culture project on the image of the city of Novi Sad—The perception of the local community. *Turizam 25*(2), Article 2.

Sharma, P., Charak, N. S., & Kumar, R. (2018). Sustainable tourism development and peace: A local community approach. *Journal of Hospitality Application and Research 13*(1), 36–56.

Vasiljević, Đ. A., Vujičić, M. D., Stankov, U., & Dragović, N. (2023). Visitor motivation and perceived value of periurban parks—Case study of Kamenica park, Serbia. *Journal of Outdoor Recreation and Tourism 42*, 100625.

Vujičić, M. D., Stankov, U., Pavluković, V., Štajner-Papuga, I., Kovačić, S., Čikić, J., Milenković, N., & Zelenović Vasiljević, T. (2023). Prepare for impact! A methodological approach for comprehensive impact evaluation of European capital of culture: The case of Novi Sad 2022. *Social Indicators Research 165*(2), 715–736.

Contents

Aspects of a Sustainable Tourism

A Readable Wukang Building: A Case Study on Cultural Sustainable Tourism (CST)

Xiangchong Pan, Yinan Guo, Wei Zhong, and Fei Yin

Abstract

Old houses hidden in old trees, Shikumen in the old alley, a-century-long exotic building clusters in the Bund, which demonstrates the historical, cultural and international city of Shanghai, allures people locally and abroad. Taking the almost 100-year-old Wukang Building as a case study, this paper aims to explore a novel touring mode featured with local characteristics in response to cultural sustainable tourism. Launched by Shanghai's municipal government, the project, "a Readable Building", is set to amply revitalize the existing antiqued buildings in the city and make them Internet celebrities through the implementation of storytelling, various activities and eye-catching souvenirs with the logos of the buildings to attract tourists as well as boost the local economy. Based on qualitative and quantitative analyses, this paper clarifies the project's process and its favorable results. Hopefully, this paper will help introduce this successfully proven, reproducible, and applicable new touring mode to the rest of the world to further the efforts of Cultural Sustainable Tourism.

Keywords

A readable building • Wukang building • New touring mode • Cultural sustainable tourism

1 Introduction

A popular saying among the local folks goes: "If you want to learn about a ten-year-old Shanghai, please visit Pudong located in the Eastern part of Shanghai. If you want to learn a-hundred-year-old Shanghai, please visit the Bund located in the Western part of Shanghai. If you want to learn a-thousand-year-old Shanghai, please visit Qibao Ancient Town located in the Western part of Shanghai". When speaking of Pudong and how it is filled with rows upon rows of modern skyscrapers, such as the Shanghai Tower, the Oriental Pearl TV Tower and Shanghai Jinmao Tower, it leads you to think of Manhattan, New York, by default. The newly built 632 m Shanghai Tower is considered the highest of Pudong buildings, serving as the center of business, tourism, entertainment and hospitality. The Shanghai Tower is the tallest building in China and second tallest in the world, standing at a height just shorter than that of Burj Khalifa in Dubai, a building recognized as the tallest in the world. It takes tourists less than one minute to reach the 118th floor of Shanghai Tower, known as the observation platform, with the help of the advanced lift in order to enjoy a top panoramic view of the city of Shanghai, China (Shanghai Tower Baike.baidu, n.d.). With a height of 468 m, the Oriental Pearl TV Tower is composed of 11 differently sized balls which are arranged from top to bottom, making the tower a Shanghai landmark. The Oriental Pearl TV Tower is endowed with multiple functions including sightseeing, catering, shopping, entertainment, cruise ships, exhibitions, historical exhibitions, radio and television transmissions as well as other functions. The Oriental Pearl TV Tower is especially proud of its globally

X. Pan
College of Business Foreign Languages, Shanghai Business School, Shanghai, China

Y. Guo (✉)
Shanghai Shenergy Innovation & Development Co., Ltd, Shanghai, China
e-mail: evonnemasaki@163.com

W. Zhong
Department of Hotel Business Management, Shanghai Business School, Shanghai, China
e-mail: 21130012@sbs.edu.cn

F. Yin
Shanghai Food and Strategic Reserves Bureau, Shanghai, China
e-mail: frankyinfei@hotmai.com

© The Author(s), under exclusive license to Springer Nature Switzerland AG 2024
J. Chica-Olmo et al. (eds.), *Sustainable Tourism, Culture and Heritage Promotion*,
Advances in Science, Technology & Innovation, https://doi.org/10.1007/978-3-031-49536-6_1

well-known revolving restaurant, located in the tower's 267 m tall ball, making it the highest revolving restaurant in Asia and simultaneously providing 350 guests with an amazing view, extraordinary food and warm service. The dining guests are able to walk around the ring-shaped buffet restaurant allowing them to make a choice from 100 exquisite Chinese and Western dishes, all displayed on the dining car and surrounded by flowers, allowing them to enjoy their favorite foods and an all-round view of Shanghai. At the same time, the guests receive friendly and warm services from the welcoming ladies and the shuttle waiters, all of whom are dressed in elegant outfits, as they enjoy their meals, making the whole revolving restaurant experience outstanding (The Oriental Pear TV Tower Baike.baidu, n.d.). As high as 88 floors, the Shanghai Jinmao Tower is 420.5 m tall and serves as a business office, hotel and sightseeing location. The unique characteristic of the Shanghai Jinmao Tower is that, as the flagship hotel of Hyatt International Group in China, the top-end of Shanghai Jinmao Grand Hyatt Hotel is located on the 53rd–87th floors of this tower. This hotel is equipped with 548 luxury guest rooms featuring traditional Chinese culture in the form of calligraphy, sculptures and lacquer arts, several restaurants of different themes such as Chinese cuisine or Western flavors, as well as bars and cafes (Shanghai Jinmao Tower Baike.baidu, n. d.). The round Oriental Pearl TV Tower, accompanied with the square Shanghai Jinmao Tower, resembles Chinese philosophy of maintaining a balance between Heaven and Earth. In Chinese culture, the round shape resembles that of Heaven and the square resembles the shape of Earth. The shape of Heaven echoes the shape of Earth, symbolizing that the two towers are as steady as Heaven and Earth; standing still forever. When it comes to the Bund, which is the name card of Shanghai, it is known as the Bund International Architecture Expo Group and displays many century-old classical revival buildings with different styles such as Gothic, Baroque, Romanesque, Classicism, Renaissance, as well as mixed styles of Chinese and Western architecture. At present, those elegant and historical buildings are used for luxury hotels, banks, government offices, top-end restaurants and bars, the boutiques of world-class brands and so on and so forth. The Bund sits on the West side of Huangpu River and Pudong is on the East side of Huangpu River, sitting across from each other. In addition, the Bund represents the old Shanghai, while Pudong represents the new Shanghai. At night, when the colorful decorating lights of the Bund and Pudong are flashing, the sight takes tourists' breaths away. Especially when tourists take the Huangpu River cruise, they witness the amazing views of the two sides along the river, soaking in the harmony of the old and new, as well as the classic and modern, immediately taking their troubles, stress and sorrows away, and allowing them to enjoy life. As for the Qibao Ancient Town, located in western Shanghai, it is

famous for its street food with a local flavor, its Buddhist temple and the natural beauty of the water town. Qibao Old Street is the heart of Qibao Ancient Town and it retains the characteristics of the Northern Song Dynasty, which refers to the period between the years of 960 to 1127 and also consists of the South Street and the North Street. The South Street is featured with various local snacks, alluring tourists to taste because of the irresistible and exquisitely shaped foods and peculiarly smelling food. The North Street is known for its creative and novel handcrafts, antiques and artworks such as the calligraphy and paintings, attracting the crowds to enjoy and buy. Legend has it that in 1008 AD, the emperor bestowed the name "Qibao Temple" on the Buddhist temple, and therefore Qibao Town, that is, Qibao Ancient Town, was officially named. The Qibao Temple, rebuilt in 2000, features the architectural style of the Han Dynasty (202 BC–220 AD) and Tang Dynasty (618–907) and reflects its simplicity, generosity and solemnity. It consists of many halls such as scripture halls and Buddha halls. It also has one pagoda with a height of more than 40 m and one garden with delicate stone carvings (Qibao Ancient Town Baike.baidu, n.d.). The attractive sports of Shanghai are far beyond just these three spots where there are 130 A-level tourist attractions, 11 national historical and cultural towns and 3449 immovable cultural relics. It also has 1058 outstanding historical buildings, 44 historical feature protection zones, 153 museums, 94 art galleries, 217 theaters and new performing arts spaces and 385 cinemas in Shanghai (Yangtze River Delta Daily, 2021). The overwhelming charm of Shanghai calls for people from home and abroad to visit Shanghai all year long whether it is wet spring, hot summer, cool autumn or cold winter. According to data (Zhiyan Consulting, n.d.), 340 million domestic tourists and 8.9371 million foreign tourists visited Shanghai in the whole of 2018, demonstrating how popular Shanghai was. But the sudden outbreak of covid-19 imposed severe side-effects on global tourism and Shanghai was no exception. The hustle and bustle of the city of Shanghai turned into quiet and empty place when covid-19 was prevalent. Although the epidemic prevention and control measures such as lockdowns, dynamitic zero–covid-19 policy, and at-home quarantines effectively prevented the epidemic from spreading further, invisible walls and barriers were also set and stopped people from traveling. As human beings, we want to live, we want a life and we want to see the world. In light of the recent epidemic era, what can we do to make tourism sustainable? What can we do to establish more culturally sustainable tourism? And what can we do to keep cultural tourism sustainable? These tough questions urge us to take prompt actions to respond to the epidemic's impacts on tourism. In hopes of making tourism lively and booming again, this paper takes "a Readable Wukang Building", a touring program launched by Shanghai Municipal government, as an example to illustrate novel touring modes. It will analyze the results of carrying out those touring modes based on collected data, as well as provide suggestions and answer questions of: what we can do? and how we can achieve culturally sustainable tourism?

2 Touring Mode of a Readable Wukang Building

2.1 Background

As mentioned in the introduction, Shanghai contains historically rich buildings, which tells the stories of Shanghai's history while presenting its cultural diversities such as the Red culture, Jiangnan culture and Shanghai culture. The Red culture is an advanced culture with Chinese characteristics that was jointly created by the Chinese communists, advanced activists and the masses during the revolutionary war. It contains a rich revolutionary spirit and profound historical and cultural connotations, covering tangible and intangible culture (Red Culture Baike.baidu, n. d.). The tangible Red culture can be seen in many places such as in the Sihang Warehouse Anti-Japanese Invasion War Memorial Hall, Lu Xun's Former Residence, Songhu Anti-Japanese Invasion War Memorial Hall and more. The Sihang Warehouse Anti-Japanese Invasion War Memorial Hall, which was a war site during the Anti-Japanese invasion war, contains six exhibition halls, bringing forth a moving and tragic history through various sculptures and paintings. At that time, many ancestors sacrificed their lives guarding this warehouse in order to fight against the Japanese invasion. Lu Xun's Former Residence was a three-story building with red bricks and red tiles, located at No. 9, Lane 132, Shanyin Road, Hongkou District, Shanghai, where Lu Xin spent his final years from the year of 1933 to 1936 before passing away in Shanghai (Lu Xun's Former Residence Baike.baidu, n.d.). Lu Xin is a well-known figure for both the young and old Chinese. He was a famous writer, thinker, revolutionist, educator, a democratic warrior, an important participant in the New Culture Movement and one of the founders of modern Chinese literature. He was born on September 25, 1881, and died on October 19, 1936 (Lu Xun Baike.baidu, n.d.). He created numerous works including the True Story of Ah Q and Diary of a Madman, some of his most famous works. Moreover, the Memorial Hall displayed a large number of books, videos and other materials which recorded real history during the war, allowing tourists to experience that history. As for the intangible Red culture, they are presented in art such as film, ballet and acrobatics to demonstrate the Communist Party of China's struggles and successes. The Jiangnan culture, which refers to East

Chinese culture, reflects the spirit of pioneering, openness, inclusivism, perfectionism, education-priority and practicality (Jiangnan Culture Baike.baidu, n.d.). There are many ancient Jiangnan towns in or near Shanghai for tourists to see as well as experience Jiangnan culture through ancient towns such as Zhujiajiao, Xitang, Zhouzhuang and so on. They are all historical and cultural water towns, but they possess different characteristics. In the ancient town of Zhujiajiao, there is a vintage post-office built during the Qing dynasty (1636–1912), the most attractive sight in Zhujiajiao. Mr. Sheng's House makes the ancient town of Zhouzhuang unique. It is said that Mr. Sheng was extremely rich in ancient China and that his splendid house, namely, Mr. Sheng's House, would bring a chance for tourists to become wealthy if they visited it. The ancient town of Xitang is famous for its well-preserved old stone bridges, making it special. Moreover, the Shanghai culture is based on the Jiangnan culture and has been impacted by the modern industrial civilization from Europe and the United States, after the opening of the port of Shanghai. It has a unique style that respects diversity, individuality, and takes into account personal and social interests. Therefore, it is a rational, easy-going and mature business culture dominated by the spirit of contract (Shanghai Culture Baike.baidu, n.d.). Some of the Shanghai culture is like air, people do not see it nor are they told about it, but people cannot live well without knowing it if they live or work in Shanghai. They have to figure it out by themselves after encountering numerous cultural shocks. However, some of the Shanghai culture can be seen where the typical places presenting Shanghai culture cover 35 attractive spots such as the Bund, the Moller Villa, and Shikumen. Similarly, the Bund has also been briefly introduced. The Moller Villa used to be Moller's private garden. Moller is a British Jewish who came to Shanghai in 1919 and became a wealthy businessman by betting on horses and serving as the Shanghai racecourse executive. He spent nine years building this villa and completed it in 1936 to make his dearest daughter's dream come true (The Moller Villa Baike.baidu, n.d.). It is said that Moller's daughter had a dream one night of a big house like a fairytale castle, and so Moller built this chic Norwegian house based on what his daughter saw in her dream. As for the Shikumen, they are houses located in the lane of Shanghai where the local residents dwell. They were built in the nineteenth century with the style of traditional Jiangnan courtyards at the very beginning. The whole building was a closed independent unit with a two-story brick and wood structure. At that time, Shikumen was an ideal residence for large families. With the economy developing and business booming, outsiders flocked into the city to find jobs, therefore resulting in a higher demand for more houses. To meet the people's needs, the traditional large courtyard was replaced with the new style of Shikumen with one bedroom or two bedrooms made of concrete and novel building materials. Meanwhile, the western elements of architecture were incorporated with the new style of Shikumen, making it look like a private villa garden (Shanghai-style architecture: Eight styles of traditional Chinese architecture Baijiahao. baidu, n.d.). By turning these historical and cultural buildings into the touring program, the Readable Buildings, which was initiated by Shanghai municipal government in 2017, it not only offered foreign tourists access to understanding the history and culture of Shanghai, but also provided local citizens with an immersive cultural experience. With the Shanghai municipal government promoting the touring program, the Readable Buildings has gone through three stages since 2018: the 1.0, 2.0, and 3.0 versions. The 1.0 version meant that tourists scanned the QR code on the exterior wall of the building to learn about the historical and humanistic stories of the buildings. The 2.0 version allowed tourists to visit more historical buildings including ones that used to be closed. The 3.0 version offered tourists the opportunity to experience a digital tour of historical buildings. From 2018 to 2021, the touring program, the Readable Buildings (RBs), reached each district of Shanghai, covering 16 districts. Moreover, 1037 historical buildings were opened and 2437 QR codes were attached to those historical buildings, making those historical buildings readable, audible, viewable and tourable (The version 3.0 of the Readable Buildings makes people feel the warm of people-centered city of Shanghai in an all-round way, n.d.). In 2021, the Shanghai government updated the Readable Buildings into the 3.0 version, the digital version, through big data, AI, Internet of Things and cloud platforms to make this touring program achieve remarkable results in digital experience, systematic service, citizen satisfaction, social participation and cross-border integration, as well as resulting in further building the brand of the Shanghai Culture. For instance, by cooperating with social media APPs and online e-bookstores, the Shanghai government launched a movement to encourage the public to comment on the RBs and post them on the APPs to offer the data for the local government to improve services. The Shanghai government also invited local citizens and tourists to act as the storytellers of the RBs to create a database of architectural stories. Moreover, the Shanghai government collected the RBs pictures taken by the public to build a database of the RBs images, laying the foundation for the development of the RBs photographic identification functions in the next stage. Finally, the Shanghai government held the first RBs Souvenirs Creative Design Competition, selected the top ten among the participants' works and virtually displayed them on the Tik-Tok platform (Build the digital version 3.0 of the Readable Buildings with ten actions, n.d.).

All these measures and activities brought life to those antique architectures, improved tourism businesses and gave an answer to reducing the epidemic impacts on the tourism industry.

2.2 The Wukang Building

The Wukang building, located at the fork in the road of the Middle Huaihai, Wukang, Xingguo, Tianping and Yuqing, is the most renowned building among Shanghai's historical buildings. Shaped like a huge ship, the Wukang building stands with the height of 30 m and construction area of 9755 m^2 as well as eight floors. It was designed by the famous Hungarian architect Laszlo Hudec who lived in Shanghai at that time and was built in 1924 thanks to the funds provided by the International Savings Society, a French financial institution in China in the 1920s (The International Savings Society Baike.baidu, n.d.). At that time, Laszlo Hudec referenced the style of Renaissance buildings in Paris and cleverly used the naturally existing 30° street corner on the plot to design it, making it look like "a big ship that travels the waves". After the building was completed, it was named Normandy Apartments and was rented out as high-end apartments by some Shanghai-nationals and foreign businessmen. After the liberation of Shanghai in 1953, the building was renamed Wukang Building. Some Chinese movie stars lived there after the year of 1953 including Dan Zhao, Renmei Wang, Yi Qin, Daolin Sun, Junli Zheng, Wenjuan Wang, etc. In 2019, the Wukang Building was updated based on the principle of "architectural archeology" and "Chinese medicine conditioning", which is based on maintaining the original flavor of the Wukang Building. The renovation project not only repaired the aging brick walls, doors and windows to beautify its "face", but also cleaned up its damaged pipes and electric lines to improve the quality of life of more than 100 residents who currently live in the building. Moreover, intelligent sensors were installed inside the building, they can carry out all-weather, full-time, full-coverage intelligent and refined management (Wukang Building Baike.baidu, n. d.). The picture of the renovated Wukang Building is shown below (see Fig. 1).

Now, the newly repaired Wukang building is an internet celebrity of sorts, attracting a mass of tourists to visit due to the tourism program, the RBs. On weekends in 2021, the Wukang Building greeted more than 40,000 people per day (Fang, n.d.). (During that time, the successful domestic epidemic control made China a safe place to travel across the nation). In addition, in the past, tourists took the Bund as their first stop to visit after arriving in Shanghai, but now, they enjoy the Wukang Building first and then visit the Bund. There are many ways for tourists to experience

Fig. 1 Wukang Building (Wukang Building Nowadays, n.d.)

visiting the Wukang Building. Firstly, tourists can click the tourism APP made by the Shanghai government to read the brief introduction on this building, which includes the construction features, the history and some of its residents who are famous Chinese artists, as a way of virtually traveling the Wukang Building. Secondly, tourists can buy the pocket travel book, "Meet Wukang Building", to follow the book's guidance to easily walk around the Wukang Building and its neighborhood by themselves for a site tour. Thirdly, tourists can choose to read books to know the stories, culture and read about the individuals' lives that are related to this building such as "Live in Wukang Building", "Wukang Road" and "Heritage Architectures under Plane Trees", instead of doing a site tour. Generally speaking, the Readable Wukang Building touring program created a novel model to lead tourists to enjoy the architecture itself, the local culture and the lives of the locals. Besides enjoying the building itself, there are bookstores, shops, cafes and restaurants on the first floor of this building where tourists can get their vision, hearing, taste and so on, satisfied. For instance, the Dayin Bookstore, featured with an elegant Chinese rhyme, is right there to offer readers a quiet and spacious place to read books, most of which are about the culture and history of Shanghai as well as Wukang Building. They can also have a pot of tea and enjoy some delicate snacks. In addition, there is an ice cream shop where tourists can have ice cream shaped like the Wukang building with the logo of Readable Buildings as well as special flavors including salty cheese flavor and hawthorn flavor (the red color of hawthorn is the same as the color of Wukang Building). Among those shops, the most attractive is the Violet Hair Salon which debuted in the Wukang Building in 1936, it then disappeared from the eyes of the public for more than 20 years but is now back. The Violet Hair Salon

used to be the fashion center to lead hair trends in Shanghai in the 1980s and 1990s. Unfortunately, it closed in 2006 due to losing its economic advantages. Now it has reopened with advanced equipment and skilled and creative hairstylists to restore its former glory (Violet Hair Salon, n.d.). Tourists can walk into the Violet Hair Salon to have their hair cut or trimmed to experience the trendy and personalized service provided by an experienced hairstylist. Moreover, when tourists walk around Wukang building, they will meet voluntary storytellers narrating stories of the building, helping them get a deeper understanding of the building's culture. If any of the tourists step into the building's art gallery, they will be drawn to appreciate amazing artwork. If they go to the building's music shop, they will be able to enjoy music. What is worth mentioning is that there are many wonderful places near Wukang Building that tourists cannot miss. For instance, tourists can wander into the streets or lanes near the Wukang building to enjoy the historical and cultural neighborhood, including Soong Ching-Ling's Former Residence (Shanghai), China Post Cafe and so on. It is about a two-minute walk from Wukang Building to Soong Ching-Ling's Former Residence (Shanghai). Madame Soong Ching-Ling, was born on January 27, 1893, and died on May 29, 1981, and was one of the founders of the People's Republic of China, was an honorary president of China, a great patriot, democrat, internationalist and communist, and was acknowledged as a great and world-famous woman in the twentieth century (Soong Ching-Ling Baike.baidu, n.d.). She used to live in, hold important state activities and meet distinguished foreign guests in her residence. The main building of Soong Ching-Ling's Former Residence (Shanghai) is a two-story western-style house with a brick and wood structure. The house is boat-shaped with sailboats and iron anchors decorated on its green wooden windows, and its roof's chimney resembles that of a ship's chimney. In the spring of 1948, Soong Ching-Ling moved into this house and spent fifteen years living in it. At present, the display of Soong Ching-Ling's Former Residence (Shanghai) is completely in accordance with the layout of how Soong Ching-Ling's house used to be and the exhibits are all originals (Soong Ching-Ling's Former Residence (Shanghai) Baike.baidu, n.d.). Moreover, it takes several minutes for tourists to get to the China Post Cafe from Wukang Building on foot to have a cup of coffee from a post-office themed cafe, enjoy precious memorial stamps, and deal with postal business. The most interesting postal business offered by the China Post Cafe is that tourists can post letters or postcards to their future selves. In other words, a Readable Wukang Building provided an impressive solution to make cultural tourism sustainable while further stimulating local businesses and the digital economy when the epidemic blocked foreign travelers from entering the country and made local residents home bound.

3 Literature Review

The definition of sustainable tourism development is that it is all kinds of tourism developments that make a notable contribution to or, at least, do not contradict the maintenance of the principles of development in an indefinite time, without compromising the ability of future generations to satisfy their own needs and desires (Tosun, 1998, p. 596). According to an interpretation by Menon et al. (2021), culture refers to a kind of system covering spiritual value and lifestyle. Tourism is the carrier of culture. Cultural Tourism is a part of tourism by which tourists can appreciate traditional culture (such as art, architecture, religion and so on) presented in different areas and countries, trace cultural relics, and join in various cultural activities. As for the historical and cultural architecture of sustainable tourism, numerous research has been done previously, among them being Osman and Farahat (2021) who have indicated that people had a better understanding of the value of heritage buildings and their conservation through demolition and renovation. Ferwati et al. (2021) suggested the Albergo diffuso model, that promoted local culture, stimulated the local economy and considered that environmental sustainability should be the reference to implement tourism sustainability. Khodadadi et al. (2021) pointed out six obstacles which covered policy and planning, knowledge, resources, desire, marketing activities and awareness that challenged the sustainable development of small heritage sites in Shiraz and across Iran. They also emphasized that it would be the key solution to make clear and complete policies to target sustainably developing small heritage sites in Iran. Baker et al. (2021) proposed carbon emissions should be considered when it comes to the application of heritage.

4 Methodology

To examine the publicity, popularity and influence of the application of the novel touring mode, Readable Buildings, the author adopted SO JUMP APP, a professional online questionnaire platform to collect and analyze the data. First, the author designed the questions and made the questionnaire. Second, the author released the questionnaire on social media like Wechat. Finally, the author received the feedback of 392 participants (See Table 1).

5 Findings

Based on the data collected from questionnaire, Q1 (see Table 1) showed that the majority of the participants were young people, covering 72.19% of the pie chart, the minority

was the old participants, occupying 0.26% and the middle-aged held 27.55% (see Fig. 2).

Q2 (see Table 1) stated that the young participants were local students (69.64%), 23.21% of those participants were local employees/employers, only 1.02% participants came to Shanghai for travel and 6.12% did not reveal their identity.

Q1 and 2 indicate that the case study was limited to young people/students and middle-aged people/working staff. The opinions mentioned in this paper mainly presented their likes or dislikes in regards to the Readable Wukang Building (RWB).

Q3 (see Table 1) displayed that the number of people who did not know of the Readable Buildings (RBs) was 1.6 times as many as those who knew of the Readable Buildings (see Fig. 3). Q4 (see Table 1) stated that the number of people who did not know of RWB was almost twice as many as those who knew of RWB (see Fig. 4). Both Q3 and Q4 suggest that the promotion and popularization of the RBs and RWB should be enhanced, making them reach as many people as possible.

In terms of Q5 (see Table 1 and Fig. 5), 34.95% participants had access to social media to know of the RWB, 2.04% participants knew the RWB through TV, 0.51% participants knew the RWB through radios, 0.51% participants knew the RWB through brochures, 24.23% participants were informed of the RWB through their friends and 37.76% participants did not state how they knew of the RWB. Figure 5 proves that social media and word of mouth were good ways to advertise the RWB, while the traditional media like TV, radios, brochures played weak roles in propagandizing the RWB. It is necessary to address the importance of the roles that traditional media take in broadcasting the RWB because the traditional media are still welcomed by most of people.

According to the data from Table 1, Q6 (see Table 1) claims that 80.87% participants wanted to visit the RWB if they were available, only 6.38% chose not to visit it and 12.76% had not decided. This data shows that the RWB has the potential to reach its peak in the future. Q7 (see Table 1) explains that 51.02% of the participants who visited the RWB were fond of it and a few of them did not like it, accounting for 4.34%. These numbers prove that the RWB is widely enjoyed among the people who have been there. On the other hand, Q7 (see Table 1) displayed that 44.64% participants did not express their preferences because they had not been to the RWB yet. Q8 (see Table 1) conveyed that the majority of the participants (72.19%) would like to read the stories related to the Wukang Building, confirming that it would be feasible to make cultural tourism sustainable through having readable historical architecture. Q9 and 10 (see Table 1) stated that 73.47% participants were interested in the video introduction to the RWB, where 83.42% preferred a short video, suggesting that a digital tour would be a

Table 1 Questionnaire survey on tourism program of a Readable Wukang Building

Item	Subtotal	Percentage (%)
1. What is your age? [single choice]		
Under 23	283	72.19
23–65	108	27.55
Above 65	1	0.26
Valid number	392	
2. What is your ID? [single choice]		
Study in Shanghai	273	69.64
Work in Shanghai	91	23.21
Tour around Shanghai	4	1.02
Other	24	6.12
Valid number	392	
3. Do you know of the tourism program, readable buildings? [single choice]		
Yes	150	38.27
No	242	61.73
Valid number	392	
4. Do you know of the tourism program, a readable Wukang building? [single choice]		
Yes	138	35.2
No	254	64.8
Valid number	392	
5. How do you know of the tourism program, a readable Wukang building? [single choice]		
Social media	137	34.95
Word of mouth	95	24.23
TV	8	2.04
Radio	2	0.51
Brochure	2	0.51
Other	148	37.76
Valid number	392	
6. Would you like to visit Wukang building if you are available? [single choice]		
Yes	317	80.87
No	25	6.38
Indeterminate	50	12.76
Valid number	392	
7. Did you like the tourism program after visiting Wukang building? [single choice]		
Yes	200	51.02
No	17	4.34
Indeterminate	175	44.64
Valid number	392	
8. Are you interested in reading the stories about the Wukang building? [single choice]		
Yes	283	72.19
No	45	11.48
Indeterminate	64	16.33
Valid number	392	
9. Are you interested in watching an introductory video to the stories of Wukang building? [single choice]		
Yes	288	73.47
No	45	11.48

Table 1 (continued)

Item	Subtotal	Percentage (%)
Indeterminate	59	15.05
Valid number	392	
10. With respect to the stories of Wukang building, what type of video format would you like to watch? [single choice]		
Short-form video (less than 10 min)	327	83.42
Long-form video (more than 10 min)	23	5.87
Don't care	42	10.71
Valid number	392	
11. Do you like products with the Wukang building logo? [single choice]		
Yes	229	58.42
No	31	7.91
Indeterminate	132	33.67
Valid number	392	

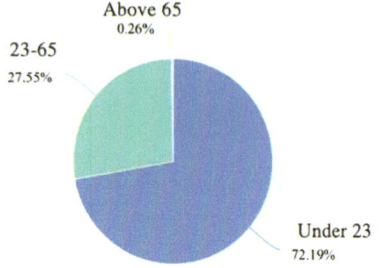

Fig. 2 Participants' ages

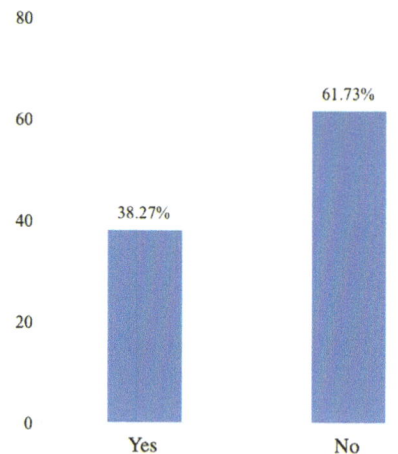

Fig. 3 The number of participants who knew/did not know of the Readable Buildings

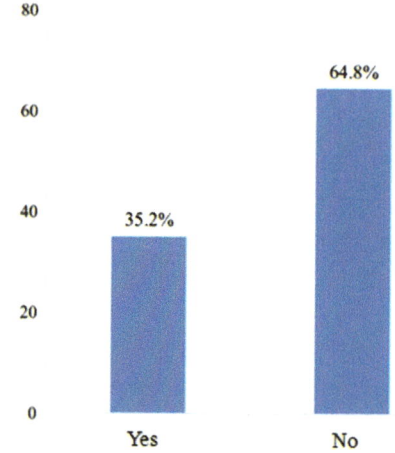

Fig. 4 The number of participants who knew/did not know a Readable Wukang Building

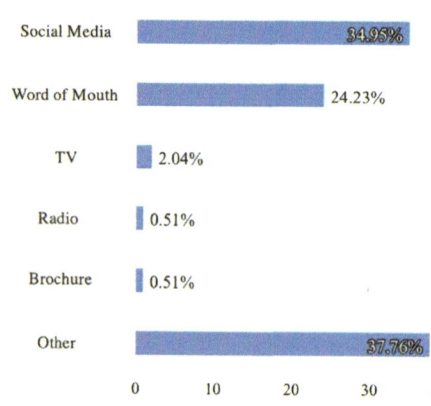

Fig. 5 The methods and numbers of how the participants knew of a Readable Wukang Building

smart approach to realize CST. In addition, based on the data for Q11 (see Table 1), the number of participants who liked the goods with the Wukang Building logo was 6 times more than those who did not like them, determining that it would be practical for the local government to use the RWB, RBs and CST to boost the local economy.

6 Discussion and Conclusion

Before the covid-19 outbreak, Shanghai used to be crowded with tourists from home and abroad regardless of it being sunny, windy or rainy. But this picture has been altered since the occurrence of the epidemic. From the sudden outbreak of covid-19 in 2019 until now (July of 2022), there have been few foreign tourists in Shanghai. It seems that it is as if the tourists have faded away. How we can revive the tourism business challenges not only the Shanghai government but also the tourism industry. Thanks to the success of the domestic control over covid-19 China was made a safe place for domestic tourists throughout 2020 and 2021. The Shanghai government launched the RBs campaign to help boost the local tourism business, and it turned out to be fruitful with the RWB waving Internet and attracting mass travelers. The RWB attracted more than 424,000 tourists from Shanghai and out of Shanghai to check in within three days during the Dragon Boat Festival in June 2021, making it join the list of popular attractions in Shanghai and even the whole country (424,000 people in 3 days! The Wukang Building is popular again. People visit it to read it rather than take a picture, which is the correct way to check in the old buildings in Shanghai, n.d.).

The RBs and RWB did temporarily thrive in the local tourism industry, but how to make this improvement sustainable is also a tough task for the Shanghai government because of the uncertainty of the epidemic. In fact, the sudden surge of the Omicron variant in Shanghai during the middle of March 2022 made tourists vanish from the streets for several months because of the lockdown, indicating that the traditional site tour cannot be sustainable. In order to advance CST as well as tackle the side-effects on CST brought on by covid-19 and its variants, the Shanghai municipal government put tremendous efforts into practice. The Shanghai government launched the tourism campaign with the theme of "architecture can be readable, urbanism can be virtually explored" in the 23rd Shanghai Tourism Festival of 2021. During this tourism festival, the method of attendance turned to a virtual one instead of a traditional site tour. With the joint efforts of the Shanghai government and a group of leading digital cultural and tourism companies in China, including Tik Tok, Meituan and Ctrip, etc., the virtual tour had a fruitful result with a total reach of 1.17 billion. In addition, the tour program (Readable Buildings) was successful. Nearly 38,000 people visited the mini program of WeChat to tour the historical buildings in Shanghai during the duration of this tourism festival (Readable Buildings, Strolling Streets, and People-centered City, n.d.). The RBs and RWB as well as digital tours went beyond the limits of traditional cultural site tours as they lessened the grievous impact brought on by the epidemic on CST.

These numbers (see references 29 and 30) witnessed the success of the new touring model, the RBs and RWB, which can be implemented through online and offline methods. The RBs not only respond to the negative impacts the tourism business is facing brought on by the epidemic, but also offer solutions to CST. The new touring model of the RBs and RWB has been quickly extended to the rest of China like Beijing, Xian, Guangzhou and so on, bringing new life to a historical building, boosting the local businesses, as well as the digital economy. In addition, the RBs can be easily copied by the rest of the world. Hopefully, this new touring model will bring the tourism industry back to how it was prior to the epidemic, and for it to even potentially prosper in the near future.

Although the RWB acted as a successful response to the challenges of the epidemic and CST, it still has room to improve its influence. Based on the results mentioned in the Findings section (part 5 of this paper), the publicity and popularity of the RWB and RBs fall short of its desired degree. First, there is quite a number of people who live in Shanghai but do not know of the RWB and RBs (see Figs. 3 and 4). Second, the advertisements lack a sense of diversity, causing them to ignore traditional paths such as TV, radios and brochures, leading to the inadequate reach of middle-aged and senior tourists. Third, the books and videos, which can systematically introduce the RWB and RBs to the public, are insufficient. Finally, most people do not know what the goods with the logo of the RWB are in the current market and where they can be bought, leaving a huge potential market for the local government and businessmen.

In order to make the new touring model of the RBs and RWB reach its potential, the following measures are suggested. First, the publicity and popularity of the RWB and RBs should be enhanced. This can be done through the use of popular social media such as WeChat, QQ, Facebook, Twitter and so on. Traditional media covering methods such as newspapers, radios, TV, brochures and so on to broadcast the new touring model of the RBs and RWB to let it reach as many people of different ages as possible, can also be used. Second, developing the products with the RWB or the RBs logo to meet people's demands for having them. Lastly, advertising the products with the RWB or the RBs logo to let people know when, where, and how they can buy these goods and how much they will pay for them.

In short, this paper, based on abundant facts and numbers, thoroughly illustrated the new touring model, the RBs, covering its intentions, contents, implementation and influence. Moreover, this paper, according to the case study on RWB, explored the strengths of this new touring model, which dismissed the passive impacts on the tourism business due to the epidemic and made cultural tourism sustainable. It also boosted the digital economy, pointed out the shortcomings of this new touring model when it comes to its promotional methods and its product development, while also suggesting ways to perfect the RBs. Hopefully, the novel touring model of the RBs and RWB will help shed a light on CST.

References

Baker, H., Moncaster, A., Remøy, H., & Wilkinson, S. (2021). Retention not demolition: How heritage thinking can inform carbon reduction. *Journal of Architectural Conservation, 27*(3), 176–194. https://doi.org/10.1080/13556207.2021.1948239

Build the digital version 3.0 of the Readable Buildings with ten actions. (n.d.). http://m.yun.jxntv.cn/p/393545.html

Fang, S. (n.d.). *Creatively open the Readable Buildings*. https://sghexport.shobserver.com/html/baijiahao/2021/09/18/541940.html

Ferwati, M. S., El-Menshawy, S., Mohamed, M. E., Ferwati, S., & Nuami, F. A. (2021). Revitalising abandoned heritage villages: the case of Tinbak, Qatar. *Cogent Social Sciences, 7*(1). https://doi.org/10.1080/23311886.2021.1973196

Jiangnan Culture Baike.baidu. (n.d.). Baike.baidu. https://baike.baidu.com/item/%E6%B1%9F%E5%8D%97%E6%96%87%E5%8C%96/55217693

Khodadadi, M., Pezeshki, F., & O'Donnell, H. (2021). Small but perfectly (in)formed? Sustainable development of small heritage sites in Iran. *Journal of Heritage Tourism, 17*(1), 74–90. https://doi.org/10.1080/1743873x.2021.1933992

Lu Xun Baike.baidu. (n.d.). Baike.baidu. https://baike.baidu.com/item/%E9%B2%81%E8%BF%85/36231

Lu Xun's Former Residence Baike.baidu. (n.d.). Baike.baidu. https://baike.baidu.com/item/%E9%B2%81%E8%BF%85%E6%95%85%E5%B1%85/5959555v

Menon, S., Bhatt, S., & Sharma, S. (2021). A study on envisioning Indian tourism—Through cultural tourism and sustainable digitalization. *Cogent Social Sciences, 7*(1). https://doi.org/10.1080/23311886.2021.1903149

Osman, K., & Farahat, B. I. (2021). The impact of living heritage approach for sustainable tourism and economics in mount lebanon. *HBRC Journal*. https://doi.org/10.1080/16874048.2021.1996062

Shanghai Culture Baike.baidu. (n.d.). Baike.baidu. https://baike.baidu.com/item/%E6%B5%B7%E6%B4%BE%E6%96%87%E5%8C%96/33264

Shanghai Jinmao Tower Baike.baidu. (n.d.). Baike.baidu. https://baike.baidu.com/item/%E4%B8%8A%E6%B5%B7%E9%87%91%E8%8C%82%E5%A4%A7%E5%8E%A6/2946581?fromtitle=%E9%87%91%E8%8C%82%E5%A4%A7%E5%8E%A6&fromid=486833&fr=aladdin

Shanghai-style architecture: Eight styles of traditional Chinese architecture Baijiahao.baidu. (n.d.). Baijiahao.baidu. https://baijiahao.baidu.com/s?id=1668552461750531876&wfr=spider&for=pc&searchword=%E6%B5%B7%E6%B4%BE%E6%96%87%E5%8C%96%E5%BB%BA%E7%AD%91%E4%BB%A3%E8%A1%A8

Shanghai Tower Baike.baidu. (n.d.). Baike.baidu. https://baike.baidu.com/item/%E4%B8%8A%E6%B5%B7%E4%B8%AD%E5%BF%83%E5%A4%A7%E5%8E%A6/5135128?fromtitle=%E4%B8%8A%E6%B5%B7%E4%B8%AD%E5%BF%83&fromid=10516840&fr=aladdin

Soong Ching-Ling Baike.baidu. (n.d.). Baike.baidu. https://baike.baidu.com/item/%E5%AE%8B%E5%BA%86%E9%BE%84/115132?fr=aladdin

Soong Ching-Ling's Former Residence (Shanghai) Baike.baidu. (n.d.). Baike.baidu. https://baike.baidu.com/item/%E4%B8%8A%E6%B5%B7%E5%AE%8B%E5%BA%86%E9%BE%84%E6%95%85%E5%B1%85/5933986?fr=aladdin

The International Savings Society Baike.baidu. (n.d.). Baike.baidu. https://baike.baidu.com/item/%E4%B8%87%E5%9B%BD%E5%82%A8%E8%93%84%E4%BC%9A/10679785

The Moller Villa Baike.baidu. (n.d.). Baike.baidu. https://baike.baidu.com/item/%E9%A9%AC%E5%8B%92%E5%88%AB%E5%A2%85/7443991

The Oriental Pear TV Tower Baike.baidu. (n.d.). Baike.baidu. https://baike.baidu.com/item/%E4%B8%8A%E6%B5%B7%E6%98%8E%E7%8F%A0%E7%94%B5%E8%A7%86%E5%A1%94/8283626

The version 3.0 of the Readable Buildings makes people feel the warm of people-centered city of Shanghai in an all-round way. (n.d.). http://www.ctnews.com.cn/paper/content/202103/18/content_55268.html

The 85-year-old "Violet Hair Salon" returns to the Wukang Building, and the Shanghai old brand tells a new story. (n.d.). https://baijiahao.baidu.com/s?id=1709672242660406122&wfr=spider&for=pc&searchword=%E7%B4%AB%E7%BD%97%E5%85%B0%E7%BE%8E%E5%8F%91%E5%8E%85%E5%8E%86%E5%8F%B2. Accessed June 12, 2023.

Tosun, C. (1998). Roots of unsustainable tourism development at the local level: The case of Urgup in Turkey. *Tourism Management, 19* (6), 595–610. https://doi.org/10.1016/S0261-5177(98)00068-5

Qibao Ancient Town Baike.baidu. (n.d.). Baike.baidu. https://baike.baidu.com/item/%E4%B8%83%E5%AE%9D%E5%8F%A4%E9%95%87/5142529?fr=aladdin

Readable buildings, strolling streets, and people-centered city. (n.d.). http://sh.xinhuanet.com/2021-10/07/c_1310229737.html

Red Culture Baike.baidu. (n.d.). Baike.baidu. https://baike.baidu.com/item/%E7%BA%A2%E8%89%B2%E6%96%87%E5%8C%96/4998455

Wukang Building Baike.baidu. (n.d.). Baike.baidu. https://baike.baidu.com/item/%E6%AD%A6%E5%BA%B7%E5%A4%A7%E6%A5%BC/4669442?bk_tashuoStyle=topLeft&bk_share=shoubai&bk_sharefr=lemma&fr=shoubai

Wukang Building Nowadays. (n.d.). https://www.meipian.cn/32k3aam1

Yangtze River Delta Daily. (2021, September 24). Enhancing the construction of a world-famous tourist city! Shanghai did these during "14th Five-Year Plan".3g.163.com. https://3g.163.com/dy/article/GKM1UL7C0550DDYR.html

Zhiyan Consulting. (n.d.). Analysis on the trend and composition of tourism revenue, tourist numbers in China in 2018. https://www.chyxx.com/industry/201908/777282.html

424,000 people in 3 days! The Wukang Building is popular again. People visit it to read it rather than take a picture, which is the correct way to check in the old buildings in Shanghai. (n.d.). https://sghexport.shobserver.com/html/baijiahao/2021/06/15/460612.html

Xiangchong Pan Conceptualization, Methodology, Software, Investigation, Formal Analysis, Data Curation, Validation, Visualization, Resources, Supervision, Writing—Original Draft; Review and Editing.

Yinan Guo Conceptualization, Resources, Software, Data Collection, Editing.

Wei Zhong Conceptualization, Resources, Data Collection.

Fei Yin Conceptualization, Resources, Data Collection.

Sustainable Tourism Development in Less Touristy Destinations; The Case of Epirus, Greece

Eleni Gimouki

Abstract

While Greece is a destination that everyone knows and maybe has visited at least once in their life, not all Greek destinations are touristy just like Santorini or Crete are. For this reason, in the last few years, there has been an attempt to develop strategies for sustainable tourism in less-known destinations in Greece. In order to achieve sustainable tourism development, Governments and Destination Management Organizations (DMOs) must take into account several parameters. For example, not all destinations provide the same natural landscapes or similar cultural heritage. That means that different strategies must be applied so as to achieve the goal of sustainable tourism development. First of all, in this essay, development strategies for sustainable tourism are going to be discussed. It is very significant to point out good practices which can lead to tourism sustainability. In this way, both touristy and less touristy destinations can increase their sustainability. Much more, the case of Epirus in Greece is a good example of how we can apply development strategies for sustainable tourism. For example, the Cultural Route of the Ancient Theaters of Epirus is a great paradigm of how a destination can improve its tourism sustainability by using its cultural resources. In this way, it can attract more tourists all year round. To conclude, by developing good strategies for sustainable tourism, countries achieve to have tourism not just for a few months of the year but almost all year round. Finally, there are many destinations that need to develop tourism so as to develop their economy, create new jobs and thus reduce unemployment.

Keywords

Development strategies • Sustainable tourism • Destination management • Destination development • Economy development

1 Introduction

Good strategies for sustainable cultural tourism development can offer countries numerous advantages. In order to achieve this, Governments and Destination Management Organizations (DMOs) must take into account several parameters. For example, not all destinations provide the same cultural heritage and resources. That means that different strategies must be applied so as to achieve the goal of sustainable cultural tourism development.

In the present article, there is an attempt to provide information through the narrative literature review methodology on cultural tourism and strategies and tools that are used in order for stakeholders to develop it in a sustainable way. Moreover, it is interesting to investigate how Greece make a try to develop its cultural tourism.

A characteristic example of this attempt is the case of Epirus, Greece. While Epirus is a less-known and popular destination in Greece, but full of cultural resources, it is interesting to find out which cultural resources are used the most in order to develop the tourism sector.

To conclude, when good strategies for sustainable tourism are being developed, countries have the opportunity to achieve to develop tourism not just for specific periods of time during the year but almost all year round. Besides, there are many destinations that need to develop tourism as, in this way, they will be able to develop their economy, in a wide range.

So, following appropriate strategies can lead less touristy destinations to become more popular and thus increase their annual tourist percentage.

E. Gimouki (✉)
Sustainable Tourism Development, Cultural Heritage, Environment, Society, Harokopio University, Athens, Greece
e-mail: elengim@gmail.com

2 Research Methodology

The methodology that is followed in the present article is the narrative literature review. In order to conduct a search, key words were used on Google Scholars and Science Direct. More specifically, the main key words of this research were "development strategies", "sustainable tourism", "destination management", "destination development" and "economy development". Other key words, such as "cultural tourism", "cultural tourism in Greece", and "cultural tourism in Epirus" were used in order to find more information on this subject. The next step was to review abstracts and articles in order to provide the relevant information and to reach the results of this review (Snyder, 2019).

3 Cultural Tourism

First of all, cultural tourism is tourism concerning the history, the arts, and the culture of a destination. The most common activities in cultural tourism are visiting museums and archeological sites, watching theatrical plays or even taking part in the destination's customs. The aim is for visitors to be educated on this new culture in a fun way (Mousavi et al., 2016). Cultural tourism is a type of tourism activity in which the essential motivation of the visitor is to learn about, discover, experience, and consume the tangible and intangible cultural attractions and/or products of a destination. At the same time, stakeholders strive to preserve local and regional cultural resources (Richards, 2011).

The recent years, Governments and Destination Management Organizations (DMOs) around the world make an effort to develop suitable strategies in order to achieve sustainable cultural tourism. In general, the need for tourism sustainability was initially stated in the 1970s by international organizations. The *World Conservation Strategy* was published by the International Union for the Conservation of Nature (IUCN) in 1980, and the *World Environment and Development Commission* report stating the six principles of sustainability was published in 1987.

In the decades of 1970s and 1980s, standardized mass production of package holidays in exotic destinations was being developed. In the 1990s, it became a necessity to develop different types of tourism in a more sustainable way. As a result, the mass market began to fragment into different niches. Cultural tourism is a characteristic paradigm of a niche tourism market (Richards, 2011).

Today, it is observed that the number of people who choose cultural tourism is increasing. Some reason why this shift toward new areas of culture, such as the intangible culture or the popular culture happens is that more and more people have access to higher education and get better job positions. As a result, they obtain a better income and a higher status level in the market. Besides, the extended use of the Internet in order to get more information about the destinations and their culture plays a significant role in the development of cultural tourism. Finally, there is a need for people to feel the co-presence and for this reason, they choose to take part in different cultural events, like festivals (Richards, 2011).

3.1 Strategies for Cultural Tourism Development

On this basis, nowadays, significant strategies for the sustainability of cultural tourism have been developed. First and foremost, in general, heritage must be accessible to people. That means that Governments must fund heritage through taxes, and public and private funds in order to preserve the cultural resources and let people visit them with low-cost tickets. However, we need to have in mind that not all cultural resources have the same significance or attractiveness, so it is not necessary for countries to fund every small archeological site but to try to figure out what people are interested in visiting and experiencing. Moreover, heritage needs to be protected by the legal system. Besides, there are several examples of trespassing on archeological sites. Finally, Governments need to find the appropriate resources; human and financial resources that will help cultural tourism to be maintained and developed (Woodside & Martin, 2007).

In order to achieve this aim, a few significant processes of management must be followed. Firstly, planning is the most significant phase as the vision of the heritage must be stated. To be more precise, both a Government and the Destination Management Organizations (DMOs) of a country have to agree to a cultural tourism development plan. In this plan, they will state which tangible and intangible cultural resources can emerge, the reasons why these resources are more important than other ones, and the guidelines that must be followed to achieve the development of cultural tourism (Richards, 2011).

The second step concerns the implementation of this plan. In other words, financial and human resources must be found. It is vital for the successful development of cultural tourism to determine the appropriate human resources—all these people who will work on this aim of developing cultural tourism. This is not only about employees in administrative positions, but even the employees who directly come in touch with the visitors such as ticket issuers, securities, and guides. As it concerns the financial resources, that have to do with the preservation of cultural heritage. So, it needs to be decided about the taxes, and the public and private funds, as it was mentioned above.

After the planning and implementation phases, monitoring is the last step for successfully developing cultural tourism in a destination. Monitoring may be the most significant step as it is the time when stakeholders watch their plan and implementation start operating. The reason why this is really important is that they will have the chance to develop strategies in order to solve any problems (Woodside & Martin, 2007).

So, what stakeholders must take into consideration during the last phase of monitoring is how many people visit the place, how much the costs and the economic results are, and how we can reduce the costs when there aren't visitors. Stakeholders also need to think about the temporary employees and the volunteers who may wish to work and if such solutions would be effective. Concerning the archeological sites and museums, it must also be decided whether the whole museum will be open or just a part of it. Finally, they need to determine where the funds come from (expenditures of tourists, partnerships, or even indirect profits, e.g., from employees who buy goods with their salaries).

Moreover, a destination has to provide transportation in order for visitors to easily approach the different archeological sites and museums. Therefore, a less popular cultural resource in an inaccessible place of the destination may not be worth including in a tourism development strategy.

In order to develop sustainable cultural tourism, each destination must upgrade its labor market, offer a certain level of qualification, and obtain stable prices. In this way, tourists will choose the place, while the destination itself will have the opportunity to develop not only its tourism sector but almost all of its economic sectors. In this way, residents will be more interested in participating in the organization and promoting the cultural heritage of their place.

Finally, a strong brand name must be created. A proper marketing strategy makes a product attractive so that it will be sold. In this context, it is important to study the profile and interests of tourists in order to develop a suitable marketing strategy that will attract more tourists interested in cultural tourism.

3.2 Tools for Cultural Tourism Development

Except for adopting suitable strategies for cultural tourism development, Governments and Destination Management Organizations (DMOs) in cooperation with the destination's stakeholders need to also use some useful tools in order to achieve the goal of cultural tourism development.

It is well known that most people use their smartphones all day long. It is therefore understood that in order to achieve the development of cultural tourism, it is important to use new technologies as our major tools. So, a useful tool

in order to evaluate the impact of cultural tourism is the Mobile Positioning Data. Using this tool helps find out the number of tourists that visit a specific destination on a daily basis. Moreover, it offers information about how many tourists stayed either at hotels, at their friends' or relatives' houses, at Airbnb, or left their destination the same day. This type of monitoring helps us analyze the daily, weekly, and seasonal rhythm of visitors (Kalvet et al., 2020). For example, tourists who stay longer in a place tend to consume more cultural products than those who stay for one or two days, and they only visit the most popular cultural sites.

Besides this, we may use a social network to link to new audiences. In other words, most people nowadays are registered users on one or more social network platforms. That is to say, in addition to utilizing social networks to associate with their friends and upload posts, they additionally use these platforms with the intention of discovering new destinations to expand theirs after the holidays. Travelers are exceptionally repeatedly impacted by the outcomes of their research while we are surfing the web regarding goals. These outcomes are straight associated with quest outcome rankings, metadata, and paid links.

Based on the information above, search engines can be seen as an information space where businesses in the tourism industry compete for the attention of online travelers because the representation of the domain is largely based upon the ranking and position of search results and will be influenced by contingent factors such as the presence of online advertisements, according to Kim and Fesenmaier (2008).

For this reason, the use of new technologies by Governments and stakeholders, in general, is an urgent need for cultural tourism development. However, in connection with the use of new technologies, stakeholders also need to create new promotional ideas and present new celebrations or new heritage merchandisers or even proceed to the development of interactive guides where visitors will have an active role during their visit in a museum or a monument (Richards, 2011).

Finally, an innovative tool that we can use to successfully develop cultural tourism is virtual reality (VR). It is a tool that is now used in several cases and allows the public to experience past historical periods as if they had been there themselves. In this way, museums or archeological sites are no longer mere exhibition spaces, they come alive and attract more young people fascinated by virtual reality.

Clearly, we have more tools at our disposal today than in the past, most of them directly related to new technologies. So, if we use these kinds of tools in cultural tourism, we will not only succeed in developing this kind of tourism, but also in keeping the cultural heritage of the destinations alive and modern and, of course, appealing to the younger audience by making the cultural heritage interesting and entertaining.

4 Cultural Tourism in Greece

Greece is a tourist destination rich in cultural resources. It is best known for its ancient history and antiquity in general. A lot of its monuments are famous around the world. Among these is the Parthenon in Athens, the monument of Poseidon in Sounion, Athens, the ancient theater of Epidaurus and the ancient Olympia in Peloponnese, the Knossos Palace in Crete, and much more.

Parthenon, for instance, according to the statistics, attracts about 2 million visitors every year. It is the 7th most visitable attraction in the world. It is impressive the fact that even during the winter months thousands of people visit the monument of the Parthenon. For example, in January 2019, 92,937 tourists visited the Acropolis of Athens (Greek Statistical Authority, 2019). As it is already known, Parthenon is the symbol of democracy, and this is the first reason why most people wish to visit it. Except for that, it provides a unique architecture that everyone admires. So, it isn't difficult to understand the reason why it became this popular even among people who aren't interested in history and culture.

However, at this point, it is significant to state the fact that the Parthenon is built in the center of Athens, the capital of Greece. As Athens is a big city, this is a great advantage to developing sustainable cultural tourism. The reason why this has already happened is that Athens provides numerous amenities such as a lot of hotels and restaurants in a wide price range, convenient transportation, and a lot of activities to do. Besides, a visitor doesn't travel to a place just to visit and admire the cultural resources, but he/she also wishes to spend his/her holidays having fun and gaining new experiences.

On the other hand, there are a lot of archeological sites such as the Knossos Palace which isn't located in a big city like Athens, but it is built on an island, Crete. Knossos Palace is a symbol of Minoan Culture, and it is also as popular as other monuments in Greece. Every year, about 700,000 people visit Knossos Palace in Crete. Again, according to the statistics, only in January 2019, 3166 tourists visited Knossos Palace (Greek Statistical Authority, 2019). So, although it is located on an island, people keep on visiting it every year. In this case, visitors combine their summer holidays with visiting archeological sites and museums. It is true that the tourism model of "sea-sun-sand" belongs to the past. Nowadays, tourists wish to spend their summer holidays not only enjoying the sun and the sea but also becoming a member of society, learning more about its customs, and expanding their knowledge of its history and culture.

However, it is important to state that in recent years, more and more small and less-known destinations are becoming more popular as regional stakeholders make try to promote these destinations and highlight their significant cultural resources, following the successful paradigm of Athens and other popular destinations.

4.1 Destination Selection and Evaluation Methodology

All of the above is the result of the strategy for tourism development in Greece. More specifically, the Greek Ministry of Tourism follows a specific strategy that contains several steps which are necessary for cultural tourism development. First, it set some significant criteria following a destination evaluation in general.

More precisely, the most important criterion seems to be accessibility to a destination. It is obvious that when a destination is easily accessible, more people are willing to visit it. The second criterion is the main tourist products that a destination provides. In order to develop tourism, stakeholders need to focus on some specific tourist products, either cultural or environmental, that each destination offers. These will be the tourist products that will help the destination build a successful brand name. In connection with the last criterion, complementary tourist products must be considered.

Next, international fame and recognition, points of interest and UNESCO Monuments play a vital role in Greek tourism development. Finally, the existing demand according to the arrivals at the accommodation and the existing supply, i.e., the number of rooms provided at each destination is also an important criterion (INSETE, 2022).

After defining the criteria, the formulation of the tourism development strategy will start as shown below. The international market was largely analyzed with a specific focus on trends (megatrends) that decide the international direction of Greek tourism products and tourism. In addition, tourism and other resources accessible in Greek destinations are recorded, which beget the basis for the advancement of tourism products. Digital upgrades are crucial as the digital age is affecting the tourism sector as tourists use tech before, in the course of, and afterward their trips. An extra crucial goal for accomplishing cultural tourism growth is environmental protection and sustainability as cultural destinations are repeatedly surrounded by affluent commonplace environments. Therefore, we must ensure the protection of this natural environment and ensure the sustainability of our cultural tourism products. Finally, it is important to develop new and innovative tourism products.

Since it is difficult to develop a tourism development strategy for each destination in a country, the Ministry of Tourism decided to select 36 indicative destinations in

Greece. For each destination, the attractiveness of the tourism products to be developed and the markets that are able to be addressed were identified and prioritized. So, the last step was to sum up all the overhead in a national action way and set objectives for both the particular regions and the general country.

5   The Case of Epirus

In recent years, intensive efforts have been made to develop low-tourist destinations and increase annual tourist numbers. Epirus is such a characteristic case. The region of Epirus is located in northwest Greece, it is a focal point among Italy, Northern Greece, and the Balkan Countries and includes four regional units (Arta, Thesprotia, Ioannina, Preveza). Although it is a destination rich in both cultural and environmental resources, only in the last few years has it achieved to promote and highlight all these resources and thus become more popular as a cultural destination.

So, it is very interesting to collect and investigate the tourist products that the stakeholders created in order to achieve this goal.

5.1   Wine Tourism

For the tourism industry, grapevine cultivation and the operation of winemaking are captivating and attractive. It could be considered also as a product but largely as an experience. It is largely connected with quality tourism and with tourists who are seeking a new credible experience. So, taking portion in this encounter provides tourists also the knowledge of grapevine cultivation and winemaking, but additionally the opportunity to come in touch with the destination's gastronomy by tasting exceptional wine varieties besides local cuisine, with its history and culture, and the way of living (Bonarou et al., 2019).

According to Hall and Macionis definition about wine tourism (1998): *"Wine tourism is a visitation to vineyards, wineries, wine festivals and wine shows for which grape wine tasting and/or experiencing the attributes of the grape wine region are the prime motivating factors for visitors"* (Hall & Macionis, 1998).

In Epirus, as well as in many other Greek destinations, a form of wine tourism has been developed due to the number of vintage wine distilleries that there are in the region in combination with the willingness of nowadays' tourists to visit such places, try the wine tasting experience, and also taste other wine products and the local cuisine (Bada, Pappas, Dalla, & Lolis).

On this basis, the Tourist Wine Route offers visitors a wide range of knowledge not only about history, culture, and

customs but also about the real living conditions of the people living in the Epirus region from the past to the present. The cultural experience of wine tasting can allow visitors to discover the economic and social conditions of a destination over a long period of time.

5.2   Architectural Heritage

Another form of cultural tourism in Epirus is the architectural heritage of the destination. In Epirus, visitors can discover hundreds of traditional stone bridges built in past decades and a lot of traditional villages with many stone-built houses. Besides, the stone building techniques constitute a past tradition of Epirus architecture. Today, a lot of people visit Epirus villages in order to enjoy this architecture.

In two surveys conducted in 2008 and 2010 by Dr. Giannakopoulou Stella and Prof. Kaliampakos Dimitris, the aim was to identify the motivation to travel to Metsovo and Sirako villages, in Epirus, both rich with architectural heritage. They also recorded information from people who have already visited these two villages about whether they would visit them again and if they would propose this trip to other people. A rate of 67.5% of the visitors in Sirako stated that they also visited other villages during their trip while a rate of 57.7% of visitors in Metsovo stated that this village was their exclusive destination. Similar rates were recorded for the year 2010. Also, most of the visitors of both villages used their own cars in order to arrive at their destinations (Giannakopoulou & Kaliampakos, 2015).

According to the survey of 2008, in the case of Sirako, visitors were attracted by the natural landscape and isolation (24%), the local architecture with the stone-built houses (22%), and the general beauty of the place that combines the natural with the architectural environment (21%). In the case of Metsovo, 35% of the visitors stated that they admired the local architectural environment and 24% liked the combination of natural and building environment. In the surveys of 2010, visitors traveled to Sirako mainly because of the natural environment (37%), the local architectural heritage (36.5%), and the local history and tradition (15%) (Giannakopoulou & Kaliampakos, 2015).

5.3   Ancient Theaters

The rise of cultural tourism in Epirus' ancient theaters serves as a last illustration. The Region of Epirus and Epirus Development, along with the Diazoma Association, planned the "Cultural Route of the Ancient Theaters of Epirus" event in November 2018 and hosted it. The "Cultural Route of the Ancient Theaters of Epirus" is a suggestion for managing cultural heritage as a comprehensive entity that perpetuates the past and present, its tangible and intangible manifestations, its economy, and its aesthetics.

This cultural route, above all, consists of 5 archeological sites (Fig. 1). The ancient theater of Amvrakia, the ancient theater of Gitana, the ancient theater of Dodona, the ancient theater of Kassope, and the ancient theater of Nikopolis. All of these theaters are located in Epirus and the wider region of Western Greece (Ancient Theaters of Epirus, 2018).

This cultural product, which connects the historical path in a special way with the local population, its goods, gastronomy, and the natural environment, is an innovative program aimed at both the domestic and foreign markets. Visitors have the opportunity to see locations rich in natural and other cultural elements, including food, festivals, customs, etc., along with archeological sites with 2500 years of history. It is a 344 km journey (Fig. 2) (Ancient Theaters of Epirus, 2018).

According to the Deputy Governor of Preveza region Mr. Ioannoy: "*From an early stage, the route became a common collective goal of multiple agencies, stakeholders, businessmen, groups, but mainly of the citizens of Epirus.*

Fig. 1 Epirus' ancient theaters (Ancient Theaters of Epirus, 2018)

Fig. 2 Map of the "Cultural Route of the Ancient Theaters of Epirus" (Ancient Theaters of Epirus, 2018)

It is that decisive link of public and institutional action with the private sector, which is the real vehicle of development. Achieving unprecedented synergies, the classic glamor of the monuments is exploited, which is re-proposed as a reason to visit in the most modern way in the modern tourist competition. Moreover, in a globalized tourist market, the culture of a place constitutes a unique element of identity and differentiation" (Chrisostomidou, 2018).

The theatrical activity was significant in ancient times. The theater, along with the agora, was a symbol of the city-state and the center of public life, where people would congregate to watch religious and dramatic acts designed not just for amusement but also for education. One of the key pillars of ancient city life is the theater as a structure. Moreover, one of the immaterial commodities is theater as an act. Visitors can now get a strong sense of antiquity while exploring and learning about some of Greece's most significant historic theaters.

6 Results of Cultural Tourism Development

The benefits of developing cultural tourism are numerous and significant for the community and the location. The growth of the local culture is one of the most important positive consequences, to be more precise. In other words, cultural heritage is preserved when tourists travel to an area and engage in local activities like visiting museums, attending plays, or even participating in local festivals. This presents the destination with an excellent opportunity to showcase and publicize less well-known cultural resources, both tangible and intangible.

Another significant result of the development of cultural tourism in a region is the protection of the natural habitat. By developing cultural tourism as the main type of tourism in a destination, visitors tend to mainly visit museums and archeological sites or take part in the destination's festivals and customs rather than visiting natural landscapes and doing sports such as trekking and rafting. In this way, the natural environment is much less affected, as tourists' visits are more balanced between cultural and natural heritage (Kasimoglu & Aydin, 2012).

Moreover, since cultural tourism is not a seasonal activity, it is advantageous that the tourist season can be prolonged. Also, spring or autumn is the best times to visit museums and archeological sites in a country like Greece, where the summers are extremely hot. As a result, both the summer and other seasons can see a sizable influx of tourists at the location.

Concerning the community, cultural tourism development can bring several positive effects. Above all, the sustaining of the population levels is the most important of the positive effects (Lozano-Oyola et al., 2012). People nowadays are increasingly leaving small towns, villages, and islands to relocate to large cities in search of work. As a result, when cultural tourism develops in these areas, people will be able to stay and work in the tourism industry. Furthermore, tourists do not only visit a destination to visit a specific museum or archeological site, but they also stay for a while and consume, e.g., they choose to have lunch at a local restaurant, enjoy a coffee there, or buy some souvenirs, among other things.

Thus, in order for towns to move forward with investments in real estate and the service industry, there has been a large increase in tourism revenue. It is crucial to see the positive outcomes here. Most importantly, these investments may contribute to the expansion and rise in the proportion of visitors who decide to stay longer in the area. Concurrently, new employment opportunities may arise not only in the hotel or restaurant industry but also in other service sectors like markets and pharmacies. Further to the aforementioned benefits, this also reduces unemployment and poverty.

The improvement of the region's infrastructure and transportation systems is the final good outcome. The journey to the destination must be simple and comfortable for tourists. For this reason, airports, motorways, and roads are built to provide simpler and faster access while tourism is increasing. Also, a place with a high volume of tourists needs well-maintained medical facilities, pharmacies, banks, ATMs, etc. The improvement of a region's infrastructure and transportation for tourism purposes also benefits the locals, making life much easier for them. This is yet another positive effect that encourages people to return to their towns and villages.

7 Conclusions

In this paper, it is analyzed what cultural tourism is and what the best practices and tools are in order for a destination to develop this alternative form of tourism. Cultural tourism is a recent form of the tourism sector, and it is related to the visit of museums and archeological sites, watching theatrical plays and taking part in a destination's customs and festivals, and also broadening the knowledge of its gastronomy and local cuisine. Tourists who visit destinations in order to enjoy cultural tourism seek to feel the co-presence and to broaden their knowledge of the culture and history of a new place.

It is no coincidence that the development of cultural tourism in the last decades when there is a clear development of the level of education of the people, which has led them to obtain better jobs and higher incomes. Moreover, the development of new technologies and the active presence of people on different social media platforms help them discover new destinations with rich cultural heritage resources.

On this basis, strategies for the development of cultural tourism are developed with the use of tools mainly coming from new technologies, such as Mobile Positioning Data, social media platforms, and Virtual Reality. These strategies contain significant economic and social resources and the making of an appropriate development and management plan.

These strategies and tools are also used by the Greek Ministry of Tourism in order to achieve the goal of the development of cultural tourism not only in popular destinations, like Athens but also in less popular destinations. Such a characteristic paradigm is the region of Epirus. As it was stated before, the stakeholders in Epirus decided to use some specific cultural heritage resources to develop sustainable cultural tourism. These resources are the very popular wine tourism, the traditional architecture, and the innovative construction of a cultural route to the ancient theaters of the whole region of Epirus.

Investigating all of the above with the use of narrative literature review methodology, it was easy to reach some significant results for the development of cultural tourism. These are the development and the strengthening of the regional culture, the protection of the natural habitat, the extension of the tourism season, the new investments that lead to new job openings and the reduction of unemployment, and the development of new transportation and infrastructures.

Taking all the above into consideration, we conclude that, nowadays, the development of sustainable cultural tourism is more than necessary. It seems that more and more people are asking for living new experiences and coming in touch with new cultures. What Governments and stakeholders, in general, must do is identify this need and proceed with planning an appropriate strategy using suitable tools in order to develop and support this new tourism era. Such a development is going to offer numerous positive effects on the destinations and communities.

So, through this paper, first and foremost, we achieve to gather some useful methods and tools regarding cultural sustainable tourism development using the narrative literature review. Collecting all this useful and most important information from the articles related to this field, it is easy to understand the level of development that cultural tourism has reached. In other words, such a process is vital on a theoretical level as we need to know the exact methods and tools that are proposed in such a new area of research.

At a practical level, it is clear that by applying these methods and using the tools suggested above, countries can develop cultural tourism even in less popular tourist destinations. Moreover, taking the specific example of Epirus, Greece, it is clear that the region, in general, will require considerable cultural, human and infrastructural resources to achieve this goal.

So, Governments and regional stakeholders who make a suitable plan and use specific tools to the aim of cultural sustainable tourism development have the opportunity to increase the annual percentage of visitors. However, it is understood that to achieve this goal, they need to decide which cultural resources they will emphasize in. In the characteristic paradigm of Epirus, the Greek Government, and the regional stakeholders decided to highlight the wine distilleries, the traditional architecture, and the ancient theaters.

Finally, we conclude that cultural tourism is not only about the big cities or popular destinations. On the contrary, with the correct use of the mentioned methods and tools, it is possible for any smaller or less popular destination to develop tourism to a satisfactory degree by promoting and highlighting its cultural resources.

References

Ancient Theaters of Epirus. (2018). Ανάκτηση από Ancient Theaters of Epirus. https://ancienttheatersofepirus.gr/en/

Bada, K., Pappas, C., Dalla, E., & Lolis, T. (χ.χ.). *Routes for tourism and culture.*

Bonarou, C., Tsartas, P., & Sarantakou, E. (2019). E-storytelling and wine tourism branding: Insights from the "Wine Roads of Northern Greece". In M. Sigala, & R. Robinson (Eds.), *Wine tourism destination management and marketing: Theory and cases* (pp. 77–98). Palgrave Macmillan.

Chrisostomidou, V. (2018). The ancient theaters of Epirus, a vehicle for development. *Kathimerini.*

Giannakopoulou, S., & Kaliampakos, D. (2015). The role of architectural heritage in regional development of mountainous areas, evidence from Epirus, Greece. In *Sustainable mountain regions: Make them work* (pp. 193–197). Fakel.

Greek Statistical Authority, G. (2019). *Visitation of museums and archaeological sites—January 2019.* Greek Satistical Authority, GSA.

Hall, C., & Macionis, N. (1998). Wine tourism in Australia and New Zealand. In R. Butler, M. Hall, & J. Jenkins (Eds.), *Tourism and recreation in rural areas* (pp. 197–224). Wiley.

INSETE. (2022). *Greek tourism 2030: Action plan, methodology.* INSETE.

Kasimoglu, M., & Aydin, H. (2012). *Strategies for tourism industry: Micro and macro perspectives.* IntechOpen.

Kim, H., & Fesenmaier, D. (2008). Persuasive design of destination Websites: An analysis of first impression. *Journal of Travel Research,* 3–13.

Lozano-Oyola, M., Blancas, F., González, M., & Caballero, R. (2012). Sustainable tourism indicators as planning tools in cultural destinations. *Ecological Indicators,* 659–675.

Mousavi, S., Doratli, N., Mousavi, S., & Moradiahari, F. (2016). Defining cultural tourism. In *International Conference on Civil, Architecture and Sustainable Development* (pp. 70–75). London (UK).

Richards, G. (2011). *Tourism trends: Tourism, culture and academia* (pp. 21–39).

Snyder, H. (2019). Literature review as a research methodology: An overview and guidelines. *Journal of Business Research,* 333–339.

Woodside, A., & Martin, D. (2007). *Tourism management: Analysis, behaviour, and strategy.* CABI.

Actual Conditions of Tourist Guides in the Covid-19 Pandemic: Ecuador Case Study

Angélica González-Sánchez, Fernanda Navas-Moscoso, Sylvia Andrade-Zurita, and Edgar Encalada-Trujillo

Abstract

This research on the labor reality of tourist guides in Ecuador is considered because with the arrival of the pandemic caused by covid-19, tourism has been one of the seriously affected sectors. Consequently, all positions related to this activity have been forced to stop providing their services. The objective of this research focuses on analyzing the current employment situation of tour guides in Ecuador, using a mixed-method methodology, through a questionnaire consisting of 25 questions with a Likert-type rating of 5 points, divided into 3 dimensions: (1) sociodemographic data, (2) employment status, and (3) glass ceiling; validated by experts and with a Cronbach's Alpha of 0.80, to a sample of 134 legally registered tour guides, the non-parametric Kolmogorov–Smirnov (K-S) test was used to identify the distribution of data within informality in tourism activity and occupational segregation affected professional development in times of pandemic. The results show that tour guides have faced economic, emotional, psychosocial, and political impacts, conditions that weaken the comprehensive well-being of all the tourism system. Also, the employment situation of tour guides has not been analyzed deeply and therefore this research explores the reality that tour guides have faced in times of pandemic. Finally, the economic impact that instability brings with it, one cannot fail to mention the changes that must be considered in the labor competencies of the guides, of tourism: protecting the mobility of small groups, respecting the allowed capacity, the management of the bio-security standards always, to promote rapid growth in proximity tourism with overall sustainability; even though the process of returning to work depends on exogenous factors.

Keywords

Tourism • Tourist guides • Labor competencies • Economy • Covid-19

1 Introduction

The tourism sector has been listed until the end of 2019 as the third most important item in the contribution of 2.2% to the direct contribution to Ecuador's GDP, registering that until December 2019 it employed more than 477,382 people, mainly related to accommodation and food and beverage establishments, representing 6.6% of the economically active population (Adame Obrador et al., 1998). However, as a result of the pandemic, 44% of tourism jobs were suspended; undoubtedly, the effects of the pandemic have several edges for analysis and in the case of the tourism system, very little has been studied about the current employment situation of tourist guides who are called the country's ambassadors and play an important role in the process to make visitors feel welcome in a tourist destination. What is more, this group of workers has been excluded in the analysis of the legal variables that their employment situation implies, a situation that has been most affected during the last year 2020—2021, who in turn have managed to generate resilience strategies to face a number of problems: economic, social, and emotional, which for many has become a very serious issue to continue or change jobs permanently.

A. González-Sánchez (✉) · S. Andrade-Zurita · E. Encalada-Trujillo
Facultad de Ciencias Humanas y de a Eduación, Universidad Técnica de Ambato, Ambato, Ecuador
e-mail: am.gonzalez@uta.edu.ec

S. Andrade-Zurita
e-mail: sylviajandradez@uta.edu.ec

E. Encalada-Trujillo
e-mail: eg.encalada@uta.edu.ec

F. Navas-Moscoso
Facultad de Turismo, Universidad Laica Eloy Alfaro de Manabí, Manta, Ecuador
e-mail: mariela.navas@uleam.edu.ec

© The Author(s), under exclusive license to Springer Nature Switzerland AG 2024
J. Chica-Olmo et al. (eds.), *Sustainable Tourism, Culture and Heritage Promotion*,
Advances in Science, Technology & Innovation, https://doi.org/10.1007/978-3-031-49536-6_3

The reduction in economic activities and the restrictions on the movement of people affect the services sector, understood as a set of operations that are directly or indirectly related to tourism or may influence in the provision of services to a tourist (Alrwajfah et al., 2020). According to the new UNWTO definition at the Conference Ottawa, 1993, tourism is understood as a set of activities carried out by people during their trips and stays instead of other than their usual environment, for a consecutive period of less than one year for leisure, business, and other reasons (Amador-Muñoz, 2016).

The tourism sector comprises a large umbrella of services such as hotels, restaurants, travel agencies tour guides, with complementary actors such as carriers, and artisans. Being one of the sectors with the greatest diversification which generate both direct and indirect jobs.

Tour guides are tourism professionals who are responsible for informing, assisting, and serving tourists during a visit or trip (Armas-Arias et al., 2021). It is the industry of professional tour guides, which has received a harsh coup due to the closure of international borders and the implementation of restrictions on spaces for mass attendance. Under this background, analyzing the current employment situation of tourism guides in Ecuador, exploring the different challenges they have faced during the first year of the pandemic caused by the presence of this virus, as long as to a distinction on the gap gender for women who decide to work as tour guides, invites to a critical reflection.

1.1 How Did Covid-19 Affect the Tourism Industry in Ecuador?

The tourism industry in Ecuador was significantly affected by the covid-19 pandemic. The Galapagos Islands economy, e.g., is highly dependent on tourism and the crisis has raised serious questions about the viability of the local fishing sector (Banerjee & Chua, 2020). The closure of national borders, the drop in tourists, and the loss of ties with the Ecuadorian mainland disrupted the food supply system of the archipelago, forcing households to reduce their consumption. The tourism sector represents an average 65.5% of Galapagos GDP, and strong linkages with other sectors of the Galapagos economy exacerbated the impact of the crisis. The overwhelming contribution of tourism to the local economy made it difficult to find another viable economic activity that could replace tourism in the short term (Blomstervik & Ottar, 2022).

The collapse of tourism in the archipelago had an amplified ripple effect due to the spillover effect, which stimulates other sectors of the economy. Likewise, the tourism industry in other parts of Ecuador was also negatively affected, with cities such as Manta and Salinas-Santa Elena being particularly affected, which receive a large number of tourists (Campos-Soria et al., 2011). The long-term impacts of the pandemic on the tourism industry in Ecuador remain to be seen. However, post-covid-19 tourists are expected to be more respectful of the environment and take more precautions, such as increased hygiene and disinfection practices in tourist establishments.

2 State of the Art

2.1 Structuring Your Paper

Due to the constant declarations of states of exception and confinement due to the presence of covid-19 and its variants in Ecuador until the first quarter of 2021, there is an unemployment rate of 5.8%, underemployment 22.6%, adequate employment 34%, and a global participation of 65.4% (Casado-Díaz & Simon, 2016). A situation that has fallen into a domino effect and that has become very "difficult to bear, especially for accommodation, recreation, tourism, food, and beverage services" (Ugarteche, 2020), a fact that in addition to the suspension and closing of business has driven migration and capital flight.

In this sense, the employment situation in tourism prior to the health crisis had the characteristics of being a great impulse coefficient and generator of many employment opportunities (CONGOPE, 2013), even two are identified typologies: (1) formal employment in tourism, which provides employees with a stable work environment, good social security, more training opportunities and, above all, the income of the entire group is relatively stable; and (2) informal employment in tourism, quite the opposite of the first and which is constantly affected by the low season and job instability, where employees are self-employed and face such problems as: low wages, long working hours, high intensity labor. and low social welfare. These social facts are precisely the characteristics of the tourist guides, who, in turn, do not feel represented by any union, college or association that safeguards their rights (Constitución de la República del Ecuador, 2008).

In relation to covid-19 and the labor reality of tourist guides in Ecuador, research is scarce, and the few that exist are inclined to the competitiveness of tourist guides in time of covid-19 (Ecuador, 2021), including in the design and content of the guide's online profile (Gössling et al., 2020).

The pandemic around the world has been a brake to tourist activity since the operators have stopped going to their destinations since the measures to avoid contagion are isolation and social distancing between people, aspect incompatible with the tourist activity which requires great social contact. The economic losses in the tourism sector are very high and the social consequences for those who work in it are very serious, since many of these activities have been temporarily or permanently closed. As tourism has been important in several countries and even the main economic activity, generating employment, by losing income, the losses have been disastrous not only for those directly involved but for society as a whole. It is an activity that stopped operating, but that needs to be re-established, adapting to the new normality, in order not to put at risk the health and lives of the people who work in this sector and the tourism consumers, as well (Grančay, 2020).

The pandemic highlighted the fragility of the tourism system in Ecuador, where the confinement followed parameters similar to those that occurred globally. This can be largely explained by the fact that, as it was an unprecedented event in recent history, there were no enough protocols to be followed (Hall & Brown, 2022).

The effects generated by covid-19 showed that there are territories that are better prepared than others for the development of this activity. In this sense, the territories in which this sector had been promoted, prioritizing the well-being of its inhabitants, managed to rely on complementary economic structures. Meanwhile, where there was predominantly enclave tourism, foreign investors abandoned those places, "tourist destinations" from which they had benefited (INEC, 2021).

2.2 Political and Territorial Management of Tourism in Ecuador

The planning of tourist activities in the territories allows for defining the course that must be followed by the different actors that are part of the sector, in this way, the real and potential resources (natural and cultural) for the tourist operation can be shown (INEC, 2020). Then, a brief tour of the territorial management of tourism in Ecuador is presented, from the first actions to the most solid current planning structures.

2.3 The Evolution of the Profile and Number of Tourists

During the last two decades, the growth in the number of tourists from abroad has grown steadily. However, and despite this encouraging data, questions arise as to whether the income, registered by immigration control, corresponds entirely to it. Since the country receives constant migratory flows of people fleeing from armed conflicts or others who are looking for work options. Despite this questioning, it is necessary to point out that, in recent years, the information regarding the profile of tourists arriving in Ecuador has improved quantitatively and qualitatively. Additionally, the Ministry of Tourism has more and better spaces for disseminating information, as is the case of the Tourist Information Display (Ioannides & Gyimóthy, 2020).

The year 2018 was the year with the highest growth, even significantly exceeding the figures corresponding to the high season months. However, the year and the figures coincide with the Venezuelan migration process, a situation that can be evidenced by the decrease that occurs in 2019. During the month of March 2020, due to the processes of confinement and closure of borders and airports, begins a dramatic drop in the number of tourists arriving in the country. Although the figures for October show a recovery, they are well below the historical numbers (Jomo & Chowdhury, 2021).

3 Methodology

The research adopted a mixed study approach, descriptive-exploratory, which allowed the integration of quantitative and qualitative results that were obtained by reviewing the literature and applying a structured questionnaire which was validated by experts in the area, in addition to the Cronbach Alpha statistic, obtaining a reliability of 0.80, an instrument composed of 25 questions with qualification 5-point Likert scale, divided into 3 dimensions that helped to identify the effects of the pandemic: (1) sociodemographic data, (2) employment status, and (3) glass ceiling; understood as the unwritten normative difficulties within a group that prevent women from occupying positions of high institutional hierarchy.

The scarce participation that women have in the operation and management of tour is services demanded by long-term trips (from 8 to 12 days), considering the difficulty of meeting certain requirements such as mastery of a third and fourth language; in addition to the concentration involved in reconciling family needs and restrictions as long as the working hours and social stereotypes are within the main reasons for this low female participation.

Before the presence of covid-19, a 50% of those surveyed expressed that the activity of tourist guidance was very important in the personal and family economy of this group of professionals, not so for the 25.37% who think it was important, a 17.91% neutral, while the 5.22% classify it as not very important, just as for a 1.49%, it was considered as not important.

4 Results

According to the sociodemographic characteristics of the 134 professionals surveyed in Ecuador, it was identified that 76.12% are men and 23.88% are women, who in turn are registered in the SEVEN system of the Ministry of Tourism, as national guides 60.45%, local 5.22%, specialized in adventure 28.36%, and in heritage 5.97%, whose origin is established in the cities of Quito 64%, Manta 8.95%, Cuenca 19.84%, Latacunga 6.72%, Baños 23% and in the Galapagos Islands 5.97%.

It is identified that the values of the three variables are very significant in the development of the research, as a result, the alternative hypothesis is accepted where it is concluded that the pandemic affects the employment situation of professional tourism guides. According to the results obtained, the covid-19 pandemic caused high impact effects on tourism specialists, and which brought with it a crisis: 70.15% economic, 15.67% emotional, 13.43% psychosocial, and 0.75% political; economic problems prevail, due to the fact that a large part of the tourist workers were left without employment due to the cancelations and/or postponement of trips to Ecuador, a situation that left 61.19% of the surveyed professionals immobile. While 20.15% of the professionals decided to replace their profession with an undertaking, such as trading, buying and/or selling bio-security products, 13.43% opted for service activities through mobile markets, cashiers, moto parcels, online trainers, among others, and 5.22% carried out manufacturing, evidencing their work in the preparation and marketing of food and handicrafts.

Prior to the health crisis, tour guides already faced problems and challenges related to seasonal work, a condition that negatively affects the number of jobs, social security, and disposable income, therefore, the current situation aggravated all professionals related to this area. Considering that 71.64% of those surveyed affirm that the activity is carried out for fees, commonly called "freelance", who are not obliged to comply with a specific schedule, therefore their work is carried out independently from an agreement express contractual between the employer and the specialist.

On the other hand, 18.66% have a temporary document, 5.22% work seasonally hired and 4.48% have a permanent contract. However, the latter group has had to take up new positions in order not to cease functions, considering this location, workers, despite the employability risks they face, have various types of personal insurance, as follows: life insurance (3.73%), accident (3.73%), private health (41.04%), while 50.75% do not have any kind of insurance, a situation that makes work precarious, transforming the worker into a vulnerable citizen within tourism development.

Another variable studied is the invisible barriers that women face in the labor field, the social role theory affirms that the stereotypes established for the two sexes are replicated by cultural traditions where the woman assumes roles within the home while the man is the family provider and therefore, the tourist guide is not the exception, at the national level. Despite the notable intervention gap where 72.93% of tourist activity is occupied by men and 27.07% by women, a 32.09% of the surveyed guides totally agree

that women enjoy the same legal rights as men in the exercise of tourist guidance. However, when hiring a tour guide, they prefer to bet on a man, in the same way 35.58% think they agree, while 18.66% neither agree nor disagree because it does not exist in the collective unconscious the thought of equality between men and women; unlike 11.94%, think they disagree, thus a 3.73% totally disagree.

Nevertheless, as a result of the declaration of the pandemic, tourism and with special attention, the workers have endured multiple impacts that according to a 78.36% of the participants did not receive any help from the Government, while the 13.43% indicate that in the first wave of confinement (March–May), they were trained on biosafety regulations, marketing, and other areas of tourism related to the virus prevention, on the other hand, the 3.73% received help to have an approved economic loan, a 2.99% had tax exceptions and finally, a 1.49% received food and economic aid.

The future of the profession is uncertain, however, the strategies generated to contribute to the economic reactivation of this activity, according to the guides surveyed, 75.37% consider that it depends on the alliances between tourist servers, while 16.42% favor the ease in access credit for entrepreneurship, and 8.21% request urgent changes in tourism governance, mainly in drastic sanctions for informality, reduction in rates and taxes and advertising campaigns with special emphasis on sustainable rural tourism.

It should be noted that the information gathering was carried out in the months of confinement (from April to June 2020), to a sample of 134 tourist guides, out of a total of 2926 professionals registered in the SIETE system of the Ministry of Tourism of Ecuador. The survey was applied through the non-probabilistic snowball sampling technique, considering a confidence level of 93% and a probability of success of 70%. In turn, the nonparametric Kolmogorov–Smirnov test was determined using the SPSS statistical program (Table 1).

5 Discussion

The employment situation of tourist guides in Ecuador and the rest of the countries, despite being classified as a cornerstone in the experience and satisfaction of visitors in the tourist destination is catastrophic because professionals worldwide were forced to suspend their activities, to the point that their income fell by more than 80% compared to the same period in 2019 (MINTUR, 2019).

In Ecuador, the situation is disproportionate in several areas such as formality, stability, remuneration, and equality of economic conditions in relation to gender which causes an imbalance in their well-being. In this order of ideas, the sectorization of the employed population according to the (Ministerio del Turismo, 2022) is divided

Table 1 Kolmogorov–Smirnov test

		Informality affects professional development	Women enjoy the same rights	Economic activities carried out in the first year of the pandemic
N		134	134	134
Normal parameters[a,b]	Half (mean)	4.38	3.78	3.93
	Typical deviation	1.010	1.133	1.087
More extreme differences	Absolute	0.372	0.232	0.249
	Positive	0.270	0.141	0.162
	Negative	− 0.372	− 0.232	− 249
Z Kolmogorov–Smirnov		4.306	2.691	2.882
Sig. Asintot (bilateral)		0.000	0.000	0.000

[a]The contrast distribution is Normal
[b]They have been calculated from the data

into the formal sector (45.9%); domestic work (1.9%); unqualified (3.6); informal sector (48.6%), in the last group are classified the fixed self-employed (55%), itinerant (43%), and occasional (2%).

The increase in informal employment worsened throughout 2020, particularly the statistics described are directly related to the results of this research, showing that in the first year of the pandemic, tourism specialists face the cancelation of trips and therefore their services were forced to look for different activity options from those they were used to, being classified into three groups: (1) commerce (20.15%), who made the purchase and sale of bio-security products; (2) service (13.43%) dedicated to being itinerant merchants, cashiers, parcel motorcyclists, online trainers, and (3) manufacturing (5.22%) people who through their skills dedicated themselves to prepare food and making handicrafts, that is, informal work until the first quarter of 2021.

Activity that prevails in an environment of social, political, and economic instability, with constitutional rights violated by the perfect and imperfect suspension of work in the state of confinement and has even violated the right to work continuity (Amador-Muñoz, 2016). Many businessmen in the tourism sector availed themselves of the Organic Law of Humanitarian Support (2020) (Montaner, 1998) for their benefit who chose to terminate employment contracts or reduce the working day to 50%, undermining the rights of the worker in reducing their remuneration, contribution to social security and there were even cases in which employees did not receive any compensation regardless of the number of years worked; becoming more acute the fact that in Ecuador only 9.70% of guides appear as dependent workers while 91.70% are freelance or independent specialists.

Undoubtedly, the economic situation in tourism has affected all human capital due to the actions described in the previous paragraph; in this sense, for those who work under dependency, the article 328 of the Constitución de la República del Ecuador (2008) states: "The remuneration will be fair, with a decent salary that covers at least the basic needs of the worker ...", a condition in which particularly the tour guides were the first to be disengaged from any labor agenda, leaving aside any humanitarian right that they could have embraced. A year has passed, and few guides have returned to their main activity—guiding—with certain variables related to changes in the tourism market based on bio-security regulations and allowed capacity considering that internal demand recovers faster than international (Organización para la Cooperación y el Desarrollo Económicos, 2020).

Another aspect of discussion is the glass ceiling and how these invisible barriers prevent women from accessing personal and professional growth, the average salary of male specialists in the tourism industry is 6.7% higher than that of women workers (Ormaza-González et al., 2021).

Although tourism employs more men than women, it is known for providing low-paid, low-status, low-skilled, precarious and seasonal work with few development opportunities for the second ones (Santos & Varejão, 2007). The results obtained from this research show an inequality which is fed by the social stigma "the negative consideration, the inferior level and the relative impotence that society collectively grants to workers who possess a particular characteristic" (Tian & Guo, 2021).

This classification is useful to understand the rationale behind one of the variables in this study, the glass ceiling, a metaphor that has been widely used in the literature on gender and management and in the pronounced increase in the specified differences in the payment that men and women gain. The participation of women in operational areas is evident in such areas as receptionists or waitresses of an accommodation establishment, and/or counter of an airline or travel agency because it requires limited experience or skill, reasons whereby the employer offers a seasonal job, low wages, and little training for professional development (Ugarteche, 2020). Therefore, it is well known that a limited number of job positions are held by male representatives, who are mainly responsible for tourism management or administrative and technical professional positions (Universidad Central del Ecuador, 2020; Viteri et al., 2022).

6 Discussion

The employment situation of tourist guides in Ecuador and the rest of the countries, despite being classified as a cornerstone in the experience and satisfaction of visitors in the tourist destination is catastrophic because professionals worldwide were forced to suspend their activities, to the point that their income fell by more than 80% compared to the same period in 2019 (INEC, 2020).

7 Conclusion

From the rapid spread of the covid-19 pandemic, and the closure of borders and airports, the suspension and/or cancelations of trips increased, affecting the employment situation of all participants of the tourism system, identifying that among the guides surveyed more than 61.19% were immobile, without economic income.

It is confirmed that the effects of the pandemic are multiple and among these, discrimination is an interesting factor that affects employment and the income gap between male and female tour guides, who were affected mainly in the following aspects: economic (70,15%), emotional (15.67%), psychosocial (13.43%), and political (0.75%), a condition that is detrimental to the overall well-being of the entire conglomerate, and therefore, it is not shared with the encouragement from some researchers who conclude that at the end of the health emergency, tourism will be more encouraging (Gössling et al., 2020).

Furthermore, the unfortunate labor, economic, and cultural reality does not contribute to the sustainable development goals, becoming a turning point for the 2030 agenda; likewise, being in disagreement with "unfair labor practices in the tourism sector" (Ioannides & Gyimothy, 2020), when a series of abuses are observed that are against any human being; despite the fact that this situation could become in an opportunity for change in favor of all citizens.

The women guides are the ones who have faced the worst part, due to the scarce integration in the formal work environment and the implicit fulfillment of their social role regarding the care of the family and the patients of covid-19,

in this space is where the close relationship that exists between education, culture and the economy is manifested because it is evident that not all professionals who are educated, break cultural barriers, and acquire an integral development consistent with the years of study and the effort made.

On the other hand, it is necessary to highlight that resilience and innovation have played a preponderant role in all those who have sought to overcome this economic, emotional, and psychosocial crisis, despite the fact that this means an increase in the informal employment rate throughout Ecuador, divided into self-employment and underemployment, in addition to widening the gaps in job instability causing greater uncertainty, not counting that the impact of informal tourism on income is negative, which is why 37.03% of those surveyed have decided take measures of temporary suspension of their tour guide license and/or the change of activity until the international arena returns to its normal course.

In addition to the economic impact that instability brings with it, one cannot fail to mention the changes that must be considered in the labor competencies of the tourism guides:

- Protecting the mobility of small groups, respecting the allowed capacity,
- Management of the bio-security standards at all times, in order to
- Promoting rapid growth in proximity tourism with sustainability prevailing.

A special factor on which a future research could be proposed would be the psychological aspect of the tourism professional as a latent sequel due to the effect of social isolation, the difficult economy and its relationship with the high items generated by the payment of house, private health, the high value of specific medicines to treat the collateral effects produced by covid-19, the death of close relatives and the hopelessness of days where tourism activity is strengthened for the benefit of all.

Acknowledgements The authors express their gratitude to all the tour guides who participated in the survey. Also, to the Technical University of Ambato, to the Research and Development Department (DIDE-UTA) for supporting the project "Impact of digital marketing in the reactivation of tourism in the province of Tungurahua post-covid-19" approved under resolution CONIN-2020-0322-R.

References

Adame Obrador, M. T., Colom Bauzà, J., Iglesia Mayol, B. D. L., Muntaner Guasp, J. J., Rodríguez Rodríguez, R. I., Rosselló Ramon, M. R., & Sureda García, I. (1998). Pràctica 1 i 2 de la llicenciatura de Psicopedagogia. Bienni 1998–2000.

Alrwajfah, M. M., Almeida-García, F., & Cortés-Macías, R. (2020). Females' perspectives on tourism's impact and their employment in the sector: The case of Petra, Jordan. *Tourism Management, 78*, 104069.

Amador-Muñoz, L. (2016). Mujer y medio ambiente: Una aproximación desde la acción socioeducativa. *Collectivus, 6*(1), 247–270.

Armas-Arias, S., Páez-Quinde, C., Ballesteros-Lopez, L., & López. Pérez, S. (2021). Decision trees for the analysis of digital marketing in the tourism industry: Tungurahua case study. In *Multidisciplinary International Congress on Science and Technology, CIT 2021*. *Quito*.

Banerjee, S., & Chua, A. Y. (2020). How alluring is the online profile of tour guides? *Annals of Tourism Research, 81*, 102887.

Blomstervik, I., & Ottar, S. (2022). Progress on novelty in tourism: An integration of personality, attitudinal and emotional theoretical foundations. *Tourism Management, 93*, 104574. https://doi.org/10.1016/j.tourman.2022.104574

Campos-Soria, J., Marchante-Mera, A., & Ropero-García, M. (2011). Patterns of occupational segregation by gender in the hospitality industry. *International Journal of Hospitality Management, 30*(1), 91–102.

Casado-Díaz, J. M., & Simon, H. (2016). Wage differences in the hospitality sector. *Tourism Management, 52*, 96–109.

CONGOPE. (2013). Caja de herramientas de apoyo a la gestión territorial del turismo. http://www.congope.gob.ec/wp-content/uploads/2014/08/Se%C3%B1aleticaTuristica-en-areas-rurales-17-07-2013.pdf

Constitución de la República del Ecuador. (2008, octubre 20). Disponible en: https://www.oas.org/juridico/pdfs/mesicic4_ecu_const.pdf

Ecuador. Ministerio de Turismo de Ecuador. Catastro Nacional de Guías de Turismo (2021).

Gössling, S., Scott, D., & Hall, C. M. (2020). Pandemics, tourism and global change: A rapid assessment of COVID-19. *Journal of Sustainable Tourism, Early Access*. https://doi.org/10.1080/09669582.2020.1758708

Grančay, M. (2020). COVID-19 and Central European tourism: The competitiveness of Slovak tourist guides. *Central European Business Review, 2020*(5), 81–98.

Hall, J., & Brown, K. (2022). Creating feelings of inclusion in adventure tourism: Lessons from the gendered sensory and affective politics of professional mountaineering. *Annals of Tourism Research, 97*, 103505. https://doi.org/10.1016/j.annals.2022.103505

INEC. (2020). Clasificación del empleo en Ecuador. Disponible en: https://www.ecuadorencifras.gob.ec/documentos/web-inec/EMPLEO/2020/ENEMDU_telefonica/Principales_Resultados_Mercado_Laboral.pdf

INEC. (2021). Indicadores nacionales en porcentaje (%) respecto a la PEA trimestral. Disponible en: https://www.ecuadorencifras.gob.ec/enemdu-trimestral-enero-marzo-2021/

Ioannides, D., & Gyimóthy, S. (2020). The COVID-19 crisis as an opportunity for escaping the unsustainable global tourism path. *Tourism Geographies, Early Access*. https://doi.org/10.1080/14616688.2020.1763445

Jomo, K. S., & Chowdhury, A. (2021). COVIDD-19 pandemic recession and recovery. *Development, 63*, 226–237.

Ministerio del Turismo. (30 de 03 de 2022). Ecuador. (Visualizador de Información Turística). https://servicios.turismo.gob.ec/visualizador

MINTUR. (2019). Ecuador. (Ministerio de Turismo del Ecuador). https://servicios.turismo.gob.ec/index.php/anuario-de-estadisticas-turisticas

Montaner, J., Antich, J., Corgos, A., & Arcarons, R. (1998). *Diccionario de Turismo*. Síntesis.

Organización para la Cooperación y el Desarrollo Económicos. (2020). Se vislumbra una recuperación económica mundial. Disponible en: https://www.oecd.org/perspectivas-economicas/

Ormaza-González, F., Castro-Rodas, D., & Statham, P. (2021). COVID-19 impacts on beaches and coastal water pollution at selected sites in Ecuador, and management proposals post-pandemic. *Frontiers in Marine Science, 8*.

Santos, L., & Varejão, J. (2007). Employment, pay and discrimination in the tourism industry. *Tourism Economics, 13*(2), 225–240.

Tian, J., & Guo, W. (2021). A study of the income difference between tourism formal and informal employment in China. *Journal of Hospitality and Tourism Management, 46*, 414–422.

Ugarteche, O. (2020). Covid-19: el comienzo del efecto dominó. Observatorio económico latinoamericano. Disponible en: http://www.obela.org/analisis/covid19-el-comienzo-del-efecto-domino

Universidad Central del Ecuador. (2020, 25 de junio). Debate sobre la situación del empleo turístico en Ecuador. [Comunicado de prensa]. https://repositorio.uce.edu.ec/archivos/jmsalazara/Boletines/Boletines2020/300/224.pdf

Viteri, C., Rodríguez, G., Tanner, M., Ramírez-González, J., Moity, N., Andrade, S., & Pittman, J. (2022). Fishing during the "new normality": Social and economic changes in Galapagos small-scale fisheries due to the COVID-19 pandemic. *Maritime Studies, 21*, 193–208.

Accessible Tourism Businesses as a Means of Promoting Sustainable Cultural Tourism in Thessaloniki, Greece

Dimitris Kourkouridis and Asimenia Salepaki

Abstract

Thessaloniki is a city with an important cultural heritage, which is recognized internationally. In addition to the cultural monuments which must be accessible to all, the accessibility of the city's tourism businesses is also important for the sustainable development of the cultural destination. Accessible tourism has been intensively researched in recent decades, and its relationship to the long-term sustainability of tourism development is well established. In an effort to promote the city's cultural heritage, a combination of accessible tourist facilities and cultural heritage sites should be offered. The purpose of the present research is to examine the accessibility of tourism businesses, specifically hotels and food and beverage businesses, in Thessaloniki. A quantitative survey was conducted in the area ($N = 74$), where it was found that the tourism businesses are in average condition in terms of their accessibility (Mean score = 2.77 out of 4) and staff training in serving people with disabilities (Mean score = 2.86 out of 5). The areas that need to be improved in order to achieve the goal of accessible cultural development in the city were also identified. The present research is considered to contribute significantly to the development of cultural sustainable tourism in Thessaloniki, as the accessibility of the city's tourism businesses has not been studied in the past.

Keywords

Accessible tourism • Cultural tourism • Sustainable tourism

D. Kourkouridis (✉) · A. Salepaki
School of Spatial Planning and Development, Aristotle University of Thessaloniki, Thessaloniki, Greece
e-mail: kourkouridis@plandevel.auth.gr

A. Salepaki
e-mail: salepaki@plandevel.auth.gr

1 Introduction

Disability is a phenomenon very prevalent in society and, in one way or another, familiar to all people, an element of their direct and indirect social environment. It is a fact that globally over a billion individuals are disabled in some way. This equates to around 15% of the global population, with up to 190 million (3.8%) people aged 15 and older experiencing substantial difficulties in functioning (W.H.O., 2021). Accessible tourism, i.e., the offer of tourism products for all, is a sensitive social issue of human rights, but at the same time, also a factor that contributes to the economy. Accessibility in tourism offers significant advantages as it attracts people with disabilities, but also other important segments of the market, such as the elderly. These two markets (accessible tourism and senior tourism) present significant prospects (Alén et al., 2012; Bowtell, 2015; Michopoulou et al., 2015).

The offer of accessible tourist products and services is an important pull factor for these groups of the population, while the cultural heritage of a destination is a significant push factor for senior travelers (Urbonavicius et al., 2017). Although this segment of the market is characterized by heterogeneity, research has shown that older travelers (65 +) show a great interest in educational and informational activities on vacation (Backman et al., 1999). It should be noted that older travelers often have mobility difficulties (German Federal Ministry of Economics and Technology, 2004).

The purpose of the present paper is to present the results of a research conducted to investigate the accessibility of tourism businesses in the city of Thessaloniki, as a contributing factor to the city's sustainable cultural tourism. The research specifically focuses on the physical accessibility of the city's hotels, restaurants-cafes and a small indicative sample of sports facilities. The focus on people with physical disabilities is mainly because they constitute the majority of the disabled population, including the elderly. According to research conducted by the German Federal Ministry of Economics and Technology (2004) on tourists with disabilities, more than half of the participants were over 65 years old, while the majority (more than 60%) had mobility difficulties.

2 Literature Review

2.1 Accessible Tourism

Accessible tourism refers to collaborative processes between stakeholders involved in the provision of services that enable people with disabilities to function independently and with dignity in the tourism services offered (Darcy et al., 2011). It is defined as (W.H.O., 2016, p. 18): '*tourism and travel that is accessible to all people, with disabilities or not, including those with mobility, hearing, sight, cognitive, or intellectual and psychosocial disabilities, older persons and those with temporary disabilities*'. Accessible tourism is of great concern to the scientific community and also to decision-makers, both because of the sensitivity of the issue regarding human rights, and because of the significant contribution of this particular segment of tourism to the economy (Ambrose et al., 2012).

Moreover, accessible tourism is an issue that has been extensively researched in recent decades, and its connection to the sustainability of tourism development is well known (Darcy et al., 2010; Qiao et al., 2021). The inclusion of people with disabilities in tourism is a key factor of equality and equity, two important factors of sustainable tourism development (Lwoga & Mapunda, 2017; Richards et al., 2010).

Sustainability and inclusiveness are strongly connected in a variety of ways (Heylighen, 2008), while the concept of social sustainability has arisen as a term that encompasses both inclusive and sustainable design (Ostroff, 2011).

Early approaches to disability followed the medical model, seeing it as the result of some physical impairment. However, today the perception that disability is a social construct is widespread. In fact, disability is created due to barriers that exist in the environment, and therefore, social action is needed to make the environment accessible to all people. This social action includes, among other things, the application of universal design to the built environment (Darcy et al., 2011).

Accessible tourism has grown from the idea of providing accommodations or adaptations so that persons with disabilities can participate in tourism, to a concept of quality tourism for all, with the knowledge that accessibility is an integral aspect of that quality. An accessible environment must take into account the needs of all categories of people and ensure that each of them can move and live as autonomously as possible, adapted to their capabilities to compensate for any weaknesses. Despite the fact that people with disabilities have the highest need for accessibility, accessible tourism caters to a wide range of customers with varying access requirements (not always visible), which can be caused by impairment, illness, injury, age, stature, foreign language proficiency, or culture (W.H.O., 2016).

The markets of accessible tourism and senior tourism present significant prospects, as can be seen in Table 1. It is a potentially major and profitable market (Bowtell, 2015), that

Table 1 Overall demand for people with disabilities and senior citizens (estimations)

Location	Estimate	% of the population
Overall demand for people with disabilities		
Global	From 600 to 859 million people	From 9 to 13
USA	Almost 54 million people	21
	From 50 to 80 million people	From 16 to 26
Canada	More than 2.7 million people	15.5
Australia	More than 3 million people	18
Europe	Around 45 million people in the EU-25 countries (aged 16–64 years)	15.7
	From 45 to 90 million people who have some type of disability	From 10 to 20
	50 million people in extended Europe	Aprox. 11
	From 69 to 92 million people	From 15 to 20
	From 60 to 80 million disabled and reduced mobility people	From 13 to 17
	From 92 to 115 million people	From 20 to 25
Senior citizens		
Global	About 600 million people (in 2025 it is estimated that it will be 1.6 billion people)	10
USA	About 31.88 million people	12.4
Europe	More than 85 million people	16.6

Source Alén et al. (2012, p. 148–149)

is expected to increase in the coming years due to the increase in aging population (Bloom & Luca, 2016), while it is also expected to have an impact on the future competitiveness of tourism destinations (Michopoulou et al., 2015). Research shows that accessible tourism can make a significant contribution to reducing the seasonality of tourism and to increasing occupancy (Machado, 2020). Research also shows that even in the case of small tourist markets, accessibility contributes positively to the economic development and demographic resilience of the regions (Ibanescu et al., 2020).

In many countries, there are currently specific guidelines for serving people with disabilities in the tourism sector. The guidelines apply in Greece are the designing instructions 'Designing for Everyone' of the Ministry of Environment, Energy and Climate Change. The main goal of these instructions is to remove architectural barriers and anything that prevents the autonomous movement and living of people with disabilities—and people with reduced mobility in general—inside buildings or outdoors. It should be noted that these guidelines are consistent with findings in the literature on external barriers faced by people with disabilities (Turco et al., 1998). However, there are still some barriers and challenges in the tourism sector, such as inadequate and unsuitable facilities, but also insufficiently prepared and trained staff (Darcy, 1998; Ozturk et al., 2008). The lack of physical access to tourism infrastructure is the biggest deterrent to the decision of people with disabilities to travel, but also the staff attitudes, as they are the direct contacts of people with disabilities in their tourist experience (Avis et al., 2005; Bi et al., 2007; Darcy, 1998).

2.2 The Relationship Between Accessible and Cultural Tourism

Culture and tourism are two completely related concepts. The cultural elements of a destination are key motivations for traveling, while the travel itself is a generative cause of culture (Richards, 2018). Cultural tourism is characterized as experiential, as it is based on visiting but also engaging with the cultural heritage of the destination (Lwoga & Mapunda, 2017).

Older visitors seem to be more interested in visiting cultural attractions (Backman et al., 1999), while cultural sites are among the most important attractions they wish to visit in a destination (Zielińska-Szczepkowska, 2021), which makes the accessibility of both the attractions and the entire tourist experience (accommodation, restaurants, etc.) important. Although the senior traveler market is highly heterogeneous (Alén et al., 2017), many surveys report a segment of this market that is attracted by the archeological/historical/cultural attractions in a destination (Huang & Tsai,

2003; Norman et al., 2001; Sangpikul, 2008). Moreover, it should be noted that older visitors are those who value the place attributes of a destination as the most important obstacles when they travel (Nyaupane & Andereck, 2008), therefore physical accessibility is a factor that contributes to sustainable cultural tourism development of a destination. For the tourism industry, and cultural tourism in particular, to have a more sustainable future, exclusion and discrimination against people with disabilities should be reduced, attracting them into the industry in a way that benefits them (Darcy et al., 2020). Thus, there is a constant need to address the lack of accessibility of the sector that creates social inequality (Gillovic & McIntosh, 2020).

In the literature, there are researches that investigate the physical accessibility of cultural areas, as a pillar of sustainable cultural tourism development. For example, Lwoga and Mapunda (2017) studied the accessibility of the Village Museum site in Tanzania, using the barrier approach. Among the factors, they considered are parking spaces, entrance, reception, corridors, toilets, rest areas and other physical aspects of the site. They also examined the service provided to people with disabilities and especially by the site's interpreters. The results showed that cultural tourism sites have significant difficulties in serving people with disabilities as visitors. Accordingly, museums have been studied in relation to their accessibility, as, e.g., in Wiastuti et al. (2018) survey, where among others, parking areas, signage, wheelchair access, elevator accessibility, toilets, etc., were investigated. The results showed the need to improve the accessibility of the museums examined. Of particular interest is the research by Naniopoulos and Tsalis (2015), regarding the accessibility of the monuments of Thessaloniki. The researchers identified important accessibility problems and formulated specific proposals to improve the accessibility of the monuments of Thessaloniki.

3 Methodology

3.1 Study Area

The study area of the research is the city of Thessaloniki, which has a metropolitan character as the second largest city of Greece (Frangopoulos et al., 2009) after the country's capital (Athens), with a population of 1,091,424 (Greek Statistical Authority, 2022). It is a city with significant cultural heritage and history. In the area of today's Thessaloniki, there was an organized human presence, which developed settlements and small towns from prehistoric times. Also, the Roman monuments of the city are important. In 305 AD the Caesar Galerius Valerius Maximianus settles in the city and the most important public Roman buildings are built in it, such as the famous Galerian Complex which included the palace of Galerius (Navarino square), the Rotunda, which was connected to the palace with the triumphal arch (Camara), the Hippodrome, the Ancient Market, etc. The Galerian Complex, the most important monumental complex of Thessaloniki, was built at the crossroads of two worlds, the Roman and the Byzantine, while defining the urban fabric of modern Thessaloniki in this area. During the Byzantine period, Thessaloniki reached the peak of its glory with significant economic, military, political, cultural and also religious importance for the Byzantine Empire. In the Byzantine era, the city developed further and became even larger, becoming the second most important city of the empire after Constantinople itself. Magnificent churches and other buildings were built in several areas, while extensive defensive construction took place. Also, the history of the city is parallel and connected to its walls. Quite large sections of the city walls are preserved to this day in many places and their size testifies to the greatness of their Byzantine splendor. Eptapyrgio, which dominates the north-eastern end of the walls, is one of the most emblematic fortress complexes of the Balkans. In 1430, the Ottoman period began for Thessaloniki. The city now exudes the air of an Eastern city with mosques, covered markets and hammams. The most emblematic monument of the city, the White Tower, was built by the Ottomans. In the course of about five centuries, the history, society, architecture of the city was profoundly affected, preserving to this day the signs of the Ottoman presence in its streets (Papagiannopoulos, 1995).

Today Thessaloniki is an established and highly developed tourism destination, as it is one of the most competitive and popular City Break destinations at national level due to its significant resources, infrastructure and points of interest (second largest city in the country, climate, air and road connectivity, proximity to the Balkan countries, urban culture, landscape that combines mountain and sea, commercial market, etc.). Also, in recent years, Thessaloniki has actively participated in the organization of large international conferences and exhibitions (Kourkouridis et al., 2019; Tsiftelidou et al., 2017). Its strategic location, ease of access by all means of transport, quick movement within the city, conference and hotel infrastructure and rich cultural heritage make it a force to be reckoned with in the conference tourism industry. The city

offers a total of 8217 rooms in 145 hotels (Table 2) of various categories as well as guesthouses, while it has the most 5* hotels of any other destination in Greece. In order to meet the global demand for MICE tourism, conference rooms and exhibition halls have been integrated into many hotels in Thessaloniki (Hellenic Chamber of Hotels, 2021; INSETE, 2022; Kourkouridis et al., 2018).

In 2017, 15.2% of the country's food and beverage service activities are located in the Region of Central Macedonia. This percentage ranks it in second place behind the Attica Region (22.0%). Compared to 2013, the number of food and beverage businesses increased by 5.7% (INSETE, 2022).

3.2 Data Collection and Analysis

The research sample consists of tourism businesses operating in the study area and specifically, it was chosen to include hotels, food and beverage service activities (restaurants—cafes) and a small indicative sample of sports facilities. Random sampling was used using an electronic questionnaire, which was distributed to tourism businesses in the area via e-mail. The number of businesses included in the survey was 74 (34 accommodation, 34 food and beverage, 4 sports facilities and 2 a combination of the above).

The research tool, i.e., the research questionnaire, was designed based on designing instructions 'Designing for Everyone' of the Ministry of Environment, Energy and Climate Change. Tourism businesses are included in Chapter: 'Buildings used for the public' of the instructions. Based on these guidelines, 21 questions were formulated, which aim to collect information about the extent to which tourism businesses in Thessaloniki are in compliance with these guidelines. Specifically, the main variables examined are those presented in Table 3.

The tools of descriptive statistics were used to capture the results of the research, followed by inductive statistics tests. In particular, the Chi-square test of independence statistical hypothesis test was applied, which is nonparametric and is used to control two categorical (nominal) variables (Kourkouridis, 2021). The p-value was calculated using the statistical package IBM SPSS Statistics.

Table 2 Hotels in the study area

Region	Region unit		5*	4*	3*	2*	1*	Total
Central Macedonia	Thessaloniki	Units	19	32	42	23	29	145
		Rooms	2314	2166	2202	775	760	8217
		Beds	4342	4060	4293	1469	1571	15.735

Source Hellenic Chamber of Hotels (2021)

Table 3 Variables

Variables facilities	
Accommodation businesses	
1	Rooms for People with Disabilities (PwD)
Food and beverage (Restaurants–cafes)	
2	Accessibility table–seats
All categories	
3	Special provision at the entrance
4	Wheelchair free movement
5	Parking for People with Disabilities (PwD)
6	Accessible elevator
7	Toilets for People with Disabilities (PwD)
8	Accessible website
Variables serving People with Disabilities (PwD)	
All categories	
9	Legislation awareness
10	Staff knowledge
11	Staff experience

4 Results

4.1 General Information

Table 4 presents the general information on the tourism businesses that participated in the survey. It appears that many of the businesses employ a large number of staff (43.2% more than 15 employees), they have been operating for several years (51.4% for more than 15 years), and their clients come from both Greece and abroad (67.6%). 45.9% of businesses belong to the accommodation category, 45.9% to the food and beverage category, 5.4% to sports activities and 2.7% is a combination of the above.

Table 4 General information

		Frequency	Percent		Cumulative percent
Accessibility					
Number of employees	Less than 5	20	27.0		27.0
	From 5 to 10	12	16.2		43.2
	From 11 to 15	10	13.5		56.8
	More than 15	32	43.2		100.0
	Total	74	100.0		
Years of operation	Less than 5	12	16.2		16.2
	From 5 to 10	18	24.3		40.5
	From 11 to 15	6	8.1		48.6
	More than 15	38	51.4		100.0
	Total	74	100.0		
Clients' origin	Greece	22	29.7		29.7
	Abroad	2	2.7		32.4
	Both	50	67.6		100.0
	Total	74	100.0		
Type of business	Accommodation	34	45.9		45.9
	Food and beverage	34	45.9		91.9
	Sports activities	4	5.4		97.3
	Combination	2	2.7		100.0
	Total	74	100.0		

4.2 Descriptive Statistics

The factors that we used in order to assess the accessibility of the tourism businesses are the following: (a) accessibility of the entrances, (b) wheelchair free movement inside the business, (c) special parking spot(s) for people with disabilities, (d) accessible elevators, (e) toilets for people with disabilities, and (f) website accessibility. The 'Accessibility' factor emerged from these factors. The 'Accessibility' score is on a scale from 0 (not accessible) to 4 (accessible). The factors that we used in order to assess the staff training in serving people with disabilities are the following: (a) legislation awareness, (b) staff knowledge, and (c) staff experience. The 'Staff' factor emerged from these factors. The 'Staff' is on a scale from 1 (not at all good) to 5 (excellent). Table 5 gives the descriptive statistics results.

Regarding accommodation businesses, we found that the majority (76.5%) have at least one room specially configured for disabled people, although there is a notable percentage (23.5%) that do not have any disabled room. In food and beverage businesses, 64.7% stated that they took into account the accessibility of the table–seats, 11.8% that they did not take it into account, while the percentage that does not know is significant (23.5%).

The majority of businesses (75.7%) stated that the business entrance is accessible, although there is a significant percentage of 24.3% that is not accessible. Inside the majority of businesses (82.4%), wheelchairs can move freely, while in 73.9% of businesses that do have a parking there is at least one parking space for people with disabilities. 69.7% of businesses that do have an elevator are accessible, while the percentage that is not accessible is quite significant (30.3%). It is noteworthy that in more than half of the businesses examined (56.8%), there is no toilet accessible to disabled people. Finally, the majority of businesses (64.7%) do not have an accessible website. The Accessibility Factor resulting from the above variables has an average score of 2.77 (out of 4).

As for business staff, 64.8% know the legislation for disabled people, even a little, while their knowledge of serving disabled people is at a moderate level. Their experience in serving people with disabilities also seems to be at a moderate level. The Staff Factor resulting from the above variables has an average score of 2.86 (out of 5).

4.3 Inductive Statistics

Table 6 gives the results obtained from the Chi-square test of independence statistical hypothesis for the factors: 'Accessibility' and 'Staff'. The analysis is related to the association of these factors with the following three variables:

(a) number of employees, (b) years of operation and (c) type of business.

We see that there is a statistically significant relationship ($p < 0.05$) between the variables: (a) Accessibility–Years of operation, (b) Accessibility–Type of business and (c) Staff–Type of business (bold in Table 6). Regarding the Accessibility Factor, 78.3% of businesses that scored 3 and 52.2% that scored 4 have been operating for more than 15 years, while 66.7% that scored 1 have been operating for less than 5 years (Fig. 1). Also, 69.6% of businesses that scored 4 and 52.2% that scored 3 are accommodation businesses, while all businesses with a score of 1 are food and beverage businesses (Fig. 2). However, regarding the Staff Factor, 66.7% of businesses that scored 4 are food and beverage businesses, while 75% of those scored 2 are accommodation businesses (Fig. 3).

5 Discussion

The results of the present research showed that the situation of the tourism businesses of Thessaloniki, in terms of their accessibility, is at a moderate level. In terms of physical accessibility, improvements should be made in relation to disabled toilets, elevator accessibility, entrance accessibility, while businesses also seem to be lagging behind on the accessibility of their websites. It should be noted that the lack of accessible toilets, entrances and elevators unfortunately leads to the complete exclusion of people with mobility disabilities from the tourism businesses of Thessaloniki. Also, the staff factor seems to be at moderate levels, in terms of serving people with disabilities. Examining the relation of the Accessibility Factor and other variables of the survey, we found that the more accessible tourism businesses are operating more than 15 years and they are accommodation businesses. However, regarding the Staff Factor, food and beverage businesses seem to achieve higher scores.

Addressing the sustainable development goals in tourism for all, including people with disabilities, is a major challenge (Darcy et al., 2020), while of particular importance is the need to provide clear information about accessible tourist facilities, and the training of employees in the special needs of these individuals (Loi & Kong, 2015). The findings of the present study are in line with the literature that highlights the need for more accessible cultural tourism (Lwoga & Mapunda, 2017; Naniopoulos & Tsalis, 2015; Wiastuti et al., 2018). As Gillovic and McIntosh (2020, p. 9722) point out: 'we still have some way to go'.

In order to achieve the sustainable cultural tourism development of Thessaloniki, the specific barriers that the present research revealed should be completely eliminated. Of course, a comprehensive accessible cultural tourism experience for visitors should be ensured. Therefore, the

Table 5 Descriptive statistics results

		Frequency	Percent	Valid percent	Cumulative percent
Accessibility					
Accommodation businesses					
Rooms for people with disabilities	Not accessible	8	10.8	23.5	23.5
	Accessible (at least 1 room)	26	35.1	76.5	100.0
	Total	34	46.0	100.0	
Food and beverage businesses					
Accessibility table–seats	Not accessible	4	5.4	11.8	76.5
	Accessible	22	29.7	64.7	64.7
	Don't know	8	10.8	23.5	100.0
	Total	34	46.0	100.0	
All categories					
Accessible entrance	Not accessible	18	24.3	24.3	24.3
	Accessible	56	75.7	75.7	100.0
	Total	74	100.0	100.0	
Wheelchair free movement	Not accessible	12	16.2	17.6	17.6
	Accessible	56	75.7	82.4	100.0
	Total	68	91.9	100.0	
Parking for PwD	Not accessible	6	8.1	26.1	26.1
	Accessible	17	23.0	73.9	100.0
	Total	23	31.1	100.0	
Accessible elevator	Not accessible	10	13.5	30.3	30.3
	Accessible	23	31.1	69.7	100.0
	Total	33	44.6	100.0	
Toilet for PwD	Not accessible	42	56.8	56.8	56.8
	Accessible	32	43.2	43.2	100.0
	Total	74	100.0	100.0	
Website accessibility	Not accessible	22	29.7	64.7	64.7
	Accessible	12	16.2	35.3	100.0
	Total	34	45.9	100.0	
Accessibility factor	*N*	**Range**	**Minimum**	**Maximum**	**Mean**
	74	3	1	4	2.77
Staff					
All categories					
Legislation awareness	I do not know if anyone knows the law	16	21.6	21.6	21.6
	Yes, very well	22	29.7	29.7	51.4
	Yes, moderate	14	18.9	18.9	70.3
	Yes, a little	12	16.2	16.2	86.5
	No	10	13.5	13.5	100.0
	Total	74	100.0	100.0	

(continued)

Table 5 (continued)

		Frequency	Percent	Valid percent	Cumulative percent
Staff knowledge	Excellent	0	0.0	0.0	0.0
	Very good	8	10.8	10.8	10.8
	Moderate	16	21.6	21.6	32.4
	A little good	24	32.4	32.4	64.9
	Not at all good	26	35.1	35.1	100.0
	Total	74	100.0	100.0	
Staff experience	Excellent	6	8.1	8.1	8.1
	Very good	10	13.5	13.5	21.6
	Moderate	26	35.1	35.1	56.8
	A little good	26	35.1	35.1	91.9
	Not at all good	6	8.1	8.1	100.0
	Total	74	100.0	100.0	
Staff factor	N	**Range**	**Minimum**	**Maximum**	**Mean**
	74	3	1	4	2.86

Table 6 Chi-square test results

Pearson chi-square	Value	df	Asymp. Sig. (2-sided)
Accessibility			
Accessibility * Number of employees	18.989	9	0.125
Accessibility * Years of operation	37.942	9	**< 0.001**
Accessibility * Type of business	30.151	9	**< 0.001**
Staff			
Staff * Number of employees	12.323	9	0.196
Staff * Years of operation	25.435	9	**0.003**
Staff * Type of business	15.241	9	0.085

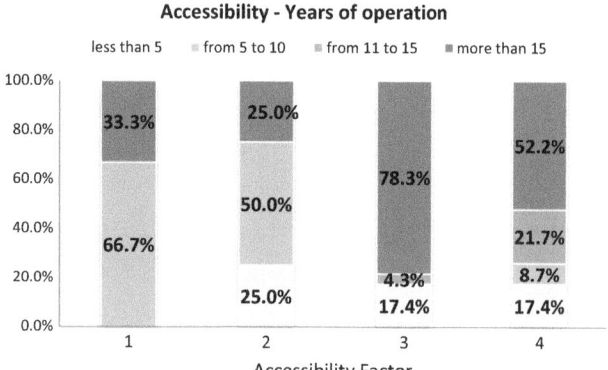

Fig. 1 Crosstab accessibility factor–years of operation

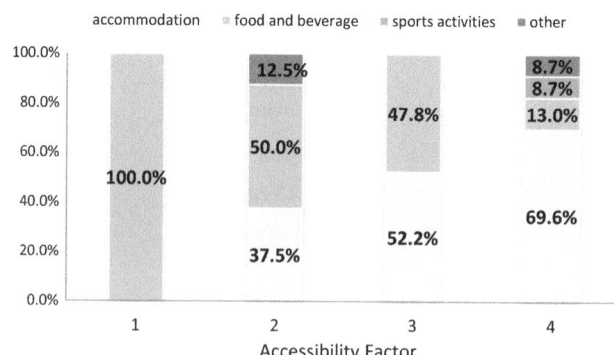

Fig. 2 Crosstab accessibility factor–type of business

scope of this research, combined with the findings of the research of Naniopoulos and Tsalis (2015) on the monuments of Thessaloniki, can contribute significantly to this integrated approach of sustainable cultural tourism. It is a fact that there are many important cultural monuments and

sites in Thessaloniki and the overall accessibility experience of visitors is of great importance in order for all to reap the benefits of culture without discrimination. Naniopoulos and Tsalis (2015) found difficulties in the movement of people with limited mobility in many of the city's monuments, a fact

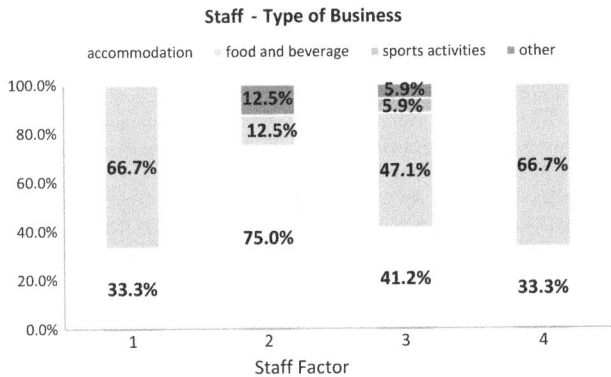

Fig. 3 Crosstab staff factor–type of business

which, combined with the shortcomings we identified in the present research, leads to the conclusion that significant efforts are needed to ensure the cultural sustainable tourism development of the city.

However, it should be pointed out that accessible tourism does not refer only to the physical accessibility issues studied in this research, nor only to people with mobility difficulties. Research based on the experience of tourists with disabilities has shown that tourism accessibility is a multifactorial issue (Yau et al., 2004), while some researchers argue that each person is unique in their abilities and preferences (Buhalis & Michopoulou, 2011). But providing a barrier-free experience regarding physical accessibility is definitely one of the most important parameters of this multifaceted relationship between people with disabilities and their satisfaction with the tourist experience.

6 Conclusion

The research on the accessibility of tourism businesses in Thessaloniki revealed significant room for improvement. These improvements, combined with improvements in the accessibility of the city's cultural monuments, can contribute to the sustainable cultural tourism development of the city. However, the limitations of the research, i.e., the focus only on the physical accessibility of people with mobility problems, creates the need for further investigation of the subject, so as to include all dimensions of accessibility and the varied needs of people with disabilities.

References

Alén, E., Domínguez, T., & Losada, N. (2012). New opportunities for the tourism market: Senior tourism and accessible tourism. In M. Kasimoglu (Ed.), *Visions for global tourism industry: Creating and sustaining competitive strategies* (pp. 139–166). BoD–Books on Demand.

Alén, E., Losada, N., & de Carlos, P. (2017). Profiling the segments of senior tourists throughout motivation and travel characteristics. *Current Issues in Tourism, 20*(14), 1454–1469. https://doi.org/10.1080/13683500.2015.1007927

Ambrose, I., Darcy, S., & Buhalis, D. (2012). Introduction. In D. Buhalis, Dimitrios, S. Darcy, & I. Ambrose (Eds.), *Best practice in accessible tourism: Inclusion, disability, ageing population and tourism* (pp. 1–15). Channel View Publications.

Avis, A. H., Card, J. A., & Cole, S. T. (2005). Accessibility and attitudinal barriers encountered by travelers with physical disabilities. *Tourism Review International, 8*(3), 239–248. https://doi.org/10.3727/154427205774791591

Backman, K. F., Backman, S. J., & Silverberg, K. E. (1999). An investigation into the psychographics of senior nature-based travelers. *Tourism Recreation Research, 24*(1), 13–22. https://doi.org/10.1080/02508281.1999.11014853

Bi, Y., Card, J. A., & Cole, S. T. (2007). Accessibility and attitudinal barriers encountered by Chinese travelers with physical disabilities. *International Journal of Tourism Research, 9*(3), 205–216. https://doi.org/10.1002/jtr.603

Bloom, D. E., & Luca, D. L. (2016). The global demography of aging: Facts, explanations, future. In J. Piggott & A. Woodland (Eds.), *Handbook of the economics of population aging* (Vol. 1, pp. 3–56). Elsevier.

Bowtell, J. (2015). Assessing the value and market attractiveness of the accessible tourism industry in Europe: A focus on major travel and leisure companies. *Journal of Tourism Futures, 1*(3), 203–222. https://doi.org/10.1108/JTF-03-2015-0012

Buhalis, D., & Michopoulou, E. (2011). Information-enabled tourism destination marketing: Addressing the accessibility market. *Current Issues in Tourism, 14*(2), 145–168. https://doi.org/10.1080/13683501003653361

Darcy, S. (1998). *Anxiety to access: Tourism patterns and experiences of New South Wales people with a physical disability.* Tourism New South Wales.

Darcy, S., Cameron, B., & Pegg, S. (2010). Accessible tourism and sustainability: A discussion and case study. *Journal of Sustainable Tourism, 18*(4), 515–537. https://doi.org/10.1080/09669581003690668

Darcy, S., Ambrose, I., Schweinsberg, S., & Buhalis, D. (2011). Conclusion: Universal approaches to accessible tourism. In D. Buhalis & S. Darcy (Eds.), *Accessible tourism, concepts and issues* (pp. 300–316). Channel View Publications.

Darcy, S., McKercher, B., & Schweinsberg, S. (2020). From tourism and disability to accessible tourism: A perspective article. *Tourism Review, 75*, 140–144. https://doi.org/10.1108/TR-07-2019-0323

Frangopoulos, I., Dalakis, N., & Kourkouridis, D. (2009). Urban structure and mobility in a modern city: The example of the submerged tunnel of Thessaloniki through the empirical investigation of the citizens opinion. *International Journal of Sustainable Development and Planning, 4*(4), 333–344. https://doi.org/10.2495/SDP-V4-N4-333-344

German Federal Ministry of Economics and Technology. (2004). *Economic impulses of accessible tourism for all.*

Gillovic, B., & McIntosh, A. (2020). Accessibility and inclusive tourism development: Current state and future agenda. *Sustainability, 12*(22), 9722. https://doi.org/10.3390/su12229722

Greek Statistical Authority. (2022). Results of ELSTAT 2021 population—households census (in Greek). https://elstat-outsourcers.statistics.gr/Census2022_GR.pdf

Hellenic Chamber of Hotels. (2021). Hotels in Greece 2021—by region—prefecture—Island. Available at: https://www.grhotels.gr/ksenodocheiako-dynamiko-elladas-2021-ana-perifereia-nomo-nisi/. Last visited 22 February, 2023 (in Greek).

Heylighen, A. (2008). Sustainable and inclusive design: A matter of knowledge? *Local Environment, 13*(6), 531–540. https://doi.org/10.1080/13549830802259938

Huang, L., & Tsai, H. (2003). The study of senior traveler behavior in Taiwan. *Tourism Management, 24*(5), 561–574. https://doi.org/10.1016/S0261-5177(03)00008-6

Ibanescu, B. C., Eva, M., & Gheorghiu, A. (2020). Questioning the role of tourism as an engine for resilience: The role of accessibility and economic performance. *Sustainability, 12*(14), 5527.

INSETE. (2022). *Greek tourism 2030, action plans. Region of Central Macedonia.* Institute of the Association of Greek Tourism Enterprises (INSETE) (in Greek).

Kourkouridis, D. (2021). *The socio-spatial dimension of Exhibitions, the host cities and their interactive dynamics—The case study of Thessaloniki* [Dissertation, School of Spatial Planning and Development, Aristotle University of Thessaloniki] (in Greek).

Kourkouridis, D., Dalkrani, V., Pozrikidis, K., & Frangopoulos, I. (2018). Hosted Buyers Program (HBP)—Tourism development and the city TIF-HELEXPO HBP for the period 2014–2016. In *Innovative approaches to tourism and leisure* (pp. 537–551). Springer.

Kourkouridis, D., Dalkrani, V., Pozrikidis, K., & Frangopoulos, Y. (2019). Trade fairs, tourism and city: Thessaloniki international fair and the concept of honored countries. *Tourismos, 14*(2), 30–56. https://doi.org/10.26215/tourismos.v14i2.571

Loi, K. I., & Kong, W. H. (2015). People with disabilities (PwD) in the tourism industry-concepts and issues. In *Critical tourism studies conference VI "10 years CTS: Reflections on the road less travelled and the journey ahead", Opatija, Croatia, 26–30 June 2015.* Critical Tourism Studies.

Lwoga, N. B., & Mapunda, B. B. (2017). Challenges facing accessible tourism in cultural heritage sites: The case of Village Museum in Tanzania. *Revista de Turism-Studii di Cercetari in Turism, 24,* 45–54.

Machado, P. (2020). Accessible and inclusive tourism: Why it is so important for destination branding? *Worldwide Hospitality and Tourism Themes, 12*(6), 719–723. https://doi.org/10.1108/WHATT-07-2020-0069

Michopoulou, E., Darcy, S., Ambrose, I., & Buhalis, D. (2015). Accessible tourism futures: The world we dream to live in and the opportunities we hope to have. *Journal of Tourism Futures, 1*(3), 179–188. https://doi.org/10.1108/JTF-08-2015-0043

Ministry of Environment, Energy and Climate Change. *Designing instructions 'designing for everyone'.* Athens (in Greek).

Naniopoulos, A., & Tsalis, P. (2015). A methodology for facing the accessibility of monuments developed and realised in Thessaloniki, Greece. *Journal of Tourism Futures, 1*(3), 240–253. https://doi.org/10.1108/JTF-03-2015-0007

Norman, W. C., Daniels, M. J., Mcguire, F., & Norman, C. A. (2001). Wither the mature market: An empirical examination of the travel motivations of neo-mature and veteran-mature markets. *Journal of Hospitality & Leisure Marketing, 8*(3/4), 113–130. https://doi.org/10.1300/J150v08n03_08

Nyaupane, G. P., & Andereck, K. L. (2008). Understanding travel constraints: Application and extension of a leisure constraints model. *Journal of Travel Research, 46*(4), 433–439. https://doi.org/10.1177/0047287507308325

Ostroff, E. (2011). Universal design: An evolving paradigm. In W. Preiser & K. H. Smith (Eds.), *Universal design handbook* (2nd ed., pp. 1.3–1.11). McGraw-Hill.

Ozturk, Y., Yayli, A., & Yesiltas, M. (2008). Is the Turkish tourism industry ready for a disabled customer's market?: The views of hotel and travel agency managers. *Tourism Management, 29*(2), 382–389. https://doi.org/10.1016/j.tourman.2007.03.011

Papagiannopoulos, A. (1995). *History of Thessaloniki.* Rekkos Publications.

Qiao, G., Ding, L., Zhang, L., & Yan, H. (2021). Accessible tourism: A bibliometric review (2008–2020). *Tourism Review, 77*(3), 713–730. https://doi.org/10.1108/TR-12-2020-0619

Richards, G. (2018). Cultural tourism: A review of recent research and trends. *Journal of Hospitality and Tourism Management, 36,* 12–21. https://doi.org/10.1016/j.jhtm.2018.03.005

Richards, V., Pritchard, A., & Morgan, N. (2010). (Re)Envisioning tourism and visual impairment. *Annals of Tourism Research, 37,* 1097–1116. https://doi.org/10.1016/j.annals.2010.04.011

Sangpikul, A. (2008). Travel motivations of Japanese senior travellers to Thailand. *International Journal of Tourism Research, 10,* 81–94. https://doi.org/10.1002/jtr.643

Tsiftelidou, S., Kourkouridis, D., & Xanthopoulou-Tsitsoni, V. (2017). Assessment of impact-contribution of cultural festival in the tourism development of Thessaloniki. In *Tourism, culture and heritage in a smart economy* (pp. 411–424). Springer.

Turco, D. M., Stumbo, N., & Garncarz, J. (1998). Tourism constraints for people with disabilities. *Parks and Recreation-West Virginia, 33,* 78–85.

Urbonavicius, S., Palaima, T., Radaviciene, I., & Cherian, J. (2017). Push and pull factors of senior travelers: The lingering influence of past restrictions. *Trziste/Market, 29*(1), 93–108. https://doi.org/10.22598/mt/2017.29.1.93

W.H.O. (2016). *Manual on accessible tourism for all: Principles, tools and best practices—Module I: Accessible tourism—Definition and context.* World Health Organization.

W.H.O. (2021). *Disability and health.* World Health Organization.

Wiastuti, R. D., Adiati, M. P., & Lestari, N. S. (2018). Implementation of accessible tourism concept at museums in Jakarta. In *IOP conference series: Earth and environmental science* (Vol. 126, No. 1, p. 012061). https://doi.org/10.1088/1755-1315/126/1/012061

Yau, M. K. S., McKercher, B., & Packer, T. L. (2004). Traveling with a disability: More than an access issue. *Annals of Tourism Research, 31*(4), 946–960. https://doi.org/10.1016/j.annals.2004.03.007

Zielińska-Szczepkowska, J. (2021). What are the needs of senior tourists? Evidence from remote regions of Europe. *Economies, 9*(4), 148. https://doi.org/10.3390/economies9040148

A Bottom-Up Approach for Sustainable Cultural Tourism in Ladakh: An Initiative Taken by Women and Homestays

Skalzang Dolma and Ashwani Kumar

Abstract

Tourism industry is considered an important sector for economic development. The notion of tourism is not entirely new; it has been popular since the dawn of civilization. Earlier the movement of people was either for pilgrimage or in the quest to know the geographical landscapes. During the 1960s and 1970s, this industry started developing at a rapid rate and the primary reason for the movement of tourism was to visit the different regions to experience different cultures and nature and also to visit historical sites, religions, and places. A similar case is also seen in Ladakh, a Trans-Himalayan region in northwestern India. The region is known for drawing tourist attractions (both international and domestic). Tourism in Ladakh has seen such a surge in the last two decades that it has become a cause of concern for the ecology, society, and culture. However, the region has adopted a sustainable cultural tourism approach to mitigate potential challenges posed by the recent surge. The present paper aims to study the working of homestays with particular reference to the women's contribution to community-based rural tourism or maintaining sustainable cultural tourism. For this study, the village of Phyang from Ladakh has been selected as the area has the presence of large numbers of homestays with active women participation. The methods adopted for the study are literature review, participatory observation, and open-ended and close-ended interviews. This study revealed that homestays with women's participation as a bottom-up approach have been playing a significant role in maintaining sustainable cultural tourism.

S. Dolma (✉) · A. Kumar
Department of History, Central University of Punjab, Bathinda, India
e-mail: skalzang.sd@gmail.com

A. Kumar
e-mail: ashwani.kumar@cup.edu.in

Keywords

Ladakh • Challenges • Sustainable cultural tourism • Homestay • Women • Phyang

1 Introduction

Among many other industries in the world, tourism is contended to be one of the major industries which accelerates at a rapid rate and has enhanced the economic operation of many countries all over the world (Amerta et al., 2018). Tourism expanded the possibilities to find employment opportunities which curtailed the number of job seekers. The significance of this industry is not just limited to economic activity but is also observed in various sociocultural phenomena affecting people in multiple ways (Kumar, 2017). The historical background of this lucrative sector can be traced back to the beginning of humankind's civilization which was initially marked by the migration of people for pilgrimages and other religious travel. This industry is seeing rapid development in the present day (Table 1), and the driving factor for tourists to visit new places is to have a personal experience of the lifestyles as well as the customs and practices of the places they visit (Fig. 1). These visits lead to the direct connection between two societies with different cultures resulting in the speeding up of the process of cultural exchange. Thus, it plays a key role as a cultural phenomenon; it also has a notable factor in the economic sector where its multiplier effect can bring immediate profits (Carter et al., 2015). Due to its economic significance, the industry does not fail to garner respect among the business community (Ramaswamy & Sathis Kumar, 2010).

The economic advantages of tourism have widely overshadowed its impact on the environmental and sociocultural aspects, especially in developing countries (Salazar, 2012). Tourists or visitors intend to visit places with a clean and serene natural environment with rich cultural and traditional

Table 1 International tourist arrivals

Year	ITAs (in million)			Percentage (%) share and rank of India in world		Percentage (%) share and rank of India in Asia and the Pacific	
	World	Asia and the Pacific	India	% share	Rank	% share	Rank
2009	883.0	181.1	5.17	0.59	41st	2.85	11th
2010	948.0	204.9	5.78	0.61	42nd	2.82	11th
2011	994.0	218.5	6.31	0.63	38th	2.89	9th
2012	1039.0	233.6	6.58	0.63	41st	2.82	11th
2013	1087.0	249.7	6.97	0.64	41st	2.79	11th
2014	1137.0	269.5	13.11	1.15	24th	4.86	8th
2015	1195.0	284.1	13.77	1.15	24th	4.84	7th
2016	1241.0	306.6	15.02	1.21	26th	4.90	8th
2017	1333.0	324.1	16.81	1.26	26th	5.19	7th
2018	1413.0	346.5	17.42	1.23	23rd	5.03	7th
2019	1466.0	360.1	17.91	1.22	25th	4.97	8th
2020	405.0 (P)	59.1(P)	6.33	1.56	19th	10.71	3rd
2021	427.0 (P)	20.7 (P)	7.00	1.64	–	33.82	–

Source Indian Tourism Statistics, 2022, *P* Provisional, *ITA* International Tourist Arrival

Fig. 1 Tourist arrival's main purpose. *Source* UNWTO (2022)

activities, which are considered the core component of the tourism industry (Neto, 2003). However, it is unfortunate that these characteristics which tourists seek get distorted during their visits (Angelevska-Najdeska & Rakicevik, 2012). The destructive effect of unregulated tourism includes damaging the environment, creating pollution, exploitation of resources, overcrowding, haphazard construction of infrastructural facilities, harming the tangible as well as intangible cultural heritages, importing dangerous viruses, drinking water shortages, and the production of litter (Singh & Mishra, 2004). The imprudent operation of tourism even facilitates instability in the local economy; Istoc states that "this risk occurs when the tourism facilities in the area are owned by foreign companies that gain the profit and take it out from the area without making any investments" (2012). Despite limitations, tourism is considered a major economic force, and hence economic development due to tourism cannot be stopped. Therefore, a sustainable approach to tourism is needed in order to minimize human impact on the environment, expand employment opportunities, and appreciate traditional–cultural values.

1.1 Sustainable Tourism Development

Tourism, seen as a growing and relevant sector in many localities incurred a negative impact on environmental as well as sociocultural practices (which have been overshadowed) (Manzoor et al., 2019). It has been identified as one of the five main industries in need of achieving sustainable development (Dangi & Jamal, 2016). Therefore, the sustainable development of tourism is considered a fundamental process to balance tourism's negative impacts and community well-being (Durovic & Lovrentjev, 2014; Rocha, 1997). Clarke (1997) differentiates "sustainable tourism" from the "mass tourism" concept, its purpose is not just to maintain the country's economic growth but also to maintain the natural resources. In order to increase the feasibility and longevity of tourism development plans, sustainable principles are indispensable (Okazaki, 2008). "Sustainable tourism development meets the needs of present tourists and host regions while protecting and enhancing opportunity for the future" (UNWTO, 1994). WTO (World Tourism Organisation) defined "sustainable tourism" as "tourism that takes full account of its current and future economic, social and environmental impacts, addressing the needs of visitors, the industry, the environment and host community" (WTO, 2005). It is an approach to developing tourism in a structured and organized way which ensures sociocultural sustainability and inhibits the degradation or altering of the environment (Istoc, 2012). Sustainable tourism development, also called "green tourism" or "community-based rural tourism," was developed mainly as a response to the adverse effects of industrialization and globalization (Singh & Mishra, 2004).

Sustainable tourism paves a path to the judicious use of natural resources, eco-friendly development of a region, locals' participation in the tourism industry, provision of livelihood for the local communities, and maintaining harmony among the people, economy, and environment. A study done in Cambodia showed that tourism is the country's main source of economic growth, and as economies increase, so does environmental deterioration. As a result, the long-term growth of tourism in Cambodia is anticipated to be strongly influenced by sustainable tourism development (Irawan & Nara, 2020). Sustainable agrotourism in the Bhaktapur district of Nepal has facilitated not only economic growth but also the promotion and preservation of sociocultural practices that have allowed tourists to interact closely with the local population (Pandey & Pandey, 2011). Another study (Pjerotic et al., 2017) deals with the indicators of sustainable tourism development in the rural areas of countries of Hungary, Romania, Lithuania, Poland, and Bulgaria. The empirical study conducted shows that tourism's contribution to employment, its share of GDP, expenditure and capital investments are the major indications of sustainable rural tourism development. One of the primary purposes of sustainable tourism is to contribute to the society and economy of the host community. It contributes to the local people's lives, improves their living standards, and promotes local cultures. It also plays a primary role in preserving society's customs and culture (Amerta et al., 2018). The existing top-down centrally planned approaches have failed to accomplish a well-integrated and organized tourism development as they do not consider many stakeholders' views, i.e., the views of tourists, citizens, local communities, and tourist agencies (Ioan-Franc & Istoc, 2007; Chiabai et al., 2013). Therefore, community-based rural tourism, often associated with alternative tourism or a sustainable approach, is adopted.

1.2 Community-Based Tourism (CBT) or Rural Tourism

The traditional knowledge that the locals hold is of prime significance in structuring or restructuring the community to enhance regulated tourism and planned socioeconomic development (Agarwal & Mehra, 2019; Clouse, 2020). Thus, in promoting sustainable tourism, the local community plays an important role. Community-based tourism or rural tourism is defined as having a positive outlook and contribution to the economic and social structure of rural society, and it focuses on causing minimal damage to the social and the physical environment (Swarbrooke, 1996). It generates employment in rural areas, contributes to rural development, and is economically viable and ecologically sustainable. It is designed to promote the interaction between the host community, visitors, and public and private sectors. It is a form of tourism that provide opportunities for people who are not directly involved in tourism (Amerta, 2017). The first of CBT's three main goals is to enhance the local area's natural and cultural resources. Secondly, to advance the area's social and economic facets. Thirdly, the local community must be in charge of planning and tourism management (Manyara & Jones, 2007).

CBT, or rural tourism, enhances the sustainable living pattern of rural areas by increasing farmers' income and giving importance to farm-related economic opportunities or restructuring the rural industry while meeting sustainability needs (Gao et al., 2009). Community-based tourism draws a broad distinction from the traditional or conventional style of tourism in its operation, structure of management, people institution, direct benefits to the host community, and sustaining of the environment (Murphy & Murphy, 2004). The community-based tourism of a region results in its economic, psychological, social, and political empowerment (Dangi & Jamal, 2016) see (Appendix 1).

According to a study done on the island of Santiago (Cape Verda), community-based tourism has a significant impact on the socioeconomic growth of the area because it offers tourists services like lodging, dining, entertainment, and transportation, all of which generate income. Additionally, community-based tourism has prompted the construction of public facilities in the area (López-Guzmán et al., 2011). A study on community-based tourism sites in South Korea (Bukchon Hanok Village, Ewha Mural Village, Seorae´ Village, Achasan Mountain Ecological Park, Seoul Forest Park, and Jeonju Hanok Village) has proven that it is immensely beneficial in increasing the sustainability of the socio-ecosystem, preserving local traditions, conserving natural resources, reducing poverty, and protecting cultural authenticity (Han et al., 2019). It is also regarded as an important reason for poverty reduction in most developing countries. The study (Demkova et al., 2022) shows that community-based tourism in Pastangna, Sikkim, India, is a pro-poor tourism policy as it enhances regional economic growth. In Zimbabwe, community tourism development is viewed as a measure of economic growth and poverty eradication. However, the locals of the region were unable to reap the benefits as they played a negligible role (Chiutsi & Mudzengi, 2012). CBT thus contributes its substantial share in supporting the development of rural infrastructure and communication, containing the rural population within the village (Anand et al., 2012), creating jobs as well as additional income opportunities like opening a market for handicrafts and organic farm produce (traditional food and drinks) (Swarbrooke, 1996).

CBT or rural tourism plays a significant role by empowering rural communities and generating employment for women and youths who otherwise are abandoning the villages in search of jobs (McAreavey & McDonagh, 2011). It has much potential for empowering women as women have progressed from passive involvement to active participation in culture-related tourism ventures (Moswete & Lacey, 2015). Empowerment emphasizes individual inclusion in an organization and its organizational decision-making, (Rocha, 1997). The community-based tourism initiatives aim for the holistic goal of empowering underprivileged groups sustainably (Dangi & Jamal, 2016). It has influenced the participation of women in owning and managing homestays and preparing food for tourists to making crafts, showing cultural dances, and exhibiting local foods (Irawan & Nara, 2020). The study conducted in the Dalla community (in Nepal) witnessed that the rural women have benefitted from the homestays (community-based tourism) because they have been able to obtain financial resources while caring for their families (Woli, 2022). The

success of sustainable development, women's active participation in homestays, and their empowerment in terms of income generation and decision-making are all demonstrated in a case study conducted in Kasari Devi, India (Chakraborty, 2019). According to a study conducted in the Jammu (Surinsar and Shamachak villages) region of India, with 60 women respondents, it was observed that rural tourism enhanced women's empowerment in socioeconomic upliftment, household decision-making, strengthen participation in educational activities within and outside the state, and supplement the family income (Slathia et al., 2015).

1.3 Sustainable Cultural Tourism in Ladakh

Ladakh, a Trans-Himalayan remote region, had a self-sufficient existence that remained undisturbed for centuries (Hodge, 2013). The economic condition of pre-independence Ladakh was largely dependent on trade, and the nineteenth century Ladakh was considered an important trading route and center of Central Asian trade and the vast majority of the Ladakhis lived by subsistence farming in small village communities (Kaul, 1998; Mann, 2002). "In the pre-independence time, there were caravans travelling and the only 'tourists' were the occasional army officer and the only foreign resident were the Moravian missionaries" (Rizvi, 1999). It was only from 1974 onwards that the region's economic condition took a significant shift as Ladakh was opened for tourism. Since then, tourism started as a substitute for agriculture, contributing a major share of Ladakh's GDP (Dar et al., 2019). The exotic Buddhist cultural practices, its breathtaking, serene landscape, and the warmth and hospitality of the locals became the unique selling point of Ladakh tourism (Chaudhary & Angmo, 2020). In 2010, around 77,800 tourists visited the Leh district (22,115 international and 55,685 domestic). By 2019, tourist arrival in Leh district has seen substantial growth to 279,937 (241,285 domestic and 38,652 international) (Pelliciardi, 2013) (Fig. 2).

Tourism, no wonder, has led to economic development and improvement in the livelihood status of the region as it contributes nearly 50% to the GDP of Ladakh (Loram, 2004) and it creates a significant number of low-skill and medium-skill jobs along with some high-skill jobs and entrepreneurial opportunities for local residents in its associated industries, including hotels, guesthouses, restaurants, catering services, transportation, guides, and others and attracts a large number of migrant workers and business people during the summer season (Leh Vision, 2030). Expanding tourism brought in investments and creates economic opportunities for local participation. An increase in income, access, and connectivity, additionally, as a rise in the number of tourist accommodations have also become the key drivers of tourism growth.

Fig. 2 Tourist arrival in Leh district (2010–2022). *Source* Tourism Department Leh District

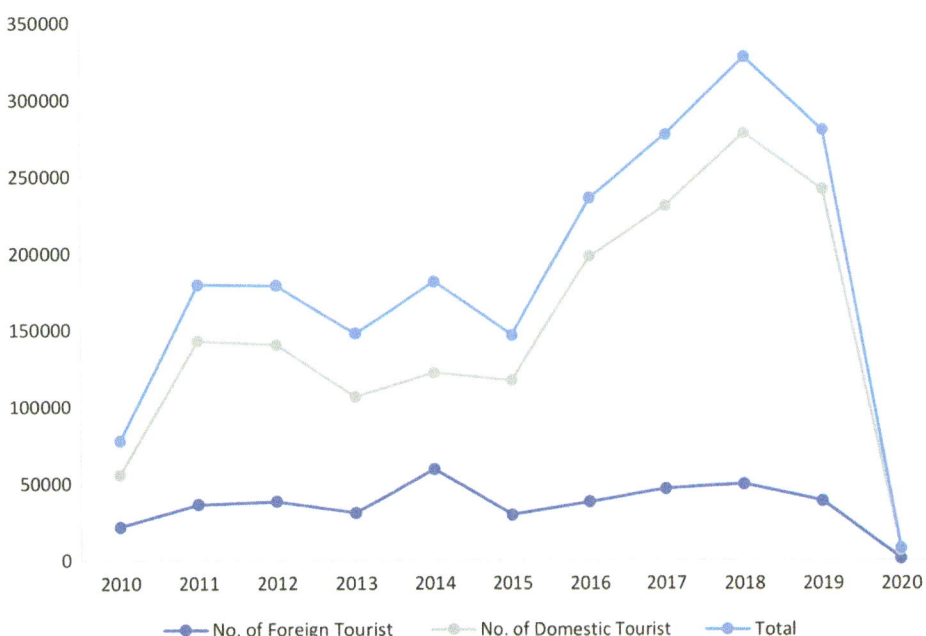

Tourism, interconnectivity, and development resulted in the economic expansion of Ladakh but at the cost of social and environmental change (Michaud, 1996). The changes have taken a heavy toll on the society, environment and traditional–cultural values of the Ladakhi people, resulting in the loss of its communal cooperative spirit (Cyr, 2018). The lack of a proper sewage disposal system in Ladakh is leading to sewage overflow and pollution of local water, and expanding infrastructure to accommodate rapidly growing modern tourism consumes more resources than can be supported by local ecosystems (Rajashekariah & Chandan, 2013). The studies conducted by ICIMOD suggest that glaciers in Ladakh are shrinking at a rapid rate and about 35% of them will disappear within two decades (Menon, 2011). Brian (2021) views that the increase in wastage of water due to the introduction of non-traditional flush toilets in hotels, waste disposal, and soil degradation due to off-road driving and the impact of pack animals (used by trekking groups) are some of the negative impacts of tourism in the region. Considering the current scenario, it becomes imperative to question whether further expansion in tourism is possible or whether will it be able to comfortably sustain the additional tourist arrival. Thus, the importance of sustainable or responsible tourism is acknowledged as a better solution for the unregulated tourist arrival in Ladakh, and community-based rural tourism is considered a medium for maintaining sustainable tourism in which homestays play a significant role as a bottom-up approach for responsible tourism.

Homestays are an attractive sector of the CBT which is significant in maintaining sustainable cultural tourism

(Acharya & Halpenny, 2013). It is typically defined as a type of low-cost accommodation in which visitors experience a homely stay by interacting and building relationships with the host family, learning the locals' cultural and traditional practices and taking part in their daily chores (Janjua et al., 2021). Homestays play a prominent role in building and achieving the threefold sustainable community-based tourism goals of environmental, cultural, and local employment protection (Reimer & Walter, 2013). "Variants of the rural homestay concept include cultural homestay, farm stay, heritage homestay, agricultural homestay, leisure homestay and cottage homestay" (Hamzah, 2007). These are seen as an effort to retain the uniqueness of traditions (Pradana & Arcana, 2020). Homestay's program, which forms a part of CBT, provides tourists with a unique opportunity to experience the atmosphere, lifestyle practices, and activities of rural communities in the countryside (Mapjabil et al., 2015). According to Dutta (2012), the purposes of the homestay program are as follows:

- To lessen the impacts of tourism on the traditional–cultural values of the local indigenous communities.
- Advancement of economic empowerment in the remote locations.
- Preservation and reinforcement of local cultures and values.
- Empowering villagers to make their own decisions.
- Income generation for women and other disadvantaged members of the village.
- To give the visitors the experience of authentic lives and the host community's culture.

Fig. 3 Major homestay villages
in Leh district

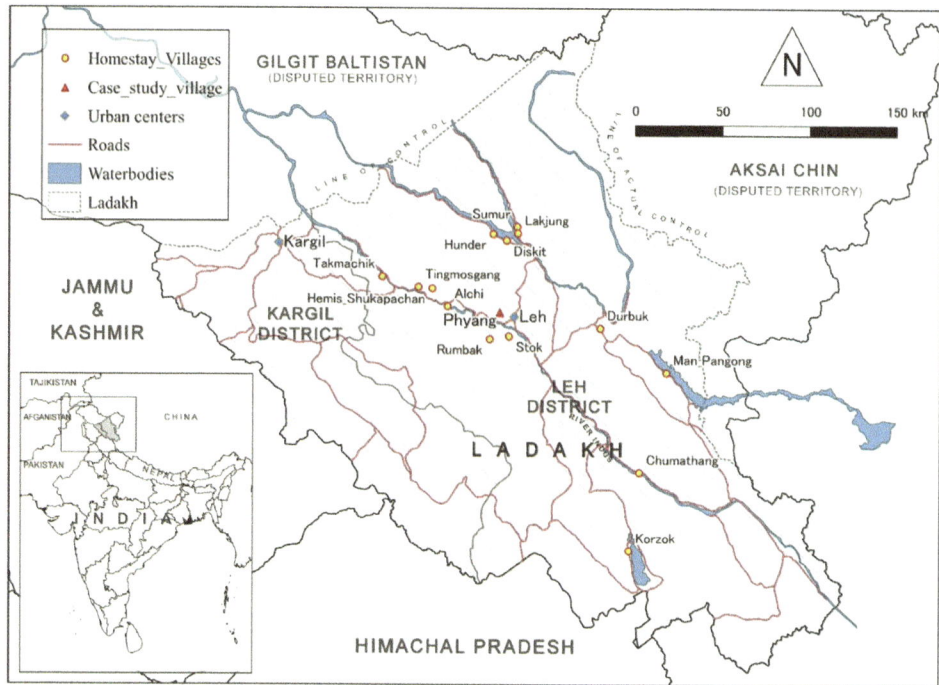

The concept of "Homestay in Ladakh"is as old as its culture itself as men from each village would leave to sell and buy goods and rest at the homes of those villages that come on the way (Satterfield, 2009). Furthermore, this concept was given a formalized character when there prevailed intense episodes of human–wildlife conflict and unregulated tourism. The wildlife protection and tourism department responded to these situations as an opportunity to plan and develop a system or policies to regulate tourism, especially in eco-sensitive regions. In addition to this, the Snow Leopard Conservation Trust of India (SLC-IT) projected Ladakh homestays in 2000 as an opportunity to generate income by introducing a snow leopard expedition, which would help the farmers to see these "highly endangered cats as a source of income rather than a pest needed to be eradicated" (Peaty, 2009). This idea was first incepted in the village of Rumbak, a village with nine homes located at an altitude of 4050 m along the famous Markha Valley trekking route in Hemis National Park, the home of India's largest population of snow leopards (Pelliciardi, 2010).

Homestays in Korzok, the region in eastern Ladakh, were introduced by World Wide Fund for Nature (WWF-India); this drive initiated by the said organization integrated community-based rural tourism, which protects the lake Tsomoriri in Korzok and provides supplementary income and generates employment opportunities to the Changpa population (Anand et al., 2012). Similarly, the Global Himalayan Expedition team, who were installing sustainable solar-based microgrids in 20 odd remote villages of Ladakh, realized that these electricity-ridden villages could be

attractive stay options for trekkers (Saini & Singh, 2022). Moreover, in 2012, the Department of Tourism initiated its homestay program in most of the villages of Leh district. Since then, the homestay modal has been adopted by both public and private entities in many villages of Ladakh. Figure 3 shows the major homestay villages in the Leh district.

2 Methods

While burgeoning literature on many aspects of tourism in Ladakh is readily available, there still exists a deficit in research into rural tourism or sustainable tourism development in Ladakh. The present paper thus aims to study the bottom-up approach to maintaining sustainable or responsible tourism in the Leh district with the help of the case study of Phyang village.

2.1 Study Area

The village is situated 20 km away from Leh town. The total household in the village is 352, with a population of 2036 (Amenity 2018–2019). The primary purpose for selecting this area for the study is that the village experiences a presence of tourism in large numbers in May–August and it is a typical example where rural tourism or community-based tourism prevails. Apart from that it is also in close proximity to other tourist sight-seeing areas of Leh town.

One of the most popular tourist destinations for both domestic and foreign visitors in the village is Phyang Gompa (Monastery), the museum of the monastery contains 900-yearold scriptures, idols, Tibetan and Mongolian firearms and weapons. Guru Lakhang of Phyang village is another tourist attraction, with its original wall paintings and architectural structure from the fourteenth century, this temple has exceptional historic value. The village monastery festival known as "Phyang Tseruk" celebrated annually in the month of 2nd or 3rd week of July is another important tourist attraction. A historic trade route linking the Indus and Nubra valleys, which were a part of the old Silk Road network extending from India to Central Asia, is followed by the hikers travelling from Phyang to Hunder through the Nubra valley. The ice stupa in the village is an important winter tourist spot. Apart from this, the Himalayan Institute of Alternative Learning is also a major tourist attraction. Its Director, Sonam Wangchuk, an environmental engineer and a social reformer, is known for his innovative approach to dealing with environmental challenges. Homestays in the village are also an important field of attraction for tourists. There are around 30 households with homestays (both registered and unregistered). However, in this study, only 27 homestay owners have been interviewed as the other three owners had stopped engaging in homestay sectors.

This study was conducted in January–March and July–August in 2022. Primary, as well as secondary data, have been collected for the study. The primary data collection is done through close-ended (Keller & Conradin, 2020) and open-ended interviews (Demkova et al., 2022). The close-ended interviews were conducted with purposefully selected samples (Kunjurman & Hussin, 2016) of ($n = 27$) homestay owners (women) of Phyang village because they

have experience in managing homestays. The descriptive statistics analysis of the close-ended interview has been conducted. Furthermore, open-ended interviews were carried out with key informants (Chiutsi & Mudzengi, 2012) village headman ($n = 1$), the founding director of HIAL ($n = 1$), and the tourism officer ($n = 1$) to gather information regarding their attitudes and perceptions about sustainable community-based rural tourism development and women's participation. The responses have been analyzed using text analysis. Thus, the primary data for the study have been collected from a total of ($N = 30$) respondents. The data collected from the government reports and research papers published in journals on the concerned theme supplement the primary data.

3 Results

Table 2 represents the age profile, gender, and number of years of experience of respondents in the homestay sector. A sample of 27 homestays of the village with 100% of women respondents (as the homestays were managed by women) was selected. About 7.4% of the respondents were in the age group 20–30; 29.6% in between 31 and 40; 25.9% in between 41 and 50 and 51 and 60, and 11.1% of the respondents were 65 and above. The data show that the majority of women managing homestays are in the age group of thirty-one to forty, which is then followed by the forty-one to fifty and fifty-one to sixty, and these women were mostly housewives with no independent source of income. The year's experience table further outlines that 66.7% of the sample had six to ten years of experience in managing homestays, which provides us with the

Table 2 Demographic profile of the respondents ($N = 27$) of Phyang homestays

		Frequency	Percent
Age			
Valid	20–30	2	7.4
	31–40	8	29.6
	41–50	7	25.9
	51–60	7	25.9
	> 65	3	11.1
	Total	27	100.0
Gender			
Valid	Female	27	100.0
Year's experience in homestays business			
Valid	1–5 yrs	5	18.5
	6–10 yrs	18	66.7
	11–15 yrs	4	14.8
	Total	27	100.0

Fig. 4 Homestay: an important measure for sustainable tourism

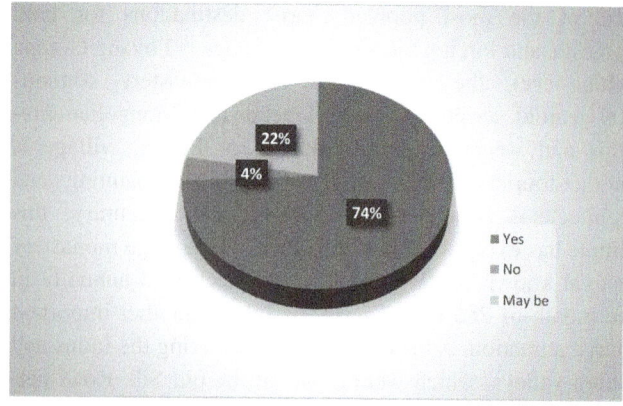

Fig. 5 Homestays as an additional source of family income. *Source* Field Visit

information that most of the families started running homestays from the year 2016–17 onwards. The other 18.5% had an experience of one to five years, and only 14.8% had the experience of eleven to fifteen years.

Figure 4 illustrates how homestays are considered an important measure for sustainable tourism. The interview conducted by the author in the Ladakhi language with the homestay owners has been translated and presented below:

> Homestays have made village life economically viable, people made their houses a source of income. (Respondent 1, personal communication, July 29, 2022)
>
> The young people who have left for Leh because there is not enough income from the farm can depend on tourism through farm stays and homestays. (Respondent 1, personal communication, July 29, 2022)
>
> The economy in Ladakh has changed, earlier it was self-reliant people consumed what they produced, now there is a need for cash to buy things and for children's education and farm production is not sufficient, in fact, it takes more cash than gives, so it is no longer sustainable, we can't live off barley field so as a supplement we see tourism. (Respondent 2, personal communication, July 29, 2022)

The narratives and the word cloud representation in Fig. 4 (which has been obtained using the text analysis method) shows that homestay in the village is considered to be an important measure for responsible tourism. Homestays have an impact on both the visitors and the villagers. The visitors or tourists experience the local lifestyle and culture, whereas the host community or the villagers earns livelihood from homestays. The respondents viewed that most of the homestays have their own organic kitchen gardens which become a prominent tourist attraction. The number of households running homestays in the village is rather small when compared to the overall number of households, but the advantages that the homestay offers by increasing income are substantial. The Tourism Department of Leh district and the HIAL (Himalayan Institute of Alternative Learning: a non-government body situated in the Phyang village) also acknowledged the homestays as an excellent source of income in rural areas, especially in the border or remote areas.

It is shown with the help of Fig. 5 that homestays generate additional income in the village by providing accommodation to the tourists, 74% of the respondents hold the opinion that homestays have become an additional source of income which has helped increase family income. Local communities derive income not only from providing accommodation but also from guiding services, campsites, food, and handicrafts (hats, socks and gloves) (Peaty, 2009). The majority of the income from tourism, however, in this village is achieved only through accommodation. Homestay businesses in the village have also become a crucial aspect of restoring cultural values and traditions for the future generation. The practices and rituals in the village are still followed as it is an important tourist attraction, thus becoming the main component of sustainable rural tourism.

The word cloud presentation (Fig. 6) illustrates that the homestays have benefitted the village's women to a great extent. Women play a primary role in the working of the homestays; as the male members of the family go out to earn income, it is the women/mothers *(Amalays)* who do all the work in the household and look after the proper functioning of the homestay. However, it is also true that the family also supports them. It was viewed that many of them earlier used to go to Leh market to sell their vegetables which took much hard work but did not reap good income, and homestays have helped them in earning easy money. The homestay earnings have contributed to their self-sufficiency and given them an equal opportunity to support their family financially. A noticeable percentage of women in the village are unemployed, and homestays have helped these women to generate income and gain self-confidence as they get to interact with people from different countries and regions.

The empowerment framework of Scheyven categorized the empowerment of women on the basis of social empowerment, economic empowerment, psychological empowerment, and political empowerment (Kunjurman & Hussin, 2016). Based on the study of the Scheyven empowerment

Fig. 6 Word cloud presentation on advantages of homestays to women of the village

Figure 7b shows that 93% of the respondents believe that women's engagement in tourism activities has boosted their self-esteem. Most of the respondents viewed that they feel proud as they could contribute to the development of their family. About 96% of the women in Fig. 7c illustrates that homestays have become an additional source of income. Women who were earlier unemployed could earn additional income for supporting their families. According to the respondents, they feel independent as they do not have to ask for money from the male members of the family as they were doing before. Around 89% of women in Fig. 7d hold the opinion that the additional income that they earn helps them in enhancing their decision-making power, thus making them feel politically empowered. None of the respondents in any of the figures responded with "no," rather a small percentage of 19% in Fig. 7a, 7% in Fig. 7b, 4% in Fig. 7c, and 11% in Fig. 7d responded with the option "maybe" which shows that the respondent who believes that they are not socially, politically, economically, and psychologically empowered due to homestays are quite less in number.

framework (Fig. 7), this study also shows that the women associated with homestay in the village feel socially, politically, economically, and psychologically empowered.

Figure 7a shows that 81% of the sample population witnessed that they were empowered socially as their participation, dedication, and achievement have been recognized and acknowledged by the community members.

Fig. 7 Scheyven women empowerment framework:
a social empowerment,
b psychological empowerment,
c economic empowerment,
d political empowerment

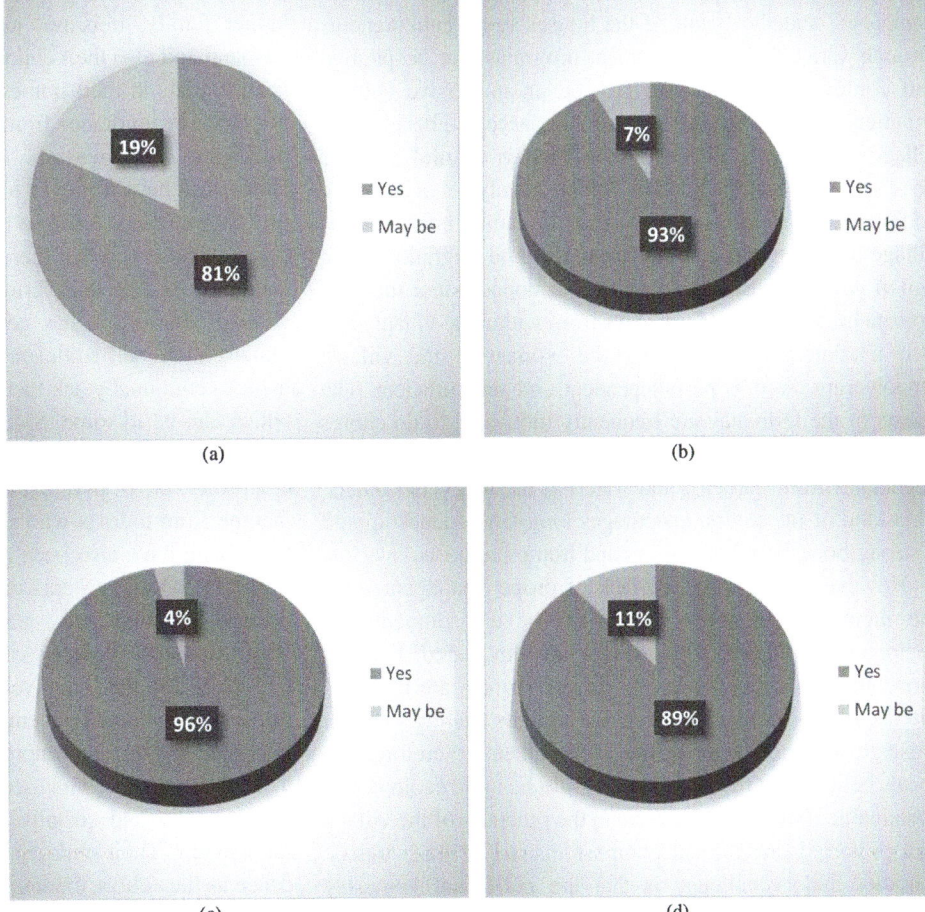

4 Discussion

The results indicated that community-based rural tourism, which establishes a bottom-up approach to sustainable tourism development, is considered an important means for generating and increasing income in the village. It also contributes significantly to the preservation of identity. Furthermore, homestays as a tool of community-based tourism have helped in empowering women of the village. However, the fundamental shortcomings are a lack of management and marketing of homestays which can be overcome with the support of government authorities.

4.1 Contribution of Homestays in Phyang

Homestays in Phyang village have been operating since the year 2017. It was first started by the Himalayan Institute of Alternative Learning (HIAL) and was widely known as the "Himalayan Farm Stays." Since its establishment, homestays have contributed to generating family income, empowering women by strengthening their role in social, economic, and political aspects. The Himalayan Farm Stay seeks to create village livelihood opportunities by training the mothers "*Amalays*" and the youth of the houses to be entrepreneurs through various skill development programs like hospitality, and communications and to provide an immersive cultural experience to the travelers. It is a widely accepted belief in the village that merely depending on the agricultural produce does not increase the income in the family. Hence, as an additional source of earning income, the family houses of the village switched to homestay business. The migration problem of youths to nearby towns has also stopped since they can operate homestays and farm stays from within the village. It is also a source of income for the women in the village, empowering them to be independent and self-sufficient. The guests of the farm stay are frequently motivated to do chores around the farm, which often involve cooking, gathering vegetables from gardens, and watering the fields and others. This kind of interaction encourages long-term friendship and a strong bond between visitors and homestay hosts.

However, because of the lack of proper management of the increasing number of tourists, the environment and sociocultural pattern of the village are threatened. Furthermore, the challenge that the homestay owners are facing is the use of flush-water toilets by the tourists leading to water wastage and pollution to the environment; therefore, the locals believe that the homestays would be more ecologically sustainable if the tourists adhere to the practices of the village and preferred using local compost instead of flush toilets. Another major challenge is that the individual homestay owners are responsible for branding and marketing their establishments. Some have websites, but in other cases, travelers inquire about homestays and locals give recommendations. As a result, homestay operations are not properly managed, and only those with websites or links to tour operators reap the benefits, leaving others without that opportunity. The HIAL and the individual homestay website currently stand as the only source of promotion for these homestays. In order to help the locals participate in the tourist sector and attract tourists' interest in experiencing the cultural beliefs and lifestyle of the people, the government should establish partnerships with travel firms for broad advertising and marketing of these homestays. Another critical measure to consider is spreading awareness to the locals and mainly to the tourist and educating them that the village culture and practices are sacred and need to be preserved.

4.2 Women's Participation in the Homestays

The findings of this study on the empowerment of women indicate that women feel they are empowered in every aspect. They run most of the homestay businesses with the help of their family members. As one of the respondents noted, "family support is essential in the working of a homestay, and in her family, it is not just *Amalay* (the mother) but her husband and also their children who work together when they have tourists in their homestay." This shows that women do not face any hindrance from the family or community in their managing of homestays, thus socially empowering them. Homestays have achieved its goal: to provide employment for the unemployed *Amalays* and young people. One of the *Amalays* observed that there are many families with no other source of income than farming, in such a scenario, the operation of homestays has generated pocket money for them, making them self-sufficient and has eliminated the need for them to continually ask the male family members for money, which can be irksome. Since they are economically empowered, they feel independent and self-confident which in turn empowers them psychologically. In addition to receiving income from tourists who stay in their homes, the women of the village have also been able to create new jobs for themselves by producing apricot juice, jam, bread, weaving, and tailored clothing as part of organized groups like Self-Employed Women Association (SEWA) and Self-Help Group (SHG), the initiatives by self-help groups (SHG) have proved monumental in empowering women. These groups pool their resources to improve their economic conditions and their communities thus seeking to empower women both economically and socially. Homestays are also believed to assist them. Their children learn new skills and gain confidence as they allow them to interact with people from various cultures and countries.

5 Conclusion

The sustainability of tourism has been the subject of numerous studies and the present study also contributes to the plethora of such research. However, this study has a bottom-up perspective that stresses the significance of community-based rural tourism in formulating sustainable tourism development and its contribution to women's empowerment. The Phyang village of Leh district reflects an ideal example of sustainable tourism in which the local community plays a prominent role. The homestays functioning in the village are considered an important initiative of rural tourism for maintaining sustainable tourism, including prudent use of resources, conservation of culture, and bottom-up participation for better decision-making at the local level. It significantly contributes to the village economy's development and has generated employment for the marginalized section of society and has empowered them. The study revealed that the homestay program provides tourists with a first-hand experience of the village atmosphere, customs, and activities of rural communities and their lifestyles. However, several issues have emerged, which initially did not seem severe and can have detrimental effects in the long run. It is witnessed that the sole purpose of homestays, i.e., sustainable development, is hampered as there is no proper management in keeping a check on its working in the village or making aware the people of its purpose, and as a result, the proper guidelines of the homestay are not adhered by the host village as well as the tourists.

Furthermore, there have been irregularities in tourist arrival since it lacks a proper marketing strategy. This issue can be mitigated by emphasizing and promoting homestay's actual purpose and significance. The stakeholders associated with a homestay in the village should issue proper guidelines which must be adhered to and maintained. There is a need for creating training programs and awareness of homestay guidelines for tourists and the host community. The conclusion drawn from the study can be applied as a framework to establish well-managed community-based tourism in the other rural areas of Ladakh.

6 Limitation

The current study emphasizes the significance of homestays and the bottom-up involvement of women in sustainable cultural tourism. Although the research provides a very comprehensive picture of the advantages of rural tourism, this paper has a few limitations, such as the small sample size collected presents the view of those engaged in the tourism industry or homestay, and it falls short of presenting such a detailed overview of the adverse impacts of rural tourism. Based on the current study's findings, future studies could concentrate on new regions and communities while advancing the standards of sustainable or rural tourism.

Acknowledgements The authors would like to express their gratitude to all the respondents of the village who answered and participated in the interview and without their support this paper would not have been possible. Additionally, we want to express acknowledgement to the Sonam Wangchuk, Founding Director of HIAL (Himalayan Institute of Alternative Learning), for his contribution to this paper.

Appendix 1

(A1): Integrated Approach to "Sustainable Community-Based Tourism" (Dangi & Jamal, 2016).

Dimensions of community empowerment	Factors
Economic empowerment	• Economic benefits • Local ownership of business • Entrepreneurship/skill development • Marketing • Rural area development
Psychological empowerment	• Participation • Information access • Knowledge gain • Educational activities
Social empowerment	• Improvement of social skills • Improvement in quality of life • Improvement in women status • Respect for local culture and tradition
Political empowerment	• Support from local/national government • Involvement/collaboration • Passionate leaders

References

Acharya, B. P., & Halpenny, E. A. (2013). Homestays as an alternative tourism product for sustainable community development: A case study of women-managed tourism product in rural Nepal. *Tourism Planning & Development, 10*(4), 367–387. https://doi.org/10.1080/21568316.2013.779313

Agarwal, S., & Mehra, S. (2019, June). Socio-economic contributions of homestays: A case of tirthan valley in Himachal Pradesh (India). In *Tourism International Scientific Conference Vrnjačka Banja-TISC* (Vol. 4, No. 1, pp. 183–201). http://www.tisc.rs/proceedings/index.php/hitmc/article/view/251

Amerta, I. M. S. (2017). Community based tourism development. *International Journal of Social Sciences and Humanities, 1*(3), 97–107.

Amerta, I. M. S., Sara, I. M., & Bagiada, K. (2018). Sustainable tourism development. *International Research Journal of Management, IT and Social Sciences, 5*(2), 248–254.

Anand, A., Chandan, P., & Singh, R. B. (2012). Homestays at Korzok: Supplementing rural livelihoods and supporting green tourism in the Indian Himalayas. *Mountain Research and Development, 32*(2), 126–136. https://doi.org/10.1659/MRD-JOURNAL-D-11-00109.1

Angelevska-Najdeska, K., & Rakicevik, G. (2012). Planning of sustainable tourism development. *Procedia-Social and Behavioral Sciences, 44*, 210–220. https://doi.org/10.1016/j.sbspro.2012.05.022

Brian, H. (2021). The impact of a tourism boom in an environmentally-sensitive region: A case study of Ladakh (Kashmir, India). *Japanese Journal of Policy and Culture, 29*, 21–41.

Carter, R. W., Thok, S., O'Rourke, V., & Pearce, T. (2015). Sustainable tourism and its use as a development strategy in Cambodia: A systematic literature review. *Journal of Sustainable Tourism, 23*(5), 797–818. https://doi.org/10.1080/09669582.2014.978787

Chakraborty, B. (2019, June). Homestay and women empowerment: A case study of women managed tourism product in Kasar Devi, Uttarakhand, India. In *Tourism International Scientific Conference Vrnjačka Banja-TISC* (Vol. 4, No. 1, pp. 202–216).

Chaudhary, M., & Angmo, S. (2020). The economic, socio-cultural and environmental impacts of tourism on the residents of Leh. In R. Jacob, T. K. Thomas, S.R. Babu & Anita (Eds.), *Asian tourism research, interpreting the landscape of Asian tourism* (Vol. 1, Issue. 1).

Chiabai, A., Paskaleva, K., & Lombardi, P. (2013). E-participation model for sustainable cultural tourism management: A bottom-up approach. *International Journal of Tourism Research, 15*(1), 35–51. https://doi.org/10.1002/jtr.871

Chiutsi, S., & Mudzengi, B. K. (2012). Community tourism entrepreneurship for sustainable tourism management in Southern Africa: Lessons from Zimbabwe. *International Journal of Academic Research in Business and Social Sciences, 2*(8), 127.

Clarke, J. (1997). A framework of approaches to sustainable tourism. *Journal of Sustainable Tourism, 5*(3), 224–233.

Clouse, C. (2020). *Climate-adaptive design in high mountain villages: Ladakh in transition.* Routledge.

Cyr, E. (2018). *Preserving identity through discourse in a changing Ladakh.*

Dangi, T. B., & Jamal, T. (2016). An integrated approach to "sustainable community-based tourism." *Sustainability, 8*(5), 475.

Dar, S. N., Shah, S. A., Khatun, A., & Batool, N. (2019). Role of tourism in poverty alleviation and socio-economic development: A study of Leh (Ladakh). *History Research Journal, 5*(5).

Demkova, M., Sharma, S., Mishra, P. K., Dahal, D. R., Pachura, A., Herman, G. V., & Matlovicova, K. (2022). Potential for sustainable development of rural communities by community-based eco-tourism. A case study of rural village Pastanga, Sikkim Himalaya, India. *GeoJournal of Tourism and Geosites, 43*(3), 964–975.

Durovic, M., & Lovrentjev, S. (2014). Indicators of sustainability in cultural tourism. *The Macrotheme Review, 3*(7), 180–189.

Dutta, P. K. (2012). Guidelines for the promotion of homestays in Arunachal Pradesh. *Western Arunachal Landscape programme, WWE.*

Gao, S., Huang, S., & Huang, Y. (2009). Rural tourism development in China. *International Journal of Tourism Research, 11*(5), 439–450. https://doi.org/10.1002/jtr.712

Hamzah, A. (2007). Malaysian homestays from the perspective of young Japanese tourists: The quest for Furusato. In *Asian tourism: Growth and change* (pp. 213–228). Routledge.

Han, H., Eom, T., Al-Ansi, A., Ryu, H. B., & Kim, W. (2019). Community-based tourism as a sustainable direction in destination development: An empirical examination of visitor behaviours. *Sustainability, 11*(10), 2864.

Hodge, H. N. (2013). *Ancient futures: Learning from Ladakh.* Random House.

Ioan-Franc, V., & Iştoc, E. M. (2007). Cultural tourism and sustainable development. *Romanian Journal of Economic Forecasting, 1* (2007), 89.

Irawan, N., & Nara, V. (2020). Managing women empowerment through participation in sustainable tourism development in Kampong Phluk, Siem Reap, Cambodia. *International Journal of Economics, Business and Accounting Research (IJEBAR), 4*(02). https://jurnal.stie-aas.ac.id/index.php/IJEBAR

Istoc, E. M. (2012). Urban cultural tourism and sustainable development. *International Journal for Responsible Tourism, 1*(1), 38–57.

Janjua, Z. U. A., Krishnapillai, G., & Rahman, M. (2021). A systematic literature review of rural homestays and sustainability in tourism. *SAGE Open, 11*(2). https://doi.org/10.1177/21582440211007117

Kaul, H. N. (1998). *Rediscovery of Ladakh.* Indus Publishing.

Keller, S., & Conradin, K. Semi-structured interviews (2020) |SSWM —Find tools for sustainable sanitation and water management! *Sustainable Sanitation and Water Management Toolbox.*

Kumar, A. (2017). Cultural and heritage tourism: A tool for sustainable development. *Global Journal of Commerce & Management Perspective, 6*(6), 56–59.

Kunjurman, V., & Hussin, R. (2016). Women participation in ecotourism development: Are they empowered. *World Applied Sciences Journal, 34*(12), 1652–1658.

LAHDC (Village Amenity Directory) 2018–2019.

Leh Vision 2030: www.lehvision2030.com

López-Guzmán, T., Borges, O., & Cerezo, J. M. (2011). Community-based tourism and local socio-economic development: A case study in Cape Verde. *African Journal of Business Management, 5*(5), 1608.

Loram, C. (2004). *Trekking in Ladakh* (2nd ed.) Trailblazer Publications.

Mann, R. S. (2002). *Ladakh then and now: Cultural, ecological, and political.* Mittal Publications.

Manyara, G., & Jones, E. (2007). Community-based tourism enterprises development in Kenya: An exploration of their potential as avenues of poverty reduction. *Journal of Sustainable Tourism, 15* (6), 628–644.

Manzoor, F., Wei, L., Asif, M., Haq, M. Z. U., & Rehman, H. U. (2019). The contribution of sustainable tourism to economic growth and employment in Pakistan. *International Journal of Environmental Research and Public Health, 16*(19), 3785.

Mapjabil, J., Ismail, S. C., Ab Rahman, B., Masron, T., Ismail, R., & Zainol, R. M. (2015). Homestays-community programme or alternative accommodation? A re-evaluation of concept and execution. *Geografia, 11*(12).

McAreavey, R., & McDonagh, J. (2011). Sustainable rural tourism: Lessons for rural development. *Sociologia Ruralis, 51*(2), 175–194. https://doi.org/10.1111/j.1467-9523.2010.00529

Menon, S. (2011). Two sides to Ladakh tourism. *The Hindu Business Line.*

Michaud, J. (1996). A historical account of modern social change in Ladakh (Indian Kashmir) with special attention paid to tourism. *International Journal of Comparative Sociology, 37*(3–4), 286–301.

Moswete, N., & Lacey, G. (2015). "Women cannot lead": Empowering women through cultural tourism in Botswana. *Journal of Sustainable Tourism, 23*(4), 600–617. https://doi.org/10.1080/09669582.2014.986488

Murphy, P. E., & Murphy, A. E. (2004). *Strategic management for tourism communities.* Channel View Publications.

Neto, F. (2003, August). A new approach to sustainable tourism development: Moving beyond environmental protection. In *Natural resources forum* (Vol. 27, No. 3, pp. 212–222). Blackwell Publishing Ltd. https://doi.org/10.1111/1477-8947.00056

Okazaki, E. (2008). A community-based tourism model: Its conception and use. *Journal of Sustainable Tourism, 16*(5), 511–529. https://doi.org/10.1080/09669580802159594

Pandey, H., & Pandey, P. R. (2011). Socio-economic development through agro-tourism: A case study of Bhaktapur, Nepal. *Journal of Agriculture and Environment, 12*, 59–66.

Peaty, D. (2009). Community-based tourism in the Indian Himalaya: Homestays and lodges. *Journal of Ritsumeikan Social Sciences and Humanities, 2*, 25–44.

Pelliciardi, V. (2010). Tourism traffic volumes in Leh district: An overview. *Ladakh Studies, 26*, 14–23.

Pelliciardi, V. (2013). Estimating total receipts for 2011 from Tourism in Leh District. *Ladakh Studies, 29*, 6–12.

Pjerotic, L., Delibasic, M., Jokšienė, I., Griesienė, I., & Georgeta, C. P. (2017). Sustainable tourism development in the rural areas. *Transformations in Business & Economics, 16*, 21–30.

Pradana, G. Y. K., & Arcana, K. T. P. (2020). Balinese traditional Homestay in a sustainable tourism entering millennial era. *Journal of Xi'an University of Architecture & Technology, 12*(3), 4208–4217.

Rajashekariah, K., & Chandan, P. (2013). Value chain mapping of tourism in Ladakh. *WWF-India: New Delhi, India.*

Ramaswamy, S., & Sathis Kumar, G. (2010). Tourism and environment: Pave the way for sustainable eco-tourism (March 5, 2010). https://doi.org/10.2139/ssrn.1565366

Reimer, J. K., & Walter, P. (2013). How do you know it when you see it? Community-based ecotourism in the Cardamom Mountains of southwestern Cambodia. *Tourism Management, 34*, 122–132. https://doi.org/10.1016/j.tourman.2012.04.002

Rizvi, J. (1999). Ladakh: Crossroads of high Asia. *OUP Catalogue.*

Rocha, E. M. (1997). A ladder of empowerment. *Journal of Planning Education and Research, 17*(1), 31–44.

Saini, N., & Singh, K. (2022). A study of determinants affecting homestay tourism in Kullu district of Himachal Pradesh. *Specialusis Ugdymas, 1*(43), 2219–2232.

Salazar, N. B. (2012). Community-based cultural tourism: Issues, threats and opportunities. *Journal of Sustainable Tourism, 20*(1), 9–22. https://doi.org/10.1080/09669582.2011.596279

Satterfield, L. (2009). Trailing the snow leopard: Sustainable wildlife conservation in Ladakh (India). *Mount Holyoke College.*

Singh, R. B., & Mishra, D. K. (2004). Green tourism in mountain regions-reducing vulnerability and promoting people and place centric development in the Himalayas. *Journal of Mountain Science, 1*(1), 57–64.

Slathia, P. S., Paul, N., & Nain, M. S. (2015). Socio-economic empowerment of rural women through rural tourism projects in Jammu Region of J&K state in India. *Indian Journal of Extension Education, 51*(3 and 4), 40–43.

Swarbrooke, J. (1996). Towards the development of sustainable rural tourism in Eastern Europe. *Tourism in Central and Eastern Europe: Educating for Quality, ATLAS, Tilburg*, 137–163.

United Nations Environment Program, (United Nations) World Tourism Organization (UNEP-UNWTO). (2005). *Making tourism more sustainable—A guide for policy makers* (pp. 9 and 20). UNWTO. Available online: http://www.unep.fr/shared/publications/pdf/DTIx0592xPA

United Nations World Tourism Organization (UNWTO). (1994). *Agenda 21 for travel and tourism: Towards environmentally sustainable tourism.* WTO, WTTC and the Earth Council.

United Nations World Tourism Organization (UNWTO). (2022). http://www.unwto.org/tourism-data/global-and-regional-tourism-performance

Woli, L. (2022). Impact of homestays on socio-economic opportunities of the local community. *KMC Journal, 4*(2), 212–223.

Promotion of World Heritage Sites in Kyoto, Japan

Rodolfo Delgado

Abstract

This paper reports on a study of perceptions and challenges in promoting less popular World Heritage Sites in the City of Kyoto, Japan. According to the Kyoto Promotion Office, national and international tourists visit the most common World Heritage Sites of Kinkaku-ji, Ginkaku-ji, Kiyomizu-dera, and Nijo-jo castle but many miss the opportunity of visiting and getting to know other World Heritage Sites within the city. However, attracting tourists to less visited cultural sites is key to the sustainability, development, and survival of all cultural World Heritages Sites in the city. Prior to the covid-19 pandemic, overtourism was a challenge for the middle sized but compact City of Kyoto as most people that visited the city seemed to be concentrated in the same places and attractions. Meanwhile, the lack of knowledge, promotion, and marketing of less popular World Heritages Sites in Kyoto limited the sustainability, conservation, and preservation of those cultural attractions. Therefore, promoting, branding, and marketing of less popular World Heritage Sites will boost visitation, contribute to their conservation, survival and will increase repeat visits to the City of Kyoto overall. Meanwhile, the collaboration of several actors is key to the successful dissemination of information about these sites.

Keywords

World • Heritage • Sites • Sustainable • Conservation • Marketing • Promotion • Branding

R. Delgado (✉)
Department of English, Faculty of Foreign Studies,
University of Kitakyushu, Kitakyushu City,
Fukuoka Prefecture, Japan
e-mail: delgado@kitakyu-u.ac.jp

1 Introduction

This manuscript analyzes the need for policy, management, and marketing strategies to solve the problem of overtourism in the city of Kyoto. Local tourism in Kyoto has been affected by the covid-19 pandemic and understanding the impact of return to "normal" is needed in terms of providing tourism products and services that will meet the needs of national and international tourists, meanwhile, it be able to manage overcrowding in the most visited cultural sites in the city.

The City of Kyoto is highly depended on the tourism industry. Nevertheless, the Tourism Promotion Office located at the Kyoto Station can help manage overtourism in well-known World Heritage Sites by promoting the less visited World Heritage Sites and local communities instead of promoting and attracting tourists to the most visited cultural attractions. Less visited World Heritage Sties can benefit from a marketing diversification strategy to boost the economy in the Kyoto area. For example, instead of continuing to promote the Kinkaku-ji, Ginkaku-ji, Kiyomizu-dera, and Nijo castle sites in Kyoto, the office could start promoting less visited World Heritage Sites to boost local economies in the region and revitalize the economy and the destinations. Among the less visited World Heritage Sites of Kyoto are those distant from the city center, e.g.,: Byodoin, Enryakuji, and Ujigami-Shrine.

The problem that this research addresses is that the well-known World Heritage Sites in Kyoto mentioned above are constantly overcrowded by national tourists and international residents and on the other hand, tourists miss the opportunity of visiting less known World Heritage Sites because they do not receive enough advertising, promotion and marketing assistance from the city tourism office.

The general objective of this research project is to define how local tourism offices in charge of promoting tourism can generate traffic to less visited World Heritage Sites as a strategy of destination diversification to overcome the problem of overtourism in the most visited World Heritage Sites. The specific objectives are: (a) to explain how overtourism put into risk sensitive tourists' cultural attractions and destinations (b) to demonstrate the importance of tourism for local tourists' destinations (c) to provide viable solutions to overcome the challenge of overtourism; and (d) to create diversification strategies to attract tourists to less visited World Heritage Sites.

2 Literature Review

2.1 The Importance of Tourism

According to the United Nations World Tourism Organization (2021), "Tourism is one of the world's major economic sectors. It is the third largest export category (after fuels and chemicals) and in 2019 accounted for 7% of global trade. For some countries, it can represent over 20% of their GDP and, overall, it is the third largest export sector of the global economy".

Faulkner and Tideswell (1997) argue that there are specific factors that affect a destination: (1) stage of tourism development, (2) tourist/resident ratio, (3) types of tourists, and (4) seasonality. The tourism industry plays a significant role in promoting sustainable development, communities are dependent on for their jobs, for support for their families, their economy and their culture in their City of Kyoto. Therefore, diversity and inclusiveness of community members is essential, and their voices should be taken into consideration when attempting to act and create policies that will increase the sustainable development of tourist's destinations, achieve economic prosperity by attracting tourists that are aware of the overtourism problem, and that are also willing to promote other tourist attractions to contribute to tourism diversification and marketing of heritage attractions with less traffic within the city after the pandemic.

2.2 Heritage Tourism

In previous research studies, heritage tourism has been defined in two categories: tangible and intangible. Tangible heritage is characterized by physical and cultural values (UNESCO, 2000). McKercher and du Cros (2002) define intangible heritage as "traditional culture, folklore, or popular culture that is performed or practiced with close ties to 'place' and with little complex technological accompaniment".

Heritage tourism needs to take into consideration two aspects, first the attraction itself and the existing support for the attraction. This support to the attraction is based on the demand of heritage destination, the perception, motivation, and experience of tourists visiting the attraction (Chhabra

et al., 2003; Moscardo, 2001; Richards, 1996; Silberberg, 1995). Moscardo (2001) analyzed how important experiences and interactions in heritage tourism were for the attractions of visitors. On the other hand, Richards (1996) discusses the need for changes in supply and demand to boost visits to heritage sites. Meanwhile, Timothy (2000) argues that some challenges in the demand of heritage tourism like: "inaccessibility, lack of education, and psychological barriers" will require marketing strategies to attract visitors to the heritage sites. On the other hand, Poria et al. (2001) discuss the importance of perception during experiential visits to heritage tourism sites because this is strongly related to the expectations and motivations of tourists. Some of the motivations described during the experience of heritage tourism are: nostalgia, social distinction, and desire for an 'authentic' (Poria et al., 2001) experience. According to Apostolakis (2003), authenticity is a key motivation aspect for attracting tourist to World Heritage Sites, and it can be adjusted to attract diverse segment of a market according to tourist motivations.

McCann an Ortega-Argilés (2013) describes how the economies of growth have a significant role in accelerating the process of innovation to create and design new products and services to boost local economies and increase sustainable development in a region. Innovation policies have the role of promoting cluster development in local and regional destinations to boost traffic and visit to the tourist sites. Several factors of Regional Innovation Systems (RSIs) have been analyzed in diverse occasions and places in the last decades to illustrate these impacts, but nowadays a regional innovation policy approach is emerging in the field. There is a need to build cooperation with companies, not-for-profit organizations and other participants in the local communities to create networks, synergies, and exchange information among the actors involved.

2.3 Tourism Perceptions

Tourists' perceptions are important because these will facilitate destination marketers to target places, cities, and local communities. Understanding tourism products and services consumers will facilitate the creation of niche markets and target new destinations that need to be discovered and promote places that promise authenticity after the covid-19 pandemic.

National and International tourists look for authenticity in the tourism experience because as tourists they are interested in finding new destinations and uniqueness in the places that they visit, therefore, promoting diverse cultural attractions tourists' destinations that will allow the discovery, innovation, and understanding of tourist destinations in local communities. Therefore, it is important to develop destinations that will bring a diversity of tourists, while using the Internet to promote and market these destinations, especially the remote World Heritage Sites and local communities.

Government agencies could help the development of tourism in local communities by organizing the communities and members of these areas by providing training and the understanding of foreign cultures, build expectations for tourist visits, and show at the same time how to protect the natural and cultural resources in tourist destinations (Delgado, 2019).

Several scholars have defined tourism as a consumer activity (Moutinho, 1987; Roselius, 1997; Zhang et al., 2004). As a result, there is often the risk that the tourist products and services will not satisfy the needs of customers during their experiences in the destinations (An & Fu, 2005; Oliver, 1980). The concept of "risk" in the tourism industry (Jones & Boer, 2003; United Nations, 1989) is defined as "the possibility that tourists subject to various unfortunate on a trip or a tourist destination" and "tourists cannot determine the consequences or negative results after making travel decisions" (Chen et al., 2009; Cui et al., 2016; Schiffman & Kanuk, 2000).

Tourists can rapidly make changes when they see they have made the wrong decisions with negative results. Broadly, tourists are looking for authenticity and visiting places where they feel accepted and welcome. Therefore, it is important to develop tourist destinations that will bring a diversity of tourists to each community. In addition, it is now necessary to involve the Internet to promote and market the destinations, especially for remote local communities.

The Internet plays an important role in facilitating the access to new destinations to international tourists (Faulkner & Tideswell, 1997). On the other hand, Allen et al. (1988) analyzed how retail sales contributed to the tourism development in several communities. This demonstrates how different business activities and industries can contribute to tourism development.

2.4 Tourisms Economic Benefits

Tourism is a multisector activity, affecting the economic, social, and environmental aspects of local communities. The development of tourism needs to consider several players in the industry, and to consider the cooperation between government organizations, not-for-profit organizations, private organizations, and the communities. The development and planning of tourism should not rely on the private sector, and it requires the close collaboration of local government working together with private organizations and not-for-profit organizations.

2.5 Overtourism

According to studies conducted by the UNWTO (2021), the covid-19 pandemic has affected many people dependent on the tourism industry for their livelihoods, but its redevelopment also needs to be closely controlled. Organizations such as the UNWTO can help manage overtourism with the collaboration of national and local government offices after the pandemic by helping to promote new local community destinations unknown to tourists, or emerging destinations instead of attracting tourists to the most visited World Heritage Sites and destinations.

For example, the World Heritage Organizations promoted by the United Nations Educational, Scientific and Cultural Organization (UNESCO) has a vital role in attracting national and international tourists to World Heritage Sites, and future heritages sites can be chosen based on the potential participation of tourism development in local communities. There is a need to diversify urban tourism to regional and local tourism cultural attractions, and to reduce the agglomeration of people in city centers. Therefore, analyzing and continuing research tourism in destinations like Kyoto is key to determining how cultural attractions and local communities can benefit from attracting tourists to its local destinations to revitalize the tourism industry in the city and region.

According to the UNWTO, overtourism or tourism congestion refers to "the number of visitors and the capacity to manage them." (UNWTO, 2021). There are many tourist destinations that struggle with overcrowding and there are other destinations that receive very few visitors. UNWTO has defined tourism' carrying capacity as "the maximum number of people that may visit a tourist destination at the same time, without causing destruction of the physical, economic, and sociocultural environment and an unacceptable decrease in the quality of visitors' satisfaction" (UNWTO, 2021).

There are four factors in "overtourism" that requires solutions and management:

(a) Managing overtourism requires creating capacity limit strategies to control the excessive number of tourists' visiting natural and cultural attractions in various destinations. One of the strategies is limiting the number of tourists entering a natural and/or cultural destination.

(b) Managing overtourism requires better control of the excessive number of tour buses creating congestion on the roads, generating street noise and disturbance in local communities;

(c) Managing overtourism will also need to control the proliferation of hotels, hostels, and services offering products to international and national tourists visiting crowded tourist destinations; and

(d) Managing and planning Tour Guide Programs to less visited World Heritage Sites will facilitate the visit of national and international tourists and residents (UNWTO, 2021).

Providing understanding, information, and feedback on past overtourism in the Kyoto World Heritage Sites is key to promotion of diversification of destinations in less visited World Heritage Sites. There is also the need to determine how tourists visiting well-known World Heritage Sites can be attracted to less visited World Heritage Sites. Therefore, it is necessary to conduct surveys constantly to find out tourists' desires to promote tourism diversification. NGOs, government organizations and businesses can operate with local governments to achieve these goals.

2.6 Sustainable Development of the City of Kyoto

Local communities in Kyoto need to do research to determine how they can actively participate in the process of sustainable development and contribute to its sustainability. Pigram (1990) discusses the importance of available policy strategies support for businesses and organizations and the efforts that exit to adapt to local, national, and global levels.

According to the UWTO, "Tourism will only be sustainable if developed and managed considering both visitors and local communities. This can be achieved through community engagement, congestion management, reduction of seasonality, careful planning that respects the limits of capacity and the specific nature of each destination, and while promoting product diversification" (UNWTO, 2021). Thus, "The United Nations New Urban Agenda and the 17 Sustainable Development Goals, namely Goal 11 "Make cities and human settlements inclusive, safe, resilient, and sustainable", are priorities for all" (UNWTO, 2021).

2.7 Marketing Diversification of World Heritage Sites

Accessibility is an essential aspect to take into consideration when Marketing World Heritage Sites, because national and international tourists are not interested in commuting for long periods of time due to time limitation, access to infrastructure and managing traffic congestion are other aspects that need attention.

Marketing strategies will be key in the development and promotion of world heritage tourist destinations because they can boost the sustainable development of a destination and deliver messages and communications that are going to promote the values of the local communities, economy and culture. The United National World Tourism Organization (UNWTO) aims to work together with destination managers to promote local tourist destinations, history, heritage sites and the development of niche tourism markets to attract international and national tourists. Therefore, the collaboration of NGOs, businesses, and local government offices is the key to the attraction of the tourists that are conscious of meeting the Sustainable Development Goals (SDGs) (UNWTO, 2021).

Schulz (2001) mentioned that a marketing concept should be flexible enough to adapt to its mission and goals. Nowadays, the marketing of tourist destinations should include community purposes and stakeholder interests. Arnould and Thompson (2005) suggest that marketing should also consider customer experiences, culture, and social aspects to maximize satisfaction of their needs.

3 Methodology

The methodology of this manuscript is based on qualitative research, involving interviews with the Tourism Office at Kyoto station and observations conducted while visiting the various World Heritage Sites in Kyoto.

The research hypothesis states that if there is more promotion for less visited World Heritage Sites in Kyoto, this will allow more tourists to visit them and contribute to their sustainability, and conservation, and facilitate the financial costs of building maintenance and renovations of the cultural attractions for its sustainable development and conservation.

The theoretical framework of this study is based in the sustainable development and marketing of heritage tourism. One of the challenges in this form of sustainable development is the need to plan for heritage tourism, (a) heritage tourism should be developed based on the location and characteristics of the products and services provided; (b) effective management is the key to the economic and sustainable development of such tourist sites.

4 Discussion

The personnel of the Tourism Office at Kyoto station were interviewed and recommended the most visited World Heritage Sites (Kinkaku-ji, Ginkaku-ji, Kiyomizu-dera, and Nijo-jo) for this research. If the same staff members were directing national and international tourists to less visited World Heritage Sites with the same vigor, these recommendations would increase the visitations at these cultural destinations. Traveling to these less visited World Heritage Sites can be described as going to a remote site and seeking the permission of the cultural destination for a scheduled visit as in the case of Saihoji.

Apart from increasing recommendations to other cultural heritage sites, it is necessary to consider if these recommendations will increase the traffic of these alternative destinations to the point of overcrowding them as well as the main ones. Therefore, it is also necessary to consider the availability of public transportation, and how far in advance a letter of intent to visit the cultural heritage site is required to avoid disappointment when they visit without knowing the cultural destination requirements. According to the tourism office somewhere between a day and a week is required to visit Saihoji, a World Heritage site located about an hour by bus from the Kyoto City center. This is the only World Heritage Site with a reservation system to manage overcrowding in the cultural attraction and yet it is not one to the sites recommended for visitation. Saihoji counts the number of tourists on daily basis, and it manages the number of visitors that visit the cultural attraction at the same time.

Many World Heritage Sites in Kyoto should follow the example of Saihoji and manage an overtourism plan with a reservation system to visit the cultural attractions. The tourist capacity for visiting a tourist destination in Kyoto needs to be calculated to determine the number of tourists that can stay in each cultural destination at the same time.

World Heritage Sites have an opportunity to capitalize on marketing and branding authenticity because national and international tourists have a demand for it. Therefore, directing tourism to other sites will be a competitive advantage for the less visited cultural attractions. If transportation is an issue, cultural attractions like: Byodoin, Enryakuji, Ujigami-Shrine, and Saihoji could arrange a bus tour with a travel agency and facilitate tourists access to some of these destinations. These tours are required to be developed to attract visitors to cultural attractions that struggle to attract visitors. These cultural attractions sites services can be developed on a sustainable development basis to control the number of visitors and avoid overtourism in these destinations.

According to a study conducted by the United Nations World Heritage Sites (UNWTO, 2021), there are five areas that need to be considered when trying to improve the well-being of the local communities with the help of government support, focusing on health, safety, travelers, host communities, and roadmaps that will contribute to the improvement of the tourism industry. Government agencies could help the development of tourism in local communities by organizing the communities and members of these areas through providing training and understanding of foreign cultures, expectations of the tourists' visits, and how to protect the natural and cultural tourists' destinations (Delgado, 2019).

Diversification of tourist products and services is the key to attracting national and international tourists to less visited World Heritage Sites because it contributes to the sustainable development, management, cultural education and exchange of knowledge about these tourist destinations. There is a perception that some World Heritage Sites are visited less frequently due to their relatively remote distance from the city center, tourists may need to use several transportation systems to access cultural sites and may also need to apply for a visit either by letter or online to visit the attraction (Saihoji).

On the other hand, inclusiveness of communities near tourists' destinations is important to increase the effectiveness of promoting and marketing, because its people will support the growth and capacity building of these sites. If residents in Kyoto are included in the tourism process, they will contribute to the development of the tourism industry in the destination, by providing guided tours in local areas, accommodation, organize cultural events like local dances and music, and promote emerging tourist' destinations. Communities should be the center of the tourist activities and, this will boost the economy, social well-being, and cultural development of the regions and promote the diversification of cultural attractions, entertainment, and tourist site visits.

5 Conclusions

Objective 1 of this research stated: "To define how local tourism offices in charge of promoting tourism can generate traffic to less visited World Heritage Sites as a strategy of destination diversification to overcome the problem of overtourism in the most visited World Heritage Sites". In conclusion, the Kyoto tourism office personnel are aware of the importance of promoting all the 17 World Heritage Sites in Kyoto and that there is a need to give equal opportunity and importance to the promotion and marketing is essential to increase the tourist traffic in these cultural destinations.

Objective 2 was: "To explain how overtourism put into risk sensitive tourists' cultural attractions and destinations". In conclusion, overtourism needs to be managed to preserve World Heritage Sites in Kyoto, to avoid traffic congestions in the city and to avoid overcrowding the most visited World Heritage Sites in the city.

Objective 3 stated: "To demonstrate the importance of tourism for local tourists' destinations". In conclusion, the tourism is a key industry in the City of Kyoto and sustainable development of tourism is important to continue to attract national and international tourists to the city and enforce the sustainable development of the cultural attractions with those tourists' visits.

Objective 4 was defined as: "To provide viable solutions to overcome the challenge of overtourism". In conclusion, the City of Kyoto needs to develop an overtourism management program in collaboration with the World Heritage sites in Kyoto to create a reservation system to the cultural sites to better manage the number of tourists' that visit these cultural attractions at the same time. Therefore, this system will contribute to reduce the problem of overcrowding in the tourist destination.

Objective 5 was explained as: "To create diversification strategies to attract tourists to less visited World Heritage sites". In summary, creating diversification strategies to attract tourists to less visited World Heritage sites will increase the number of national and international tourists visiting the cultural attractions and their visits will contribute to the sustainable development of the cultural destinations and to the financing of the renovations required to maintain these attractions for future generations.

It is necessary to understand the tourists' perceptions because this knowledge will allow destination marketers to target places, cities, and local communities. Understanding tourism products and services consumers will facilitate the creation of niche markets in new destinations that need to be discovered and the promotion of places that promise authenticity. After the covid-19 pandemic, there is going to be growing demand for authenticity because tourists are interested in finding new destinations and uniqueness in the places that they will visit but are not interested in overcrowding.

References

Allen, L., Long, R., Perdue, R. R., & Kieselbach, S. (1988). The impact of tourism development on residents' perceptions of community life. *Journal of Travel Research, 27*, 16–21. https://doi.org/10.1177/00472875880270010

An, H., & Fu, R. (2005). The subjective factors influence tourists risk perception and implications for tourism crisis management. *Zhejiang Academic Journal, 1*, 196–200. (in Chinese).

Apostolakis, A. (2003). The convergence process in heritage tourism. *Annals of Tourism Research, 30*(4), 795–812. https://doi.org/10.1016/S0160-7383(03)00057-4

Arnould, E., & Thompson, C. (2005). Consumer Culture Theory (CCT): Twenty years of research. *Journal of Consumer Research, 31*(4), 868–882. https://doi.org/10.1086/426626

Chen, N., Qiao, G. H., & Liu, L. (2009). Tourism association studies risk perception and outbound tourists' travel preferences—Tourists in Beijing. *Geography, 6*, 97–102. (in Chinese).

Chhabra, D., Healy, R., & Sills, E. (2003). Staged authenticity and heritage tourism. *Annals of Tourism Research, 30*, 702–719. https://doi.org/10.1016/S0160-7383(03)00044-6

Cui, F., Liu, Y., Chang, Y., Duan, J., & Li, J. (2016). An overview of tourism risk perception. *Natural Hazards, 82*, 643–658. https://doi.org/10.1007/s11069-016-2208-1

Delgado, R. (2019). The role of tourism businesses to promote local destinations in Japan to attract international tourists. *Ritsumeikan Journal of Asian Pacific Studies, 37*(1), 46–57. https://doi.org/10.34382/00011508

Faulkner, B., & Tideswell, C. (1997). A framework for monitoring community impacts of tourism. *Journal of Sustainable Tourism, 5*(1), 3–28. https://doi.org/10.1080/09669589708667273

Jones, R., & Boer, R. (2003). Assessing current climate risks adaptation policy framework: A guide for policies to facilitate adaptation to climate change [EB/OL]. *UNDP*. http://www.undp.org/cc/apf-outline.htm

McCann, P., & Ortega-Argilés, R. (2013). Modern regional innovation policy. *Cambridge Journal of Regions, Economy and Society, 6*, 187–216. https://doi.org/10.1093/cjres/rst007

McKercher, B., & du Cros, H. (2002). *Cultural tourism: The partnership between tourism and cultural heritage management.* Haworth.

Moscardo, G. (2001). Cultural and heritage tourism: The great debates. In B. Faulkner, G. Moscardo, & E. Laws (Eds.), *Tourism in the 21st century* (pp. 3–17). Continuum.

Moutinho, L. (1987). Consumer behavior in tourism. *European Journal of Marketing, 21*(10), 5–44. https://doi.org/10.1108/EUM0000000004718

Oliver, R. L. (1980). A cognitive model of the antecedents and consequences of satisfaction decisions. *Journal of Marketing Research, 17*(4), 460–469. https://doi.org/10.1177/00222437800170040

Pigram, J. (1990). Sustainable development policy considerations. *The Journal of Tourism Studies, 1*(2), 3–9.

Poria, Y., Butler, R., & Airey, D. (2001). Clarifying heritage tourism research. *Annals of Tourism Research, 28*, 1047–1049. https://doi.org/10.1016/S0160-7383(00)00069-4

Richards, G. (1996). Production and consumption of European cultural tourism. *Annals of Tourism Research, 23*, 261–283. https://doi.org/10.1016/0160-7383(95)00063-1

Roselius, E. (1997). Consumer rankings of risk reduction methods. *Journal of Marketing, 35*(1), 56–61. https://doi.org/10.1177/00222429710350011

Schiffman, L. G., & Kanuk, L. L. (2000). *Consumer behavior.* Prentice Hall.

Schulz, D. (2001). Is it now time to change marketing's name. *Marketing News, 35*(22), 8.

Silberberg, T. (1995). Cultural tourism and business opportunities for museums and heritage sites. *Tourism Management, 16*, 361–365. https://doi.org/10.1016/0261-5177(95)00039-Q

Timothy, D. (2000). Building community awareness of tourism in a developing country destination. *Tourism Recreation Research, 25*(2), 111–116. https://doi.org/10.1080/02508281.2000.11014916

UNESCO World Heritage Center. (2000). *World heritage convention.* http://www.unesco.org/whc

United Nation. (1989). *International decade for natural disaster reduction, resolution 44/236 adopted at the 44th session of the United Nations General Assembly* [EB/OL]. http://www.un.org/ru/documents/ods.asp?m=A/RES/44/236

UNWTO. (2021, June 8th). *'Overtourism'?—Understanding and managing urban tourism growth beyond perceptions.* https://doi.org/10.18111/9789284420070

Zhang, S. Y., Yi, W. C., & Wang, E. P. (2004). Consumer psychology of risk perception. *Advances in Psychological Science, 12*(2), 256–263. (in Chinese).

CittaSlow: Hospitality and Sustainable Urban Tourism Development. The Case of Vizela (Northern Portugal)

Eduardo Cordeiro Gonçalves⬤, António J. D. V. T. Melo, and Ricardo Jorge da Costa Guerra

Abstract

This study focuses on the Slow Cities or CittaSlow movement, which advocates a sustainable and green model of urban social development. Based on the example of the city of Vizela (Northern Portugal), we propose research on how the ideology of slowness can be applied in these urban settings. Thus, the study results in the search and documented identification of how, through education, they permeate the principles of good practices in sustainable tourism. It is, therefore, a proposal for a bottom-up model of strategic management of the sustainable cultural development. Furthermore, we propose to evaluate how environmental conservation practices and identity strengthening contribute to the tourist valorization of cities with population density according to the CittaSlow model. Based on the case study method and using open techniques (qualitative/deductive), pre-existing data collection, focus groups, and participatory research were used. Based on the information provided by the CittaSlow Education project, now complemented by the indicated procedures, it was possible to substantiate the key role of "education" in the implementation of CittaSlow values with a view to sustainable urban development.

Keywords

CittaSlow • Governance models • Tourism and hospitality • Sustainability • Urban tourism

E. C. Gonçalves (✉)
University of Maia and CEGOT, Maia, Portugal
e-mail: egoncalves@umaia.pt

A. J. D. V. T.Melo
University of Maia and CEDTUR, Maia, Portugal
e-mail: amello@umaia.pt

R. J. da Costa Guerra
Polytechnic Institute of Guarda and CEGOT, Guarda, Portugal
e-mail: ricardoguerra@ipg.pt

1 Introduction

The dynamics of local and endogenous development are increasingly articulated with those of its geographic surroundings, whose territorial thickness today tends toward the scale of the region or subregion (Angeles, 2000). The resulting model should associate the creation of collaborative networks (May & Knox, 2006; Bekar, 2015) with institutional cooperation, especially at municipal and inter-municipal community levels. This cooperation between organizations should include the sharing of resources with more integrated approaches at the territorial level, such as eminently social concerns, in order to highlight the policy proposals aimed at territorial cohesion. In this light, we can understand the so-called "creative cities," or "sustainable and entrepreneurial," or even digital and other territorial mentions that are associated with the particularities in the governance of the territory (Mendes, 2014). With roots in the ideology of the "Slow Movement," CittaSlow assume a figure of urban social development embodied in local governance oriented toward sustainability (Ferreira et al., 2014). These small- and medium-sized cities (less than 50 thousand inhabitants) form an international network called slow cities and configure an alternative approach to urban development, whose objectives focus on improving the quality of life of local populations, guaranteeing the good be providing to its visitors, as well as preserving the environment and reinforcing the identity of the place (Heitmann et al., 2011; Pérez-Mongiovi & Cardoso, 2015; Pink, 2008; Rete Urbana delle Rappresentanze, 2012).

The CittaSlow movement emerged in Italy, in 1999, and was founded by Paolo Saturnini, at the time mayor of Greve, in Tuscany. His ideas were quickly adopted and expanded by the mayors of the cities of Bra (Francesco Guida), Orvieto (Stefano Cimicchi), and Positano (Domenico Marrone), with the support of the president of "Slow Food," Carlo Petrini. In fact, slow cities also propagate the philosophy of slow food in local communities, assuming eco-gastronomy concepts in everyday practices. Currently, the slow city concept has matured and is present in more than 200 cities, spread over 30 countries on 5 continents. In Portugal, there are six cities associated with the network: Lagos, São Brás de Alportel, Silves and Tavira, in the south, as well as Vizela and Viana do Castelo, in the north of the country. To be part of the CittaSlow international network, candidate cities must meet criteria, covering six essential lines of action: environmental policies, infrastructure policies, technologies and facilities for urban quality, safeguarding indigenous production, hospitality, and awareness raising. Once the status of CittaSlow has been obtained, regular assessments are carried out to ensure compliance with the established guidelines (Petrini, 2001).

From the diagnosis of the priority areas of action, a vision and the operational objectives, it was possible to base the action program through education in the CittaSlow values. Based on the "CittaSlow Education" project, developed in the city of Vizela, our study has the general objective of analyzing how, through education, one can raise awareness and promote a CittaSlow culture. As a specific objective, we intend to assess the way in which the local population is involved in the preservation and dissemination of natural–cultural resources. It is also intended to understand how stakeholders participate in governance principles in a bottom-up sense.

2 Theoretical Framework and Literature Review

To find the origins of the slow movement, it is important to go back to the 70s and 80s of the twentieth century. With the marks of a protest movement, that movement assumes the reaction against a type of society that is excessively consumerist and unconcerned with the preservation of natural resources and the environment (Davis et al., 2005; Emmendoerfer et al., 2020). The concept of slowness values the experience of finding a relationship with nature and endogenousness (Carvalho, 2014). Before that, we had the organization "Slow Food," whose motivation was, even then, to oppose the "fast lifestyle," a movement that was associated with the rescue of local food traditions, these articulated with the principles of careful nutrition (Petrini, 2001). The slow philosophy is shaped in several areas such as education, the economy or tourism (Ekinci, 2014). In fact, the so-called "slow tourism" overlaps with the principles of slow cities, constituting another relevant tool for local development. Slow tourism thus promotes a quality, low-impact stay in host communities (Gardner, 2009; Holden, 2008; Miele, 2008). Of an international character, the CittaSlow movement proposes to change mentalities with regard to the government of small towns (Dickinson & Lumsdon, 2010). It

is important to encourage the strengthening of citizenship actions and the participation of all players in thinking about the city, in its valorization and in safeguarding the sustainable and responsible use of its resources (Bauer, 2016; Ekinci, 2014; UN United Nations General Assembly, 2015) .

The current research guideline on the topic of CittaSlow has had a special focus on urban dynamics around the principles of sustainability (Carvalho, 2014; Radstrom, 2014) and on local development (Ildiko, 2013). The issues of image, city promotion, and education for the theme have also been addressed (Bekar, 2015). More recently, the literature has reinforced the issue of tourism, articulated with the slow city movement, mainly due to the focus on local identity issues, culture, and tradition (Bauer, 2016). In turn, the need to educate for CittaSlow values and action has been demonstrated (Emmendoerf, 2019). The certification of slow cities (Table 1) has contributed to encouraging responsible behavior in the use of cultural/natural resources and, simultaneously, to the qualification of territories under a more collaborative prism, prioritizing the endogenous and the differentiating marks of the place (Knox et al., 2009).

The sense of territorial qualification proposed by the slow city has a collaborative character in proposing the involvement of multiple stakeholders, public and private agents and civil society (Emmendoerfer, 2019). In a transdisciplinary view, Castells, in his study *The network society*, from 2009, emphasizes that "grassroots movements" (Castells, 2009) shape the dynamics of cities, and small cities need a specific model of "urban social movement" (Pink, 2009), which greatly contributes to articulating and strengthening local dynamics in the CittaSlow (Knox & Meyer, 2009).

3 Methodology

Based on the case study method, as it is "descriptive" of the reality studied and its context (Yin, 1994). It is, at the same time, "exploratory" because it combines lesser-known problems and points out prepositions that remain open for further investigations; but also "explanatory" because it intends to explain "cause/effect relationships" from a theory (Quivy et al., 2008; Yin, 1994). We started, therefore, with a literature review followed by the collection of pre-existing data, with emphasis on the data provided by the "CittaSlow Education" project concerning Vizela.

We use deductive qualitative techniques. Thus, when collecting data, we carried out a systematic consultation and analysis of documentary sources to substantiate the objectives proposed for the study. We use serial data that characterize the territory, to which we add other empirical data collected in the field. Thus, the following data collection techniques stand out: (i) consultation of documents of a technical and political nature, such as territorial management

Table 1 CittaSlow certification criteria

Lines of action	Examples
Environmental policies	"Control of the quality of air, soil and water; waste management; light pollution control; alternative energy sources; energy saving plans; prohibition of the use of O.G.M. in the farming."
Infrastructure policies	"Urban planning for safe transport and mobility; restoration and improvement of historic buildings and parks; construction of an urban circuit of cycle paths, ecopaths, sports, social and leisure centers and medical assistance centers; construction of infrastructure aimed at people with reduced mobility."
Technologies and facilities for urban quality	"Supply of optical fiber and wireless to the city; promotion of bioarchitecture; noise control plan
Safeguarding indigenous production	"Planning for the development of organic agriculture; promotion of markets with local products; safeguarding traditional professions and cultural events; preservation of local enogastronomy; educational programs on food; defense of native flora and fauna species."
Hospitality	"Training courses for tourist information and quality in hospitality; guided tourist itineraries with international indications; policies to facilitate visitor contact with the city; support for local conviviality through cultural events and gastronomic traditions."
Awareness	"Awareness programs for residents and visitors; application of the slow philosophy in educational programs (e.g., school gardens); sharing skills in cultivation techniques, food preparation and crafts."

Source Adapted from Heitmann et al. (2011), Rete Urbana delle Rappresentanze—Rur (2012)

instruments at national, regional, and local levels; studies and articles, scientific journals, legislation, regulations, statistical data series, as well as some cartography; (ii) use of direct observation, which allowed a better understanding of the characteristics of the territory and its management model; (iii) use of the focus groups technique, used as a primary source of qualitative information, naturally articulated with other tools.

4 A "Slow" Destination: The Case of Vizela

Located in the Northern region of Portugal, the municipality of Vizela has an area of 24.7 km^2 and is inscribed in the district of Braga, being an integral part of NUT III/CIM Ave (Fig. 1). The population density is about 961 people per km^2, with a resident population of 23,898 individuals and 66% of the active population (MEM, 2020). Per capita purchasing power is 83% in relation to national indicators and 93% if we consider indicators for the North region. Among the largest employers in the county, we find the textile sector (confection and clothing), soon followed by the manufacture of footwear. The division by sectors of activity finds the tertiary sector in first place, followed very closely by the secondary sector and, lastly, the primary sector, although with significant expression (INE, 2020). Regarding tourism indicators, the municipality only has a capacity of 214 beds, out of the 4300 of the entire NUT III Ave, with a net occupancy rate of about 17.6%, corresponding to 7116

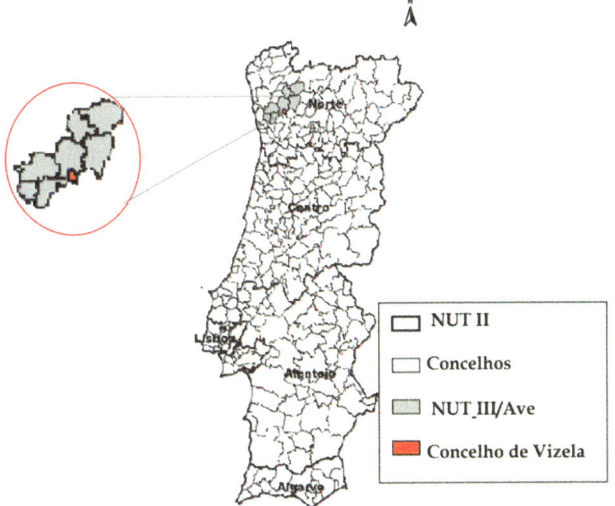

Fig. 1 Municipality of Vizela in its regional surroundings. *Source* Municipality of Vizela/IGP 0–100 km

overnight stays (INE, 2020). It is also worth mentioning its thermal vocation, which has now been rehabilitated for the scope of health and wellness tourism. Often referred to as the "queen of spas in Portugal," its thermal waters and thermal practice dates back to a remote past, today offering renewed services and a cross-selling environment that permeates gastronomy, accommodation and tourist entertainment. Elevated to city status on September 1, 1998, by law 63 of the Assembly of the Republic (Portugal, 1998), Vizela became the first slow city, established in the North of Portugal, in 2011, after a candidacy to the CittaSlow international network, accompanied by the development of multiple initiatives in the field of gastronomy and education. Vizela's accession to the CittaSlow network meant a reconfiguration in its relationship with the territorial environment, but also with the natural and cultural framework, with a view to connecting the city to its historical roots, its people and its identity ethos (Pérez-Mongiovi & Cardoso, 2015). The inclusion of the city of Vizela in the CittaSlow network enhances a vision of governance that integrates multiple actors, whether from the municipal, associative, business, and civil society contexts (Ferreira et al., 2014).

5 Through Education

As highlighted in the ENDS 2015—*Implementation Plan of the National Strategy for Sustainable Development* (Portugal, 2007), sustainable development policies embody an integrative vision of development, combining economy, society, and the environment (Pink, 2008). From here derives the meaning respect for biodiversity and natural resources, solidarity between generations, as well as cooperation between territories. Based on these assumptions, the municipality of Vizelle supported the "CittaSlow Education" project, which developed a diagnosis of the municipality's potential and identified priorities for its development (Pérez-Mongiovi & Cardoso, 2015). Supported by the dynamics and structure of the international movement, the governing bodies of the Agrupamento de Escolas de Infias, in Vizela, promoted the implementation of the project with the school community in order to reflect on students the basic concepts of CittaSlow, with the aim of providing a basis for education in the values of the movement. Through the young students, it is intended to inculcate an attitude, an identity of the place. As part of the project, curricular programs were developed with the aim of providing students with the understanding and formulation of attitudes and personal skills from a pedagogical practice. However, sustainable tourism meets the development of space and expresses the relationship of people with their culture, traditions, and history. Thus, tourist activities end up transforming the space and assume the role of transforming agents of reality, changing the relationships that people establish with the environment in which they are inserted. The paradigm is no longer that of socioeconomic growth and starts to include respect for natural and endogenous resources that foster balanced and harmonious development, in a synthesis between the community and the environment in which it operates. Thus, this theme was also included in the school context, passing the concept of sustainable tourism to be approached as a pedagogical practice and to assume an alternative place of

environmental, social, and cultural education. Furthermore, it has become an environmental education tool with the particularity of being able to be experienced by students in their contact with nature, experiencing tasks, activities, and traditions. In this way, they acquire new knowledge and information about the rural area, interacting with the tourist attractions/resources visited. With these dynamics, it was intended to continue learning related to curricular contents, ethical and esthetic values, in addition to training attitudes toward sustainability. The commitment to education and training as a way of promoting a CittaSlow culture is possible, as demonstrated by the results of the "CittaSlow Education" (Fig. 2) project developed by the Infias, Vizela.

School Group and which derived from the strategic vision of the international CittaSlow movement. The CittaSlow concept was started to be studied in the context of the school's grouped curriculum with approximately 1333 students (Carvalho, 2022).

The "CittaSlow Education" program started with the identification of priority areas of action, passing through the presentation of a vision and continuing with the operational objectives in order to draw up an action program (Fig. 3). Following the holding of two focus groups, the priority areas of action were listed, namely the economy and employment, the social area, the community and culture, environment/ mobility and spatial planning, as well as education and training.

Through the Focus Group, a vision of the local and school community was drafted, which was asked about the aspirations for the city of Vizela for the 2020s (Fig. 4). The results are leaked in a vision that permeates by: (i) assumption of the CittaSlow philosophy; (ii) incentive to education/training for the citizenship of the future; (iii) participation in local governance; (iv) appreciation of the endogenous and the city's resources; (v) cultural and economic enhancement of the local heritage; (vi) qualification of urban and natural spaces; (vii) investment and promotion of tourism; (ix) stimulation of co-creation and tourist

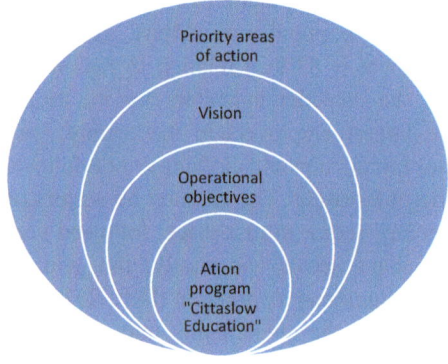

Fig. 2 Sustainability strategy of the "CittaSlow Education" program. *Source* Projeto "CittaSlow Education"

Fig. 3 Priority areas of action.
Source Focus groups; "CittaSlow Education."

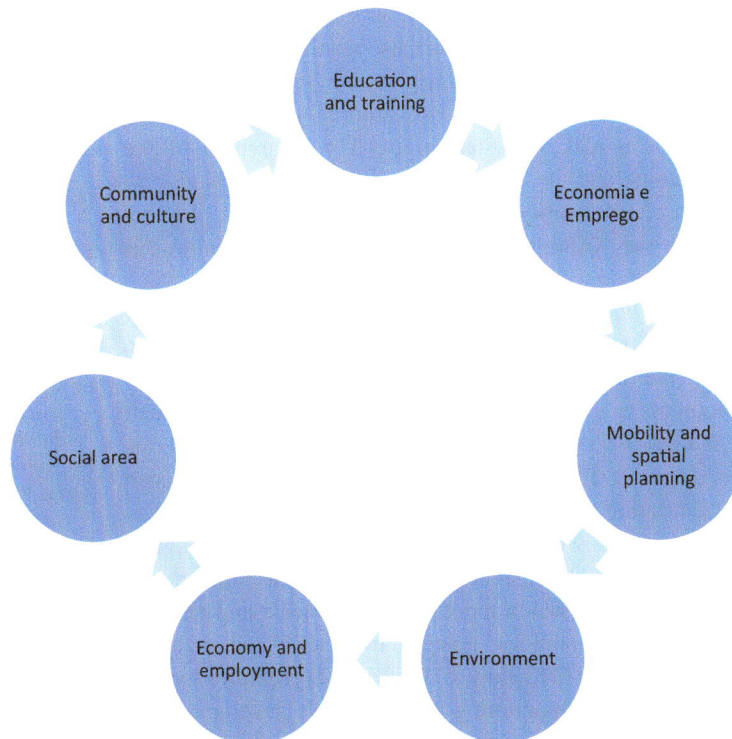

Fig. 4 CittaSlow vision keywords for Vizela. *Source* Grupos focais; "CittaSlow Education."

entertainment; (x) encouragement of quality standards for local production and marketing; (xi) investment in the social area; (xii) investment in infrastructure that guarantees sustainability circuits; (xiii) the city's notoriety for the "bem-viver" brand inspired by the CittaSlow model. Thus, considered the "priority areas of action" having as ballast the philosophy of CittaSlow, the city of Vizelle has in the School the starting point for training in the values of citizenship for conscious participation in the face of the principles of sustainability based on the use of water and energy resources, as well as natural and cultural heritage. Its economy should focus on the specificities of the territory, the environment, culture, and tourism, enhancing social cohesion. In summary, the priority areas of action thus focus on "Education and training," "Community and culture," and "Social area." From these priority areas of action, 3 essential operational objectives derive: (i) to create an inclusive society, mainly by satisfying basic needs such as health,

education, and culture, promoting participatory governance; (ii) invest in the natural and cultural heritage, as well as in all territorial specificities, namely tourist management; (iii) establish an articulation between the school and the economic agents, in order to create synergies and enhance the pursuit of common objectives in the subject of sustainability and the local economy.

6 Conclusion

The present study was intended to investigate local development strategies in low-density territorial areas, focusing on experience and strengthening local skills focused on "people." Based on the use of the CittaSlow "brand" of urban social development, with a demographic expression of less than 50 thousand inhabitants, the search was made on how the resources of the place can be trained. In this light, territorial development is based on a bottom-up approach, that is, it should be crossed with the contribution of the private sector and the community. It is in this sense that more inclusive, less hierarchical modes of regional development should be understood, considering the connection with networks and partnerships. Development goes beyond economic exclusivism to incorporate a process in which historical–geographic, cultural, political, socioeconomic and educational components intertwine. Based on the analysis of the "CittaSlow Education" project and the promotion of a diagnosis of the potential of the municipality of Vizelle, priorities were sought for its development. This also resulted in a vision that enshrines the CittaSlow principles and that points to the assumption of environmental and cultural–natural values that lead to a model of sustainability. The awareness of the community and its mobilization for a civic behavior in respect of the identity values of the place are principles that regulate the "CittaSlow Education Program." Meaning this the deepening of the ability to "build the future" which should go through the involvement and "bottom-up" strategy of the actors with relevance in the decision and in the action. This line of development is supported by endogenous resources, in a change that is envisaged from a management articulated with the local agents that constitute the core of this development. Finally, the set of priority action areas listed —"Education and training," "Community and culture," and "Social area"—with a view to local development are aligned with the UN 2030 Agenda for sustainable development. Of note is SDG 11, which reports on "sustainable and resilient cities and communities," particularly articulated with the guidelines of the CittaSlow movement with regard to support for planning and monitoring the sustainable development of this territorial thickness.

References

Angeles, L. C. (2000). *Using gender-aware participatory research methods in community poverty profiles, project planning and policy assessment.* UBC Women's Studies.

Bauer. (2016). *Movimento Slow Travel no contexto cultural do turismo no Brasil: desafios e perspectivas.* E ACH/USP.

Bekar, A. (2015). The concept of CittaSlow as a marketing tool for destination development: The case of Mugla, Turkey. *American International Journal of Social Science, 4*(3), 54–64.

Carvalho, R. M. A. F. (org.). (2022). *Projeto Educativo: Agrupamento de Escolas de Infias.* AEI.

Carvalho, R. M. R. (2014). Lentidão, território e bem-estar: O movimento da cidade lenta e a sustentabilidade do lugar. *Periódico Técnico e Científico Cidades Verdes, 2,* 73–89.

Castells, M. (2009). *A sociedade em rede.* Paz e Terra.

Davis, A. K., Knox, P., & Meyer, H. (2005). Urban vitality through the Cittàslow charter movement. In *Book of Abstracts of the International Conference for Integrating Urban Knowledge 6 Practice* (pp. 94–95).

Dickinson, J., & Lumsdon, L. (2010). *Slow travel and tourism.* Earthscan.

Ekinci, M. (2014). The Cittaslow philosophy in the context of sustainable tourism development; The case of Turkey. *Tourism Management, 41,* 178–189. https://doi.org/10.1016/j.tourman.2013.08.013

Emmendoerfer, M. L. (2019). *Inovação e empreendedorismo no setor público.* ENAP.

Emmendoerfer, M. L., Fraga, B. O., Costa, V. N. G., & Ferreira, M. A. M. (2020). Análise da inserção das cidades no movimento slow city como diretriz de qualificação e inovação territorial. *Desenvolvimento Regional em Debate, 10*(núm. Esp.), 171–194.

Ferreira, P. F., Seabra, C., & Paiva, O. (2014). Slow cities (Cittaslow): Os espaços urbanos do movimento slow. *Revista Turismo & Desenvolvimento, 21–22,* 191–192.

Gardner, N. (2009). A manifesto for slow travel. *Hidden Europe Magazine, 25,* 10–14.

Heitmann, S., Robinson, P., & Povey, G. (2011). Slow food, slow cities and slow tourism. In P. S. Robinson, S. Heitmann, & P. Dieke. (Eds.), *Research themes for tourism* (pp.114–127). CAB International. https://doi.org/10.1079/9781845936846.0114

Holden, A. (2008). *Environment and tourism* (2nd ed.). Routledge.

Ildiko, G. H. (2013). Regional and town development in hodmezo-vasarhely from the aspect of CittaSlow. *Analele Universităţii Din Oradea, 1*(20), 197–204.

INE Instituto Nacional de Estatística. (2020). *Anuário Estatístico.* INE.

Knox, P., & Meyer, H. (2009). *Small town sustainability: Economic, social and environmental innovation.* Birkhauser Verlag AG.

Mendes, J. F. G. (2014). *O futuro das cidades.* Interciência.

Miele, M. (2008). Cittaslow: Producing slowness against the fast life. *Space and Polity, 1*(12), 135–156. https://doi.org/10.1080/13562570801969572

Ministério da Economia e do Mar. (2020). *Sínteses estatísticas.* Lisboa: GEE/ MEM.

Pérez-Mongiovi, D., & Cardoso, A. (2015). O turismo no modelo *Cittaslow* de desenvolvimento urbano. O caso de Vizela no Norte de Portugal. In A. Santana Talavera, E. Gonçalves, & X. Pereiro Pérez (Eds.), *Governança e Turismo* (pp. 193–209). Ed. ISMAI e CEDTUR.

Petrini, C. (2001). *Slow food: The case for taste.* Columbia University Press.

Pink, S. (2008). Sense and sustainability: The case of the slow city movement. Local environment. *The International Journal of Justice and Sustainability, 13*(2), 95–106.

Pink, S. (2009). Urban social movements and small places: Slow cities as sites of activism. *City, 13*(4), 451–465. https://doi.org/10.1080/13604810903298557

Portugal. (1998). Lei 63/98, de 1 de setembro. Criação do município de Vizela e elevação a cidade. *Diário da República—I Série A, 201,* 4526–4527.

Portugal. Presidência do Conselho de Ministros (2007). *ENDS 2015 PIENDS Plano de Implementação da Estratégia Nacional de Desenvolvimento Sustentável.* PCM.

Quivy, R., & Campenhoudt, L. V. (2008). *Manual de investigação em ciências sociais* (5ª). Gradiva.

Radstrom, S. A. (2014). Place sustaining framework for local urban identity: An introduction and history of Cittaslow. *Italian Journal of Planning Practice, 1*(1), 90–113.

Rete Urbana delle Rappresentanze—Rur (2012). Cittaslow: From Italy to the world. International network of cities where living is easy. *FrancoAngeli, s.r.l., Milano, Italy.*

UN United Nations, General Assembly. (2015). Resolution adopted by the General Assembly on 25 September 2015. Transforming our world: the 2030 Agenda for Sustainable Development. *United Nations, Treaty Series, 2302*(41032), 1–35.

Yin, R. (1994). *Case study research: Design and methods.* Sage Publications.

Future Prospects and Trends of Tourism

Contemporary Digital Age Pilgrimage in Chichibu in Japan

Chieko Nakabasami

Abstract

This article focuses on what makes contemporary pilgrimage sustainable in Chichibu, northwest of Saitama Prefecture, located north of Tokyo, Japan. The 34 Chichibu pilgrimage temples are dedicated to Kannon bodhisattva, the Goddess of Mercy, who is very popular in Japan. The Chichibu pilgrimage originated in the thirteenth century, and it endures to this day. Especially in the Edo period, between 1603 and 1867, Chichibu was a bustling place for many Edo people for the pilgrimage to the Kannon faith. Today, there are efforts to use gamification to adapt to the current digital age to make the Chichibu pilgrimage attractive to tourists of all generations. Gamification is the use of elements of game design in non-game contexts and is designed for social activities and services. A field survey has been conducted in Chichibu to explore what tourists want to get from contemporary pilgrimage. Results have indicated that one of the key factors in the success of contemporary sustainable pilgrimage is gamificated cultural communication between local stakeholders and tourists. In this article, the future use of emotional mapping on cultural routes is also proposed to attract more tourists to destinations based on a contemporary digital age pilgrimage.

Keywords

Contemporary pilgrimage · Gamification · Cultural communication · Emotional mapping · Chichibu · Saitama

C. Nakabasami (✉)
Faculty of International Tourism Management, Toyo University, Tokyo, Japan
e-mail: chiekon@toyo.jp

1 Introduction

In this article, we focus on what makes contemporary pilgrimage sustainable in Chichibu, in northwest Saitama Prefecture, located north of Tokyo, Japan (Fig. 1). With mountainous natural beauty and a rich traditional culture, Chichibu has attracted many tourists. Some research papers can be found on tourist resources in Chichibu. For example, "Chichibu Night Festival" features gorgeous fireworks that color the night sky and local folk arts vividly performed (Akaike, 1976; Alaszewska, 2012), and, on the "Geo Tour," tourists admire the natural heritage by the "Geopark Chichibu" (Sakaguchi, 2014). According to the Japan National Tourism Organization (2018), Chichibu was visited by more than 9.8 million people, including 130,000 overseas tourists in 2017. After enduring the Covid-19 pandemic, tourists have been gradually returning to Chichibu today.

The 34 Chichibu pilgrimage temples are dedicated to Kannon bodhisattva, the Goddess of Mercy, who is very popular in Japan. Many reference books about the Chichibu pilgrimage, including guidebooks, have been published (e.g., Sato, 2004). Tourists visit the temples on foot, by bicycle, or by car. Especially in Japan, which benefits from the traditional numbering of the temples, we expect that several gamificated cultural communications using the numbers could be well established between hosts and guests on pilgrimage. On the Chichibu pilgrimage, a digitalized *stamp rally* is offered for tourists to circle the temples with a gaming feeling. A s*tamp rally* is a game almost unique to Japan, in which one visits different locations to collect stamps on a card. In the rally, one can win prizes according to the number of stamps collected. In Chichibu, tourists collect digitalized stamps by scanning a two-dimensional code on a smartphone. In addition to such a digitalized stamp rally, for future use, we propose *emotional mapping* for tourists to discover various interesting spots between the temples. According to Pánek and Benediktsson (2017), emotional mapping is a method by which people display

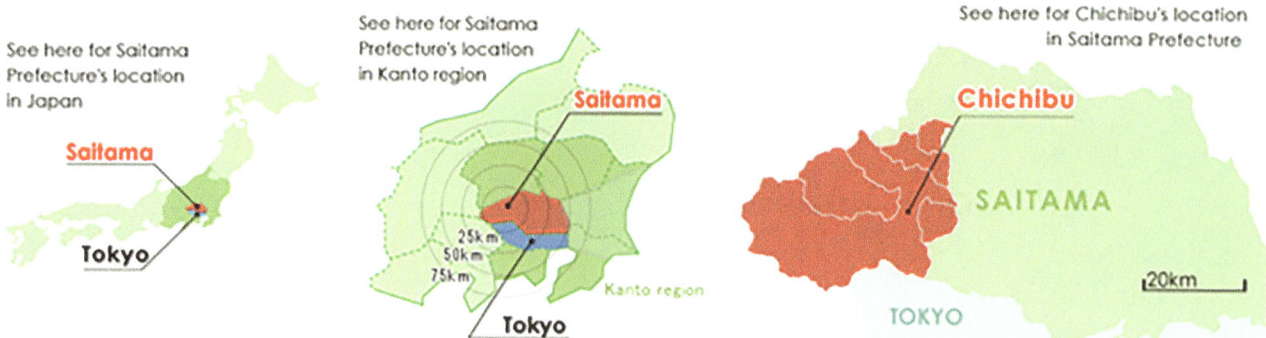

Fig. 1 Map of the Chichibu region of Saitama prefecture, north of Tokyo. *Source* https://www.chichibu-omotenashi.com/en/ (accessed on August 13, 2022)

subjective, qualitative, and bottom-up spatial information on the map. Using emotional mapping, tourists can share various information regarding the places they visit. We think that by developing cultural routes linking the temples with a social networking service (SNS), we can discover more local experiences than by only visiting each temple. SNS is a powerful tool for achieving this.

Turning global research activities, the interest of tourism researchers worldwide has been drawn to religious tourism for one of the significant tourism fields, as shown by the many contributions in ATLAS Special Interest Group on religious tourism and pilgrimage (ATLAS, 2016).

Subsequent sections are as follows. In Sect. 2, contemporary secular pilgrimage in Japan is described with reference to traditional religious pilgrimages. In Sect. 3, the Chichibu pilgrimage is introduced in more detail, and interviews with two stakeholders in the Chichibu pilgrimage are highlighted. In Sect. 4, we discuss what makes digital age pilgrimage sustainable based upon an on-site survey and analyze the results. In Sect. 5, I conclude and use the results to propose the future use of emotional mapping on cultural routes. In this article, three terms—pilgrims, tourists, and guests—are used interchangeably. In particular, "guests" will be mainly used in conjunction with "hosts."

2   Contemporary Pilgrimage in Japan

We think that there is no remarkable distinction between pilgrims and tourists in contemporary pilgrimage in Japan. In contrast to a contemporary pilgrimage, traditional religious pilgrimages were the mainstream of pilgrimages in the premodern era. In Japan, pilgrimage was purely a religious journey. In Buddhism, originally in India, sacred places were born in the course of the dissemination of Buddhism, and people started making pilgrimages to the sacred places. India's four biggest pilgrimages relating to the Buddha are: Lumbinī (Nepal), Bodhgaya, Sarnath, and Kushinagar. From

India, Buddhism arrived in Japan in 538 A.D. via the Korean Peninsula, and pilgrimage originated in the Heian period (794–1185 A.D.). In the Heian period, pilgrimage flourished in the Kinki region (around Kyoto, Osaka, and Nara in western Japan). People wished for peace and rest from daily life by circling the temples and shrines in the Kinki region. After the Kinki region, spiritual places for the Kannon bodhisattva faith were developed in Shikoku (the smallest of the four main islands of Japan located in the southwestern part of the Japanese archipelago) and Bando (Tokyo and its neighboring area) until now.

In the present age, pilgrimages in Japan have very diverse facets. Though the truly religious pilgrimage has been retained, many contemporary pilgrimages are secular, with the main goal of recreation. Before, pilgrimages were primarily travel for elderly and middle-aged people; however, they are currently popular even among younger generations. Young people are visiting hot spots, which mean miracle-working places, seeking for spiritual experiences. Their purposes are also diverse: self-discovery, relaxation, stunning experiences, etc. They are especially eager to take Instagrammable pictures to contest Instagrammability of the picture. Looking at the recent research on pilgrimage in Japan, most papers talk about "pilgrimage tourism" and treat it as a touristic commodification of pilgrimage. Kimura (2007) discussed this influence on sacred sites and how to manage such touristic trends, while Nanchi (2022) measured pilgrimage experiences through questionnaires for tourists. Kadota (2010) described the commodification of folk religion by contemporary pilgrimages from ethnographical perspectives. In this article, in contrasted to these perspectives, we discuss the sustainability of pilgrimage tourism from the viewpoint of tourism marketing strategies that maximize the Japaneseness of pilgrimage.

In addition, we would like to explain about *Seichi Junrei* (which means "pilgrimage to sacred places" in Japanese). *Seichi Junrei* is one type of travel in which fans visit the places featured in films, manga, animations, novels, and in relation to famous people. In this sense, *Seichi Junrei* can be considered both fan tourism and content tourism. *Seichi Junrei* has become more popular in the Internet age, and fans publish and share information about the target places on the Internet. Recently, more than a few local governments have used *Seichi Junrei* places for regional vivification. In fact, in 2011, one animation film set in Chichibu, titled "Anohana: The Flower We Saw That Day," was a great success and increased the number of tourists. Amano (2012) and Okamoto (2015) introduced Chichibu content tourism, and Jang (2015) discussed the contribution of content tourism to the traditional local culture. While *Seichi Junrei* is certainly a new type of pilgrimage, and many papers have been published recently on this theme, we have a different focus than *Seichi Junrei*. In this article, we consider the original pilgrimage to temples and shrines.

One remarkable characteristic of the pilgrimage in Japan is that the routes encircle a certain area (large or small), and pilgrims can traverse the route however they want. In the pilgrimage in Japan, sacred places are numbered sequentially. In Chichibu, there are 34 numbered temples; in other places, there are 88 temples and 33 temples. No one knows who numbered these temples; the numbers existed before our contemporary age. On their own, the numbers have no special meaning. Today, in many tourism guidebooks and maps, each tourist site is often numbered for the purpose of information and convenience. In comparison, pilgrimage numbering in Japan is unique in the sense that the numbers existed historically without the intention of contemporary tourism stakeholders.

Concerning pilgrimage itineraries, many pilgrimage routes in the world have a final destination, and pilgrimage routes have been developed for arriving at the destination. In Japan, there is no final destination in the pilgrimage. In the case of Chichibu, as it is close to the Tokyo metropolitan area, it is convenient for people from Tokyo to visit Chichibu. They visit as many temples as they can in a day, return home, and come back again to Chichibu whenever they would like to visit temples they have not visited yet. They can choose which temples to visit as they want. This is one peculiar feature of Japanese pilgrimage.

Smith (2003) mentioned that "the current trend towards spiritual tourism focuses on the quest for the enhancement of self through physical, mental and creative activities". In Smith (2003), religious tourism is referred as a sort of spiritual tourism like yoga tourism, ashram visits. Richards (2006) summarized some of the main dimensions about contemporary religious tourism, as shown in Fig. 2. Richards (2006) described the four dimensions of religious tourism. Religious tourism is considered to be a continuum

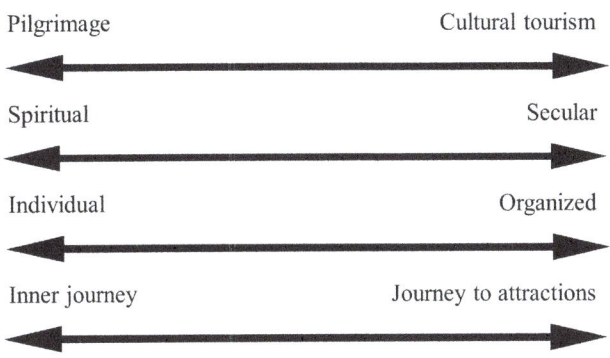

Fig. 2 Dimensions of religious tourism. Made by the author referring to Richards (2006)

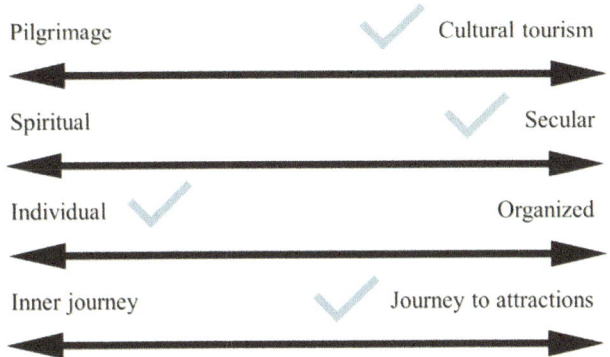

Fig. 3 Applying four dimensions of Japanese pilgrimage. Made by the author

ranging from pilgrimage to cultural tourism; however, the boundaries between them continue blur. Also, researchers note that religious tourism has changed in the course of time. Japanese pilgrimage can be adapted as Fig. 3 to the four dimensions of Fig. 2. In Fig. 3, one alternative is checked for the four dimensions. Here, we do not claim that the checked item fits all Japanese tourists, but the majority of people can be said to have such a tendency.

Though various pilgrimage tours are organized by tourism agents, we have the impression that the majority of pilgrims traverse the sacred places individually or in small groups in Japanese pilgrimage. Many secular tourists do not think of themselves as pure pilgrims; however, we would like to say that so-called eclecticism is working in the attitudes of Japanese pilgrims. In other words, in the pilgrimage, tourists seek to enjoy both spiritual experiences and secular attractions. At the same time, many temples warmly welcome secular tourists. In this article, pilgrimage is used in the same meaning as religious tourism from the situation that in Japan mainly contemporary pilgrimage includes cultural tourism.

3 The 34 Temples of the Chichibu Pilgrimage

3.1 History of the Chichibu Pilgrimage

As mentioned in Sect. 1, the 34 temples of the Chichibu pilgrimage are dedicated to Kannon bodhisattva, the Goddess of Mercy, who is very popular in Japan. The Chichibu pilgrimage originated in the thirteenth century, and it has endured until the present. Especially in the Edo period, between 1603 and 1867, Chichibu was a bustling place where many Edo people made pilgrimage to the Kannon faith. The total distance of the Chichibu pilgrimage is only 100 km. These 34 sacred Kannon temples are dotted across one city and three towns: Chichibu City, and Yokoze, Minano, and Ogano, respectively, in the Chichibu region (Fig. 4).

In Japan, as mentioned earlier, the temples on pilgrimage routes are numbered and identified by their numbers. The number has become so familiar to local people as well as tourists that each temple is often simply called by its number. Temples are also called *fudasho*, the place where pilgrims dedicate and receive *fuda* (a wooden plate) to certify their visit. A temple can be called the numbered *fudasho*. For example, *Jigenji*, a temple of "eyes" dedicated to the medicine Buddha who cures suffering using the sacred medicine (Fig. 5), is the "13th *fudasho*," or is called simply "13." Pilgrims make a resolution on *fuda* and nail the *fuda* to the wall of the temple. *Fuda* may be considered to be a lucky charm. Today, the wooden *fuda* has been replaced with one made of paper. In Chichibu, the temples are numbered from 1 to 34 and identified by their numbers, and pilgrims can visit the 34 temples in any order. The most important thing for pilgrims is to visit all of the temples on the pilgrimage route to fulfill their wish.

3.2 Gamification of the Chichibu Pilgrimage

Gamification is defined as the use of elements such as game thinking and designed for social activities and services (Deterding et al., 2011). The word *gamification* was coined around 2002 by Nick Pelling, a British-born computer programmer. By applying gaming elements to activities and services originally not games, a customer's incentive and loyalty are supposed to be strengthened. Gamification has

Fig. 4 Chichibu pilgrimage map (34 temples with their numbers in red small circles). Retrieved from https://navi.city.chichibu.lg.jp/wp/wp-content/uploads/2014/09/edojunrei_11.pdf

Fig. 5 The 13th temple *Jigenji*. Photograph taken by the author

Fig. 6 Cycle stamp rally signboard with QR code located at the 29th temple *Chosen-in*. Photograph taken by the author

been studied and implemented in various real-world domains (Hamari et al., 2014; Seaborn & Fels, 2015).

In Japan, pilgrims receive *shuin* (seal stamps) with a signature from the temple on a sheet of paper. *Shuin* are often collected in a special notebook called *shuin-cho*. Originally *shuin* were received as proof that pilgrims dedicated a copied sutra by themselves to the temple. Today, pilgrims can get *shuin* with pay at the temple without offering the sutra, while some temples set rules for receiving *shuin*, for example, one should bring copied sutra(s) as before or should wear white out of respect for the gods or need a reservation in advance. After people have collected *shuin* on the *shuin-cho*, it is treated as sacred and put on a sacred shelf in their house for pray every day. With *shuin*, people are satisfied with their sacred activity. We dare to say that collecting *shuin* might be a sort of stamp rally of the contemporary age in Japan.

In the Chichibu pilgrimage, a digitalized *stamp rally* cycle is offered for tourists to travel around the temples with a gaming feeling. The stamp rally is said to be a sort of gamification that may be unique to Japan, in which a person visits different locations to collect stamps on a card. In the rally, one can win prizes according to the number of stamps collected. Prizes that can be won in Chichibu include local money exchangeable only in Chichibu, local hotel coupons, or something useful in Chichibu. Cyclists collect digitalized stamps by scanning two-dimensional QR codes on a smartphone (Fig. 6).

Stamp rallies are a promising method of attracting tourists. In our ancestors' arbitrary numbering of the temples, we find a gamification element that contributes to tourism promotion such as the stamp rally. It is said that in the latter half of the seventeenth century of the Edo period, some pilgrimage guidebooks were published and became very popular among ordinary people. At that time, they were not allowed to travel as high-ranking people did, and pilgrimage was their only occasion to travel. It is believed that the authors of the guidebooks numbered the temples arbitrarily, and these numbers remain today. We imagine that for pilgrims, the numbers might be a kind of itinerary record that helps them estimate how long it should take for them to complete their travel and receive their wish. Psychologically, gamification has the potential to encourage completion of the pilgrimage, as people tend to concentrate on achieving the end goal; in other words, they want to win the game. It is interesting to speculate whether people in the Edo period more than 300 years ago had the same game feeling as people of today.

3.3 Interview with Local Stakeholders

Interview with the Chief Officer of the Chichibu Omotenashi Tourism Organization

We conducted an interview with the Chief Operations Officer of the Chichibu *Omotenashi* Tourism Organization. The Japanese word *Omotenashi* means "hospitality." This organization is a Destination Management (or Marketing) Organization (hereafter DMO) of the Chichibu region (Chichibu Omotenashi Tourism Organization, 2014). One of the DMO's missions is to plan strategies for realizing the regional development of tourism on regionally oriented concepts in collaboration with various stakeholders. DMOs of Japan have various governed areas, for example, one city, one prefecture, and between cities. The Chichibu *Omotenashi* Tourism Organization is of the type between cities. This organization is composed of the members of Chichibu City and four neighboring small towns of western Saitama Prefecture: Yokoze, Nagatoro, Minano, and Ogano. These five towns make up the Chichibu region. The Chichibu region has an area of approximately 900 km², and its population is around 94,000.

The officer of the Chichibu *Omotenashi* Tourism Organization is the very person who invented the Chichibu pilgrimage's cycle stamp rally. Through the interview, we gained much useful information about the Chichibu pilgrimage. The brief summary of the interview is written below.

- *How was the cycle stamp rally invented?*
  Saitama Prefecture is said to be the origin of the bicycle in the beginning of the eighteenth century, in spite of the fact that various origins are claimed around the world. Due to the verisimilitude of this tradition, we planned a stamp rally by bicycle. On the current wave of smart media, we used a digital method of QR code scanning instead of a paper card.
- *Who is using the cycle stamp rally most?*
  Middle-aged to elderly males are the most frequent users. This is because of the bicycle, but it may change to a younger generation when overseas tourists return in the future as before Covid-19. We expect that ecology-intensive tourists will tend to use bicycles to move among the temples. In addition, people can move faster on bicycles than on foot, and parking space is not necessary as it is for cars.
- *Have you discovered something new recently in the Chichibu pilgrimage?*
  I have found that the people who make the pilgrimage have been increasingly younger. I think that one reason was some promotional commercial message (CM) on TV by one railway company. In the CM, a pair of young ladies visit Chichibu. It presented such an appealing image that the younger generation wanted to visit Chichibu. The other reason was thanks to the animated film "Anohana: The Flower We Saw That Day." The film was set in Chichibu City, and we saw many local places of Chichibu in the film's scenes. In fact, the 17th temple, *Jorinji*, was chosen as one of the main film scenes. Many film fans have visited *Jorinji* after watching the film.
- *What is the driving force for promoting the Chichibu pilgrimage?*
  We are proud to say that sound communication has been established among our organization and the temples. All 34 temples are very cooperative with local sustainable vivification projects. In the cycle stamp rally, we were afraid that not all 34 temples would agree to put up QR code signboards. Fortunately, our worry was in vain. All temples were willing to accept the stamp rally project. For example, speaking of the film previously mentioned

"*Anohana*," animation scenes drawn on *ema* are sold in the 17th temple. *Ema* is a pictured small wooden plate for making a resolution, which is hung on a special shelf in the temple.

- *We heard that there are plural Buddhist religious schools in the 34 temples. Are there any problems among them?*
  I have not found any problems. There are three different Buddhist schools—*Soto*, *Rinzai*, and *Shingon*—in the 34 temples. Each has the teaching of all three schools. In the united committee composed of the 34 temples, we discuss how to make Chichibu more appealing to tourists. In the discussion, the difference among the schools has not been raised so far.

Interview with the Temple Master of the 13th fudasho Jigenji

We interviewed the Temple Master of *Jigenji*, already mentioned in Sect. 3.1. He is also the chief of the association of the 34 Chichibu temples. Through a field survey of the Chichibu pilgrimage, we were pleasantly surprised to find that each temple seems well united in hospitality toward tourists. This is because the cycle stamp rally project cannot be achieved without the cooperation of all 34 temples. We would like to know how the association manages the cooperation of the temples.

- *What do you think of the eclecticism of spirituality and secularity?*
  I think that it is OK. Appealing to more people who are not interested in the 34 Chichibu pilgrimage temples is very important. Anyway, more people must come to know the Chichibu pilgrimage. The entrance should be opened wide as much as possible, and secular promotions play a vital role in making religious tourism sustainable.
- *How do you manage well the association of the 34 temples?*
  Like any associations, there are many complicated human relationships. When some association members are against major decisions, we must listen to them to persuade them. It can be said that we are linked by the ancestors who conducted religious festivals for a long time. Failing to cooperate for the festivals would negate that link, so we must cooperate. On the other hand, recent secular projects do not harm their faith essentially; therefore, they are willing to accept them.
- *What do you expect from researchers of tourism studies?*
  If possible, I would like to hear as many opinions of young generations as possible as a key to future sustainable tourism. I would like the researchers to collect the opinions of the young and to organize events such as dialog between young people and the temples.

4 Discussion

In this chapter, we will discuss what makes contemporary pilgrimage sustainable in Japan and explore what tourists want to get through contemporary digital age pilgrimage. Taking the field survey in Chichibu into consideration, we propose a transition toward successful sustainable contemporary pilgrimage in Japan as shown in Fig. 7. From the point of view of sustainable tourism, in Fig. 7, the two axes are set as flexibility and communication. The horizontal axis expresses the strength of communication between hosts and guests. Here, the hosts are stakeholders of the pilgrimage, including local organizations like DMO and the temples as target sacred places. Communication means interaction between hosts and guests. Communication cannot be built when only hosts offer something to guests. When guests take some action in response to hosts, communication is established in both directions. The vertical axis expresses the level of flexibility in accepting both spirituality and secularity. In other words, it means eclecticism: the higher the accepted flexibility, the greater the opportunity for tourists to obtain both spiritual and secular experiences in the pilgrimage. We consider three statuses in this two-dimensional plane: A, B, and C. The arrows on the plane represent the recommended transition from one status to another, and necessary actions will be taken to move forward in the direction of the arrow.

We try to explain each transition process toward sustainable tourism using Fig. 7. In this context, we think that the most desirable status is C—strong communication and high flexibility. We start from A, in which sufficient communication is not built between hosts and guests, and spirituality and secularity are not offered flexibly enough to guests. We explain the transitions in Fig. 7 exemplifying the Chichibu pilgrimage as follows. Currently, the Chichibu pilgrimage is thought to be in status C.

- $A \rightarrow B \rightarrow C$

  As enough communication has not been built between hosts and guests, the hosts make an effort to offer guests more flexibly secular services in connection with pilgrimage. Today, *Zen* experience is a popular spiritual activity in Japanese temples. After a *Zen* experience, some temples provide various secular activities: *Kyudo* (Japanese archery), and bathing in a sauna near the temple. In Chichibu, diverse secular activities have been planned in collaboration with the temples, the Chichibu *Omotenashi* Tourism Organization, and the Chichibu Chambers of Commerce. The cycle stamp rally introduced in Sect. 3.2 is one such activity. Another past activity was a marathon race that ran through the 34 temples. Such adjoined secular events always intended to inform about the 34 Chichibu temples. Even if cyclists and amateur marathon runners are not interested in the pilgrimage itself for the first time, these secular activities might trigger their interest in the 34 Chichibu temples sufficiently to cause them to visit them the next time. Then, to make such a visit sustainable (status C), the local stakeholders should always interact with all guests. The sociability of the SNS allows guests to write comments, and the hosts—as well as other guests—react to the comments. In fact, there are several SNS for the Chichibu pilgrimage, and many famous pilgrimage places in the world have their own Facebook or other SNS.

- $A \rightarrow C$

  It would be possible to move directly from status A to C. As one tool for achieving this, gamification elements can be applied for better interaction to plan secular activities combined with spiritual ones. Basically, gamification is a secular strategy. For example, using their passion for succeeding in the game encourages tourists to visit all of the temples; at the same time, they receive gamificated cultural information as well as information about the temples, and they typically react by visiting the places. Such behavior may well be described as an interaction between hosts and guests. Though gamification is not the only strategy, it has the power to potentially achieve both flexibility and communication simultaneously. Also, considering our current digital age, as young generations especially can easily accept the synthesis between spirituality and secularity with digital devices, the effective installation of digital arts proposed by Dominguez et al. (2014) might be another promising method for accommodating rapid interaction during traditional religious festivals.

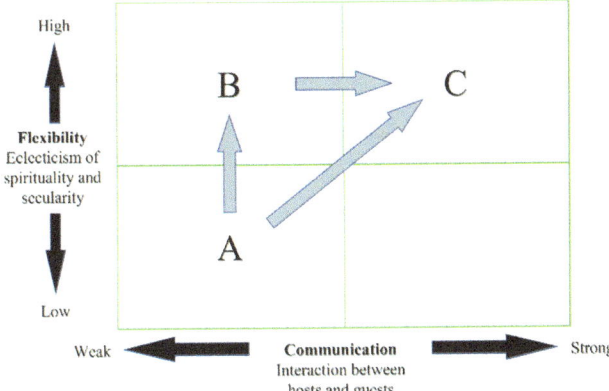

Fig. 7 Transition toward successful sustainable contemporary pilgrimage in Japan

5 Conclusions

In this article, we discussed what makes contemporary pilgrimage sustainable in Chichibu. The results from several field surveys were also shown, and the sustainability of Japanese pilgrimage was analyzed considering its unique features. In fact, the author has traveled to all 34 temples to research the Chichibu pilgrimage. I conclude that high flexibility in accepting both spirituality and secularity and strong interaction between local stakeholders and tourists play key roles in sustainable contemporary pilgrimage. Also, we would like to assert that gamificated cultural communication can be a promising tool for improving interaction.

We introduced Japanese contemporary pilgrimage and then tried to describe aspects of pilgrimage peculiar to Japan using the dimensions of religious tourism as a theoretical approach. Especially in Japan, which benefits from the temples being numbered traditionally, we expect several gamificated cultural communications using the numbers could well be established between hosts and guests during the pilgrimage. Efforts are being made to use gamification—adapted to the current digital age—to attract tourists of all generations to the Chichibu pilgrimage. We summarized interviews with the Chief Operations Officer of the Chichibu *Omotenashi* Tourism Organization and the Temple Master of the 13th of the 34 temples. Based on the field survey in Chichibu, we discussed successful sustainable contemporary pilgrimage in Japan and proposed a two-dimensional plane based on the levels of communication and flexibility.

In a future work, I plan to focus on emotional mapping of cultural routes as a way to attract more tourists to the areas between the temples. From a tourism perspective, cultural routes are very important by-products of contemporary pilgrimage. Many famous cultural routes have been developed in the world. One of the most famous and successful cultural routes is Santiago de Compostela in Spain. Santos (2002) described some factors behind rising interest in the cultural routes accompanying the pilgrimage of Santiago de Compostela as follows: a recovery of religious spirit at the turn of the century, recovery of the European spirit, tourist strategy of Galicia, and global events in Spain in 1992 such as the Barcelona Olympics and the Seville Expo. On the other hand, from the perspective of touristic behavior in the pilgrimage, cultural routes can offer special tourist resources as compared with point movement on the route traversing the temples. Tourists might discover unexpected cultural places between the temples. In the current digital age, we can use map applications on social media to obtain practical information in real time. Considering the retrieval force of real-time spatial information on social media enables tourists to visit previously unknown places. Various local tourism promotions should be conducted by using the influential force of social media. According to Pánek and Benediktsson (2017), emotional mapping is a method by which people display subjective, qualitative, and bottom-up special information and emotions on a map. Bakas et al. (2021) introduced emotional mapping into the development process of creative activity. Using emotional mapping, tourists can share various emotions they feel at a place. The information from emotional mapping is relatively self-centered. However, through emotional mapping, it is essential to produce empathy among tourists. Emotions are very much connected to spaces and places, as emotions provide a sense of place and identity to specific locales (Longley & Duxbury, 2016). Davidson et al. (2012) mentioned that emotional mapping can show ways in which meaning and values may be grounded in embodied experiences. This information can allow tourism planning designers to build in opportunities for tourists to have these embodied experiences, moving closer to reaching the "authentic everyday experience" that many tourists really desire. We would like to propose digitalized emotional mapping on a cultural route using map applications of earned media like SNS. Using a map application like Google Maps for the current digital age, the sharing of real-time information (not published in guidebooks) is needed, and the sociability aspects evoke empathy and co-creation among tourists. From the point of view of communication, the sharing of real-time information can also be a very strong interaction between hosts and guests.

As for something needing improvement, language internationalization in the gamification offering will be needed to increase accessibility for every tourist. While many Chichibu Web sites accommodate several languages, the cycle stamp rally has only the Japanese version. In future, after the Covid-19 pandemic, many inbound tourists are expected to come to Japan, so more hospitality must be prepared in multiple languages for exchanging cultural communication.

Acknowledgements Mr. Masayuki Inoue, the Chief Operations Officer of the Chichibu *Omotenashi* Tourism Organization, and Mr. Yukiyasu Shibahara, the Temple Master of the 13th temple *Jigenji*, pleasantly provided interviews for this research. I would like to express my gratitude for their kind cooperation and very useful comments.

References

Akaike, N. (1976). Festival and neighborhood association: A case study of the Kamimachi neighborhood in Chichibu. *Japanese Journal of Religious Studies*, 127–174. (in Japanese).
Alaszewska, J. (2012). Promoting and preserving the Chichibu Night Festival: The impact of cultural policy on the transmission of Japanese folk performing arts. In *Music as intangible cultural heritage: Policy, ideology, and practice in the preservation of East Asian traditions* (pp. 197–212).

Amano, K. (2012). Sightseeing promotion practice which utilized contents—The case of the Chichibu anime-tourism. In *Proceedings of the General Meeting of the Association of Japanese Geographers, Spring* (p. 100260). The Association of Japanese Geographers. (in Japanese).

ATLAS. (2016). *ATLAS special interest group: Religious tourism and pilgrimage research group.* Accessed 15 August, 2022 from http://www.atlas-euro.org/sig_religous.aspx#2013

Bakas, F. E., de Castro, T. V., & Osredkar, A. (2021). User-centred design for creative tourism prototyping: The Maribor experience. In *Creative tourism: Activating cultural resources and engaging creative travelers* (pp. 93–104). CABI.

Chichibu Omotenashi Tourism Organization. (2014). *About Chichibu.* Retrieved August 13, 2022, from https://www.chichibu-omotenashi.com/en/

Davidson, J., Smith, M. M., & Bondi, L. (Eds.). (2012). *Emotional geographies.* Ashgate Publishing, Ltd.

Deterding, S., Dixon, D., Khaled, R., & Nacke, L. (2011). From game design elements to gamefulness: Defining" gamification". In *Proceedings of the 15th International Academic MindTrek Conference: Envisioning Future Media Environments* (pp. 9–15).

Dominguez, M., Paulino, F. F., & Silva, B. M. (2014). Between the sacred and the profane in the S. João d'Arga's festivities: A digital art installation. *International Journal of Creative Interfaces and Computer Graphics (IJCICG), 5*(1), 1–20.

Hamari, J., Koivisto, J., & Sarsa, H. (2014). Does gamification work? —A literature review of empirical studies on gamification. In *47th Hawaii International Conference on System Sciences* (pp. 3025–3034). IEEE.

Jang, K. (2015). The Anohana Rocket at the Ryūsei Festival and Menma's wish: Contents tourism and local tradition. In *The theory and practice of contents tourism* (pp. 51–56). Hokkaido University.

Japan National Tourism Organization. (2018). *Inbound research report: Case study in the Chichibu Omotenashi Tourism Organization.* Retrieved August 13, 2022, from https://action.jnto.go.jp/wp-content/uploads/2019/06/chichibu_inbound_2.pdf

Kadota, T. (2010). Commodification of folk religion and its re-embedding in everyday life: An experimental ethnography on pilgrimage tourism in contemporary Japan. *Bulletin of the National Museum of Japanese History, 156,* 201–243. (in Japanese).

Kimura, T. (2007). Pilgrimage to roman catholic churches and tourism in Nagasaki. *Nagasaki International University Review, 7,* 123–133. (in Japanese).

Longley, A., & Duxbury, N. (2016). Introduction: Mapping cultural intangibles. *City, Culture and Society, 7*(1), 1–7.

Nanchi, N. (2022). An investigation on the measurement of pilgrimage tourism experiences—Based on the questionnaire survey for Saikoku Sanjusansho pilgrimage tourists. *Journal of Japan Institute of Tourism Research, 33*(1), 89–105. (in Japanese).

Okamoto, T. (2015). Otaku tourism and the anime pilgrimage phenomenon in Japan. *Japan Forum, 27*(1), 12–36.

Pánek, J., & Benediktsson, K. (2017). Emotional mapping and its participatory potential: Opinions about cycling conditions in Reykjavík, Iceland. *Cities, 61,* 65–73.

Richards, G. (2006). Religious tourism in northern Portugal. In *Cultural tourism* (pp. 227–250). Routledge.

Sakaguchi, S. (2014). Restructuring of regional tourism with the development of the Geopark and its sustainability; A case study of geopark—Chichibu, Saitama. In *Proceedings of the General Meeting of the Association of Japanese Geographers, Spring 2014* (p. 100137). The Association of Japanese Geographers. (in Japanese).

Santos, X. M. (2002). Pilgrimage and tourism at Santiago de Compostela. *Tourism Recreation Research, 27*(2), 41–50.

Sato, H. (2004). *Henro to junrei no shakaigaku.* Jinmon Shoin. (in Japanese).

Seaborn, K., & Fels, D. I. (2015). Gamification in theory and action: A survey. *International Journal of Human-Computer Studies, 74,* 14–31.

Smith, M. (2003). Holistic holidays: Tourism and the reconciliation of body, mind and spirit. *Tourism Recreation Research, 28*(1), 103–108.

Digital Interpretation as a Visitor Management Strategy: The Case of Côa Valley Archeological Park and Museum

Gorete Dinis, Maria João Carneiro, Michelle Maiurro, Maria Mota, and Rita Abrunhosa

Abstract

The context of the health emergency caused by the Covid-19 pandemic led museums to adopt visitor management measures and interpretation techniques, which would lead to a closer relationship with the public and encourage visits. The heritage interpretation is more than transmitting information, and it is an educational activity, which can contribute to improve the visitor's experience and to the sustainable development of local territories. The development and growing use of information technologies and multimedia techniques has boosted the use of digital media as a way of interpreting heritage. These techniques are crucial tools for museum management. However, digital interpretation is more than just a selection of technologies. It requires an integrated interpretive framework. This paper intends to understand how digital interpretation may be implemented in the scope of archeological heritage. In order to achieve the research objectives, besides a brief literature review, the Côa Valley Archeological Park and Museum is analyzed as a case study. This attraction was selected since the Côa Valley Archeological Park is classified as a UNESCO World Heritage Site and is an open-air gallery of Paleolithic carvings, which was born from the strength and perseverance of civil society, revolting against the construction of the dam in the Baixo Côa, and which became the main theme of the Côa Museum. The case study was analyzed through observation and an interview carried out with the representative of the Archeological Park under analysis. It was found that hard and soft strategies are being adopted for visitor management, and that various kinds of digital interpretation techniques are used, allowing great interactivity with the visitor. However, after the diagnosis, suggestions for improvement were proposed at this level.

Keywords

Digital interpretation • Heritage • Museum • Tourism • Côa Valley Archeological Park • Visitor management

G. Dinis (✉)
Research Units on Governance, Competitiveness and Public Policies (GOVCOPP), Polytechnic Institute of Portalegre, Aveiro, Portugal
e-mail: gdinis@ipportalegre.pt

G. Dinis
CITUR Algarve, Leiria, Portugal

M. J. Carneiro
Research Unit on Governance, Competitiveness and Public Policies (GOVCOPP), University of Aveiro, Aveiro, Portugal
e-mail: mjcarneiro@ua.pt

M. Maiurro · M. Mota · R. Abrunhosa
University of Aveiro, Aveiro, Portugal
e-mail: michellemaiurro@live.ua.pt

M. Mota
e-mail: mariamota@live.ua.pt

R. Abrunhosa
e-mail: regalorita@ua.pt

1 Introduction

Archeology, encompassing some of the material remains of culture, may provide people important information. This information may range from history to fields such as sociology and economy (MacWhite, 1956). Various archeological heritage is also appreciated by visitors in the scope of tourism trips. However, the characteristics and relevance of some archeological remains are not easily understood by visitors, decreasing thus the value of these artifacts. Moreover, some of these remains are fragile and care must be taken when exploiting this kind of heritage to tourism. Interpretation has a crucial role in this scope, due to its ability to assign significance and value to touristic resources (Tilden, 1977), while simultaneously contributing to protect resources (Bramwell & Lane, 1993).

Interpretation can be implemented through several means such as publications and some personal interpretations (Murta & Goodey, 2002). Nevertheless, technologies opened a wide range of opportunities to new forms of interpretation. Digital interpretation, making use of computers and of Internet, encompasses a wide range of techniques such as searching additional information in computers, to virtual reality (VR), augmented reality (AR) (Puig et al., 2020), among many others. This kind of interpretation revolutionized the way information was provided and extended the information that may be presented to visitors.

Despite the important role of interpretation in the case of archeological resources and of the high potential of technologies in the field of interpretation, there is scarce research on how the digital interpretation may be implemented in archeological sites. Moreover, to the best of our knowledge, there is no research providing an overview on the range of digital interpretation techniques that may be adopted in archeological heritage nor on how these techniques may be implemented. This paper aims to identify the digital interpretation techniques that may be used in archeological resources and analyze which kind of these digital techniques were implemented in an archeological site in Portugal, to better understand the process of implementation of these techniques. A case study approach involving interviews, among other techniques, was used to carry out the empirical research.

This paper is structured in five sections. The first section corresponds to the introduction, where the relevance and objectives of the research are presented. In the second section, a literature review on interpretation, digital interpretation, and digital interpretation of archeological heritage is discussed. Next, in the third section, the methodology adopted in the empirical research is described. The findings of the empirical research are analyzed and discussed in the fourth section. The paper finishes, in the fifth section, with the most important conclusions and implications. Some limitations and suggestions for further research are also presented.

2 Literature Review

2.1 Heritage Interpretation

Freeman Tilden, pioneer of interpretation, says in his book that numerous naturalists, historians, archaeologists, and other experts are working to disclose, to tourists who request the service, some of the beauty and wonder, inspiration, and spiritual significance that lay beyond what the visitor can experience physically. This process is designed as interpretation (Tilden, 1977, p. 3). As stated by Beck and Cable (2011), the purpose of interpretation is to shed light on the significance of our cultural and natural resources. Interpretation improves our understanding, admiration, and conservation of historical monuments and natural wonders through a variety of media, including presentations, narrated tours, and displays. In parks, forests, wildlife refuges, zoos, museums, and cultural places, interpretation is a process that provides educational and inspirational content. Even if these locations are inspiring themselves, interpretation can help us appreciate their beauty and significance in greater detail while also preserving their integrity (Beck & Cable, 2011).

Tilden (1977) identified six principles that should be considered in interpretation, according to which the interpretation should: (i) be more than information, revealing something based on information; (ii) establish a connection between what is displayed and the personality or experience of the visitor; (iii) encompass many arts, and each of them may be taught; (iv) provoke visitors; (v) provide a holistic presentation of the heritage, not only of a part; (vi) be different for children and for adults. Taking into account these principles, as Tilden states, it may be concluded that interpretation must stimulate the visitors' interest and curiosity, as well as provide visitors with enduring information, which is relevant to them. A lot of other researchers developed their own sets of guidelines for heritage interpretation.

According to ICOMOS, the term "interpretation" describes the entire spectrum of activities aimed at raising public awareness and enhancing knowledge of cultural heritage sites - print and electronic publications, public lectures, on-site installations, and those related to them off-site, of which educational programs, neighborhood activities, continuous research, training, and interpretation process evaluation are only a few examples (ICOMOS, 2008).

Off-site and on-site interpretation can be provided in either a personal (guided walks) or a nonpersonal (brochures) format. Tilden (1977) first identified the molds of guided media (with the presence of a guide) and self-guided media (e.g., without a guide, using maps, boards, folders). Self-guided (or indirect) means can be used in conjunction with guided (or direct) means to complement the information transmitted to visitors and thus provide a better understanding of the cultural and/or natural heritage found on the site (Tilden, 1977).

Murta and Goodey (2002) use a different categorization approach, establishing that interpretation techniques can be classified into three types: live interpretation, texts and publications, and design-based interpretation. Live interpretation, also known as personal interpretation, requires a person, such as a guide or an actor, to explain the theme to visitors. This technique can be represented through demonstration, representation, and performances, walking excursions, bicycle or motorized excursions, and the use of an interpreter is a key factor in the process (Murta & Goodey,

2002). Publications, which include illustrated maps, guides and scripts, folders and postcards, are important components in the interpretation process. They supplement the information obtained in the exhibits, serve as personal guides for visitors, and occasionally function as souvenirs that can be taken home. The publications are appropriate to the specific needs of visitors and should contain interesting information, while also provoking curiosity about the site, revealing details that can cause awareness (Murta & Goodey, 2002).

Natural and cultural heritage both use design-based interpretation, which ranges from traditional information boards and panels to video and computerized equipment. These means, which continue to be classified as static media (texts, illustrations, and representations) and animated media, make public visits more appealing and enjoyable (due to the sound, light, image, and movement) (Murta & Goodey, 2002).

A good heritage interpretation program, according to Bramwell and Lane (1993), should be an effective tool for managing the movement of visitors to a heritage area. This is accomplished by constructing so-called filter centers, which divert visitor traffic around the area by redirecting them to the many different attractions and thus influencing the flow of tourist streams in both time and space. These activities seek to redirect tourists away from areas most vulnerable to tourism pressure by directing them to alternative attractions, sightseeing routes, and heritage sites (Bramwell & Lane, 1993).

2.2 Digital Interpretation

Various digital technologies, including technological devices of virtual reality (VR), augmented reality (AR), mixed reality (MR), extended reality (XR), help to disseminate information and implement storytelling in a more effective way. These digital storytelling methods are increasingly popular in cultural heritage (Liarokapis et al., 2020; Rizvić, 2017).

Okanovic et al. (2022) state that "digital storytelling is an evolving methodology where the content is multimedia." It is considered an effective tool for presenting cultural and historical heritage, especially with the shortage of budget and professionals. Moreover, digital storytelling has become more consolidated in times of the Covid-19 pandemic. Museums have been impacted with restrictions, interrupted personal visits, and implemented social distancing (Magliacani & Sorrentino, 2021). In this sense, Bradburne (2019), Samaroudi et al. (2020), and Agostino et al. (2021) say that turning to digital technologies is a resilience strategy that museums can adopt to maintain visitor access.

McCrary (2011) argues that digital technologies are cultural creations and can be used to transform the institutional cultures, methods, and interaction with the public of cultural heritage. This researcher also states that the technological challenges of cultural heritage are not of the character of the digitization process, but of the interpretation process. A growing trend is an increasingly active public, where the visitor is placed in the role of active participant, reformulating the role of the service suppliers (McCrary, 2011).

Some of the digital technologies, that go from Web sites, videos (Roche & Quinn, 2017), and audio guides (Van Winkle & Backman, 2011), to more complex technological approaches, are already explored in cultural heritage and tourist destinations, enabling greater immersion. Puig et al. (2020) argue that VR can be used to recreate virtual historical sites, permitting the visitor to see scenes of ancient societies and have a holistic understanding of that societies, which often becomes difficult only by notes and objects displayed in museums. Still within virtual technologies, there is AR, which combines virtual space and physical scenes (Carmigniani et al., 2011; Garzon et al., 2019), adding real objects with digital audio, video, and virtual information artifacts (Azuma, 1997; Sommerauer & Müller, 2014). Both VR and AR are based on experiential learning (Behrendt & Franklin, 2014; Kolb, 2014). Kolb (2014) states that experiential learning involves four interactive processes: concrete experience, reflective observation, abstract conceptualization, and active experimentation. VR makes cultural heritage digitally accessible, even though some health constraints may be imposed in its use in certain contexts, as in the case of the Covid-19 pandemic (Chotrov & Bachvarov, 2021). Mixed reality technologies give the user control of both real and virtual objects. Most AR applications use mobile devices, simple to implement and effective in their use (Haugstvedt & Krogstie, 2012; Jung & tom Dieck, 2017).

Okanovic et al. (2022) highlight the relevance of using web technologies for preserving cultural heritage and also argue that the content available on the web reaches a larger number of people, not only those who are in the cultural heritage. Thus, the technology can be used to meet the needs of diverse public, with different ages, levels of education, and technological skills. Therefore, the interaction of the user with the technology should be simple and intuitive. Moreover, new forms of exploration and visits to heritage sites are being offered and promoted through alternative itineraries, creating adaptations, responding to different demands of visitors, sometimes even dispensing the physical presence in the sites, and opening a range of possibilities.

Digitization and the Internet create opportunities for visitors to enjoy new narratives, transforming the interaction with sites, namely regarding access and cultural information available to the public (McCrary, 2011). Museums, for example, can adapt their traditional interpretation of design areas, redirecting the contents and relationships through digital technologies, and build a virtual interpretation (Magliacani & Sorrentino, 2021). Magliacani and Sorrentino (2021) also mention that after the shock of the Covid-19 pandemic, museums had to resort to digital technologies to maintain their services and co-creation, with the most common technique being used being virtual interpretation.

To make the visits more attractive, some VR applications use games to make the story more interesting and attract more people, also making the user more active in exploring cultural and historical heritage (Okanovic et al., 2022). Digital interpretation can play a major role improving access of visitors to the site before, during, and after the trip. Digital interpretation not only extends the set of information to which visitors have access while visiting attractions, as also provides them more information prior the trip, in order to plan it. Furthermore, digital interpretation creates an opportunity for users to generate content, or even make decisions about where and how they want to experience the heritage (McCrary, 2011).

Pandemic has brought the need of new ways of accessibility to monuments and cultural sites, and digitalization has proved to be useful and, in some cases, a solution to ensure people's access to these places. In this way, improving the practices and experiences of VR, AR, and XR, among other techniques, becomes paramount for the cultural heritage sector in its digital transformation (Maietti et al., 2021).

Despite the crucial role that digital media may have in the interpretation of cultural heritage, some researchers identify important issues to consider when implementing VR or AR in a site. Gatelier et al. (2022) proposed a business model for innovation methodology to implement digital interpretation experiences in cultural heritage attractions. They suggest to first select sites which have already innovated or are in process of innovating, collecting data from visitors and from managers of the sites, and then analyzing data to, after that, select the best innovation approach in terms of VR or AR for the site. In the same vein, De Ascaniis and Cantoni (2022) suggest the framework ABCDE to increase understanding on the way digital media should be used to promote a sustainable relation between tourism and heritage. This framework suggests that the digital media should be implemented in cultural heritage in a way it permits to: (i) increase access to the heritage, delivering more information; (ii) improve visitors' experiences, generating more value to them; (iii) create a stronger connection between the supplier and the visitor; (iv) offer suppliers and visitors the opportunity to directly communicate with each other; (v) train suppliers. The approaches proposed by Gatelier et al. (2022) and De Ascaniis and Cantoni (2022) suggest that digital interpretation must be implemented with special care in cultural heritage.

2.3 Digital Interpretation of Archeological Heritage

According to Ballart (1997, p. 17), "heritage, as we understand it in the sense of what we possess, appears historically when in the course of generations, an individual or group of individuals identifies as their own an object or set of objects." Therefore, heritage is not any object that is linked to the past, but objects of the past that are linked to the culture of the existing people. Silva (2000) refers that the element that defines the concept is the capacity of symbolic representation of the identity of people, which builds a bridge between the past and the present.

Archeology studies the various stages of human history, from the Paleolithic period to the present day (Zafra De La Torre, 2017). Thus, "Archeological heritage is the set of material elements, both movable and immovable, whether or not they have been extracted and whether they are found on the surface or underground, in the ground or in the water, which together with their context […] will serve as a historical source for the knowledge of humanity's past" (Manuel et al., 2011, p.72).

Through archeological heritage, it is possible to get to know the habits and experiences of our ancestors and the usefulness of the artifacts to individuals (Zafra De La Torre, 2017) to have a better interpretation and understanding of history. The artifacts found reveal, in some way, the culture and identity of the people.

Interpretation of archeological artifacts may provide a wide variety of information that ranges from history, for example concerning the development and changes of group patterns across the time, to sociology, regarding, for example behaviors of certain cultures, and even extending to economy, revealing characteristics of economic activities of archeological sites (MacWhite, 1956). In the same line, according to Zafra De La Torre (2017), its study encompasses physical, socioeconomic, and symbolic dimensions (Zafra De La Torre, 2017).

Despite the relevance of archeology, many characteristics and specificities of archeological heritage may be difficult to interpret by many visitors. Moreover, some archeological artifacts may be fragile, requiring special care by visitors who visit the sites where they are located, and some parts of archeological sites may not be very accessible. Digital interpretation can play a major role in this context, improving access of visitors to the site, while also preserving most fragile remains.

Considering the technological advances that have occurred over time, the field of archeology is no exception. One evidence of this was that, in 1990, Reilley advocated and defined the concept of virtual archeology, as a form of simulating or describing an archeological site, namely objects, which do not exist anymore or that are in a bad conservation state, by creating models or replicas using computer techniques (Reilly, 1990). The first International Congress of Archaeology held in Spain led to the need for the Spanish Society of Virtual Archaeology (SEAV) to draw up the Seville Charter, which would set out the principles and criteria for measuring the quality of virtual archeology projects (Cáceres-Criado et al., 2022).

From this concept, several studies and projects arose, not only for research, but also for attracting tourists to sites with archeological heritage. Researchers from Duke University and University of California Merced studied various ways of applying this concept. Among the various projects developed, stands out the development of a Dig@IT application for the archeological site of Çatalhöyük, in Turkey, from the Neolithic and Chalcolithic era that creates inferential models for archeological interpretation, and through which it is possible to have a 3-D visualization of the various layers and the surrounding landscape, and even virtually excavate the site (Lercari et al., 2017), which may be used to the interpretation of the site.

As well as using 3-D, VR can also be used in archeological sites. Cassidy et al. (2019) present a VR platform to analyze archeological sites of difficult access, highlighting the opportunity and relevance of using VR as a tool for archeological sites to attract tourists, but also to facilitate access to archeologists and other stakeholders. The authors underline that "The introduction of immersive technologies into the archaeological and heritage sector presents an opportunity to overcome these access problems in new ways, and for multiple stakeholders" (Cassidy et al., 2019, p. 168). The researchers conclude that through the interaction of VR and other technologies, it is possible to obtain new perspectives of the site that in real life is not possible, with archeologists being able, for example, to separate the various layers of painting for better interpretation of the data, which can open doors to a new way of attracting visitors (Cassidy et al., 2019).

In the Old Zuoying City, in Taiwan, VR has been used to provide details on the construction of the old village. Augmented reality was also used to provide information on artifacts discovered in diggings and to do virtual restorations of damaged artifacts. Technologies are also used to provide gaming opportunities (Liu, 2020). Interactive apps have also been adopted in some archeological sites to better identify the archeological remains. This is the case, for example, of the digital app implemented to present two monuments to visitors—Y Pigwn Roman marching camp and Waun Ddu Fortlet—with a guided tour using an audio-narrative and animated reconstructions based on digital mapping (Smith et al., 2022). Other digital interpretation methods have also been used in archeological sites, such as projection mapping and digital animation (Liu, 2020). Digital interpretation contributed to learning, to entertainment, and to a more engaging experience (Liu, 2020).

3 Methodology

The aim of this empirical research is to understand how digital interpretation may be implemented in the scope of archeological heritage. Given the research purpose, a case study methodology was adopted, which is appropriate to deal with a contemporary phenomenon within its real-life environment (Yin, 1994a, 1994b, 2003) and which allows the analysis of data of different nature gathered through a combination of different methods (Eisenhardt, 1989; Gillham, 2000).

The case study research strategy was designed based only on a single-case design (Gustafsson, 2017; Yin, 1994a, 1994b), and this design is considered appropriate when dealing with an unusual or revelatory case (Yin, 2018), for particularly unexplored or underexamined topics for which little or scarce empirical evidence exist (Çakar & Aykol, 2021). Thus, the research will focus on the Côa Museum, an archeological museum in the northeast of Portugal, which is going to be described in the following section. In this type of research, it is recommended to use multiple information sources that will complement themselves, as well as various researchers, so that the case study is analyzed from the perspectives of different researchers (Eisenhardt, 1989). This strategy was implemented in the data collection.

The type of case study used in this research is the exploratory case study (Yin, 1994a, 1994b), since it is a data collection approach which serves the interest of the researchers and is "often a prelude to additional research efforts" (Hancock & Algozzine, 2011, p. 37).

In the empirical research, two types of data collection methods were used. One of the methods was direct observation, through the visit to the attraction—the Côa Museum—by three researchers in the 24th of November. This method is recommended when there are no data available through other methods (Noor, 2008), similar to the case study technique. These data were complemented with data obtained through two semi-structured interviews—one to the President of the Côa Park Foundation (August 24) and another to a Senior Technician of the Côa Museum (November 24), carried out online and face-to-face, respectively. Table 1 provides information on the focus of the interviews.

Table 1 Focus of the interviews

Script	General interview focus	Focus of the questions	Informant	Date
1	General information about the museum and detailed information on interpretation techniques, specifically on digital interpretation	Visitors' profile Visitors' statistics Visitor management strategies Interpretation techniques	I1	11/24/2021
2	Detailed information about the specific topic under analysis (digital interpretation)	Digital interpretation techniques: • Implementation process and evaluation • Future strategies	I2	08/24/2022

4   Findings and Discussion

4.1   Côa Valley Archeological Park and Museum: A Brief Characterization

In 1996, the Côa Valley Archeological Park was created with the mission of managing, protecting, researching, and showing to the public the rock art, encompassing more than 80 rock art sites and about 1200 panels engraved on schist rocks, almost all of the following four species: aurochs (wild bulls), horses, deer, and mountain goats (Comissão Nacional da UNESCO, 2022). This extensive set of open-air rock art testimonies, discovered in the Côa River Valley, in the Alto Douro region (Northern Portugal), dated between ± 25,000 BC and 10,000 BC, was one of the most important archeological discoveries of the Paleolithic throughout Europe, recognized and classified, in 1998, as a UNESCO World Heritage Site. "The museum is the only cross-border material asset in Europe to be inscribed on UNESCO's World Heritage List (I2)."

"The Upper Palaeolithic Rock Art of the Côa Valley is an exceptional illustration of the fast development of the creative genius of man/woman at the dawn of their cultural development; [...] it is an extraordinary demonstration of the social, economic and spiritual life of the first ancestor of mankind (Portugal-Patrimônios da Humanidade, 2018)." The rock art, which can be visited by the public, is structured in three main nuclei—Canada do Inferno, Penascosa, and Ribeira de Piscos (Fig. 1).

The Côa Museum, object of study of this research, is one of the largest Portuguese museums, with its headquarters in Vila Nova de Foz Côa and three museums supporting the Côa Valley Archeological Park (CVAP). Based on the idea that "Palaeolithic art in the Côa Valley" is perhaps the first manifestation of 'Land art' in the History of Humanity, it was designed by the architects Camilo Rebelo and Tiago Pimentel. Due to its strategic location, the museum constitutes a gateway that allows visitors to start the discovery of the two world heritage sites of the region: the Prehistoric Art of the Côa Valley and the Douro Wine Landscape (Côa Parque, 2020a). "The Côa Museum does not replace the visit

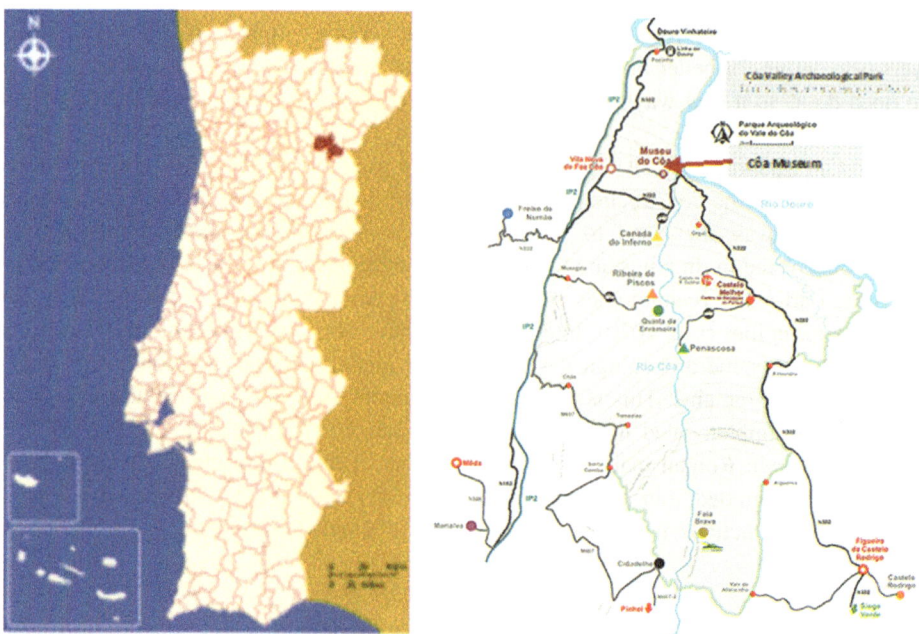

Fig. 1 Geographic location of the Côa Valley Archeological Park and the Côa museum in Portugal. *Source* Adapted from Côa Parque (2023) and Aldeias da Montanha (2023)

to the rock art sites of the Vale do Côa Archaeological Park, after all the 'true' Museum. It is the portal that will allow visitors to discover the rock art of the Côa and Douro valleys (I2)."

"The Côa Museum was founded in 2010 and is managed by Côa Parque—Foundation for the safeguarding and enhancement of the Côa Valley, also known as Fundação Côa Parque, inaugurated in 2011 (I2)." "The Côa Parque Foundation was created to manage the Archaeological Park of the Côa Valley and the Côa Museum and its main purposes are to protect, preserve, investigate and disseminate the heritage of the Côa Valley, combining the attraction capacity of the Museum with public visits to rock art" (I2). "The Foundation's main objective is, through the ongoing cultural archeology project, to promote the integrated development of the region, bringing together partners and private economic agents, highlighting the importance of the culture economy and its contribution to the well-being of the country (I2)."

According to the Strategic Plan defined for the period 2018–2022, the action of the Côa Parque Foundation should focus on the definition of a management model that integrates its activity in the main national and international cultural and tourism circuits. To do so, it must "reformulate the tourist experience in the museum and in the prehistoric art nuclei open to public fruition, promoting the updating of contents and the creation of new discursive and expository tools, more interactive and accessible which, without any compromise in the demand for scientific rigor, respond to the effective expectations of the visitor profile sensitive to cultural tourism (Côa Parque, 2018, p. 20)."

The Côa Museum, in its headquarters, includes a permanent exhibition, which develops in a continuous path, on the same floor, according to the following circuit: Room A—World Heritage of Humanity; Room B—The Territory, Man, and Côa Time; Room C—Geographical and Cultural Contextualization of the Côa Art; Room D—The Archaic Sanctuary, Room E—The Paleolithic in Everyday Life; Room F—The Endless History of Côa; and Room G—A Timeless Art. The museum also has other resources and physical infrastructures, including an auditorium with multipurpose room, a restaurant, a cafeteria, an auroque room, a meeting room, atriums, the main access, parking, viewpoints, documentation center/a library specialized in rock art, a shop, and three temporary exhibition rooms, where photography, painting, sculpture, engraving, or drawing that pay homage to or evoke the art of prehistoric engravers are exhibited (Côa Parque, 2020b).

"The Côa Parque Foundation has a policy of itinerant exhibitions, investing in national and international institutions. In the early 2000s, we had an exhibition in the cloisters of the Jerónimos Monastery and in the Gardens of the Belém Cultural Centre. In international spaces the Museum has had exhibitions in several national museums, such as: National Museum of Prehistory, in France, Museum of Human Evolution, in Burgos, and the National Archaeological Museum of Croatia. Along its route, the visitors will find original pieces, replicas and reconstitutions, as well as diverse textual and graphic information, the latter appearing in the form of drawings, photographs, videos and even holograms. The exhibition is also punctuated by several interactive exhibition elements that aim to contribute to a better understanding of the contents that are intended to be displayed (I2)."

"The Museum is also a welcoming center for researchers who wish to study the Côa, taking advantage of the largest national library dedicated to rock art (I2)."

4.2 General Characterization of Interpretation Techniques

The Côa Museum provides various types of interpretation techniques, highlighting the following "Guides/cultural mediators, Exhibitions, Guided tours, Audioguides, Mobile application, Interactive whiteboard, Physical games/tablet, Digital games, Mock-ups/replicas, Virtual and Augmented Reality (I2)."

Face-to-face interpretation techniques require direct contact between the person providing the interpretation and the visitor, and, in this sense, the museum can be visited freely or guided by a technician specialized in rock art, by appointment, at the three time periods available for this purpose. Guided tours are also available for organized groups, as well as special tours, where visitors can take advantage of customized tourist packages.

Regarding the number of visitors inside the museum, there is a limit of five people per room. "The group guided tours to the permanent exhibition have a limit of 10 people. Exceptionally, when groups have a large dimension, they are divided among the guides to facilitate the dynamics of the visit (I1)."

The museum's educational services "develop activities to welcome both the school public and the general public" (I2) and provide visits to the museum (with the option of a backstage visit), experimental and family archeology workshops, the workshop Through the Memory Stem, a puppet theater (the magic valley), and games. These pedagogical and interactive activities enable visitors to learn about cave art and the prehistoric period in a fun way, for example discovering about the techniques of making fire, making utensils, and exploring the various uses of aromatic, medicinal, and condiment plants found in the area of the Côa Museum.

In an informal context, the Côa team talks with visitors after the visits for some clarification, or even to deepen some specific theme (I1).

The Côa Museum has a wide target market, namely national and foreign visitors, consisting of families, young people, seniors, and schools. In the period from 2019 to 2021, even with the effects of the pandemic crisis, the museum received a total of 180,000 visitors, 80,000 nationals, and 100,000 foreigners. In order to promote visits to the region's museums, the Foundation has established partnerships, through protocols, with some entities—Siega Verde, Casa Grande Museum (ACDR of Freixo de Numão), Douro Foundation, and Serralves Foundation—promoting the purchase of joint tickets.

The Côa Parque Foundation, within the scope of its activities, organizes seminars and conferences where various themes are addressed, focusing on the cultural and scientific heritage of the region. Due to the pandemic, the number of seminars and conferences was reduced. However, the organization of the "European Researchers' Night" in 2021 stands out. The event aimed to promote knowledge sharing through demonstrations, lectures, conversations, and workshops, among others, and was designed to the general public and school community, bringing students closer to the valorization and interpretation of the region's heritage (I1).

In fulfilling its educational mission, the museum provides on-site training sessions open to the whole community (I2).

4.3 Digital Interpretation Techniques

Digital interpretation techniques encompass a variety of practices. In the interpretation of a site, various techniques are usually used, which complement each other (Graduate School of the Environment, 2003 *cit in* Gonçalves et al. n.d)

In order to facilitate and provide museum visitors with an in-depth understanding of all Paleolithic art in the Côa Valley and its region, and to make the visit more interactive and visual, facilitating its interpretation, the Côa Museum has technological resources and multimedia applications, namely interactive whiteboards, video mapping on rock replicas, virtual reality and augmented reality supports, audio guides, digital games, and the mobile application (APP—yourpodcast.pt/your-museum-app/) (I2). The focus on the use of technologies for museum interpretation has been evident in the Foundation's strategy, as a complement to other interpretation techniques, and intensified with the pandemic, having favored the use of individual equipment by visitors, such as the mobile application (I2). The choice and selection of interpretation techniques has been subject to evaluation by the visitors. In some pilot tests carried out, the mobile application was chosen by visitors (I2). The Côa Museum is thus accessible through the YourMuseum app,

where visitors can download the app and freely visit the museum. This app incorporates a survey that allows to evaluate user satisfaction (I2). The cultural guides/mediators, games, and interactive board are also techniques appreciated by visitors, which are adapted to all types of visitors (I2).

At the entrance of the museum, on the left side of the ticket office, you can find the TOMI, an interactive totem, user friendly, which provides information to the visitor, in Portuguese and English, about the museum, namely on the permanent and temporary exhibitions and other events taking place in the museum, as well as tourist information about the region (accommodation, restaurants, and activities).

The visitors can, since April 2018, if they wish, make the visit using the audio guide system, whose contents have been validated by the research team of the Côa Park Foundation (I1), allowing the visitor to make the visit freely. "This technique has not been used much by visitors since the Covid-19 pandemic began (I2)."

In terms of external communication, the museum is present on the Internet through its own Web site, providing plenty of information about the Côa Foundation, the Côa Valley Region, the museum and the archeological park, and also allowing visitors to purchase tickets in advance for the various types of visits. The Web site provides the functionality to connect to the social networks of the museum and park, namely Instagram, Facebook, YouTube, Twitter, and WhatsApp, and connection to the travelers' comments Web site TripAdvisor. Travelers' comments from this platform have been analyzed and considered in a museum management matrix (I2).

The Web site, the social networks, and the mobile application are thus digital interpretation techniques that can be used from home, helping to promote the museum, but also in its commercialization, thus facilitating access to the museum and avoiding the use of intermediaries.

Regarding audiovisual media, since December 2018, touch screens have been installed in the museum in rooms A, D, E, and F, allowing visitors to interact and get more information about the attraction. In room E, there is the augmented reality totem, through which the visitor can learn about the daily life of the Paleolithic people. The characters approach the visitor and show how food was prepared or how tools for hunting were made. Visitors can thus learn about the behavior of that time and even take selfies with the Paleolithic family.

In room F, we can find the virtual reality glasses, which are currently not working due to the pandemic situation. These glasses show a virtual visit to the rock engraving nuclei, to give the visitors, even those with special needs at physical level, the possibility to know the three sites of the rock engravings. In the various rooms of the museum, we can find explanatory videos (containing sequences of 2-D and 3-D animation), which illustrate, in a more enjoyable and understandable way, the main aspects related to the rock carvings and help in their interpretation. Video mapping is a tool that can be used by the guide and/or the visitor, allowing to better explain/interpret, for example, the reason why Foz Côa has such a high concentration of outdoor rock art, or to present/explore the different engravings and get to know some of the animations that were engraved in the rock at the time (byAir, 2021).

Table 2 shows a summary of the multimedia applications available for visitors to enjoy in the museum rooms.

The work developed by the Coa Parque Foundation in terms of communication and technological innovation, mainly due to the creation of the new Web site and the augmented reality experience, has been recognized by external entities in the area; it may be highlighted, for example, the first places obtained in 2019 and 2020 in the Online Communication and Management and Multimedia Application awards, respectively, attributed by the

Portuguese Museology Association, and the honorable mention in the Digital Signage Awards in 2021, in the Museums and Historic Sites category (Fernandes et al., 2021).

There is a concern in the design and creation of the museum contents to be integrated in the digital interpretation techniques, prioritizing not only the transmission of knowledge, but also the creation of a narrative that arouses emotions (I2). To this end, the contents are created and validated by a scientific committee and storytelling techniques, design, and 3-D modeling of the contents are used (I2). The digital media used in the interpretation of the heritage seek to link the message to the knowledge and dynamics with visitors, residents, and communities, in a reciprocal logic of knowledge, enjoyment, reflection, and knowledge sharing. In the near future, it is intended to introduce new technologies, such as robot guides, which may assist the museum's educational services (I2).

Table 2 Summary of the multimedia applications available in the museum

Rooms of the museum	Description	Multimedia applications
Room A—World Heritage Site	It highlights the Côa World Heritage site, offering a panoramic view of the valley and its rock art as well as the long geological process of landscape formation	Touch screens Interactive video mapping
Room B—The territory, the Man, and the Time Côa	It details the ancestral way of life in the Côa Valley, exhibiting some of the utensils recovered from the excavation of Paleolithic human occupation sites in the region	Videos (2-D and 3-D)
Room C—Geographic and Cultural Contextualization of Côa Art	It discusses the theories explaining rock art in light of the specific characteristics of the Côa Valley rock art	Videos (2-D and 3-D)
Room D—The Archaic Sanctuary	It offers an interactive exploration of the archaic sanctuary located in the large natural amphitheater comprising the Penascosa and Quinta da Barca sites	Touch screens Videos (2-D and 3-D) Two Totems
Room E—The Paleolithic in Everyday Life	It exhibits a replica of Rock 1 of Fariseu (one of the most important panels of the Côa with more than 80 engraved motifs) and the mobile art plates also from the Upper Paleolithic recovered during the excavation of this site	Augmented reality totem pole Touch screens Videos (2-D and 3-D) Video mapping
Room F—The Never ending Story of the Côa	In addition to a comparison between the art of the various Paleolithic periods, it also deals with more recent periods, such as the Iron Age. Besides rock 26 of the Vale de José Esteves, there are more replicas of rock art panels, chosen for their importance, but also because they are inaccessible	Touch screens Videos (2-D and 3-D) Virtual reality applications and glasses
Room G—A Timeless Art	This room also pays tribute to those who fought for the preservation of rock art, threatened in the 1990s by the construction of a dam on the Côa River	Videos (2-D and 3-D) Adapted documentary film (Jean-Luc Bouvert) "The Coa Battle—A Portuguese lesson" (10 min)

Source Own elaboration using data from Côa Parque (2020c), Fernandes et al. (2021), and interviews

5 Conclusion

Museums, in recent years, have sought to approach or capture younger market segments and, especially in the case of archeological museums, to show the heritage that is often inaccessible, for preservation issues, or whose understanding is more difficult and therefore also less attractive to visitors. The digital interpretation techniques are one of the most appropriate resources for that purpose, and, therefore, the aim of this article was to show, through the case study methodology, how these techniques have been used in the context of the archeological heritage, namely in the Côa Museum. To this end, two interviews were conducted with people responsible for the museum and fieldwork was carried out, which allowed direct observation of the attraction.

The results obtained led to the conclusion that the Côa Museum intends to be a gateway to the visitation of the three nuclei of rock art with the help of several digital interpretation tools. In this sense, to complement the visit and to help the guides, interpretation techniques are provided using technology that allow the visitor to enjoy the space, in a fun and dynamic way, approaching the reality of rock art. The Museum is divided into seven rooms, showing the rock art of the Côa and raising the visitor's awareness to the importance of this world heritage and the path of struggle necessary to achieve its preservation. Since 2010, the Côa Foundation, responsible for the management of the Museum, has sought to introduce technologies and multimedia applications, with a special emphasis on augmented and virtual reality, which allow the visitors to enjoy an immersive, educational, and interactive experience, giving them a greater sense of the reality of the Côa and rock art. The introduction of technologies was combined with an updating and improvement of the interpretation of the existing contents, supervised, and validated by the scientific commission. The growing number of visitors in recent years, with the exception of the decrease seen with the Covid-19 pandemic, is a reflection of the increased attractiveness of the museum, resulting from the incorporation of technology.

The study contributes to a wider knowledge of digital interpretation techniques at the level of archeological heritage in Portugal, through the presentation of the Côa Museum, which can serve as a benchmark for other museums. The paper enabled to identify a considerable range of digital interpretation techniques that may be used in archeological heritage and to better understand how they may be implemented.

This study has some limitations, such as being based only on the reality of one tourist attraction of archeological heritage—the Côa Museum—and greatly based on the perspective of the Côa Foundation. For future research, it would be interesting to analyze the usefulness of technologies in the interpretation of this type of heritage from the point of view of the visitors of the Côa Museum and extend this kind of study to other archeological museums or other archeological sites, to obtain a broader and more substantiated view of digital interpretation in the context of this type of heritage.

Acknowledgements This work was financially support by the research unit on Governance, Competitiveness and Public Policy (UIDB/04058/2020), funded by national funds through FCT—Fundação para a Ciência e a Tecnologia.

References

Agostino, D., Arnaboldi, M., & Lema, M. D. (2021). New development: COVID-19 as an accelerator of digital transformation in public service delivery. *Public Money & Management, 41*(1), 69–72.

Aldeias da Montanha. (2023). *Descobrir e visitar Vila Nova de Foz Côa*. Retrieved February 11, from http://aldeiasdemontanha.com/vila-nova-de-foz-coa

Azuma, R. T. (1997). A survey of augmented reality. *Presence: Teleoperators and Virtual Environments, 6*(4), 355–385. https://doi.org/10.1162/pres.1997.6.4.355

Ballart, J. (1997). *El Patrimonio Histórico y Arquológico: Valor y Uso*. Ariel Patrimonio.

Beck, L., & Cable, T. T. (2011). *The gifts of interpretation: Fifteen guiding principles for interpreting nature and culture* (3rd ed.). Sagamore Publishing.

Behrendt, M., & Franklin, T. (2014). A review of research on school field trips and their value in education. *International Journal of Environmental & Science Education, 9*(3), 235–245. https://doi.org/10.12973/ijese.2014.213a

Bradburne, J. M. (2019). Editorial. *Museum Management and Curatorship, 34*(1), 1. https://doi.org/10.1080/09647775.2019.1560952

Bramwell, B., & Lane, B. (1993). Interpretation and sustainable tourism: The potential and the pitfalls. *Journal of Sustainable Tourism, 1*(2), 71–80.

byAir. (2021). *Museu do Côa*. Retrieved August 10, from https://www.byar.pt/projecto/120-museu-do-coa/

Cáceres-Criado, I., Triviño-Tarradas, P., Valderrama-Zafra, J. M., & García-Molina, D. F. (2022). Digital preservation and virtual 3D reconstruction of "The Baker's house" in the archaeological site of Torreparedones (Baena, Cordoba-Spain). *Digital Applications in Archaeology and Cultural Heritage, 24*.https://doi.org/10.1016/j.daach.2022.e00218

Çakar, K., & Aykol, S. (2021). Case study as a research method in hospitality and tourism research: A systematic literature review (1974–2020). *Cornell Hospitality Quarterly, 62*(1), 21–31. https://doi.org/10.1177/1938965520971281

Carmigniani, J., Furht, B., Anisetti, M., Ceravolo, P., Damiani, E., & Ivkovic, M. (2011). Augmented reality technologies, systems and applications. *Multimedia Tools and Applications, 51*(1), 341–377. https://doi.org/10.1007/s11042-010-0660-6

Cassidy, B., Sim, G., Robinson, D. W., & Gandy, D. (2019). A virtual reality platform for analyzing remote archaeological sites. *Interacting with Computers, 31*(2), 167–176. https://doi.org/10.1093/iwc/iwz011

Chotrov, D., & Bachvarov, A. (2021). A flexible framework for web-based virtual reality presentation of cultural heritage. *AIP Conference Proceedings, 2333*(March), 1–7. https://doi.org/10.1063/5.0042542

Côa Parque. (2018). *Plano Estratégico 2018–2022*. Retrieved July 15, from https://arte-coa.pt/wp-content/uploads/2018/09/Plano-Estrate%CC%81gico-2018-2022.pdf

Côa Parque. (2020a). *O Côa—Museu*. Retrieved July 15, from https://arte-coa.pt/museu/

Côa Parque. (2020b). *Regulamento do Museu do Côa*. Retrieved July 13, from https://arte-coa.pt/wp-content/uploads/2018/09/Regulamento-do-Museu-do-Co%CC%82a.pdf

Côa Parque. (2020c). *Visita ao Museu—Planear Visita*. Retrieved August 10, from https://arte-coa.pt/visitas/visita-ao-museu/

Côa Parque. (2023). *A Região—O Côa*. Retrieved February 11, from https://arte-coa.pt/a-regiao-2/

Comissão Nacional da UNESCO. (2022). *Côa e Siega Verde*. Retrieved July 03, from https://unescoportugal.mne.gov.pt/pt/temas/proteger-o-nosso-patrimonio-e-promover-acriatividade/patrimonio-mundial-em-portugal/coa-e-siega-verde

De Ascaniis, S., & Cantoni, L. (2022). *Handbook on heritage, sustainable tourism and digital media*. Edward Elgar Publishing Limited.

Eisenhardt, K. M. (1989). Building theories from case study research. *The Academy of Management Review, 14*(4), 532–550. https://doi.org/10.2307/258557

Fernandes, A. B., Pereira, P. D., Aubry, T., & Santos, A. T. (2021). "Qual é o teu legado?" A renovação digital do Museu do Côa como instrumento de aproximação às suas comunidades. In memoriam Bruno José Navarro Marça. In P. M. Homem (Ed.), *Museus e Formação: Novas Competências para a Transformação Digital* (pp. 58–83). FLUP/DCTP. https://doi.org/10.21747/978-989-9082-07-6/musa5

Garzón, J., Pavón, J., & Baldiris, S. (2019). Systematic review and meta-analysis of augmented reality in educational settings. *Virtual Reality, 23*(4), 447–459. https://doi.org/10.1007/s10055-019-00379-9

Gatelier, E., Ross, D., Phillips, L., & Suquet, J.-B. (2022). A business model innovation methodology for implementing digital interpretation experiences in European cultural heritage attractions. *Journal of Heritage Tourism, 17*(4), 391–408.

Gillham, B. (2000). *Case study research methods*. Continuum.

Gonçalves, A., Costa, J., & Martins, P. (n.d). O Algarve: A Interpretação do Seu Património Arqueológico. *Revista dos Algarves, 13*, 14–20.

Gustafsson, J. (2017). *Single case studies versus multiple case studies: A comparative study*. Academy of Business, Engineering and Science, Halmstad University. https://www.diva-portal.org/smash/get/diva2:1064378/FULLTEXT01.pdf%20(10

Hancock, D., & Algozzine, B. (2011). *Designing case study research: A practical guide for beginning researchers* (2nd ed). Teachers College Press.

Haugstvedt, A. C., & Krogstie, J. (2012). Mobile augmented reality for cultural heritage: A technology acceptance study. In *Proceedings of the 2012 IEEE International Symposium on Mixed and Augmented Reality (ISMAR), Atlanta, GA, USA* (pp. 247–255).

ICOMOS. (2008). *ICOMOS_Interpretation_Charter_ENG_04_10_08*.

Jung, T. H., & tom Dieck, M. C. (2017). Augmented reality, virtual reality and 3D printing for the co-creation of value for the visitor experience at cultural heritage places. *Journal of Place Management and Development, 10*(2), 140–151.https://doi.org/10.1108/JPMD-07-2016-0045

Kolb, D. A. (2014). *Experiential learning: Experience as the source of learning and development* (2nd ed.). Pearson Education, Inc.

Lercari, A., Shiferaw, N., & Forte, E. (2017). Immersive visualization and curation of archaeological heritage data: Çatalhöyük and the Dig@IT App P. UC Merced.https://doi.org/10.6075/J0CN71VP

Liarokapis, F., Voulodimos, A., Doulamis, N., & Doulamis, A. (Eds.). (2020). *Visual computing for cultural heritage*. https://doi.org/10.1007/978-3-030-37191-3

Liu, Y. (2020). Evaluating visitor experience of digital interpretation and presentation technologies at cultural heritage sites: A case study of the old town, Zuoying. *Built Heritage, 4*(1), 14.

MacWhite, E. (1956). On the interpretation of archeological evidence in historical and sociological terms. *American Anthropologist, 58*(1), 3–25.

Magliacani, M., & Sorrentino, D. (2021). Reinterpreting museums' intended experience during the COVID-19 pandemic: Insights from Italian University Museums. *Museum Management and Curatorship*, 1–15.https://doi.org/10.1080/09647775.2021.1954984

Maietti, F., Medici, M., & Ferrari, F. (2021). From semantic-aware digital models to augmented reality applications for architectural heritage conservation and restoration. *Disegnarecon, 14*(26). https://doi.org/10.20365/disegnarecon.26.2021.17

Manuel, V., Bendicho, L.-M., & Grande, A. (2011). *Hacia una Carta Internacional de Arqueología Virtual*. El Borrador SEAV. http://www.londoncharter.org

McCrary, Q. (2011). The political nature of digital cultural heritage. *LIBER Quarterly, 20*(3–4), 357–368. https://doi.org/10.18352/lq.8000

Murta, S. M., & Goodey, B. (2002) Interpretação do Patrimônio para Visitantes: um quadro conceitual. In Murta, S. M., & Albano, C. (org.), *Interpretar o patrimônio: um exercício do olhar*. UFMG.

Noor, K. (2008). Case study: A strategic research methodology. *American Journal of Applied Sciences, 5*(11), 1602–1604.

Okanovic, V., Ivkovic-Kihic, I., Boskovic, D., Mijatovic, B., Prazina, I., Skaljo, E., & Rizvic, S. (2022). Interaction in eXtended reality applications for cultural heritage. *Applied Sciences* (Switzerland), *12*(3). https://doi.org/10.3390/app12031241

Portugal-Patrimônios da Humanidade. (2018). *Museu do Côa*. Retrieved July 03, from: https://portugalpatrimonios.com/2018/05/31/museu-do-coa-2/

Puig, A., Rodríguez, I., Arcos, J. L., Rodríguez-Aguilar, J. A., Cebrián, S., Bogdanovych, A., & Piqué, R. (2020). Lessons learned from supplementing archaeological museum exhibitions with virtual reality. *Virtual Reality: The Journal of the Virtual Reality Society, 24*(2), 343–358. https://doi.org/10.1007/s10055-019-00391-z

Reilly, P. (1990). *Towards a virtual archaeology scenario analysis for strategic digital archaeological knowledge management view project*. https://www.researchgate.net/publication/235902686

Rizvić, S. (2017). How to breathe life into cultural heritage 3D reconstructions. *European Review, 25*(1), 39–50.

Roche, D., & Quinn, B. (2017). Heritage sites and schoolchildren: Insights from the Battle of the Boyne. *Journal of Heritage Tourism, 12*(1), 7–20.

Samaroudi, M., Echavarria, K. R., & Perry, L. (2020). Heritage in lockdown: Digital provision of memory institutionsin the UK and US of America during the COVID-19 pandemic. *Museum Management and Curatorship, 35*(4), 337–361.

Silva, E. (2000). Património e Identidade. Os Desafios do Turismo Cultural. *Antropológicas*, 218–224.

Smith, T. A., Dunkley, R. A., & Jones, S. (2022). Storying wild landscapes: Multimodal interactions with digital app-based heritage. *International Journal of Heritage Studies, 28*(7), 803–819.

Sommerauer, P., & Müller, O. (2014). Augmented reality in informal learning environments: A field experiment in a mathematics exhibition. *Computers & Education, 79*, 59–68. https://doi.org/10.1016/j.compedu.2014.07.013

Tilden, F. (1977). *Interpreting our heritage* (3rd ed.). The University Of North Carolina Press.

Van Winkle, C. M., & Backman, K. (2011). Designing interpretive audio tours to enhance meaningful learning transfer at a historic site. *Journal of Heritage Tourism, 6*(1), 29–43.

Yin, R. K. (1994a). *Case study research. Design and methods* (2nd ed.). Sage Publications.

Yin, R. K. (1994b). *Case study research: Design and methods* (3rd ed.). Sage Publications.

Yin, R. K. (2003). *Case study research and applications: Design and methods* (6th ed.). SAGE.

Yin, R. (2018). *Case study research and applications: Design and methods* (6th ed.). Sage

Zafra De La Torre, N. (2017). El registro arqueológico como patrimonio histórico. *Complutum, 28*(1), 23–35. https://doi.org/10.5209/CMPL.58421

The Role of Social Media in the Conservation and Safeguard of Gastronomy as Intangible Cultural Heritage

Makhabbat Ramazanova, Raquel Santos Cardoso, and Isabel Vaz de Freitas

Abstract

Since 2003, when UNESCO launched the Convention for the Safeguarding of Intangible Cultural Heritage (ICH) and gastronomy was included in the World Heritage List as ICH in 2016, the importance of this topic has increased. However, globalization and the mixing of different cultures means that gastronomy as a heritage and related cultural traditions may be lost, requiring urgent action toward its preservation. Nowadays, social media platforms and their importance in today's society in various fields can be a useful tool in protecting intangible cultural heritage through its dissemination. This study was carried out in the framework of the EURICA project aimed at preserving ritual cuisine as a manifestation of heritage and folklore, making it easily accessible and usable through digitising ritual recipes. In this context, this study aims to identify and analyze already existing international platforms and networks related to gastronomy as cultural heritage, as well as the presence of these platforms in social media. As a result, three international platforms related to the theme of gastronomy as heritage were found and analyzed in terms of their amplitude and spread in social media. It has been found that many countries in different parts of the world have, relatively recently, shown a trend toward preserving their cuisine as a cultural heritage.

M. Ramazanova (✉)
REMIT/Research on Economics, Management and Information Technologies; Department of Tourism, Heritage and Culture, Portucalense University, Porto, Portugal
e-mail: ramazanova@upt.pt

R. S. Cardoso
REMIT/Research on Economics, Management and Information Technologies, Portucalense University, Porto, Portugal
e-mail: raquelsantostc@gmail.com

I. V. de Freitas
Department of Tourism, Heritage and Culture, Portucalense University, Porto, Portugal
e-mail: ifc@upt.pt

Keywords

Ritual gastronomy • Intangible cultural heritage • Heritage • Social media • Preservation

1 Introduction

Since the Convention for World Heritage was launched in 1972 (UNESCO, 1972), a focus on intangible cultural heritage (ICH) has been noticed. Especially in 2001, with the creation of the Section on ICH, it has started to be seen as a priority to the enrichment of cultural diversity and human creativity. Reading the definition of ICH, according to UNESCO (n.d.b.), it encompasses living traditions inherited from our ancestors, such as social practices, rituals, and festivals. Among them, gastronomy as heritage on the section of ICH of UNESCO was fully recognized only in 2010. And even though there are currently 631 ICH elements worldwide, corresponding to 140 countries, and there are only 27 elements present in 35 countries, related to gastronomy (UNESCO, 2022). It can be understandable, since, not all food, gastronomy or cuisine is considered as a heritage. As stated by Molina et al. (2016), *"Gastronomy does not always imply heritage"* (p. 296). (2016, Intangible heritage itself is a relatively new concept and has been subject to the evolution and development over the past half-century (Molina et al., 2016). It is important to keep in mind that not all food has heritage characteristics. Given this, there is no consensus among researchers on the definition of cultural, traditional, or ritual food. In fact, there is, still, a scarcity of literature on this topic (Ramazanova et al., 2022). Whatever the concept, the globalization process and mixture of different cultures occurring from the global phenomenon puts at risk the loss of original gastronomy. This enhances the importance of finding ways to recognize and preserve gastronomic heritage. In this context, social media can be as a facilitator for preservation and dissemination of ICH. Somehow it seems contradictory, given globalization is responsible for the mixing of cultures, but from another side, globalization is a possibility to share cultural diversity all over the world and to cultivate the sharing of authenticity and originality of countries.

In this context, social media is the quickest and best way to share and disseminate any kind of information. As can be seen, social media allows the creation and exchange of user-generated content and creates a way to promote traditional costumes and keep them alive more easily. The aim of this article is to understand the social media role in the conservation and safeguard of gastronomy as intangible cultural heritage, thus the importance of international platforms on gastronomy to raise awareness among stakeholders to preserve and promote it. As for the structure of this article, after this introduction, the literature review is presented. Section three demonstrates the methodology used. The results and the discussion are presented in section four. The last section is the conclusion, which summarizes the main results.

2 Literature Review

2.1 Gastronomy as Intangible Cultural Heritage

The focus on ICH follows the publication of the Convention for World Heritage in 1972, which expressed the concern in the preservation and protection of relevant heritage assets (UNESCO, 2003). In 2001, UNESCO created the section on intangible cultural heritage, which triggered it to be considered a priority to the enrichment of cultural diversity and human creativity, resulting in the publication of the Convention for the Safeguarding of Intangible Cultural Heritage in 2003. In this convention, the concept of intangible cultural heritage is defined as "*(…) the practices, representations, expressions, knowledge and skills—as well as the instruments, objects, artefacts and cultural spaces associated with them—that communities, groups and, as the case may be, individuals recognize it as being an integral part of their cultural heritage. This intangible cultural heritage, passed on from generation to generation, is constantly recreated by communities and groups according to their environment, their interaction with nature and their history, instilling a sense of identity and continuity, thus contributing, to promote respect for cultural diversity and human creativity* (p. 4)." The intangible cultural heritage is defined in the areas such as oral traditions and expressions, including language as the vector of intangible cultural heritage, performing arts, social practices, rituals and festive acts, knowledge and practices related to nature and the universe and traditional craft technique.

According to UNESCO, ICH encompasses living traditions inherited from our ancestors, such as social practices, rituals, and festivals. It is a key element in helping intercultural dialogue and mutual respect for different lifestyles. Even though UNESCO created the section on Intangible Cultural Heritage in 2001, gastronomy was not fully recognized as intangible heritage until 2010. It happened when UNESCO included the traditional Mexican cuisine and the gastronomic meal of the French. As can be seen, for gastronomy, it took longer to be recognized by UNESCO in a specific category. Before 2010, the first food-related nominations proposed were rejected by UNESCO. It was the persistence of countries like Mexico and Spain who enhanced the importance of gastronomic heritage recognition, for example, through an international and scientific meeting in 2008 organized by Mexico and a similar initiative made in 2009 in Barcelona (Romagnoli, 2019).

Meantime, academic community investigated this topic, being remarked the recent study of the authors Lin et al. (2021), who conducted a Bibliometric Analysis from 2001 to 2020 on gastronomy as a sign of the identity and cultural heritage of tourist destinations and stressed the scarcity of the literature on this topic. Based on this limited amount of literature, it was attempted to find definitions of traditional gastronomy as cultural heritage. Traditional gastronomy is defined by several authors, according to the summary by Guerrero et al. (2009), as: the representation of a group, linked to a territory and used by the community, being transmitted between generations. Ćirković (2016) further introduced the idea that traditional cuisine can be associated with a ritual, reinforcing that the preservation of ritual food preparation makes one community distinct from another. Vanhonacker et al. (2010, p. 453), for example, gave the following definition of traditional gastronomy: "*A traditional food product is a product frequently consumed or associated to specific celebrations and/or seasons, transmitted from one generation to another, made in a specific way according to gastronomic heritage, naturally processed, and distinguished and known because of its sensory properties and associated to a certain local area, region or country.*"

A ritual is composed of four elements, according to Ratcliffe et al. (2019), these being: artifacts or objects, a script that specifies when and how the constituent actions of the ritual are performed, the role of those who participate in the ritual, and for whom the ritual is done. Marshall (2006) adds that rituals are not only practiced in religious contexts, defining ritual as a symbolic activity, consisting of behaviors that occur on a particular occasion and are repeated on that same occasion over time. Some rituals and celebrations involve a specific food association, where food is an important part of the celebration. For example, during celebrations of special occasions such as Christmas or Easter, a different food is used and cooked for such occasions. As stated by Molina et al. (2016), a ritual cuisine, being considered intangible cultural heritage, should be passed on so that the authenticity of different communities is preserved in view of globalization, which could lead to the loss of the diversity of the distinct existing cultures.

There are many common characteristics given by different authors between ritual food and traditional gastronomy, which lead to the definition of ritual gastronomy or food as recognized by communities, groups, or individuals as part of their cultural heritage, forming part of rituals, celebrations, and festive events, involving traditional and artisanal culinary techniques, resulting from the history of the nation, symbolic value, and memory of a given region, being at risk of disappearing, and transmitted from generation to generation.

As can be verified from the discussion above, there are various perspectives regarding the characterization of gastronomy and diverse terms are used, having some of them similar contents. Thus, probably it was several attempts to recognize it as intangible heritage and took some time for UNESCO to do so. Currently, there are 631 intangible cultural heritage elements worldwide corresponding to 140 countries, recognized by UNESCO and there are only 27 elements relate to gastronomy (distributed in 35 countries), as seen in Fig. 1.

Attempts to include a specific dish in the UNESCO list demonstrate the risk of losing these dishes and the need for their protection as intangible cultural heritage. Skublewska-Paszkowska et al. (2022) define preserving as the deliberate act of keeping cultural heritage from the past for the future and can encompass documentation, protection, reconstruction, restoration, conservation, dissemination, and spreading. However, intangible cultural heritage is more difficult to preserve than tangible heritage, being intangible heritage often lost in transmission to younger generations.

Thus, the value of the projects such as EURICA is worthwhile, aiming to contribute to the preservation of this heritage at the same time taking into account the modern trends such as digital representation of the recipes and its dissemination in global context through using social media.

Moreover, a more recent adherence of several countries all over the world in sharing their gastronomic culture was noticed, using the Web and social media to do so. In this sense, social media can be seen as a useful tool for the dissemination and preservation of gastronomy and can be used as a way to promote and differentiate a specific country.

2.2 Social Media as a Tool for Intangible Cultural Heritage Preservation

Given the globalization and wide use of the technologies nowadays, social media's role is crucial in this context, according to UNESCO (2003), and is also a way to preserve and even strengthen the traditions of communities as a whole, such as with the presence of videos, images, and audios.

According to Statista Research Department (2022a, 2022b), in April 2022, there were 4.65 billion active users on social media, which represents 58.7% of the global population. Figure 2 demonstrates the statement of Statista Research Department (2022a, 2022b) that in January 2022, the three most popular social media globally were Facebook, YouTube, and WhatsApp.

Social media are applications involving the use of the Internet that, according to Kaplan (2015), integrate Web 2.0 and allow the creation and exchange of user-generated content. According to Hammou et al. (2020), social media will be a tool for the communication of intangible cultural heritage that enables its transmission, safeguarding, and promotion. For example, in the study done by the authors on the promotion of handicrafts from Marrakech through social media, stated that communication through social media is of utmost importance for the promotion of products, especially to the younger generation (Hammou et al., 2020). This type of applications have made it easier to create and share content, which is also interactive, and therefore, its growth is noted among the general population.

In accordance with Vassiliadis and Belenioti (2017), social media promotes dialogue, real-time communication,

Fig. 1 Gastronomic ICH recognized by UNESCO per continent and year. *Source* Own elaboration, based on the UNESCO's List of ICH (2022)

Fig. 2 Most popular social networks worldwide (january 2022), ranked by number of monthly active users. *Source* Own elaboration, based on the data of Statista Research Department (2022a, 2022b)

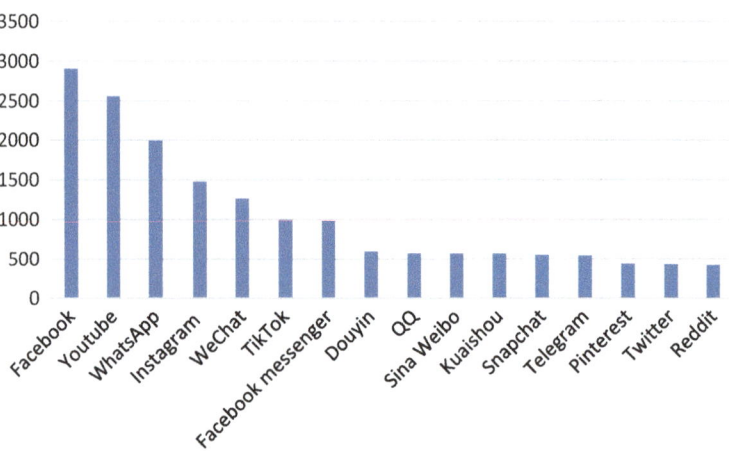

engagement with tourists and facilitates the interpretation of cultural experience. This way, social media enables a participating learning process. For example, YouTube is reported as a very dynamic tool, using as an example the Indianapolis Museum and the ZKM—Center for Art and Media, where the employees have to upload one video per day regarding these museums.

3 Methodology

In order to identify and analyze the existing international platforms and networks concerned with gastronomy as intangible cultural heritage, and the presence of these platforms in social media, an internet-based research was conducted during the months of July and September of 2022. The following keywords were used: "cultural gastronomy platforms," "gastronomy platform," "gastronomy network," and "gastronomy heritage platform." This research aimed to identify existing platforms dedicated to share gastronomy as intangible cultural heritage worldwide. Different platforms, groups, networks were found at city, country, region levels, being concerned with gastronomy as a food, promoting gastronomy for tourism development, restaurants promoting their diverse plates, projects about gastronomy. After deep analysis of the content of each Web site to understand activities undertaken by each organization and Web site, we focus on international platforms, recognizing gastronomy as cultural heritage and aiming to preserve it, such as IGCAT—International Institute of Gastronomy, Culture, Arts and Tourism, UNESCO creative cities Network, and UNWTO #TravelTomorrow. Further content analysis method was used aiming to analyze the purpose, content, meanings, directions, messages, and contribution of the platform for gastronomy preservation and aiming to raise awareness among the communities. Furthermore, analysis of the presence on social media of each platform was conducted,

aiming to understand what are the popular social media used by international organizations to promote the gastronomy as heritage. The detailed results and discussion are presented in the next section.

4 Results and Discussion

Three international platforms identified referring to gastronomy demonstrate the worldwide cover of the topic and at the same time variability within world regions, countries, and cities. The first platform is IGCAT—International Institute of Gastronomy, Culture, Arts and Tourism. The platform IGCAT is a non-profit institute established in 2012 that works with stakeholders in the areas of gastronomy, culture, arts, and tourism. IGCAT was founded the World/European Region of Gastronomy Award, intended to stimulate gastronomic creativity and innovation, educating for better nutrition, improving sustainable tourism standards, highlighting food cultures, and strengthening the well-being of the community (IGCAT, 2022). It also developed the European Young Chef Award, the World Food Gift Challenge, the Top Web sites for Foodie Travelers Award, and the Food Film Menu. It is possible to find a platform with the IGCAT member regions (currently 12 regions), called the Platform of European Gastronomy Regions, available in https://www.europeanregionofgastronomy.org/. The main goal is to increase international visibility and support knowledge share. It is worth to mention the importance of this platform in increasing awareness on gastronomy as cultural heritage and its uniqueness in a global context. Table 1 presents each regional Web site, brief description, and the social media used by the organization to disseminate its visibility in a global context. This information is expressed on the Web site of IGCAT.

The mapping of the countries involved in this platform is restricted to Europe with the collaboration of countries

Table 1 Regions of the platform of European gastronomy regions

Region	About the region	Social media presence
Hauts-de-France (2023)	Region in the north of France. Third most populous region in France. It has 187.8 inhab./km^2	Official IGCAT Web site for the region; Facebook; Instagram; LinkedIn; Twitter
Menorca (2022)	One of the Balearic Islands of Spain. It has a history of over 4000 years. It was declared a Biosphere Reserve by UNESCO in 1993	Official IGCAT Web site for the region; Facebook; Instagram; LinkedIn; YouTube
Trondheim-Trøndelag (2022)	County in central Norway. Created in 1687. The county's economy is based on food production	Official IGCAT Web site for the region; Facebook; Instagram; YouTube
Slovenia (2021)	Eastern European country. Located at the meeting of four major European regions: Alps, Dinaric, Pannonica, Mediterranean	Official IGCAT Web site for the region; Facebook; Instagram; LinkedIn; YouTube; Twitter
Coimbra (2021)	Region in central Portugal. It is the 8th largest city in the country. It has 1054 inhab./km^2	Official IGCAT Web site for the region; Facebook; Instagram; YouTube
Kuopio (2020–2021)	Region in central Finland. The largest city in eastern Finland. Founded in 1653. Known as the cultural center of Finland	Official IGCAT Web site for the region; Facebook; Instagram; YouTube; Twitter
Sibiu (2019)	Region in southern Transylvania, Romania. Gastronomic culture is one of its main tourist attractions	Official IGCAT Web site for the region; Facebook; Instagram; YouTube
South Aegean (2019)	Region of Greece. It belongs to the Cycladic and Dodecanese Island groups. It has 84.5 inhab./km^2	Official IGCAT Web site for the region; Facebook; Instagram; YouTube
North Brabant (2018)	The second largest province in the Netherlands. South of the country	Official IGCAT Web site for the region; Facebook; Instagram
Aarhus-Central Denmark (2017)	City of Denmark (on the eastern coast of the Jutland peninsula)	Official IGCAT Web site for the region; Facebook; Instagram
Catalonia (2016)	Region in northeast Spain, with four provinces: Barcelona, Girona, Lleida, and Tarragona (UNESCO recognition)	Official IGCAT Web site for the region; Facebook; Instagram; Twitter
Minho (2016)	Region in northern Portugal	Official IGCAT Web site for the region; Facebook

Source Own elaboration, based on the information of the European Regions of Gastronomy (2022)

between the North and the Mediterranean. Internationalization is one of the objectives, but for the moment, only European countries have been reached. Figure 3 depicts on the world map the regions that are involved in the IGCAT platform.

As for social media used, the social networks used by the organizations are Facebook and Instagram, YouTube, Twitter, and LinkedIn. It seems that the organizations are in coherence with globally used social networks (Fig. 4).

The second platform identified is the UNESCO Creative Cities network. The UNESCO has created this platform in 2004, called The UNESCO Creative Cities Network (UCCN). This network aims to promote cooperation between cities that use creativity for sustainable urban development.

Currently, it has 246 cities with a common objective —"*placing creativity and cultural industries at the heart of*

their development plans at the local level and cooperating actively at the international level" (UNESCO, n.d.a).

UCCN covers seven different fields: crafts and folk arts, media arts, film, design, gastronomy, literature, and music. Within the field of gastronomy, there are 48 creative cities distributed between Asia and Pacific with the registration of 18 cities, followed by Europe and North America with 14 cities registered and Latin America with 12 cities involved in the platform. Africa and Arab States count two cities each (Fig. 5).

The map of the countries involved in the creative cities of gastronomy places Brazil and China among the countries with the most cities registered on the UNESCO Creative Cities platform, both with four cities, followed by Italy, Turkey, and USA with three registered cities, Colombia, India, Iran, Japan, Mexico, and Thailand with two cities

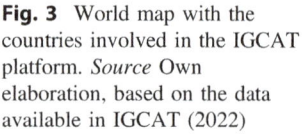

Fig. 3 World map with the countries involved in the IGCAT platform. *Source* Own elaboration, based on the data available in IGCAT (2022)

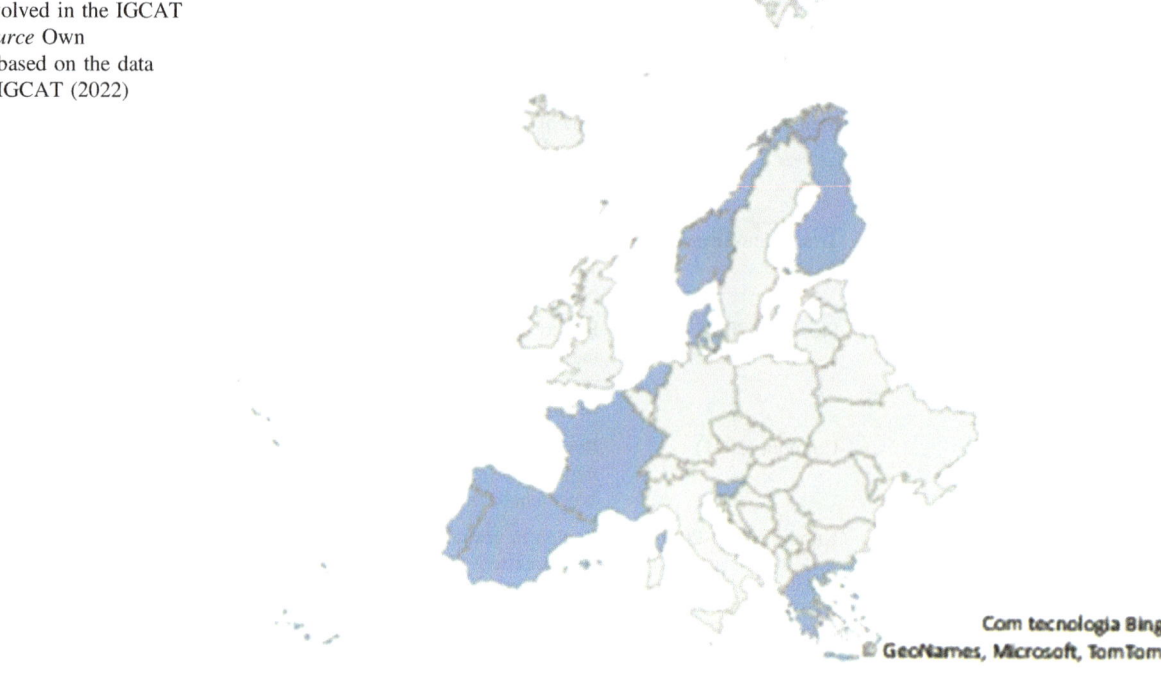

Com tecnologia Bing
© GeoNames, Microsoft, TomTom

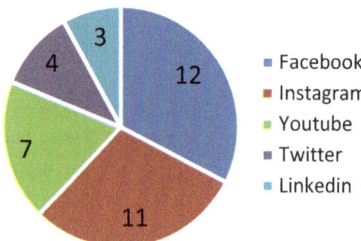

Fig. 4 Social networks used by IGCAT regions. *Source* Own elaboration, based on the data available in IGCAT

each, and 19 countries with one registered city. This platform has an international range of countries, interested in dissemination and international cooperation in the field of gastronomy (Fig. 6).

Content analysis of each regional Web site (its purpose, meaning, direction, and message) related to gastronomy was conducted. Further, key words and main terms were extracted and inserted into the WordCloud in order to create a visual representation of the most frequently used terms (Fig. 7). A WordCloud was made using keywords of the

Fig. 5 48 creative cities distribution between regions. *Source* Own elaboration, based on the data available in UNESCO (n.d.a.)

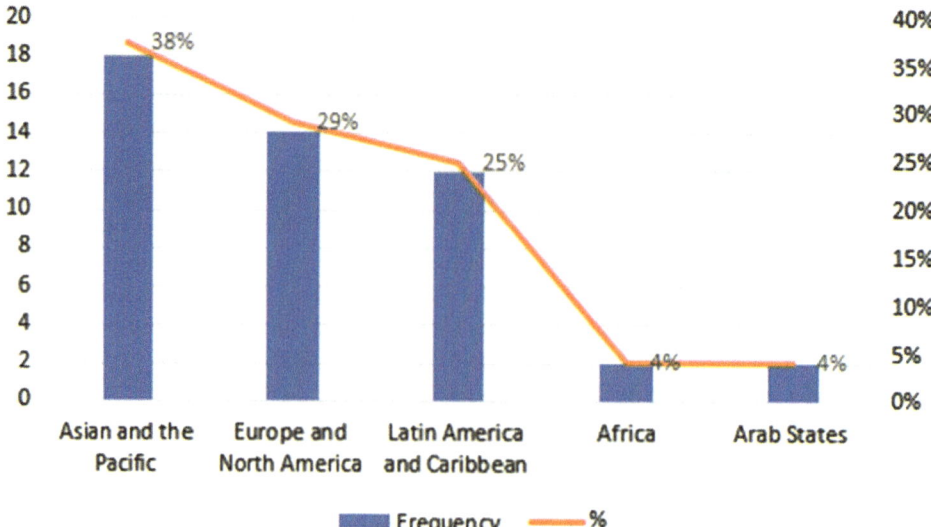

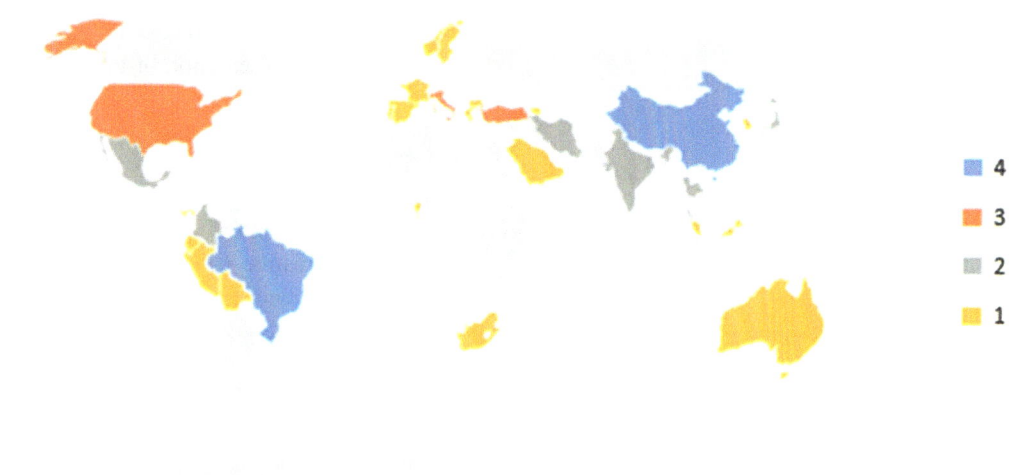

Com tecnologia Bing

© Australian Bureau of Statistics, GeoNames, Geospatial Data Edit, Microsoft, Navinfo, OpenStreetMap, TomTom, Wikipedia

Fig. 6 World map with the countries involved in UNESCO creative cities. *Source* Own elaboration, based on the data available in UNESCO (n.d. a.)

Fig. 7 Most frequently used terms in the analyzed Web sites. *Source* Wordclouds.com (2022)

"added value" presented on each Web site in the region, and it was interesting to note that some dimensions of gastronomy were important for all partners. International cooperation, dynamics with vulnerable groups, training, research, innovation, creativity and best practices and sustainability are more frequently used concepts to explain "added value." Further social, cultural, educational, academic, and economic aspects can be found as the widely mentioned terms that result from cooperation between stakeholders.

As the main objective of this study is to understand the role of media in the conservation and safeguarding of gastronomy as intangible cultural heritage, it is clear from the expressions and keywords of the creative cities of UNESCO's gastronomic partners that *preservation and protection* are important keywords, as well as *sustainability, cultural, and traditional cuisine* are keys in gastronomic fields, training, research, cooperation, or collaboration. However, it should be noted that the word "ritual" or the concept of "ritual cuisine" was never mentioned in the texts and considered as "added value" by the partners. In turn, the term of traditional cuisine is one that comes up frequently. Thus, it can be concluded that gastronomic cities are an important veil for the preservation of traditional foods, local products, and traditional methods through training, education, research, events, and in tourism activities (Richards, 2017).

The third platform is UNWTO—Travel Tomorrow, developed by the United Nations World Tourism Organization. UNWTO runs the #TravelTomorrow campaign, in which chefs from around the world show how to prepare local dishes, using YouTube as a tool to demonstrate the cooking process and thus making available worldwide. The recipes can be found at the following link: https://www. youtube.com/playlist?list=PL1J3wwM1RSVHKg-IRdRaX5tPqwxAEaOMJ. At the moment, 54 countries are represented on the Web site, spread across five regions (Fig. 9). Figure 8 shows that the most participating regions

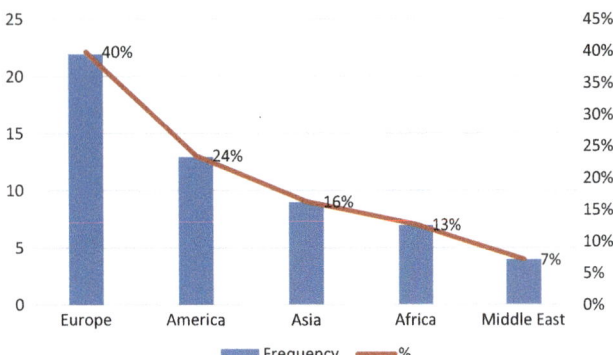

Fig. 8 UNWTO—travel tomorrow regions. *Source* Own elaboration, based on the data available in UNWTO (2022)

are Europe and America, followed by Asia, Africa, and the Middle East. Regarding social media, UNWTO is present on various social networks such as Facebook, Twitter, Instagram, LinkedIn, YouTube, and Flickr, indicating its wide social media reach.

5 Conclusion

The present study aimed to explore the role of gastronomy as intangible cultural heritage and its preservation through the analysis of international platforms and its wide presence in social media. It can be concluded that academic studies on gastronomy as ICH are very limited and can be explained by the fact that it was only recognized by UNESCO in 2010. Moreover, there is a variety of terms used in the literature, such as traditional gastronomy, ritual gastronomy, cuisine,

food, which demonstrates the lack of consensus on the definition and the differences between them. Whatever the definition, the preservation of this ICH is crucial, as ICH is harder to preserve and at risk of being lost through globalization and generational change.

Thus, we have aimed to explore the international platforms accompanying gastronomy as heritage and tried to be concise in our content analysis. Linking the analysis of the UNESCO Creative Cities Platform with the IGCAT platform, we can find common interests in preserving gastronomy as a cultural heritage and promoting creativity and innovation, education, community well-being, and sustainable development. There is an increased focus on the social and cultural aspects of gastronomy.

Based on the mapping of participants in international gastronomy-related platforms, it can be concluded that internationalization is a fact, with involvement of various countries from Asia, Africa, Australia, Europe, America (South and North).

The platforms show interesting work on international cooperation, dissemination, and exchange of gastronomy. The partner countries find in this cooperation an opportunity to develop and exchange the practices of gastronomy and to share ways related to its preservation and safeguard. As a gap, the absence of the concept of "ritual cuisine," an important aspect of gastronomic cultural heritage, that is of particular interest for future research should be noted.

Acknowledgements The authors gratefully acknowledge the support of the EU Erasmus + EURICA *Europe ritual cuisine—digital presentation and preservation* project consortium. This project has been funded with support from the European Commission. This publication reflects the views only of the authors, and the Commission cannot be

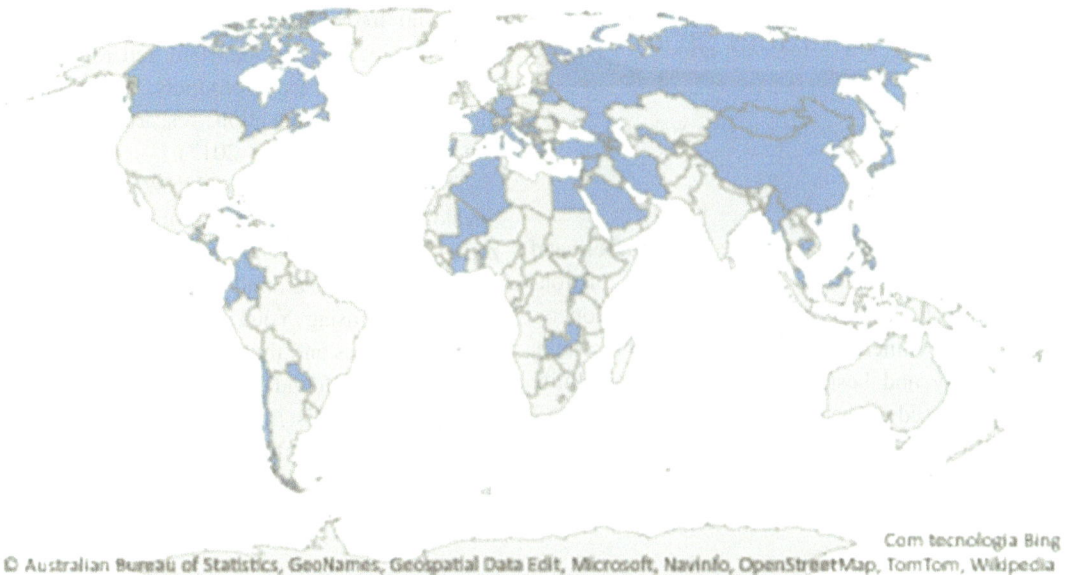

Fig. 9 Map of the UNWTO—travel tomorrow countries. *Source* Own elaboration, based on the data available in UNWTO (2022)

held responsible for any use which may be made of the information contained therein.

References

Ćirković, S. (2016). Food as intangible cultural heritage—The česnica among Serbs in Romania.

Guerrero, L., Guàrdia, M. D., Xicola, J., Verbeke, W., Vanhonacker, F., Zakowska-Biemans, S., Sajdakowska, M., Sulmont-Rossé, C., Issanchou, S., Contel, M., Scalvedi, M. L., Britt Signe Granli, B., Margrethe Hersleth, M. (2009). Consumer-driven definition of traditional food products and innovation in traditional foods. A qualitative cross-cultural study. *Appetite, 52*(2), 345–354. ISSN 0195-6663. https://doi.org/10.1016/j.appet.2008.11.008

Hammou, I., Aboudou, S., & Makloul, Y. (2020). *Social media and intangible cultural heritage for digital marketing communication: Case of Marrakech crafts.*

IGCAT (2022). *International institute of gastronomy, culture, arts and tourism.* Retrieved from: https://igcat.org/

Kaplan, A. M. (2015). Social media, the digital revolution, and the business of media. *International Journal on Media Management, 17*(4), 197–199. https://doi.org/10.1080/14241277.2015.1120014

Lin, M. P., Marine-Roig, E., & Llonch-Molina, N. (2021). Gastronomy as a sign of the identity and cultural heritage of tourist destinations: A bibliometric analysis 2001–2020. *Sustainability, 13*(22), 12531.

Marshall, D. (2006). Food as ritual, routine or convention? *Consumption, Markets and Culture, 8*(1), 65–85. https://doi.org/10.1080/10253860500069042

Molina, M. M., Molina, B. M., Campos, V. S., & Oña, M. V. S. (2016). Intangible heritage and gastronomy: The impact of UNESCO gastronomy elements. *Journal of Culinary Science & Technology, 14*(4), 293–310. https://doi.org/10.1080/15428052.2015.1129008

Ramazanova, M., Lopes, C., Albuquerque, H., de Freitas, I. V., Quintela, J., & Remelgado, P. (2022, May). Preserving ritual food as intangible cultural heritage through digitisation. The case of Portugal. In *International Conference on Tourism Research* (Vol. 15, No. 1, pp. 334–343).

Ratcliffe, E., Baxter, W. L., & Martin, N. (2019). Consumption rituals relating to food and drink: A review and research agenda. *Appetite, 134*, 86–93.

Richards, G. (2017) The role of gastronomy in tourism development. In *Proceedings of the 4th Congress of Noble Houses, Muncipality de Arcos de Valvadez* (pp. 1151–1159).

Romagnoli, M. (2019). Gastronomic heritage elements at UNESCO: Problems, reflections on and interpretations of a new heritage category. https://www.researchgate.net/publication/339487503

Skublewska-Paszkowska, M., Milosz, M., Powroznik, P., & Lukasik, E. (2022). 3D technologies for intangible cultural heritage preservation—Literature review for selected databases. *Heritage Science, 10*(1), 1–24.

Statista Research Department. (2022a). Global digital population as of April 2022. Retrieved from: https://www.statista.com/statistics/617136/digital-population-worldwide/

Statista Research Department. (2022b). *Social media—Statistics and facts.* Retrieved from: https://www.statista.com/topics/1164/social-networks/#topicHeader__wrapper

UNESCO. (1972). UNESCO World Heritage Centre—Convention concerning the protection of the world cultural and natural heritage.

UNESCO. (2003). *Convention for the safeguarding of the intangible cultural heritage.* Retrieved from: https://ich.unesco.org/en/convention

UNESCO. (2022). *Browse the lists of intangible cultural heritage.* Retrieved from: https://ich.unesco.org/en/lists

UNESCO. (n.d.a). *What is creative cities network?* Retrieved from: https://en.unesco.org/creative-cities/content/about-us

UNESCO. (n.d.b). *What is intangible cultural heritage?* Retrieved from: https://ich.unesco.org/en/what-is-intangible-heritage-00003

UNWTO. (2022). *Travel tomorrow, tourism and gastronomy.* Retrieved from: https://www.unwto.org/gastronomy

Vanhonacker, F., Verbeke, W., Guerrero, L., Claret, A., Contel, M., Scalvedi, L., Żakowska-Biemans, S., Gutkowska, K., Sulmont-Rossé, C., Raude, J., Granli, B. S., & Hersleth, M. (2010). How European consumers define the concept of traditional food: Evidence from a survey in six countries. https://doi.org/10.1002/agr.20241

Vassiliadis, C., & Belenioti, Z. C. (2017). Museums and cultural heritage via social media: An integrated literature review. *Tourismos, 12*(3), 97–132.

Sustainable Cultural Tourism and Virtual Reality the Contribution of the New Technologies Applications in the Fields of Preservation and Sustainable Tourism Management of the Cultural Heritage. The Case of Greece

Ioanna C. Chatzopoulou

Abstract

Protection and preservation of natural and cultural heritage are considered vital to the survival of human civilization. Natural environment is an integral part of the cultural heritage of a place or a people. On one hand, there are special natural environments that are characterized as sensitive ecosystems or have a unique monumental character and belong to the world's cultural heritage. On the other hand, there are components of the cultural potential that are in danger of being lost over time as their carriers have disappeared. In addition, the tangible and intangible cultural heritage is inextricably linked to the natural environment of a place and it forms the historical identity of a people. This paper examines how information and telecommunication technologies can make a decisive contribution to the protection of the natural and cultural environments. Moreover, this study analyzes how innovative and pioneering applications of new technologies can provide opportunities for the promotion of natural and cultural heritage. Besides the article demonstrates how the use of virtual and augmented reality applications contributes to the development of thematic tourism and to economic growth. It also focuses on the Greek paradigm, as a case study.

Keywords

Sustainable tourism • Cultural tourism • Virtual reality • Augmented reality

I. C. Chatzopoulou (✉)
Department of Tourism Management, University of Patras, Patras, Greece
e-mail: ichatzop@upatras.gr

1 Introduction

There is an increasing recognition of the strong interconnections between natural and cultural assets and tourism, since natural and cultural heritage offers plenty of opportunities for the promotion of thematic tourism and for sustainable economic growth (Tsartas et al., 2010). On the other hand, there is a rising awareness of the need for the protection, curation, dissemination and sustainable management of cultural and natural assets, as they are exposed to various nature and human-derived threads, that are occasionally unpredictable or unpreventable, which range from excessive exploitation by the market, vandalism, pollution or degradation of their environment, to natural disasters, like floods, fires, earthquakes, etc. (Avouri et al., 2020; European Commission, n.d.).

This article explores how the use of new technologies contributes to the preservation and restoration of natural and cultural capital, to the promotion of recreational and eco-tourism activities, and to the sustainable tourism development in the protected natural and cultural areas.

Furthermore, it identifies main relevant initiatives in Greece and suggests possible follow ups for the wider use of digital technologies (e.g. immersive, virtual and augmented reality and 3D) in the areas of digitisation, online accessibility and digital preservation of natural and cultural material.

2 Definition and Types of Environment

According to Art 2 par. 1 of the Greek Environmental Protection Law 1650/1986, the term environment refers to the complex of physical, chemical, biotic (such as climate, soil and living beings) and anthropogenic factors, that affect the ecological stability, the standard of living, but also the culture, tradition and people's values.

There are two types of environment: (1) The natural environment, which is comprised mainly by land, air, water and living organisms and (2) the human (manmade) environment, which includes the building and the cultural environment.

More specifically according to Article 1 of the Convention concerning the Protection of the World Cultural and Natural Heritage 1972, the cultural environment consists of monuments, historical and archaeological sites, architectural works, works of monumental sculpture and painting, inscriptions, cave dwellings and historic buildings (UNESCO, n.d.).

Cultural heritage also includes elements, which have no tangible, material dimension, transmitted from generation to generation, like language, aesthetics, oral traditions, performing arts, social practices, rituals, festive events, myths, customs, dance, music social organization, or knowledge and practices concerning nature and the universe [Art. 2 of the UNESCO's 2003 Convention for the Safeguarding of the Intangible Cultural Heritage (UNESCO, n.d.)].

Apart from land based heritage (both tangible and intangible), cultural heritage includes also underwater cultural material. The 2001 UNESCO Convention for the Protection of the Underwater Cultural Heritage defines this heritage as: "all traces of human existence having a cultural, historical or archaeological character which have been partially or totally under water, periodically or continuously, for at least 100 years sites, such us: (i) structures, buildings, artefacts and human remains, together with their archaeological and natural context; (ii) vessels, aircraft, other vehicles or any part thereof, their cargo or other contents, together with their archaeological and natural context; and (iii) objects of prehistoric character" (UNESCO, n.d.).

3   Interconnections Between Natural and Cultural Heritage

There is a strong interdependency between natural and cultural environment, since cultural heritage includes also natural assets and sites, like:

- caves and paleontological remains, for which there is evidence that they are related to human existence, e.g. caves with prehistorical paintings, or carved figurines dating back to the last ice age, like the six caves in the Swabian Jura in southern Germany, inscribed on UNESCO's World Heritage List, which contain art items dating from 43,000 to 33,000 years ago (including carved figurines of animals and creatures that are half animal, half human, items of personal adornment and musical instruments) (UNESCO, 2017), or cave sanctuaries and worship places, like the cave church on the Greek Island of Pátmos or the underwater caves and caverns (cenotes)

of the Mayan Riviera in the state of Yucatán, Mexico, which were once the holy places of the Mayans, since it was believed that they represent entrances to the underworld (known as Xibalba) (INCIDER, 2018).

- historical sites, meaning "areas on land or at sea or in lakes or rivers which constitute, or there is evidence that they have constituted, the site of exceptional historical or mythical events" (Art. 2 of Greek Law 3020/2002 "On the Protection of Antiquities and Cultural Heritage in General") like the Inner Ionian Sea located in the west part of Greece, where the famous battles of Actium and of Lepanto (Nafpaktos) took place, or Marathon and Thermopylae, where the famous, homonymous Battles among the Greeks and the Persians were fought in 490 and 480 BCE.
- archaeological sites, meaning areas on land or underwater, "which contain or there is evidence that they contain, ancient monuments, or which have constituted monumental urban or burial groups (Art 2 of Greek Cultural Heritage Law 3020/2002). According to the above mentioned Law archaeological sites shall also include the necessary open space "so as to allow the preserved monuments to be considered in a historical, aesthetic and functional unity" (Art 2 of Greek Cultural Heritage Law 3020/2002).
- ecosystems or natural assets of historical value or monumental character, like the island of Patmos, which has been included in the catalogue "Monuments of World Heritage of UNESCO" since 1999, because of the historical centres of the island (Chora) the Monastery of St. John the Theologian and the Sacred cave of the Apocalypse, where St. John wrote the Book of Revelation (UNESCO, n.d.).
- Agricultural Heritage Systems (GIAHS), which are agroecosystems, designated by the Food and Agriculture Organization of the United Nations around the world, which combine biodiversity, resilient ecosystems and tradition and innovation in a unique way (FAO, n.d.).
- geoparks, namely single, unified geographical areas of outstanding geological beauty, and natural and cultural significance, which are protected as part of the geological heritage.

It is worth mentioning that five Greek geoparks are included in the UNESCO'S Global Geoparks Network, which consists of 177 sites and landscapes of international geological significance in 46 countries around the world, managed through a holistic concept of protection, education and sustainable development (UNESCO, n.d.).

In particular, in Greece, the following sites are designated by UNESCO as meeting all the requirements for belonging to the Global Geoparks Network:

1. The Chelmos Vouraikos Geopark is located in Northern Peloponnese (UNESCO, n.d.).
2. The Lesvos Island UNESCO Global Geopark, which consists of an ancient forest preserved by a massive volcanic eruption 20 million years ago (UNESCO, n.d.).
3. The Psiloritis and the Sitia UNESCO Global Geoparks which are located in the island of Crete (UNESCO, n.d.; UNESCO, n.d.).
4. The Vikos Aoos UNESCO Global Geopark, which is located in the region of Epirus, Ioannina, northwestern Greece and consists of impressive gorges, rock towers, rivers, stone bridges, traditional architecture and thermal baths (UNESCO, n.d.; UNESCO, n.d.).

4 Natural and Cultural Heritage and Tourism Development

There is a strong interdependency between natural and cultural heritage and tourism, since natural landscapes are a symbol of rural identity and a factor of tourist attraction, while many tourists choose their destination for its cultural assets and because they are interested in the historical aspects of the region.

During the last years, the alienation of town dwellers from nature and their need to "return" even temporarily to nature for holiday and recreation, the rising awareness for the preservation of the natural and cultural assets for future generations and the growing number of visitors who seek to experience tourism in a more alternative manner have led to the promotion of alternative forms of tourism (rural tourism, religious tourism, natural and cultural heritage tourism, adventure tourism, etc.), which enable visitors to familiarize themselves not only with nature and the countryside, but also with traditional habits, cultural heritage and the history of each area.

Specifically, Greece, a country with remarkable cultural diversity, unique history, traditional knowledge and stunning natural landscapes, possesses a high potential for "Natural and Cultural Heritage Tourism Development".

5 Threats to Natural and Cultural Heritage

However tourism, like any other financial activity, has environmental impacts, like alienation of cultural identity, loss of authenticity, local traditions and ways of life, environmental pollution, degradation of natural assets, etc., due mainly to excess exploitation of specific tourism destinations, such as the Greek islands of Santorini and Mykonos, or the city of Venice (Almazán, 2018).

Furthermore, apart from overtourism, natural and cultural heritage face numerous natural or human-derived threats, such as improper maintenance and/or care, degradation and pollution of their environment, global warming, exploitation of the natural resources, natural disasters, like the Notre-Dame fire on 15 April 2019 (Wikipedia, n.d.), flooding and wind erosion, unauthorized activities, e.g. looting, commercial exploitation for transaction or speculation purposes, trafficking, illegal excavations, theft, illegal export or vandalism and destruction of cultural assets during wars, like the destruction of the two sixth-century monumental Buddha's statues of Bamiyan in central Afghanistan, by the Taliban during the war in 2001, or the destruction of the Temple of Bel by the ISIS in 2015—which for nearly 2000 years had been the centre of religious life in Palmyra, Syria (BBC, n.d.; Melvin et al., 2015).

For the legal protection of natural and cultural assets, from the above mentioned threats exist several international conventions, recommendations and resolutions, which complement the legal arrangements of states, such as the European Landscape Convention (2000), the European Convention on the Protection of the Archaeological Heritage (1992) or the six UNESCO Culture Conventions [Convention for the Protection of Cultural Property in the Event of Armed Conflict (1954); Convention on the Means of Prohibiting and Preventing the Illicit Import, Export and Transfer of Ownership of Cultural Property (1970); Convention concerning the Protection of the World Cultural and Natural Heritage (1972); Convention on the Protection of the Underwater Cultural Heritage (2001); Convention on the Safeguarding of the Intangible Cultural Heritage (2003) Convention on the Protection and Promotion of the Diversity of Cultural Expression (2005)] (UNESCO, 2015).

Greece's specific natural and cultural heritage legislation dates back to the period of the foundation of the modern independent Greek state in the early nineteenth century, during which the first Greek cultural heritage Law, referring also to the underwater antiquities (Law 10/22-5-1834) (Dellaporta, 2005) and the first forest Law, governing forest protection and production, but also property rights to forests, were established (Kourousopoulos, 1978).

Special constitutional provisions for the protection of the natural and cultural heritage were adopted for the first time in 1975. In particular, Art. 24 of the Greek Constitution of 1975 states that "the protection of the natural and cultural environment constitutes a duty of the State and a right of every person" and provides for the State obligation to adopt special preventive or repressive protective measures in the context of the principle of sustainable development. Furthermore, the main specific laws dedicated exclusively to the protection of natural and cultural environments are the Greek Environmental Protection Law 1650/1986 and the Greek

Cultural Heritage Law No. 3028/2002, recently codified by Law 4858/2021.

6 New Technologies Contribution to the Natural and Cultural Heritage Preservation and to Their Sustainable Tourism Development

Despite the establishment of specific international and national legal protective arrangements, the protection of natural and cultural heritage often remains incomplete, due also to the insufficient implementation of the relevant legal framework and to the ineffective integration of the environmental protection into the State's policies.

In view of the above mentioned, the role of new technologies in the preservation, protection, dissemination, research, curation, aesthetic enjoyment and sustainable tourism management of the natural and cultural property is becoming increasingly critical.

In particular, the increasing advances in digital technologies (like 3D tech, artificial intelligence and immersive, virtual/augmented reality) can play a vital role in the cultural preservation, curation and restoration, in the sustainable use and accessibility of cultural assets and in the creation of a holistic tourist product (Alivizatou-Barakou et al., 2018).

More specifically, the use of cutting-edge technologies like Artificial and Augmented Intelligence and 3D projection mapping creates possible distance education and research in the cultural heritage domain (Adamou et al., 2021).

Besides, digital technologies like the creation of next-generation 3D models of artefacts and environments, accompanied digitally by contextual technical and historical information, which users can access using virtual reality (VR) and augmented reality (AR) tools, facilitate cultural experiences and make possible time travelling and virtual visits of monuments and cultural venues (European Comission. Cordis, n.d.). Furthermore, 3D digitization is a necessity for tangible, immovable, cultural heritage at risk, but also for the preservation of the memory of destroyed monuments and cultural artefacts, like the monumental Buddha's statues of Bamiyan in central Afghanistan, destroyed by the Taliban, or the Temple of Bel in Palmyra, whose original appearance has been digitally restored and preserved.

Furthermore, the digitization of cultural heritage and the use of immersive technologies such as augmented and virtual reality can provide virtual access to cultural material, that is difficult to access or inaccessible, such as the monastic community of Mount Athos, which is completely closed to women, or underwater cultural heritage sites, which are unreachable to the general public (like the wreck of a trading ship or sunken treasures, or the ancient Roman resort of Baiae in Napoli, whose underwater, virtual exploitation and virtual visit with the use of new immersive technologies, such as augmented and virtual reality has made possible, within the EU-funded iMARECULTURE project). Moreover, augmented reality devices can enhance underwater visits of underwater cultural heritage sites, made by divers, and could also serve as a scientific tool, allowing underwater researchers to take notes and geo-position photos of new discoveries (European Comission. Cordis, n.d.; European Comission. Cordis, n.d.).

Modern technologies such as 3D modelling and augmented reality facilitate the easy access to museums, improving the experience one can undergo when visiting venues packed with cultural heritage and history (Arnold, 2005; Audunson et al., 2020; Ding, 2017; Kennedy et al., 2021; Liritzis et al., 2015; Madirova & Absalyamovaa, 2015; Turner et al., 2017).

7 Greece's Progress in the Field of Digitization, Online Accessibility and Digital Preservation of Cultural and Natural Material

According to the Digital Transformation "bible" of Greece, which was launched in December 2020 by the Greek Ministry of Digital Governance outlining a holistic digital strategy for the years 2020–2025, Greece has made during the last years significant progress in the area of digitization of cultural material (archives, libraries, art and archaeological collections) (Greek News Agenda, n.d.).

Besides, many museums and technological cultural venues have websites, which host digitized cultural material and organize virtual exhibitions, like.

7.1 The Digital Acropolis Museum

The Digital Acropolis Museum was developed under the programme "Creation of the Digital Acropolis Museum", with total budget 1,330,240,63 € excluding 24% VAT. Within the above programme, a large number of applications were developed, including the Museum Collections Database (MuseumPlus), which hosts 27,755 digital files on 10,557 objects and is enriched with new digitized cultural, research and scientific material on a daily basis, the digitization of the Museum's archival records, the 3D scanning of sixty significant exhibits and the creation of three-dimensional models and the creation of a website, that informs the public about the Museum and its temporary and permanent exhibitions, activities, events, school and family programmes, research and conservation projects (Acropolis Museum, n.d.).

7.2 The 3D Digital Reconstruction of Ancient Olympia Site

In November 2021, Microsoft implemented the project of the digital reconstruction of ancient Olympia with the collaboration of the Greek Ministry of Culture. This project enables viewers from all over the world to virtually visit and explore the site of Ancient Olympia, allowing them to travel back in time and see how it was in ancient times. This virtual journey can be achieved via an interactive mobile app, a personal computer and the exhibition of Microsoft HoloLens 2 housed in the Olympic Museum of Athens. 27 monuments in the area are digitally restored and represented, some of which are a Gymnasium in which athletes trained and a Stadium which accommodated the Olympic Games in honour of Zeus, the Temples of Zeus and Hera and the workshop of Pheidias (Keep Talking Greece, 2021).

7.3 The Centre for Historic Information at Thermopylae

The Centre is dedicated to the Battle of Thermopylae, one of the most important battles of Greek and World History, which was fought in 480 BC among the Greeks and the Persians.

The Centre provides visitors with information on the historical framework, the Battle, the protagonists, the armoury and the movements of the two opposing armies and about how the battle ended, via videos, virtual reality special glasses and special interactive tablets, which make the viewers feel as if they were present (Municipality of Lamia, n.d.).

7.4 The Underwater Museums in Magnesia of the Thessaly Region in Central Greece

The first Greece's Underwater Museum in the island of Alonissos opened to the public in August 2020. The Alonissos underwater museum enables divers to explore the fifth-century wreck off the islet of Peristera, one of the largest Classical-era shipwrecks, which was discovered in the 1990s at a depth of 25 m with more than 3000 amphorae intact (Paravantes, 2020).

In July 2022, the Greek Culture Minister inaugurated three new underwater museums in the sites of Kikinthos, Tilegrafos and Glaros in the area of Amaliapoli in Magnesia. The four sites in total are part of the diving park being created in the area of Western Pagasitic Gulf and of the Sporades islands. More specifically divers have the opportunity of visiting nine shipwrecks of the Classical, Hellenistic, Byzantine and later periods. Special tablets make

the dive to the sites even more engaging by providing information about the underwater archaeology and sea biology. Besides, a virtual reality application allows visitors, who cannot dive, to virtually visit and explore the sites in Knowledge and Awareness Centers, which operate on land. It is worth mentioning here that the underwater museums of Thessaly were awarded the first European Innovation Prize of the European Cultural Tourism Network (ECTN) in the category "Innovation and Sustainable Cultural Tourism Digitization towards smart destinations" (gdp, 2022).

Furthermore, other museums which offer virtual experiences, like virtual tours and exhibitions, website educational activities and games, videos, 3D models of objects and artefacts, are inter alia, the National Archaeological Museum, the Benaki Museums, the Archaeological Museums of Thessaloniki, Patras, Tegea and Thebes, the Goulandris Natural History Museum, the Byzantine Museum in Athens and the Technological Cultural Venue "Hellenic Cosmos", which is an ultra-modern Cultural Centre and Museum, where visitors experience Hellenic history and culture through use of state-of-the-art technology and audio visual and interactive media (Byzantine and Christian Virtual Museum, n.d.; Hellenic Cosmos Cultural Center, n.d.; Sofia Ps, 2021). The Cultural Venue includes the "Tholos", a semi-spherical Virtual Reality theatre, where digital productions are presented, like the Virtual Reality productions entitled "A Walk Through Ancient Miletus", which offers a virtual visit of the city of Miletus on the western coast of Asia Minor, one of the most important cities in Ionia. Also, "A Walk Through Ancient Olympia" which provides a complete tour of the sacred site of ancient Olympia in the second century BC and enables visitors to participate in the events and the rituals of the Olympic Games and learn about the history of the site, as well as "An Interactive Tour at the Ancient Agora of Athens" (Hellenic Cosmos Cultural Center, n.d.).

7.5 The Website VisitGreekNature

The Goulandris Museum of Natural History and the Hellenic Center for Habitats–Wetlands have recently created a website called VisitGreekNature (www.visitgreeknature.gr) whose aim is the promotion of Greece's protected areas as tourist destinations as well as the underlining of interconnections between natural and cultural heritage.

This particular website also provides information regarding 1300 protected areas of mainland and island Greece, 446 areas of the European Network Natura 2000, 51 areas protected by international treaties, 36 protected area management bodies and 2100 infrastructures and sites of special eco-tourism interest. Moreover, one can also be informed of the cultural assets found in these protected areas, such as monuments, archaeological sites and even museums. Consequently, the importance and value for recreation of Greece's both cultural and natural environment, as well as the need for their integrated management are highlighted in the website (Keep Talking Greece, 2021).

8 Conclusions

As already mentioned above, Greece has made during the last years significant progress in the area of digitization, online accessibility and digital preservation of cultural material, which has been also recognized at a European level (EKT, 2019; European Commission, n.d.).

However, the fact that the cultural heritage and the natural environment surrounding it, are increasingly threatened beyond the traditional causes of decay, by new multiple, natural and manmade threats, caused, to a wide extent, by climate change and overexploitation of natural resources and ecosystems, has highlighted the urgent need for a more holistic cultural and natural heritage protection and sustainable management policy and for the promotion of the public interest and awareness of the importance of safeguarding this unique and irreplaceable property. In these fields, the wider use of innovative and pioneering applications of new technologies can play a decisive role.

References

Acropolis Museum. (n.d.). *Digital museum.* https://www.theacropolismuseum.gr/en/digital-museum. Last accessed September 1, 2022.

Adamou, C., Polizoudi, A., Mertzimekis, N., & Mitsios, S. (2021). 3D projection mapping ως εναλλακτικός τρόπος παρουσίασης της ιστορίας των μνημείων" (=3D projection mapping.as an altenative way of narrating the monument's history. In *4th Panhellenic Conference on Digitization of Cultural Heritage, Euromed 2021, Athens 30.9.—3.10.2021, Conference Proceedings, first Edition: June 2022, Copyright "PERRAIVIA" network* (pp. 318–331).

Alivizatou-Barakou, M., Kitsikidis, A., Tsalakanidou, F., Dimitropoulos, K., Chantas, G., Nikolopoulos, S., et al. (2018). Intangible cultural heritage and new technologies: Challenges and opportunities for cultural preservation and development. In M. Ioannides, N. Magnenat-Thalmann, & G. Papagiannakis (Eds.), *Mixed reality and gamification for cultural heritage* (pp. 129–158), Springer.

Almazán, I. (2018, July 20). *Greece: New European destination victim of overtourism?* https://travindy.com/2018/07/greece-new-european-destination-victim-of-overtourism/

Arnold, D. (2005). Virtual tourism: A niche in cultural tourism. In M. Novelli (Ed.), *Niche tourism. Contemporary issues, trends and cases* (pp. 223 - 231). Elsevier Ltd.

Audunson, R., Andresen, H., Fagerlid, C., Henningsen, E., Hobohm, H. C., Jochumsen, H., & Larsen, H. (2020). Introduction–Physical places and virtual spaces: Libraries, archives and museums in a digital age. In R. Audunson, H. Andresen, C. Fagerlid, E.

Henningsen, H. C. Hobohm, H. Jochumsen, H. Larsen & T. Vold (Eds.), *Libraries, archives and museums as democratic spaces in a digital age* (pp. 1–22).

Avouri, E., Katiri, M., & Toumpouri, M. (2020). Έδρα της ΟΥΝΕΣΚΟ: Τρεις ερευνήτριες αναπτύσσουν τη σημασία της ψηφιοποίησης της πολιτιστικής κληρονομιάς στη νέα εποχή" (=*Unesco Chair: Three female researchers develop the importance of digitizing cultural heritage in the new era.* https://www.cut.ac.cy/news/article/?contentId=428380. Last accessed September 1, 2022.

BBC. (n.d.). In pictures: 3D return for Bamiyan Buddha destroyed by Taliban. https://www.bbc.com/news/world-asia-56337042. (budda statues) Last accessed September 1, 2022.

Byzantine and Christian Virtual Museum. (n.d.). *Virtual museum.* https://www.ebyzantinemuseum.gr/?i=bxm.en.virtual-museum. Last accessed September 1, 2022.

Dellaporta, K. P. (2005). "Υποβρύχια Αρχαιολογική Κληρονομιά στην Ελλάδα. Νομική Προστασία και Διαχείριση" (= "Underwater cultural heritage in Greece. Legal protection and management"). https://nomosphysis.org.gr/10093/upobruxia-arxaiologiki-klironomia-stin-ellada-nomiki-prostasia-kai-diaxeirisi-noembrios-2005/

Ding, M. (2017). *Augmented reality in museums. Arts management and technology laboratory* (pp. 1–12).

EKT. (2019, July 9). *Greece among the examples of best practice with regard to digitisation and distribution of cultural heritage.* https://www.ekt.gr/en/news/23330. Last accessed September 1,2022.

European Commission. (n.d.). *The Europeana platform.* https://digital-strategy.ec.europa.eu/en/policies/europeana. Last accessed September 1, 2022.

European Commission. (n.d.). *Sustainability and cultural heritage.* https://culture.ec.europa.eu/cultural-heritage/cultural-heritage-in-eu-policies/sustainability-and-cultural-heritage. Last accessed September 1, 2022.

European Comission. Cordis. (n.d.). *Advanced VR, iMmersive serious games and Augmented REality as tools to raise awareness and access to European underwater CULTURal heritage.* https://cordis.europa.eu/article/id/413512-new-technology-brings-europe-s-underwater-cultural-heritage-to-life. Last accessed September 1, 2022.

European Comission. Cordis (n.d.). *Inclusive cultural heritage in Europe through 3D semantic modelling.* https://cordis.europa.eu/article/id/413514-3d-models-explore-our-built-cultural-heritage-through-time-onsite-and-remotely. Last accessed September 1, 2022.

European Comission. Cordis. (n.d.). *Technical innovations help overcome access barriers to cultural spaces.* https://cordis.europa.eu/article/id/413505-technical-innovations-help-overcome-access-barriers-to-cultural-spaces. Last accessed September 1, 2022.

FAO. (n.d.). *(GIAHS) Globally important agricultural heritage systems.* https://www.fao.org/giahs/en/. Last accessed September 1, 2022.

gdp. (2022, August 2). *Central Greece adds more underwater museums to diving park.* https://news.gtp.gr/2022/08/02/central-greece-adds-more-underwater-museums-to-diving-park/. Last accessed September 1, 2022.

Greek News Agenda. (n.d.). *Digital transformation "bible" of Greece (2020–2025).* https://www.greeknewsagenda.gr/topics/business-r-d/7379-the-digital-transformation. Last accessed September 1, 2022.

Hellenic Cosmos Cultural Center. (n.d.). http://www.fhw.gr/cosmos/index.php?&lg=_en. Last accessed September 1, 2022.

Hellenic Cosmos Cultural Center. (n.d.). http://www.fhw.gr/cosmos/index.php?id=1&m=1&lg=_en. Last accessed September 1, 2022.

INCIDER. (2018, January 17). *Divers discovered a 215-mile-long underwater cave system in Mexico that's full of Mayan relics.* https://www.businessinsider.com/divers-discovered-underwater-cave-mexico-mayan-relics-215-mile-long-gran-acuifero-maya-project-2018-1. Last accessed September 1, 2022.

Keep Talking Greece. (2021, November 10). *Stunning 3D digital reconstruction of ancient Olympia site (video,pcts)*. https://www.keeptalkinggreece.com/2021/11/10/ancient-olympia-digital-reconstruction-ed-microsoft-app-greece/. Last accessed September 1, 2022.

Keep Talking Greece. (2021, September 27). *Greece launches website on protected areas reserves, mountain shelters, thermal baths, culture sites*. https://www.keeptalkinggreece.com/2021/09/27/greece-website-protected-areas-wetlands-culture/. Last accessed September 1, 2022.

Kennedy, A. A., Thacker, I., Nye, B. D., Sinatra, G. M., Swartout, W., & Lindsey, E. (2021). Promoting interest, positive emotions, and knowledge using augmented reality in a museum setting. *International Journal of Science Education, Part B: Communication and Public Engagement*, 1–17.https://doi.org/10.1080/21548455.2021.1946619

Kourousopoulos, E. (1978). Δασική Ιδιοκτησία και Διαχείριση" (= "Forest property and management").

Liritzis, I., Al-Otaibi, F. M., Volonakis, P., & Drivaliari, A. (2015). Digital technologies and trends in cultural heritage. *Mediterranean Archaeology and Archaeometry, 15*(3), 313–332.

Madirova, E., & Absalyamovaa, S. (2015). *The influence of information technologies on the availability of cultural heritage*. Elsevier Ltd.

Melvin, D., Elwazer, S., & Berlinger J., CNN. (2015, August 31). *ISIS destroys temple of Bel in Palmyra, Syria, U.N. reports*. https://edition.cnn.com/2015/08/31/middleeast/palmyra-temple-damaged/index.html

Municipality of Lamia, (n.d.). *Thermopylae's innovative centre of historical information*. https://culture.lamia.gr/en/blog/thermopylaes-innovative-centre-historical-information. Last accessed September 1, 2022.

Paravantes, M. (2020, August 4). *Greece's first underwater museum on Alonissos opens to public*. https://news.gtp.gr/2020/08/04/greeces-first-underwater-museum-alonissos-opens-public/. Last accessed September 1, 2022.

Sofia Ps. (2021, January 10). *Top-10 Greek museums you can [virtually] visit*. https://www.sofiaskaleidoscope.com/journal/top-10-greek-museums-online/. Last accessed September 1, 2022.

Tsartas, P., Stavrinoudis, T., Zagotsi, S., Kiriakaki, A., & Vassiliou, M. (2010). Τουρισμός και Περιβάλλον" (= Tourism and Enivironment"), Athens. WWF Greece.

Turner, H., Resch, G., Southwick, D., McEwen, R., Dubé, A. K., & Record, I. (2017). Using 3D printing to enhance understanding and engagement with young audiences: Lessons from workshops in a museum. *Curator: The Museum Journal, 60*(3), 311–333.

UNESCO. (2015 June 30). *Chairpersons of six UNESCO culture conventions meet*. https://whc.unesco.org/en/news/1305. Last accessed September 1, 2022.

UNESCO. (2017, July 9). *Eight new sites inscribed on UNESCO's world heritage list*. https://whc.unesco.org/en/news/1689. Last accessed September 1, 2022.

UNESCO. (n.d.). *Psiloritis Unesco Global Geopark (Greece)*. https://en.unesco.org/global-geoparks/psiloritis%20/. Last accessed September 1, 2022.

UNESCO. (n.d.). *Lesvos Island Unesco Global Geopark (Greece)*. https://en.unesco.org/global-geoparks/lesvos-island. Last accessed September 1, 2022.

UNESCO. (n.d.). *Sitia Unesco Global Geopark (Greece)*. https://en.unesco.org/global-geoparks/sitia. Last accessed September 1, 2022.

UNESCO. (n.d.). *Vikos—Aoos Unesco Global Geopark (Greece)*. https://en.unesco.org/global-geoparks/vikos-aoos. Last accessed September 1, 2022.

UNESCO. (n.d.). *List of Geoparks and regional networks*. https://en.unesco.org/global-geoparks/list. Last accessed September 1, 2022.

UNESCO. (n.d.). *Convention concerning the protection of the world cultural and natural heritage 1972*. http://portal.unesco.org/en/ev.php-URL_ID=13055&URL_DO=DO_TOPIC&URL_SECTION=201.html. Last accessed September 1, 2022.

UNESCO. (n.d.). *Text of the convention for the safeguarding of the intangible cultural heritage*. https://ich.unesco.org/en/convention. Last accessed September 1, 2022.

UNESCO. (n.d.). *Convention on the protection of the underwater cultural heritage 2001*. https://unesdoc.unesco.org/ark:/48223/pf0000126065. Last accessed September 1, 2022.

UNESCO. (n.d.). *The Historic Centre (Chorá) with the monastery of saint-john the theologian and the cave of the apocalypse on the Island of Pátmos*. https://whc.unesco.org/en/list/942/. Last accessed September 1, 2022.

UNESCO. (n.d.). *Chelmos Vouraikos Unesco Global Geopark (Greece)*. https://en.unesco.org/global-geoparks/chelmos-vouraikos. Last accessed September 1, 2022.

UNESCO. (n.d.). *UNESCO global geoparks (UGGp)*. https://en.unesco.org/global-geoparks. Last accessed September 1, 2022.

Wikipedia (n.d.). Notre-Dame fire. https://en.wikipedia.org/wiki/Notre-Dame_fire. Last accessed September 1, 2022.

Cultural Heritage and Sustainable Environment

Ecotourism and Rural Sustainable Development, Albania Case, Blezënckë Village

Juljan Veleshnja

Abstract

Ecotourism and rural sustainable development are concepts related to the tourism and are very vibrant forms of it, with clear benefits to the actors involved, to visitors, and to the local people. Rural areas offer their activity in the ecological and rural diversity, relying strongly on nature, which seems more abundant in the past. Albanian context shows two important periods of the rural development, the first during the socialist period (totalitarian regime) and the other during the long transition after 1990s with the falling of the regime. Lately, major changes in technology, cultural, political, or economic aspects, have brought a profound transformation in rural processes related to the production. The political transformation in Albania brought the increasing phenomenon of migration, which affected strongly the demographic shrinkage of the rural area, affecting the rural development. Lack of awareness, policies, and the absence of formalized planning or intervention to the rural areas, have left the possibilities for the rural community in Blezënckë to benefit from tourism development. Rural sustainable development of Albanian local communities should rely on the ecotourism and rural tourism, being a must for the contemporary rural evolution. The rural processes include a lot of natural and cultural local resources integrated with the tourist activity, evidencing its complexity and interaction with the environment and other sectors of the economy. For rural areas, the fast steps of changes have brought with it different opportunities and favorable challenges. The aim of this research is to analyze the existing relationship between rural development, tourism, and ecotourism, challenging the need for sustainable development and the preservations of the local traditions. This also rushes the economic transformation, switching from agriculture to manufacturing and services, improves welfare to farmer, and promotes an environmental conservation in rural areas.

Keywords

Rural sustainable development • Ecotourism • Rural tourism • Environment • Communities

1 Introduction

The studies on sustainable development have a particular focus on rural environment, which is a complex system that embodied culture heritage, natural resources, and most commonly agricultural and animal products. There are different international initiatives, like The Paris Agreement and the Agenda 2030 of the United Nations, but although different instruments like Local Action Group (LAG), EU Cohesion Policy, and Common Agricultural Policy (CAP), which underline the crucial importance of the rural development. This paper aims to bring the focus in the Albanian rural areas and the potential they have, for the sustainable development of the whole country.

The methodology of this research is based on a desk research, to point out the theoretical aspects and international initiatives of the sustainable rural development, which can help for a better understanding and building a sustainable frame of intervention on a study case. In addition, there have been conducted different site visits to gather reliable data of the rural area and interact, understand, and share information with local people. This methodology aims to help on building a sustainable rural development strategy for a village in Albanian country, with all the problematics of the area dictated by many aspects.

For a better structure of analyses and actions to be taken on the case study, the research will start with evidencing important initiatives which regulate the sustainable rural

J. Veleshnja (✉)
Department of Urbanism, Faculty of Architecture and Urbanism,
Polytechnic University of Tirana, Tirana, Albania
e-mail: juljan.veleshnja@fau.edu.al

development, as the starting point, evidencing briefly different goals and strategies for the sustainable rural development.

2 Sustainable Rural Development Goals

Two international initiatives that are determinative on addressing concerns for the future economic development associated with social and environmental issues are Agenda 2030 and The Paris Agreement. Similar actions must take into consideration the urban–rural gap in terms of socioeconomic conditions exposing the rural areas to social and environmental threats and pressure. The rural areas have a great role, from feeding the basic needs of the urban areas, to preserve the natural habitat and landscape, although to provide labor force, but not forgetting that the rural settlements contribute vividly to the cultural preservation and heritage of the countries.

With the late development, worldwide, the rural areas are facing problems with the poverty, demographic shrinkage, but even in a larger context with the fragmentation of the natural habitat, deforestation, climate change, and so on.

In this frame, Agenda 2030 shows sustainable development goals and specific targets to manage globally, like poverty, geographical inequality, peace, pollution, and climate change (UN G. A., 2015). Targets linked to many key aspects of human activity, and intrinsically related to the rural development.

The Paris Agreement evidences different actions to be taken in order to achieve a clean future and to build resilient communities related to climate change, for developed and developing countries too (UN, 2015).

Rural development is an important part of the Common Agricultural Policy (CAP), which contributes and supports to increase the economy and market measures, aiming the improvement of the social and environment of rural areas. Direct payments to farmers, agricultural price supports, supply controls, and border measures are some of the main instruments of the CAP, which is a locally oriented farm policy, initiated in 1962, based on three major principles: a unified market aiming for a free flow of agricultural supplies with common prices within the EU; product preference in the inner market over out-EU imports; and financial support by financing of agricultural programs. CAP priorities and implementation for the Rural Development are articulated through six priorities: knowledge transfer and innovation; farm viability and competitiveness; food chain organization and risk management; restoring, preserving, and enhancing ecosystems; resource-efficient, climate-resilient economy; social inclusion and economic development.

Within this frame, there are different approaches aiming the sustainable rural development. LEADER—Liaison Entre Actions de Développement de l'Économie Rurale, is an approach proposed by the European Commission to encourage partnerships at a sub-regional level between the public, private, and civil sectors by mobilizing the energy and resources of local people and organizations, a "bottom-up" development concept (EU-Commission, 2017). The LEADER approach is related with local empowerment through local strategy development and resource allocation, aiming at the involvement of local people in decision-making by the Local Action Group (LAG). LAG is a representative group of public, private, and civil society partners, who get together to define and implement the actions to improve their local area.

Those initiatives show that there is a huge interest on the sustainable rural development, with different objectives to be achieved. Aiming the access to proper education, the development of rural economies, and reducing socioeconomic inequalities and environmental injustice, the key strategic actions are rural resilience and circular economy. The EU is aiming for the shifting of the economic model, from linear to the circular economy framework based on the principles reduce-reuse-recycle (UN G. A., 2015). The transition to circular economy can lead to new rural economic activities which aim to promote local traditions and preserve the rural and natural landscapes based on responsible production and consumption of local natural resources (Weigend et al., 2020).

The sustainable development goals for the rural areas aim to give answers to the poverty and many other problems related to the lack of public utilities and services (mainly in developing countries) to decrease the gaps between different regions. The attention of the different projects and proposals to achieve the sustainable development goals, must be put on rural perspective for the improvement of the quality of life and to diminish the inequalities aiming a sustainable agriculture, rural resilience, and circular economy. Local government must play an important role and be reliable on supporting the linkage of urban–rural activities and the regional economies. Generally, the rural communities must be able to develop even non-farming activities and services to be combined with agriculture in order to become more resilient to the uncertain ongoing global conditions (Anderson, 2015). To help these ideas, local and central decision-makers should have continuous attention to the rural areas, to build sustainable policies of interventions, and mainly to enable projects and investments in infrastructure (Freshwater, 2015).

3 Rural Tourism

Rural areas are different from each other and carry complexities. For the economic development, the rural communities can focus on agriculture, industry, or tourism activities, depending on the local characteristics, resources, and capacities. Tourism activities are taking place more and more on rural areas, like rural tourism, agritourism, religious tourism, and ecotourism which can be good alternatives to the main economic agricultural sector, decreasing the dependency on it and ensuring the rural resilience (da Silva et al., 2017).

One of the key aspects of the tourism is ecotourism, which started in 1960, shifted the way of thinking about tourism to ecology. Ecotourism is defined as "responsible traveling to the natural areas that preserves the environment, sustains the well-being of the local people, and involves understanding and education"—the International Ecotourism Society, 2015.

Rural areas preserve a lot of natural and cultural resources which naturally can offer a lot to the tourism, and that's why the tourism sectors seem a good alternative to the sustainable rural development (Menges, 2022). Ecotourism tries to have some principles like: building cultural and environmental awareness; providing good experiences for all the actors, informative experiences for visitors aiming the raise of the sensitivity of local actors; to minimize the impact of social, behavioral, and physical aspects; to generate financial benefits for local people; to provide economic benefits for conservation and improvement of the local resources (Thompson, 2022). Rural tourism is a valuable sector which is influencing a lot the development of different rural areas, but that still needs attention and effort.

4 Albanian Context

Albania has faced different historical, cultural, social, and economic aspects which have signed the development of the country, urban and rural areas. The rural development in Albania is far from the global discourse of the sustainable rural development, exposing a problematic situation in need of attention. For a better understanding of these rural contexts, it is necessary to face historical analyses, from the structures of a fragmented economy under the occupation of the Ottoman Empire, to a centralized development during socialism leadership, to a long transitional laissez faire situation of the post-socialist system, till to the actual uncertain situation. The shift from socialist to post-socialist period presented a hostile transformation with evident traces of natural rural landscapes. The rural Albanian context presents the overlapping of different past situation, evidencing diverse realities and landscapes with signs of culture, iconography, wild nature, the post-industrial of socialist panorama, and lately, the landscape of an appropriated territory under spontaneous uncontrolled development.

The Albanian villages, before the usurpation from the Ottoman Empire, were evidencing a prosperous region with a beautiful nature. The various written documents present the Albanian territory of the time, with wide green fields and internal river valleys in the mountainous areas as natural territories with solid presence of inhabitants. The Albanian settlements were founded along the rivers, on the hills, on the slope of the mountains, or at their foot (Muka, 2007). The main economic activity of the rural areas was the cultivation of cereals, horticulture, gardening, and viticulture (Biçoku, 2005). During the Ottoman occupation, there were traces of swamping phenomenon, caused by massive deforestation and abandonment, creating so problematic situations. In 1912, Albania gained its independence and immediately after that, two world wars took place, with evident influence on the territory. During these periods, the land was managed based on the concept of manorial property and the feudal system. The feudal lords and other large landowners owned 40% of the land area.

With the end of the Second World War, Albania faced almost 50 years under a totalitarian regime. The socialist period influenced with an evident impact the rural development, under pressure of a centralized economic system. The organic relationship of the rural areas and their territories, with the regional logic that dominated the economy of the Ottoman period, changed shape during socialism, toward a centralized system on country level. The centralized economic system had expressions in all the territory through typical images of an agriculture toward industrialization. The character of different rural regions was alienated by the continued profiling of the industrialization of production. In the alienation of the rural character, it is important to understand the agricultural land management system. During socialism, agricultural land management reform faced the process of gradual collectivization of the agricultural and agrarian economy, from plain, hilly, and then mountainous terrains (Stahl, 2010), to create the cooperatives, as new economic structures.

During socialism era, the demographic movement of the people was controlled for the better use of the whole territory, organizing the urban areas, refurbishing the former rural settlements, and creating new settlements near natural resources, forests, energy sources, etc. The main elements that condition the selection of the location, for the new rural settlements, were the proximity to the labor front, being centralities of the agricultural economy, and being close to the national road network. There was a visible difference between the former rural settlements and the new ones. The former rural settlements were very organic with vernacular architecture, respecting the territory and inheriting the cultural and traditional values, with a more homogenous population. While the new socialist ones presented a more geometric spread of dwellings in the territory without any sensitivity to the territory, with a more heterogenous population in some rural areas of interest, with different cultures and way of living.

With the strong political wave of changes after 1990, with the fall of the former regime, and appearance of the democracy and free market economy, Albania faced a long transitional period (King, 2005). During this transition, from the socialist system to the post-socialist one, there was an evident degradation of the rural settlements, social, cultural, and natural sources. The country is characterized by a period of multi-faceted reforms aiming at the transformation of its economic system. During this period, there were different laws to regulate the private property turning back to the former owner, or to the continuous users of it. The reform in the privatization of agricultural land was accompanied by the diminution in production and lack of this sector to be competitive (Streule et al., 2019). On the other hand, the demographic shrinkage that accompanied the rural areas into the abandonment of the rural territory, in search of a better life, toward urban areas or developed countries, signed the decline of them.

Lately, the situation is in the process of change; there are former inhabitants turning back to their former rural settlements, investing on new economic activities, and implementing knowledge earned abroad. In parallel, there are different policies from the central government and from other organizations, which aim to help the rural development. In this frame, there was developed a project lea by the National Territorial Planning Agency (NTPA) of Albania, with the goal to build development strategies for more than 100 villages, with groups of experts, academics, and students. One of those rural area is Bezënckë village, which is taken as a case study for this research.

5 Case Study—Blezënckë Village

Blezënckë village belongs to the Municipality of Skrapar, of Berat District which is situated in the central-southern part of Albania. The area has different important natural resources, Osum river with its well-known canyon in our country and the imposing presence of the sacred mountain of Tomorr, the National Park of Albania. Blezënckë is a small rural area, but with a particular natural landscape and a very welcoming community.

For this research, there were organized different site visit and direct meetings with the local people, trying to gather information and make them aware about the actual situation and the initiatives of designing development strategies, and in this frame to invite them to be part of decision-making with a participatory process. Visitors can find hospitality of the inhabitants, in their houses and the courtyards (Fig. 1) with colorful flowers (Fig. 2). They still preserve the culture of being good hosts, with traditional food and rituals accompanied by regional folk songs.

5.1 General Description of Blezënckë

Blezënckë is a small village, immersed in nature, close to Çorovodë city. Besides other natural resources in the region, the most prominent natural elements are the Osum river—canyon and the Tomorr National Park (Qiriazi, 2017). The strategic advantage of the geographical position of the Osum Canyon, near the village of Blezënckë, creates a wonderful

Fig. 1 Blezënckë characteristic house and backyard

Fig. 2 Blezënckë characteristic house and courtyard

Fig. 3 Blezënckë natural landscape toward Osumi canyon

natural landscape with aesthetic, ecological, and cultural values (Fig. 3).

In recent decades, given the general abandonment of rural areas and the demographic movement to urban areas, the village has lost a large part of its population. The village of Blezënckë had its greatest development at the time (mainly socialist period) of the operation of agricultural cooperatives, when the whole village was active and organized in agricultural works and activities. Before 1990 in the village,

there were registered 250 inhabitants, while today, there are only 1/5 of them, 44 inhabitants, mainly older people. Its demographic shrinkage brings loss of cultural heritage and knowledge, despite creating left-over rural areas. The accessibility to the village shows a problematic infrastructure, which does not help the development of this area. The road connection with the town of Çorovoda, the nearest urban center, is quite good, while the state road Berat—Çorovodë is not in good condition. Some segments are under construction, and this doesn't help in terms of relations and connections in national scale.

The rural nature of Blezëncke village is very particular and attractive. The village has suffered for a long time from isolation that has limited its economic and social development, ranking it among the most underdeveloped villages in the region. The nearest urban area is the city of Çorovoda, which offers the village of Blezëncke administrative and public services, as well as local services such as banks, shops, transport, hospital, and social insurance. Many of its habitants have left the village and their properties moving abroad, presenting today a village with few habitants, but with the courage and will for changes. It seems difficult to talk about development of the village, taking into consideration the demographic shrinkage and partly lack of agricultural production, but the number of assets that the village and its inhabitants carries can be catalysts of several interesting initiatives in the future.

Blezëncke has several natural resources such as: water sources, which are used for basic needs in rural life; good biodiversity which directly affects agricultural production, livestock production, and product processing (Fig. 4); Osum canyon and river, which offer impressive nature and numerous attractions along the river flow, and the bridge to the village, which offers a great panorama; the picturesque natural nature, the village is surrounded by mountains and fields that provide an attractive view.

5.2 Natural Resources

Blezëncke village offers a very impressive nature, with water resources, which are a characteristic value of the area.

The village is unique, with mountainous landscape and rugged relief, offering a rich spatial heterogeneity of beautiful nature and territory. Along Osum river, close to Blezëncke, above the canyons, there are different spots with different meanings and legends that enrich the experiences while visiting the area, those are part of the regional culture and history, like the "The traces of Abas Ali" and the "The cave (whole) of the bride," stories to be discovered on the road to the village. Moving from Berat to Blezëncke, there is the impotent Tomorr Mountain with its National Park (Qiriazi & Bego, 1999), very important natural asset of our country.

The waters of Osum river, with their corrosive and dissolving power, have penetrated slowly, without interruption, into the limestone rock mass of the anticlinal structure of Çorovoda, forming the canyon of rare and attractive beauty. Osum river and its canyon are very well-known in our country and abroad. The most beautiful and picturesque part of the canyon starts from the village of Blezëncke to the end of the village of Çerenisht. The slopes of the canyon plunge over the river, from a height of 70–100 m, it is about 13 km long, 4–35 m wide, up to 70–80 m deep. The tour in the canyon can be done on foot or rafting, depending on the level of the water of the river. In the vertical walls of the canyon there are many boulders and hollows formed by karst processes (Fig. 5). The bed of Osum sometimes expands itself, swallowing the rocks below it, and sometimes it narrows under many twists, according to the currents it has had over the centuries (Fig. 6). In some segments of the canyon, large waterfalls of surprising beauty appear, two most special waterfalls named "Waterfalls of love."

These are the most important elements to be highlighted among others, which evidence the natural assets

Fig. 4 Blezëncke viticulture field (wine and raki)

Fig. 5 Osum canyon and the canyon island

Fig. 6 Osum canyon under Blezëncë

and resources which can give birth to different activities that can help on the rural development of Blezëncë.

5.3 Sustainable Development Opportunities

The development of Blezëncë rural area could be achieved by expanding the range of products and services, aiming the increase of its competitiveness at the regional level, and promoting it. Another important aspect that needs attention is increasing the accommodating capacity of Blezëncë, for visitors and tourists of various categories, mainly those who visit the Osum canyons and the surrounding areas. The principles of increasing the number of beds should be based on the "Albergo diffuso" concept (diffused accommodation), using the actual house stock of the emigrants, in order to maintain an uncontaminated rural image (Confalonieri, 2011). This aspect helps on conservation and readapting the actual building stock. Increasing the incomes of the village, another goal of the rural development could be possible by not letting it an isolated village, but by putting Blezëncë in a network with other rural areas up to Çorovodë, till to Berat. Blezëncë should be seen on a regional scale, where it can reveal its own features and characteristics, showing that it is a unique place with its own local products (Zakkak et al., 2015).

The production of local goods in the village of Blezënckë is reduced for several reasons, for the territorial extent of the village, as well as for the demographic shrinkage, the products are mainly for the families themselves, and some small sales in some market in the surrounding villages or urban areas.

This is an important aspect for the development of the village which in order to play a decisive role, the diversification of current products must be strengthened. The addition of new products and sub-products must be encouraged, but they must be closely related to the tradition and culture of the village of Blezënckë, without alienating it, thus complementing the services mainly related to gastronomic tourism and not only (Anderson, 2015).

The community of Blezënckë has a good potential in making all the rural processes. The demographic decline of this area affects the realization of these processes, as well as the spread of knowledge and traditions from generation to generation (Marot, 2022). The return of the left population and demographic increase of Blezënckë is another important aspect, in order to enable the development of all the links of the various processes.

Increasing the rural activities in parallel with the accommodation capacity for the sector of tourism is an important aspect aiming the sustainable development, but

this evidence of the need to increase social interaction as well as the training of social capital in other directions and activities that are not currently naturally present in the rural life. The training is related to the accommodation capacity, communication, as well as to the various rural activities that occur in the agricultural and livestock processes. Since the aim is to increase the experience of visitors, then hygiene, regularity of the process, and the explanatory skills must be acquired by the inhabitants.

Those activities must be coordinated and an institutional framework must be established, which must coordinate and help not only the development of all rural processes, but also informing and training of the social capital of this rural area, through training or other instruments. The institutional framework must operate on a regional scale; villages cannot work isolated for offering even other opportunities like LEADER-LAG initiatives (EU-Commission, 2017).

The institutionalization and regulation of the relations of different rural areas bring several positive aspects such as: the representation of these areas at different levels, both for promotional aspects and in engagements for economic benefits, also in finding solution for infrastructure at the regional level, as well as conversating and coordinating the activity with the relevant local institutions, thus with the aim for a sustainable rural development.

The development of Blezënckë rural area is based on the structuring, organization, and increase of rural production, as well as the ability to host tourists visiting the Osum canyons

Fig. 7 Osum canyon and river, Zabërzan bridge

and surroundings (Fig. 7). This development will be possible if local social capital is supported, made aware and empowered to carry out these processes.

5.4 Rural Tourism

Although the Tourism industry is a single sector, it currently involves various tangible and intangible elements and is affected by many other sectors and activities. The development of tourism can be useful for other economic sectors in rural areas. In the aimed economic development model of Blezëncke, the tourism industry should play an important role, which aims to promote the growth and development of the village and the region. Rural tourism, ecotourism, and agritourism are the main forms of tourism to be implemented in Blezëncke, which for their nature are needed for products and services relying on local resources and production. Those forms of tourism are supported by many other economic sectors with an emphasis on traditional activities and occupations. Rural tourism faces complex aspects and results from special interactions between the rural areas, local people, local products, and activities. The quality depends on several typical characteristics: the quality of natural resources, the structures, the ability to host visitors and tourists, the structure of the village as well as the cultural wealth.

One important natural resource of Blezëncke is Osum river and canyon, which can play a crucial role for the development of Blezëncke. There is already developed adventure tourism in the canyon of Osum by private organizations, which are not involved yet the rural area of Bleznckë in this process, for not offering enough services. Blezëncke should increase the products for tourism to be competitive and attract different actors of tourism sector and tourists.

The main aspect of rural tourism and agritourism is to provide visitors with experiences in the rural processes, with personalized contact to the nature of Blezëncke, although to allow the visitors to participate in the activities, traditions, and ways of life of the local people (Krasniqi, 2016). The community of the village with their activity can fulfill people's needs for information and participation in food production and processing and to better understand farm culture and rural heritage; fulfilling the needs for direct contact with animals, plants, and the rural environment; fulfilling human needs for experiencing the real life of a family and a rural society. Aspects that Blezëncke with its community can offer in a good way.

Consumers of rural tourism, in addition to their tendency to be active, informed, and mentally cngagcd, nccd to havc authentic experiences and hands-on experiences that will bring them back to their roots, and, in a way, will positively affect their lives. For Blezëncke a small rural area, it is important to aim not a massive tourism. It can provide necessary services for visitors, such as: reception, accommodation, sleeping, food, and health services (Buccianti, 2018). The former farm of the cooperative can be turned into the main tourist's destination, through the introduction of a restaurant, which will serve tourists traditional food with local products. It is also possible to restructure the abandoned houses as hostels to increase the accommodation capacity. The structuring, assemblage and organization of the village enable functional agritourism, guesthouses, dairies, and characteristic local foods such as: fresh dairy products and local honey. The free open areas in the surrounding offer the possibility of pitching the tents, increasing the choice of staying in the village.

In order to enable these strategies, it is necessary to organize a promotional network through social networks and other channels, thus increasing the exposure of Blezëncke and its resources.

5.5 Ecotourism in Blezëncke

The International Ecotourism Society defines ecotourism as responsible traveling to the natural areas that preserve the environment, sustains the well-being of the local people, and involves understanding and education. Ecotourism, as a nature-based tourism, has a great potential to create a good social impact of the community of Blezëncke as a positive environmental knowledge and awareness (Ballantyne & Packer, 2013), keeping in mind the rich natural environment of the Osum river with its amazing canyon, and the National Park of the sacred Tomorr Mountain. These aspects can match together research, information, knowledge, and awareness, giving many possibilities to the tourism sector and to the local people, but keeping in mind that for the small rural area of Blezëncke, the massive tourism might be damaging if not done properly. There is the risk to destroy the very environmental assets on which the rural area depends, there must be a good attention to the ecosystem and the biodiversity and particularly to the waste and pollution that in a rural area like Blezëncke, with small capacity, can be a real problem. Aiming a sustainable development of this area, there is very challenging to develop ecotourism in Blezëncke; there must be built a good partnership and cooperation between local government, inhabitants, tourism industry, and the visitors too (Wood, 2002). There is a rich natural environment along Osumi canyon and the National Park of Tomorr, which offers a perfect environment for the researchers of different disciplines. Organizing some spots which helps the researchers on their activity, but although offering them the possibility to interact with tourist and with locals too, transferring knowledge and awareness. The group of researchers can help on building up relations between all the actors involved in this process.

In this process, the community of Blezënckë must be ready and able to develop other activities and services which can be combined with agriculture, offering a rich experience and knowledge of the processes, so the village can have a positive impact on the economy and be more competitive. In this frame, an immediate answer can be done organizing agritourism and visitable processes in the rural life.

Blezënckë is a small village immersed in nature, with important natural resources and other historical, cultural, and faith aspects to be discovered in some natural itineraries, characterizing elements of slow tourism.

6 Conclusions

This research tries to show how different theoretical aspects of sustainable rural development goals of different actions and initiatives can be implemented on the Blezënckë

village. The rural areas are very different and it is not easy to have a ready answer for their future development, and the case of Blezënckë shows this in a good way. The assumption of proposals done, aiming at the sustainable development of Blezënckë, is a result of different participatory processes with the habitants of the case study. Different general theoretical objectives given into international agreements documents were shared with the local people and trying to translate them in specific objectives for the area (Fig. 8). This exchange of reciprocal information is very important on helping to build sustainable communities and to raise their awareness for different aspects related to Blezënckë development.

There is still a lot to be done with the local community and the refurbishment of the physical assets of the village in order to achieve the sustainable development goals, but it is an important starting point to raise questions and the awareness of the people. The local and central government, NGOs, and different actors in the rural development process, still have a lot to do from the legal frame, to incentives, promotions, and support to the rural community, especially in developing country like Albania, with a complex situation.

Fig. 8 Blezënckë, general map of intervention

References

Anderson, M. D. (2015). Roles of rural areas in sustainable food system transformations. *Development Journal, 58*(2–3), 256–262.

Ballantyne, R., & Packer, J. (2013). *International handbook on ecotourism.* Edward Elgar Publishing.

Biçoku, K. (2005). *Skënderbeu dhe Shqipëria në kohën e tij.* Botimpex.

Buccianti, G. L. (2018). Albergo diffuso e sostenibilità turistica. *Rivista Italiana di Diritto del Turismo, 49*, 385–433.

Confalonieri, M. (2011). A typical Italian phenomenon: The "albergo diffuso." *Tourism Management, 32*(3), 685–687.

da Silva, R. F., Rodrigues, M. D., Vieira, S. A., Batistella, M., & Farinaci, J. (2017). Perspectives for environmental conservation and ecosystem services on coupled rural-urban systems. *Perspectives in Ecology and Conservation, 15*(2), 74–81.

EU-Commission. (2017). Pathways to leader: Nities in southern-eastern Europe. *Journal of Environmental Management, 164*, 171–179.

Freshwater, D. (2015). Vulnerability and resilience: Two dimensions of rurality. *Journal of the European Society for Rural Sociology, 55*(4), 497–515.

King, R. (2005). Albania as a laboratory for the study of migration and development. *Journal of Southern Europe and the Balkans, 7*(2), 133–155.

Krasniqi, S. (2016). *Ekokultura: natyra në kulturën popullore shqiptare.* Akademia e Shkencave dhe e Arteve e Kosovës.

Marot, S. (2022). *Taking the country's side: Agriculture and architecture.* Polígrafa.

Menges, E. (2022). Ecotourism and natural areas. *Natural Areas Journal, 42*(3), 175–176.

Muka, A. (2007). *Ndërtimet tradicionale fshatare—Monografi Etimologjike.* Akademia e Shkencave e Shqipërisë.

Qiriazi, P. (2017). *Trashëgimia natyrore e Shqipërisë.* Akademia e Shkencave Shqipërisë.

Qiriazi, P., & Bego, F. (1999). *Monumentet e natyrës të Shqipërisë.*

Stahl, J. (2010). *Rent from the land: A political ecology of postsocialist rural transformation.* Anthem Press.

Streule, M., Sawyer, L., Karaman, O., & Schmid, C. (2019). Popular urbanization: Conceptualizing urbanization processes beyond informality. *International Journal of Urban and Regional Research, 44*(5), 652–672.

Thompson, B. S. (2022). Ecotourism anywhere? The lure of ecotourism and the need to scrutinize the potential competitiveness of ecotourism developments. *Tourism Management, 92*(1), 104568.

UN. (2015). *Paris agreement.* United Nations.

UN, G. A. (2015). *Transforming our world: The 2030 agenda for sustainable development.* United Nations General Assembly.

Weigend, R., Pomponi, F., Webster, K., & D'Amico, B. (2020). The future of the circular economy and the circular economy of the future. *Environment Project and Asset Management, 10*(4), 529–546.

Wood, M. E. (2002). *Ecotourism: Principles, practices and policies for sustainability.* UNEP—United Nation Publication.

Zakkak, S., Radovic, A., Nikolov, S. C., Shumka, S., Kakalis, L., & Kati, V. (2015). Assessing the effect of agricultural land abandonment on bird communities in southern-eastern Europe. *Journal of Environmental Management, 164*, 171–179.

Heritage in Socio Economic Sustainable Development: The Salzedas and São João de Tarouca Case

João Pedro Almeida Mendonça

Abstract

Heritage and culture are considered by several academic studies as forms of achieving socio economic development. This is particularly evident in the case of Salzedas and São João de Tarouca, a region near the Douro Valley in Portugal, where the lack of economic opportunities compromises his future. The recent public investments in the restoration of the heritage sites and other forms of private initiatives related to them (hotels, restaurants, active tourism enterprises, etc.) seem to change this condition. This paper discusses why heritage maintenance and preservation can contribute to population fixation and richness growth. Several social and economic data are used to prove that statement. Another aspect considered is the evaluation of the historical values in presence and its role in the cultural tourism improvement. Considering all these variables, finally, is an objective of this research to prove that in less developed regions, the heritage and culture activities can be a way to achieve sustainability.

Keywords

Heritage • Culture • Sustainable development • Salzedas • São João de Tarouca

1 Introduction

Tourism is an alternative to traditional existing ways of life for many rural areas. This is particularly notorious in the areas of lower social and economic dynamism, such as those in the interior of Portugal. One of the types of tourism that can contribute to this transformation is linked to local heritage and culture. This is the example of a small region near the Douro River—Salzedas and São João de Tarouca, in the municipality of Tarouca. It is a territory very rich in historical and heritage values, which nowadays take on new forms of use—Hotel inserted in an old Cistercian abbey; tourist houses in rural areas; municipal initiatives for cultural and tourism promotion; recovery of monasteries for museum purposes, etc.

Despite this transformation dynamics, this area of the Douro region, as well as the generality of the region has seen a demographic crises and a reconfiguration of its economy, previously very dependent on the activities of the primary sector, now replaced or complemented by the services sector. Reflecting on these examples and themes is the main objective that is proposed.

2 Cultural Tourism

Cultural tourism is one of the most complex forms of tourism, from the point of view of definition. This complexity results from the wide scope of concepts related to the word culture. Cultural manifestations as diverse as crafts, literature, music, and local popular theaters, just to mention a few examples, illustrate different ways of disseminating the tourist attractions of a region and, therefore, fostering local tourism itself. The other aspect is the growing importance that cultural tourism has in all current tourist flows. There is even one word that represents this diversity, the concept of "cultural touring," meaning not just the possibility that a person has to experience the cultural values and attractions, but also recovering the traditional concept of "touring" when the visitor makes knowledge with all the aspects of a tourist destination.

The classification of "Douro" as a UNESCO World Heritage Site on December 14, 2001, was a milestone, in which "new opportunities but also increased responsibilities" were combined, in the words of the coordinator of the application's technical team. Once this new normative document came into existence, it is pertinent to assess the degree

J. P. A. Mendonça (✉)
Universidade da Maia, Maia, Portugal
e-mail: d011336@umaia.pt

© The Author(s), under exclusive license to Springer Nature Switzerland AG 2024
J. Chica-Olmo et al. (eds.), *Sustainable Tourism, Culture and Heritage Promotion*,
Advances in Science, Technology & Innovation, https://doi.org/10.1007/978-3-031-49536-6_13

of development of the region as a tourist destination, based on a model in which natural and cultural landscape values are promoted. It will be emphasized as one of the typologies of tourist use in which the evolution of recent years has been the most striking: wine tourism.

3 Cultural Tourism and Sustainable Development

Cultural tourism has a role on the sustainable perspective of the development itself (Pereiro Pérez, 2009, 156). It is not, however, just social and economic dynamics; as a whole, this type of action can constitute an even broader development, contributing to the sustainability and future of a territory.

The development provided by investments in the cultural and property aspects of the sites has an impact on society, creating new jobs and fixing a population that would otherwise tend to abandon them. This will be shown in this paper's study case. In the case of Salzeas and São João de Tarouca, it is worth mentioning the creation of a hotel called "de Cistercian," on the site of an ancient Cistercian abbey (of which archaeological remains). In close coordination with the previous investment, there is a company in the wine sector, of national name—Caves Murganheira, whose vineyards are visible from the Hotel. This company, which began as a family investment, belongs today to the largest wine production group, national. A third level of analysis results from the consequences that this type of investment has for the territory and its sustainability. There is a greater possibility of keeping intact the local landscape aspects, both in their human elements (traditional buildings, manor houses, and the agricultural landscape), or in their physical elements (disposal of agricultural land by the slopes, preservation of watercourses, forests, and types of crops, etc.).

The main components of sustainable development are shown below, namely that it can meet the needs of current and future generations.

4 Development of Tourism in the Douro and Salzedas/São João de Tarouca

The use of the Region of Douro for tourist and leisure purposes is not a new fact. Just remember the traditional trips, and especially at the time of the harvests, of the non-resident owners, towards the farms, something that still remains today. The romantic literature of the nineteenth century attests to this tradition, being numerous, those who, essentially by rail, moved to less time-consuming stays in the region, some obeying medical prescriptions, others to accompany the viticultural labors of the harvest phase.

Resuming the analysis of what occurred in the Douro, and if we refer to some of the studies that focused on the tourist potential of this region, as well as the generality of rural areas of the interior in the chronological period between the mid-1980s and the mid-1990s, it is inferred that housing offers were reduced and little diversified. It is therefore understood that the terminology used in these assessments revealed a weak capacity for tourist reception. The authors or institutions used expressions such as "deficient hotel capacity," "reduced expressiveness in terms of number of beds" and drew attention to the small number of units with a rating equal to or greater than four stars.[1]

In recent years, this description has changed due to the growing demand from countless visitors who are beginning to show a certain fatigue regarding the forms of mass tourism of the "sun and beach" model and the improvement of access conditions to the Douro region, namely through the river and road.[2]

If we report to river cruises the trend is markedly growth. To this end, the classification of the city of Porto as a World Heritage Site and the various institutional campaigns carried out, for example by the D. Afonso Henriques Foundation, the Institute of External Trade Portuguese, Tourism Delegations, etc. can be found.

5 Tourism in the Douro and Salzedas/São João de Tarouca

The Douro Demarcated Region corresponds to an area of approximately 240,000 hectares. For the purposes of the classification were considered only about 25 thousand hectares of the total territory, corresponding to six sub-areas of the valleys of Corgo, Chanceleiros, Pinhão, Tua, Torto, and Vale da Figueira, areas of the most representative of the Demarcated Region, or by the presence of vernacular heritage that resulted in the current network of villages and farms that support the Douro wine production, identification of the most characteristic forms of disposal of the vineyards, in terraces. In addition to this area (included in 13 municipalities) where the regulations of the building and alignment of vineyards will necessarily have to be more restrictive, the rest of the Demarcated Region will also be subject to special monitoring.

According to the principles of Point 39ii of the Guidelines for the Application of the World Heritage Convention introduced in 1992, the Douro unquestionably corresponded to an example in which the landscape was synonymous with

---

[1] See among other works: Rofe, Lapworth and Lybrand Associates Ltd (1984); Cravidão and Cunha (1993), or even Dias (1995).
[2] C.f. Fontes (2000).

Fig. 1 Difficult process of preparing the land and cultivating it. *Source* The author.

a perfect combination between human intervention and the natural conditions of the region, and *it is relevant "the landscape itself remains a condition of sustainability of a territory and a community (...)"*[3]. In this respect lies possibly the biggest dilemma the region faces: the problem of progressive population loss may weaken its main productive and economic base.

The natural conditions determined the implantation of the vineyard in an area of slopes, which required the sometimes-over-human effort of construction of landfills and construction of support walls. If it was this adaptation between human action and nature that resulted in the main factor of beautification of the landscape, it is also true that the rudeness of the working conditions that these terraces originate has led many Douro inhabitants, especially the younger ones, to opt for activities other than agricultural ones. They do not always feel remunerated for the hard work that this territory of rough soils and climate of extreme temperatures, entails for those who work in the vineyards (Fig. 1).[4]

As the Douro is part of a restricted and competitive niche of quality viticulture and generous wines, this shortage of labor has required producers to take new forms of production organization, which undergo a greater mechanization of farms, assuming the provision of vineyards different forms of implementation from traditional ones. In this respect,

particular attention has been given in the application process, as the preservation of part of the traditional terraces is a key element in the allocation of classification and is obviously one of the most decisive factors for maintaining the success of the tourist operation.[5]

Farms that are most concerned with the preservation of terraces and which therefore have a negative impact in terms of production costs may be partly reimbursed for this damage, either by the direct use of the holdings, by recourse them to wine tourism, or indirectly, for example by means of disclosure on the labels of their bottles, of images or texts related to the care they place in the preservation of their heritage.

The scenic value of the river channel and the interest in knowing the details related to the viticulture of the famous Port wine are more than enough reasons to attract tourists to the Region of Douro (Fig. 2).

However, if we follow a route towards the areas more on the inland, abandoning the shale lands, even if the presence of the vineyard is diluted, the reasons of interest for the traveler are diversified. This can observe extensive orchards of almond trees or even leafy chestnut trees. Here and there, the tourist looks at viewpoints and privileged places to have another perspective of the vineyard valley, from high points, overlooking the river.[6]

---

[3] Spire; Quaternaire Portugal and Workshops De Planeamiento (1999), p. 14.
[4] C.f. Portela et al. (2001).

[5] Spire; Quaternaire Portugal and Workshops De Planeamiento (1999), p. 67.
[6] In the concrete case of Torre de Moncorvo it is worth highlighting the ridge lines of the Serra de Reboredo.

Fig. 2 Iron Museum—Moncorvo—Douro (*Source* Author)

6   Tourist, Social, and Cultural Development in Salzedas/São João de Tarouca

It is in a geographical frame of transition that the municipality of Tarouca is located on the area belonging to the Douro. One of the components of local and regional tourism is the wine-tourism routes, such as Port/Douro Wine.

The Douro, in general, and this area, can offer, as opposed to the forms of reception typical of the areas of greatest affluence[7], even if, overall, there is a demographic decline.

Due to the territorial described framework and the proposals presented, it is concluded that cultural tourism linked to the presence of a monastic community with a great historical tradition, as in the case of Salzedas/S. João de Tarouca is another way to enhance and revitalize a rural territory with a weak social and economic base. That can be seen on the population decline (Fig. 3).

The historical nature of this region (similar to the Douro Region) requires that particular attention should be paid to distinctiveness and preservation.

Another aspect mentioned above is nature tourism. Tarouca, together with the other municipalities of the Douro region, has recently bet on the promotion of natural and landscape values, which are examples of the promotion of bird watching routes belonging to the Douro International Park or the contemplation of the landscape in the Douro Demarcated Region.

A final line of reasoning should be highlighted, because, as Cravidão and Cunha refer, it will be necessary to avoid transposing the load thresholds that call into question the very survival of tourism[8], at a time when the number of visitors has been growing sustainably (Table 1). In this sense, initiatives such as the classification of the Douro as world heritage prove to be paramount.

Another aspect is related to the cultural traditions, mainly those associated to the monuments and the wine production of the Douro and Távora-Varosa Regions. These are related to the dissemination of wine-growing works, especially those concerning new forms of touristic activities.

Among several dynamics, it is worth mentioning in the case of Salzedas/São João de Tarouca:

- Presence of elements of religious and civil architectures as the mentioned monasteries.
- The Ucanha village is considered an example of a "wine-growing village."
- Existence of a fortified tower and bridge, also in Ucanha (Fig. 4).

[7] Cited by: Martins (1993).

[8] C.f. Cravidão and Cunha (1993).

Densidade populacional

Territórios		Indivíduo - Média							
		N.º médio de indivíduos por Km²							
Âmbito Geográfico	Anos	2001	2009	2010	2011	2012	2013	2014	2015
NUTS 2013	Portugal	112,5	114,6	114,7	114,5	114,0	113,4	112,8	112,3
NUTS II	Norte	173,2	174,3	173,8	173,4	172,7	171,7	170,7	169,7
NUTS III	Douro	54,6	51,7	51,3	50,8	50,3	49,7	49,2	48,7
Município	Alijó	47,8	41,6	40,7	40,0	39,5	38,8	38,1	37,5
Município	Sabrosa	44,7	41,4	40,9	40,5	40,1	39,6	39,1	38,6
Município	São João da Pesqueira	32,4	30,2	29,9	29,5	29,1	28,6	28,2	27,9
Município	Tabuaço	50,5	48,2	47,8	47,5	47,2	46,9	46,4	46,0
Município	Tarouca	82,9	81,4	80,9	80,4	79,8	79,2	78,6	78,0

Fig. 3 Population density in several territories from 2001 to 2015 *Source* INE

Table 1 Number of passengers on riverboat cruises on Douro

Years	2013	2014	2015	2016	2017	2018	2019
Passengers	545,630	615,361	721,242	946,728	1,282,241	1,296,031	1,644,937

Source APDL, 2021

Fig. 4 The Tower Bridge of Ucanha *Source* the author

7 Territory Sustainability/Conclusions

In summary, the region's own world heritage application document points to the typification of wine-growing landscapes, as being able to integrate this interactivity between man and nature, and it is also necessary to add the relevant contribution that classification can make in the maintenance and economic and social valorization of the region. Since it is an area typified as economically and socially depressed, tourism must be an integral part of the process of diversification of the economic base and be closely linked to the territorial, social, and economic viability and sustainability itself.

The conciliation of agrarian activities with others developed in rural areas (agrotourism, wine tourism, cultural, tourism and environmental activities, for example) allows them to obtain income and preserve cultural, social, and environmental values, that have consequences on the development of those areas.[9]

The tourist use of the Douro region has also been developed through various types of services and thematic tourist products. Without wanting to be exhaustive, we then list some of the typologies that currently constitute forms of tourist use in the Douro: wine route, cultural tourism related to popular traditions and handcrafts, religious heritage, boat and road touristic exploitation, etc.

---

[9] Salazar et al. (2020), UNWTO/UNESCO (2018).

In conclusion, several examples of touristic activities of the Douro region can be developed in the presented case study of Salzedas/São João de Tarouca. Based on this article, there are some that can contribute to a more sustainable economic and social development processes: wine and cultural tourism activities. We have to notice the growth in the levels of accommodations (Cister Hotel) and the importance of the wine production factory of Murganheira (Tarouca). In addition, there are more and more tourists in this region searching for the cultural and architectural elements and buildings (Monastery of Salzedas and São João de Tarouca).

References

Cravidão, F., & Cunha, J. (1993). Ambiente e práticas turísticas em Portugal. In *Inforgeo—Revista da Associação Portuguesa de Geógrafos* (No. 6, p. 90, Dez. 93). Associação Portuguesa de Geógrafos e Edições Colibri. (Cravidão, F., & Cunha, J. (1993). Environment and tourist practices in Portugal. In *Inforgeo—Revista da Associação Portuguesa de Geographographers* (No. 6, pp. 85–91, Dez. 93). Associação Portuguesa de Geógrafos e Edições Colibri.

Dias, A. (1995). *Turismo no espaço rural. A study on the rural accommodation sector in the Douro region.* The University of Trás-os-Montes and Alto Douro (final internship report).

Fontes, A. (2000). O desenvolvimento turístico no vale do Douro: um destino em fase de afirmação, uma rede institucional em discussão. In *Desenvolvimento e Ruralidades no Espaço Europeu—Actas do VIII Encontro Nacional da APDR* (Vol. I, pp. 109–124). Associação Portuguesa para o Desenvolvimento Regional. (Fontes, A (2000). The tourist development in the Douro valley: A destination in the affirmation phase, an institutional network under discussion. In *Development and Ruralities in the European Area—Minutes of the VIII National Meeting of the APDR* (Vol. I, 99. 109–124). Portuguese Association for Regional Development.

Martins, L. P. (1993). *Lazer, férias e turismo na organização do espaço no noroeste de Portugal* (p. 196). FLUP (tese de doutoramento). (Martins, L. P. (1993). *Leisure, vacation and tourism in the organization of space in the northwest of Portugal* (p. 196). FLUP (Doctoral Thesis).

Portela, J., Rebelo, V., & Vasques, C. (2001). A omnipresença da vinha e a rejeição generalizada do trabalho vitícola pelos jovens; o caso de Santa Marta de Penaguião. Comunicação apresentada ao I Congresso de Estudos Rurais, Universidade de Trás-os-Montes e Alto Douro, Vila Real, 16 a 18 de Setembro de 2001 (Portela, J., et al. (2001). The ubiquity of the vineyard and the widespread rejection of wine-growing work by young people; The case of Santa Marta de Penaguião. In Communication presented to the 1st Congress of Rural Studies, University of Trás-os-Montes and Alto Douro, Vila Real, 16–18 September 2001.

Rofe, K., Lapworth, C., & Lybrand Associates Ltd. (1984). *Estudo de desenvolvimento da Região Douro: Relatório final* (Vol. 3), s.l (Rofe, K., Lapworth, C., & Lybrand Associates Ltd. (1984). *Douro region development study: Final report* (Vol. 3), s.l.).

Salazar, D., Gonzáles, D., & Macias R. (2020). El turismo cultural y sus construcciones sociales como contribución a la gestión sostenible de los destinos turísticos. *Rosa dos Ventos Turismo e Hospitalidade, 12*(2), 406–428. https://doi.org/10.18226/21789061.v12i2p406 (viewed on August 12, 2020).

Spire; Quaternaire Portugal and Workshops de Planeamiento. (1999). *Feasibility of the douro valley's application for world heritage: Final Report.* Spidouro, Quaternaire Portugal and Workshops de Planeamiento Ed., s. l.

UNWTO/UNESCO. (2018). *World Conference on Tourism and Culture: For the Benefit of All, Istanbul, Turkey, December 3–5, 2018.* file:///C:/Users/Utilizador/Downloads/event-1464-1.pdf. (viewed on August 12, 2020).

Tourists' Perceptions of the Image of the Peneda-Gerês National Park

H. Martins and António José Pinheiro

Abstract

The image of a destination has been one of the most relevant concepts in tourism research, as it is considered one of the most important variables for market segmentation. Furthermore, image is a construct that influences, considerably, the preference, motivation and behaviour of individuals during the process of choosing a tourist destination. It is therefore important to understand which attributes tourists highlight in tourist destinations. The aim of this study is to understand which attributes (functional and psychological) of the image that tourists most identified in a particular tourist destination. The study method applied was the quantitative methodology, and data collection was carried out through the use of a questionnaire survey. The territory under analysis was the Peneda-Gerês, the only Portuguese national park which has registered a high growth in visitors, as a result of its projection and notoriety. According to the information gathered, of the functional attributes the natural landscape stood out, while the infrastructures, accommodation and restaurants stood out. Of the psychological attributes, the one that stood out the most was hospitality and welcome. Only the events and entertainment were the least considered attribute, given that most of the sample considered rest and contact with nature as the main reasons to visit this destination, and fun was the third most referenced reason. This study considers that the various stakeholders should seek to improve these attributes because although most visitors had good expectations, the degree of satisfaction does not demonstrate that this protected area can stand out from other competing destinations.

Keywords

Destination image • Functional attributes • Psychological attributes • Peneda-Gerês National Park

1 Introduction

Tourist destinations are becoming an increasingly competitive and expanding market. Currently, there is a wide range of tourist destinations, some with more specific characteristics, for a particular type of tourist, others more general to be able to reach a wider range of people. However, all these destinations multiply in advertising and promotions in order to attract new tourists to maintain market share and/or increase it. For a destination to stand out from other competing destinations, it must try to stand out for its uniqueness in order to position itself positively in the minds of consumers. It is therefore important that tourists have a distinctive and attractive perception of the destination (Martins et al., 2021a, 2021b).

The image of a destination is a fundamental variable in the process of selecting a tourist destination. However, it is difficult for destinations to gain that uniqueness, because tourists are increasingly informed and demanding and nurture high expectations about the attributes and experiences they will find in the destinations they decide to visit (Silva & Correia, 2017).

The image of a destination is therefore important since it affects consumer behaviour in the choice of the holiday destination and post-trip. The tourist, when seeking information about the next holiday destination, can be influenced by the image of the destination that is projected to him, being more likely to choose the destination that has projected a positive and appealing image. Due to technological advancement, very well used in tourism, people have at their disposal (often a click away) a volume of information available through various media (touristic guides, travel

H. Martins (✉) · A. J. Pinheiro
Centre of Studies in Geography and Spatial Planning (CEGOT),
University of Maia, Maia, Portugal
e-mail: hugomartins@umaia.pt

A. J. Pinheiro
e-mail: ajpinheiro@umaia.pt

agencies, television, internet and others). This allows individuals to perceive and create an image of the destination, previously to the visit, and that often conditions their behaviour and choice options (Chi & Qu, 2008; Echtner & Ritchie, 1991).

After the stay in that destination, after experiencing it, the tourist will (re)build the image of the destination. The comparison that the tourist will make between the two images (the projected initially, by the information he collected, and the one that remains after his experience at the destination) is a factor that will influence his level of satisfaction, the desire to return to that destination, as well as the positive or negative way he tells his experience to others, interfering, in turn, in the formation of their images (Martins, 2022a).

Some researchers have tried to measure the image of tourist destinations based on functional and psychological attributes and concluded that the receptivity of the inhabitants, the landscapes and the environment were the most mentioned attributes (Gallarza et al., 2002). It is therefore our intention to understand which attributes of the destination image are most highlighted by tourists in a particular tourist destination. The chosen territory was the Peneda-Gerês, the only Portuguese national park and which has attracted many tourists, due to its notoriety and brand image that has been consolidated both nationally and internationally.

Methodologically, we tried to answer the research question about which image attributes tourists most identify in the PGNP, based on a questionnaire survey carried out to people who stayed overnight in this tourist destination, in a pre-pandemic period. The sample was considered representative with a total of 507 respondents.

This article is organised into five chapters. After the introduction, the second chapter includes a literature review on the concept of image and its importance, focusing on the issue of attributes. Chapter three presents the study's methodological framework, characterising the procedures in terms of sample analysis and data collection. Chapter four presents the results. Finally, in the conclusion, the main inferences and recommendations of the study are presented.

2 Literature Review

The image associated with the tourist destination has its genesis in the work carried out by Hunt in 1975. Destination image is highly valued by both academics and the tourism industry; this is because the degree of consumer involvement is higher in tourism than in other services (Tavitiyaman et al., 2021).

The concept of destination image has been identified in the tourism marketing literature and has been widely studied. However, the numerous definitions of the concept "image of tourist destinations" used in several studies and empirical work are considered unclear, without a very solid conceptual structure varying from researcher to researcher (Beerli & Martín, 2004; Lv et al., 2020) (Table 1).

However, it is commonly accepted the idea of "global impression". In this context, Crompton's definition (1979) is the most widely accepted in the scientific community since it is "the sum of beliefs, ideas and impressions that tourists have about a particular destination" (Crompton, 1979) (Table 1), based on the idea that it is something constructed by the consumer. However, there are also authors who

Table 1 Definitions of image of a destination and respective authors	Image–definitions	Authors
	"Compilation of beliefs and impressions based on information processing from a variety of sources over time resulting in an internally accepted mental construct"	MacKay and Fesenmaier (1997, p. 538)
	"Perceptions or impressions of a destination engaged in by tourists in relation to the expected benefit or consumption values, including functional, social, emotional, behavioural benefits of a destination. These perceptions or impressions lead to the decision to visit a country as a holiday destination"	Tapachai and Waryszak (2000, p. 37)
	Image is the set of expectations and perceptions that a potential traveller has of a destination	Buhalis (2000)
	It is formed from the interaction of the destination's characteristics and the observer's characteristics and may contain both cognitive and affective components. The combination of these two components results in the formation of a global image	Baloglu and McCleary (1999)
	"Impression, visual or mental, of a place or product experienced by the general public"	Milman and Pizam (1995, p. 21)
	"The sum of beliefs, ideas and impressions that a person has about a destination"	Crompton (1979, p. 18)

Source Elaborated by the authors

support the idea that the image can be built by the destination (as a way of promoting the place), constructed and projected to attract the consumer/tourist, being this projection created by the destination that will influence the tourist behaviour, namely, in the decision of choosing the place to visit (Govers & Go, 2004).

Although most authors agree about the holistic nature of the image, that is, that the destination image is a global and multidimensional impression, there is still no consensus about the dimensions that form this global impression (Bigné Alcañiz et al., 2009; del Bosque & San Martín, 2008; Echtner & Ritchie, 1991; Luque-Martínez et al., 2007; Tavitiyaman et al., 2021).

Echtner and Ritchie (1991) consider that only the "impressions" or "perceptions" of a place are very vague and that it is relevant and enhanced to know which elements make up the destination image. These authors have created a referential model that supports empirical research on the image of tourist destinations. In the perspective of Echtner and Ritchie (1991), the elements that compose the destination image are visible in three dimensions: functional-psychological, common-unique and attribute-holistic (Fig. 1).

According to these authors, destination image, considering the holistic attribute dimension can be defined by the individual's perception of the attributes of the destination (e.g. warm climate and low prices), but also by the holistic impression, the overall mental image that the same individual forms about the destination (e.g. general physical characteristics such as mountainous or city). The functional-psychological dimension highlights the more functional/tangible characteristics such as climate, accommodation and attractions; and more psychological/abstract characteristics such as the atmosphere of the place, tranquillity, hospitality and reputation. This dimension is related to the previous dimension in that destination attributes and holistic impression have both functional characteristics and psychological characteristics. The common-unique

dimension is related to the common characteristics of a given place in relation to others and those that cannot be found anywhere else but, in that place, which are specific to it. This dimension is related to the two previous ones: destination image may vary from more common characteristics (functional or psychological) to more unique characteristics (functional or psychological) (Echtner & Ritchie, 1991).

Several authors affirm that the image of the tourist destination is a reflection of the rational and emotional interpretation of the consumer and, consequently, the result of two evaluations: (a) one of cognitive nature, linked to aspects of attributes and functional characteristics that the individual has about the object observed; (b) another of affective nature, associated with holistic images and psychological aspects that translate the feelings and emotions perceived by the individual when confronted with the object (White, 2004).

The perception of the destination image that the consumer has, regardless of its cognitive, affective or even mixed nature, plays a relevant role in the choice options by tourists. The more positive the perception of the destination image, the more preferred the destination will tend to be and the more likely it will be revisited, with a positive impact on consumer satisfaction (Al-Ansi & Han, 2019; Kastenholz, 2012; Zhang et al., 2018).

According to Kastenholz (2012), an image is designed by obtaining and processing information. The image generation is adjacent to the perception process, therefore, behaviour can be considered as an effect of the perceived image, and repetitive behaviour can be a consequence of a strong image. Thus, favourable destination images found on repeated visits are more likely due to prior positive experiences. This can lead to destination loyalty and the accumulation of positive images which the author has termed a positive vicious circle or virtuous circle (Fig. 2).

In this vicious circle, we believe that a positive experience, associated with a tourist destination, can influence

Fig. 1 Dimensions of destination image. *Source* Echtner and Ritchie (1991)

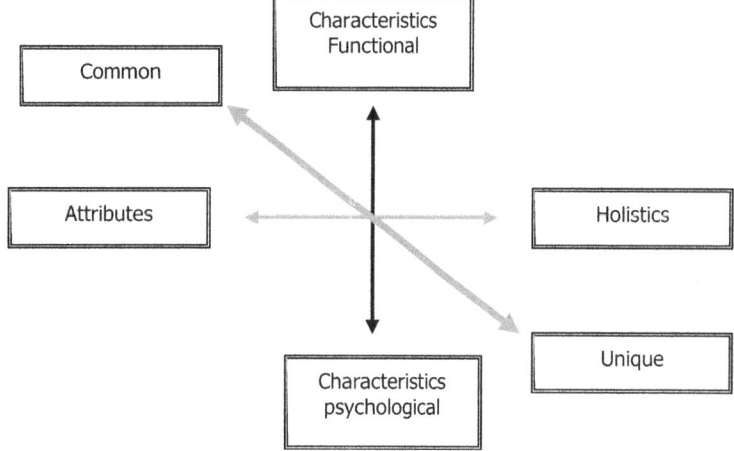

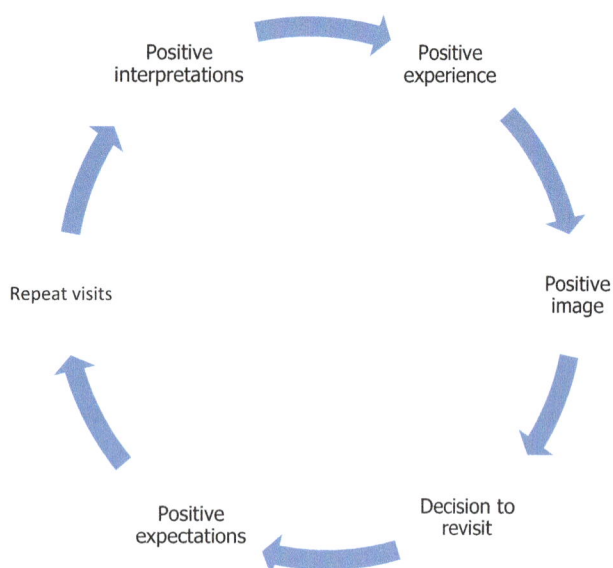

Fig. 2 Positive vicious circle/virtuous circle of destination image. *Source* Kastenholz (2012, p. 141)

future decisions, namely, revisit. As such, the image of tourist destinations is important in the organisation of a successful marketing strategy, to the extent that the perceived image influences both the behaviour and the decision-making process, and also has a relationship with the levels of satisfaction obtained regarding the tourist experience (Chi & Qu, 2008; Martins et al., 2021a, 2021b, 2023).

In relation to the tourism phenomenon, each tourist destination must develop a marketing policy, creating its brand image and being considered a product offering. According to Boo et al. (2009), destinations compete mainly on the basis of their perceived images relative to the images of their competitors in the market world. In this sense, it is necessary that a set of marketing strategies are developed in order to ensure a solid position in the competitive market with regard to attracting tourists (Beerli & Martín, 2004).

According to Buhalis (2000, p. 8), consumers have certain consumption patterns, appreciate the qualities of a destination and, as a result, if they are left with a positive impression associated with a strong degree of satisfaction, "they tend to visit certain destinations more regularly and frequently, increasing their degree of loyalty and showing willingness to pay higher prices in order to enjoy their preferred destinations".

According to Chen and Tsai (2007), attracting tourists to a destination, getting them to revisit that destination and getting them to recommend it to others are fundamental to the successful tourism development of that destination. Therefore, the branding of a tourist destination is the key piece because it is, according to the image of that destination, from there that consumers make their choices (Tan & Wu, 2016).

To make an analysis of the image of a destination it is therefore necessary to identify its attributes, which, according to Echtner and Ritchie (1991), can be distributed in attributes with more functional characteristics until typically psychological attributes. There are several studies that use lists of attributes to assess the image of a destination (Beerli & Martín, 2004; Gallarza et al., 2002; Kastenholz, 2012; Li et al., 2022; Pike, 2002; Tapachai & Waryszak, 2000).

According to Neves (2012), the attributes have not registered substantive changes, being a curious salience when it is verified that these are studies carried out by various authors based on distinct tourist destinations, regions countries or territories, and in equally distinct temporal periods (Table 2).

3 Methodology

With this study, we seek to understand the perceptions of the image of tourists who visit a tourist destination with specific characteristics, a protected area. The destination chosen was the PGNP, the only national park in Portugal. After an analysis of the articles by Echtner and Ritchie (1991), Kastenholz (2012), Gallarza et al. (2002) and Beerli and Martín (2004) who made an exhaustive analysis of attributes used in a set of vast articles, we chose to define a set of attributes that fit our object of study, the PGNP. In the present work, we will adopt the approach that destination image is formed from the interaction of destination characteristics and observer characteristics and may contain both functional and psychological attributes, with the combination of these two components resulting in the formation of a global image (Baloglu & McCleary, 1999) (Table 3).

Through the existing literature, we chose to identify which attributes would most fit the chosen tourist destination. In methodological terms, we sought to answer the following research question: "What are the attributes (physical and psychological) of the image that tourists most highlight in a tourist destination such as the PGNP?".

The method of study was the quantitative methodology. The target population was the tourists who stayed overnight in the PGNP, and the technique chosen was the questionnaire survey, made available in four languages (Portuguese, English, French and Spanish) in order to capture the opinion of national and foreign tourists who visited this tourist destination. The measurement of the constructs was done through an attitude interval scale in Likert interval format expanded to 7 points ranging from 1 = Strongly disagree to 7 = Strongly agree, where 4 = Neither agree nor disagree. Qualitative variables (nominal and ordinal) were also used namely to obtain information about the tourist and his stay.

The sample is considered significant (McDaniel & Gates, 2004) totalling 507 respondents. Being the target audience the tourists who stay overnight in the accommodation units

Table 2 Most used attributes in the evaluation of the image of tourist destinations

Author/date	Tourist destination	Attributes of the tourist destination image analysed
Prayag (2010)	Cape Town, South Africa	Quality of public transport and local infrastructure; information available at the destination; accessibility; appealing gastronomy and quality of service; activities and entertainment; climate and landscape; cultural and natural attractions
Vengesayi et al. (2009)	Zimbabwe	Historical and cultural attractions; natural attractions and recreational facilities; shopping and accessibility of attractions; touring, sports, events and outdoor activity offer
Shani et al. (2009)	South America	Natural and scenic beauty; protected landscapes and wildlife areas; places of historical and archaeological interest; experiences and personal enrichment
McCartney et al. (2008)	Macau	Gaming opportunities; nightlife/adult oriented; political stability; urban destination; cleanliness and preserved environment; unique architecture
Prebežac & Mikulić (2008)	Hawaii and Croatia	Natural beauty and landscape; adequate transport; price of transport
Woomi (2008)	New York, USA	Shopping, restaurant and entertainment; safety and hospitality; history and culture
Chi and Qu (2008)	Arkansas, USA	Natural attractions, entertainment; infrastructure, accessibility; shopping, restaurants and events; environment
James (2006)	Australia	Events, people and places; nature, art and culture; heritage and history; accommodation, thematic routes and information material

Source Neves (2012), with adaptations by the authors

Table 3 Attributes applied in the present research

Functional attributes	Psychological attributes
Activities: Outdoor activities (e.g. hiking) Entertainment activities and/or events	Gastronomy
Natural landscape	Hospitality and reception
Heritage: Historical attractions	Safety
Price	
Tourist infrastructures and/or facilities: Accessibility (to get to the park) Accessibilities (travelling in the park) Signs in the Park Infrastructures (walkways/public places/WC's) Accommodation (e.g. hotels, camping parks) Restoration (e.g. restaurants, cafés) Tourist information (tourist offices) Local Interpretation Centres Tourist entertainment agents	

Source Elaborated by the authors

within the limits of the PGNP, the sample is of the non-probabilistic type by convenience. To carry out the empirical study, the collaboration of the receptionists of the local accommodation units and tourist resorts was requested in order to deliver the questionnaire.

The empirical study was based on fieldwork, which was carried out during the months of June to October (being the most representative). The questionnaire was filled in by the respondent. After data collection, the questionnaires were coded and validated.

Statistical analysis was performed in SPSS. Continuous variables were described as means (M) and standard deviations (SD). Absolute (n) and relative (%) frequencies were calculated for categorical variables.

4 Analysis and Discussion of Results

Once the data from the sample questionnaires (*N* = 507) were entered into SPSS Statistics software (version 27), it was possible to perform a descriptive analysis of them, seeking to characterise the attributes both physical and psychological that the sample considered most relevant. The data revealed that the sample was made up mostly of male tourists (51.1%), married (58.8%), between the ages of 26–35 (30%) and 36–45 (24.9%), with a predominance of university graduates (38.9%). They are also mainly national tourists (85.8%) and repeat tourists (71%), as it is already recurrent to visit the PGNP. However, considering the sample, it is not very usual for the respondent to frequent protected areas (only 40.4% stated that they had visited protected areas).

We then sought to characterise the image with which tourists were left of the PGNP in its most diverse attributes. For the measurement, the respondents had to, on a Likert scale with seven points, rate the attributes referring to the PGNP, where 1 meant much worse than expected, 7 much better than expected and 4 neither good nor bad. However, regarding those attributes that the respondents were not aware of or did not want to answer, they could leave that indication. For this analysis, measures of central tendency (mean, median, mode and percentiles) and measures of dispersion (standard deviation) were calculated and used.

In what concerns the most functional attributes, we tried to divide them: on one side the natural landscape, the heritage and the developed activities and, on the other side, the set of infrastructures/touristic facilities.

As it is possible to observe in Table 4, in terms of average, the item that presents the highest average, standing above the value six (very good), was the natural landscape. This indication corroborates some research (Akgiş İlhan et al., 2022; Chi & Qu, 2008; Shani et al., 2009). In addition, further research indicates that the natural offer, such as landscapes, represents one of Portugal's strengths (Martins, 2022b; Turism of Portugal, 2017). All the remaining attributes ranked at value 5 (good) except for the item activities: events and entertainment.

In addition to the average, we sought to analyse the median of the items observed, since the median is the value that separates the larger and smaller half of a sample. It was an option to use the median, since the advantage of the median over the mean is that the median can give a better idea of a typical value because it is not so distorted by extremely high or low values.

Thus, comparing the average of these items with the medians, it was possible to observe that, in most items, the median is higher than the average (i.e. most of the sample gave better ratings compared to the mean terms), in the items of activities, both outdoor activities and entertainment activities. The opposite occurs in the remaining items, meaning that most of the sample is below average (Table 4).

In regard to the functional attributes related with the infrastructures, it is possible to observe that in terms of average four items are above the value five (good), namely, the accommodation (5.85), the restaurants (5.59), the tourist animation agents (5.16) and the accessibilities until reaching the park (5.00). These data corroborate other research that highlights the accommodation (Liu & Jo, 2020), restoration (Ganzaroli et al., 2017) and tour operators (Romero et al., 2020). The items that stood out the most were accommodation and restaurants. In fact, the region is known for having a great offer in terms of accommodation (Martins et al., 2021a, 2021b; Martins, 2022a), as well as in what concerns restaurants. The items that had the lowest averages, between points 4 (neither good nor bad) and five (good), were the items regarding the infrastructures (walks, public places and sanitary facilities) with an average of 4.65; followed by the item regarding the interpretative centres (4.79); item about the signage within the park (4.81); and the item about the tourism offices and their information (4.94). In terms of median and mode, only the accommodation and restaurant items are placed at value six, while all other items are placed at value five.

Regarding psychological attributes, the one that stands out most positively in terms of average is hospitality and friendliness, standing above value six (very good) (6.02) (Table 6). This indication corroborates Portugal's strategic

Table 4 Average, median, mode, standard deviation and percentiles of the image of the PGNP perceived by respondents —functional attributes—part I

Items		Landscape	Price	Heritage: historical attractions	Activities: outdoors	Activities: events and entertainment
N	Valid	499	462	447	437	407
	No answer	8	45	60	70	100
Average		6.26	5.04	5.39	5.69	4.76
Median		6	5	5	6	5
Mode		**6**	5	5	**6**	4
Standard deviation		0.76	1.04	1.02	1.03	1.25

Source Elaborated by the authors

plan and other studies when they state that the hospitality and friendliness of the local population represent some of Portugal's strengths, as they are considered the touch points with the best performance, both in terms of satisfaction and compliance with expectations (Turism of Portugal, 2017).

It is important to mention that the remaining psychological attributes also scored highly, above value five: safety (5.32) and gastronomy (5.57) (Table 6) corroborating the study of Bertan (2020).

In addition to the average and median, we tried to identify the mode in the answers of the respondents. Regarding the mode, only the events and entertainment item is in point 4 of the Likert scale (neither good nor bad). The items natural landscape, accommodation, catering, gastronomy, hospitality and welcome, outdoor activities and safety had good acceptance, with the majority considering them to be in point six (very good) on the Likert scale. The mode of the remaining items was found at point five (good) on the Likert scale (Tables 4, 5 and 6).

Given that the standard deviation is a measure of dispersion around the population mean of a variable, we also analysed this measure. The standard deviation, in all items, does not show large discrepancies, thus presenting a low standard deviation, since part of the data tends to be close to the average or expected value and, therefore, is not very dispersed throughout the Likert scale, ranging between 0.76 (natural landscape item) and 1.29 (tourist information/ tourism offices item).

It should be noted that the items interpretative centres, events and entertainment and tourist entertainment agents were those attributes that tourists did not respond to (they had a high number of unanswered questions—more than one hundred), probably due to the tourists' lack of knowledge, because it was not used or simply because they did not want to answer.

In the impossibility of covering the various attributes of the PGNP image, and according to several authors such as Echtner and Ritchie (1991), Gallarza et al. (2002),

Table 5 Average, median, mode, standard deviation and percentiles of the image of the PGNP perceived by respondents—functional attributes—tourist infrastructures and/or facilities

Items[*]		1	2	3	4	5	6	7	8	9
N	Valid	499	497	485	475	491	479	432	384	335
	No answer	8	10	22	32	16	28	75	123	172
Average		5.00	4.86	4.81	4.65	5.85	5.49	4.94	4.79	5.16
Median		5	5	5	5	6	6	5	5	5
Mode		5	5	5	5	6	6	5	5	5
Standard deviation		1.09	1.11	1.27	1.26	0.91	1.04	1.29	1.12	1.28

1. Accessibility (to get to the park)

2. Accessibilities (travelling in the park)

3. Signs in the park

4. Infrastructures

5. Accommodation

6. Restoration

7. Tourist information

8. Local interpretation centres

9. Tourist entertainment agents

Source Elaborated by the authors

Table 6 Average, median, mode, standard deviation and percentiles of the image of the PGNP perceived by respondents—psychological attributes

Items[*]		Gastronomy	Hospitality and reception	Safety
N	Valid	471	485	451
	No answer	36	22	56
Average		36	6.02	5.32
Median		5.57	6	5
Mode		6	6	6
Standard deviation		6	0.93	1.2

Source Elaborated by the authors

Kastenholz (2012) and Beerli and Martín (2004), respondents had the opportunity to add another attribute that contributed to the image that was left of the PGNP and proceed to rate it on a seven-point Likert scale. Thus, as it is possible to observe, of the eleven respondents who used this field, the maintenance, the cleanliness of the streets and the cleanliness of the tourist attraction points, the reduced number of rubbish bins, as well as the quality of the park maps were identified as aspects much worse than expected.

The car park was rated with a poor-quality attribute (point two of the interval scale). Regarding the attributes that scored 3 on the Likert scale (bad), respondents identified parking, medical support, and the toll payment for Mata da Albergaria (Table 7).

After a stay at a particular tourist destination, it is common for tourists to take an assessment of their holiday and evaluate certain aspects. Therefore, the respondents were asked to give their opinion about the overall image of the park, namely whether it met initial expectations. According to Table 8, it is possible to observe that the great majority

was quite satisfied (63.3%), as the image they had of the park corresponded to expectations (35.7% claimed to be very satisfied and 27.6% completely satisfied). Approximately 29.2% claimed to be satisfied with the image of the park, and 5.7% claimed to be neither satisfied nor dissatisfied. However, despite the small number, some tourists (1.8%) were not satisfied: 0.2% were very dissatisfied and 1.6% were dissatisfied, considering that it neither met nor exceeded their initial expectations.

Regarding the nine tourists (1.8%) who were unsatisfied with the image, they highlighted a number of situations, namely, the "feeling of abandonment and neglect by the entities that should promote and care for the park" (Q.29); "Lack of respect for nature by those who run and organize the park; Lack of signage; Lack of hygiene/garbage bins" (Q.238); "Lack of cleanliness of tourist attractions; Toll payment; There should be a monitor to explain the fauna and flora of Mata de Albergaria" (Q. 239); "Trail maintenance; Garbage bins; Information about the PGNP" (Q. 244); "PGNP should have more entertainment (Bars and pools)"

Table 7 Other attributes considered by respondents regarding the image of the PGNP

	Likert scale	Attributes	F
1.	Far worse than expected	Maintenance	1
		Cleaning-streets	1
		Cleanliness-tourist spots	1
		Waste bins	1
		Maps	1
2.	Very bad	Car park	1
3.	Bad	Medical services	1
		Car park	1
		Payment of tolls	1
4.	Neither good nor bad	–	–
5.	Good	–	–
6.	Very good	Silence	1
7.	Far better than expected	Waterfalls	1
Total			**11**

Source Elaborated by the authors

Table 8 Respondents' expectations regarding the image of the PGNP

Likert scale	f	%	
1. Completely unsatisfied	0	0	1.8
2. Very unsatisfied	1	0.2	
3. Unsatisfied	8	1.6	
4. Neither satisfied nor unsatisfied	29	5.7	5.7
5. Satisfied	148	29.2	29.2
6. Very satisfied	181	35.7	63.3
7. Completely satisfied	140	27.6	
Total	**507**	**100**	

Source Elaborated by the authors

(Q.142; Q. 314); "Garbage" (Q.72); "Lack of restoration infrastructures" (Q.295) and "Payment of tolls" (Q.254).

Trying to analyse the reasons/motives that led tourists to visit the PGNP with the image they had of the park after their stay, we crossed the data to draw some conclusions. Thus, as can be seen in Table 9, the majority was very satisfied with the image of the PGNP, falling within point six of the interval scale, e.g. rest-151; contact with nature-137; and entertainment-66.

However, we found that those who mentioned that one of the reasons for their visit was to visit monuments were completely satisfied with the park's image. On the other hand, for those who came to the park to do sports, the image was a little lower than expected, and most of them said they were satisfied with the image of the park (27/72) (Table 9).

We consider that the PGNP, despite being a territory with many tourist attractions, sport is not a very consistent and consolidated attraction. It should also be noted that there were respondents who were not satisfied with the image of the PGNP, namely those who mentioned reasons of rest and contact with nature (Table 9).

After identifying the three main reasons/motives for the respondents' visit (rest, contact with nature and entertainment), we sought to analyse how often repeat tourists (360–71% of the sample) usually visit the park.

Thus, through the analysis of Table 10, of those who revisit the PGNP ($n = 360$), the visitors who seek the park to rest (314/360), usually visit it at least once a year (109/314), followed by once in three years (70/314), several times a year (69/314) and less than once in three years (63/314).

Table 9 Motives for the stay presented by respondents distributed by expectations of the image of the PGNP post-visit

Expectations	Completely unsatisfied	Very unsatisfied	Unsatisfied	Neither satisfied nor unsatisfied	Satisfied	Very satisfied	Completely satisfied	Total
Motives	**0**	**1**	**8**	**29**	**148**	**181**	**140**	**507**
Rest	0	1	7	25	130	151	122	436
Contact with nature	0	0	7	20	109	137	115	388
Entertainment	0	0	3	10	32	66	53	164
Visit monuments	0	0	3	7	16	23	25	74
Sport	0	0	2	1	27	24	18	72
Gastronomy	0	0	1	6	14	26	24	71
Visit friends/relatives	0	0	0	2	5	12	4	23
Health	0	0	1	2	7	10	2	22
Other motive	0	0	0	0	3	4	1	8
Business	0	0	1	1	1	3	2	8
Religion	0	0	1	0	0	2	0	3

Source Elaborated by the authors

Table 10 Reasons for the stay presented by the respondents distributed by frequency of repetition ($n = 360$)

Motives	Frequency					
	1 time p/month	1 time p/year	Several times p/year	1 time in 3 years	Less than 1 in 3 years	Total
	14	**324**	**224**	**182**	**182**	**926**
Rest	3	109	69	70	63	314
Contact with nature	5	96	65	53	57	276
Entertainment	2	44	31	19	21	117
Visit monuments	1	18	20	11	13	63
Sport	1	18	18	10	5	52
Gastronomy	1	18	7	12	14	52
Visit friends/relatives	1	8	5	3	4	21
Health	0	8	5	4	2	19
Other motive	0	2	3	0	0	5
Business	0	2	0	0	2	4
Religion	0	1	1	0	1	3

Source Elaborated by the authors

Only a small number usually go to the park once a month to rest (3/314) (Table 10).

Regarding those who indicated contact with nature (276/360), the majority stated that they come mainly once a year (96/276), followed by several times a year (65/360), less than once in three years (57/276) and once in three years (53/276). A small number states that they come once a month to contact with nature (5/276) (Table 10).

With regard to entertainment, the data are similar to the previous ones: 44/117 stated that they come once a year, 31/117 several times a year, 21/117 less than once in three years, 19/360 stated once in three years; a not very significant number admitted to come once a month (2/117) (Table 10).

5 Conclusions

In this study, we sought to answer the research question about what are the main attributes that visitors highlight in a given territory with specific characteristics such as the PGNP. This research corroborates some studies conducted when they refer that the natural landscape is one of the most outstanding, namely Gallarza et al. (2002), Chi and Qu (2008), among others.

What attracts the most attention in PGNP are the functional attributes such as natural landscape, and outdoor activities. These features are unique to tourist destinations such as protected areas in general and the PGNP specifically (Martins et al., 2022; Martins, 2022a). The majority of the sample considered rest and contact with nature as the main reasons for visiting this destination, which denotes the uniqueness of this territory. It is therefore crucial to maintain and preserve this protected area in order to continue attracting and retaining tourists. In terms of infrastructures, the accommodation and restaurants stand out positively. This corroborates the existing literature, as this protected area has a well-consolidated set of infrastructures in terms of accommodation (Martins, 2022a), and is also a region that stands out for its typical and gastronomic dishes.

Of the psychological attributes, those that stood out the most were hospitality. Events and entertainment were the least identified attribute due to the specific characteristics of this tourist destination, which presupposes calmer experiences since the most stated reasons for staying are rest and contact with nature. This shows that the visitor profile of this protected area identifies more with nature and silence. However, we consider that managers and stakeholders of this protected area should try to improve these attributes, as although most visitors have a good image of the park, the percentage of satisfaction in terms of expectations (very or completely satisfied) is below 65%, which does not demonstrate that this protected area can overcome other competing destinations.

As proposals for future research, it would be interesting to understand in other protected areas, with regard to accommodation and restaurant attributes, if the same attributes are referred to or if this is just a characteristic of the PGNP because there are many accommodations.

References

Akgiş İlhan, Ö., Özoğul Balyalı, T., & Günay Aktaş, S. (2022). Demographic change and operationalization of the landscape in tourism planning: Landscape perceptions of the Generation Z. *Tourism Management Perspectives, 43*, 100988. https://doi.org/10.1016/j.tmp.2022.100988

Al-Ansi, A., & Han, H. (2019). Role of halal-friendly destination performances, value, satisfaction, and trust in generating destination image and loyalty. *Journal of Destination Marketing & Management, 13*, 51–60. https://doi.org/10.1016/j.jdmm.2019.05.007

Baloglu, S., & McCleary, K. W. (1999). A model of destination image formation. *Annals of Tourism Research, 26*(4), 868–897. https://doi.org/10.1016/S0160-7383(99)00030-4

Beerli, A., & Martín, J. D. (2004). Factors influencing destination image. *Annals of Tourism Research, 31*(3), 657–681. https://doi.org/10.1016/j.annals.2004.01.010

Bertan, S. (2020). Impact of restaurants in the development of gastronomic tourism. *International Journal of Gastronomy and Food Science, 21*, 100232. https://doi.org/10.1016/j.ijgfs.2020.100232

Bigné Alcañiz, E., Sánchez García, I., & Sanz Blas, S. (2009). The functional-psychological continuum in the cognitive image of a destination: A confirmatory analysis. *Tourism Management, 30*(5), 715–723. https://doi.org/10.1016/j.tourman.2008.10.020

Boo, S., Busser, J., & Baloglu, S. (2009). A model of customer-based brand equity and its application to multiple destinations. *Tourism Management, 30*(2), 219–231. https://doi.org/10.1016/j.tourman.2008.06.003

Buhalis, D. (2000). Marketing the competitive destination of the future. *Tourism Management, 21*(1), 97–116. https://doi.org/10.1016/S0261-5177(99)00095-3

Chen, & Tsai, D. (2007). How destination image and evaluative factors affect behavioral intentions? *Tourism Management, 28*(4), 1115–1122. https://doi.org/10.1016/j.tourman.2006.07.007

Chi, C. G. Q., & Qu, H. (2008). Examining the structural relationships of destination image, tourist satisfaction and destination loyalty: An integrated approach. *Tourism Management, 29*(4), 624–636. https://doi.org/10.1016/j.tourman.2007.06.007

Crompton, J. L. (1979). Motivations for pleasure vacation. *Annals of Tourism Research, 6*(4), 408–424. https://doi.org/10.1016/0160-7383(79)90004-5

del Bosque, I. R., & San Martín, H. (2008). Tourist satisfaction a cognitive-affective model. *Annals of Tourism Research, 35*(2), 551–573. https://doi.org/10.1016/j.annals.2008.02.006

Echtner, C., & Ritchie, J. (1991). The meaning and measurement of destination image. *Journal of Tourism Studies, 2*(2), 2–12.

Gallarza, M. G., Saura, I. G., & García, H. C. (2002). Destination image: Towards a conceptual framework. *Annals of Tourism Research, 29*(1), 56–78. https://doi.org/10.1016/S0160-7383(01)00031-7

Ganzaroli, A., De Noni, I., & van Baalen, P. (2017). Vicious advice: Analyzing the impact of TripAdvisor on the quality of restaurants as part of the cultural heritage of Venice. *Tourism Management*, *61*, 501–510. https://doi.org/10.1016/j.tourman.2017.03.019

Govers, & Go, F. (2004). Cultural identities constructed, imagined and experienced. A 3-gap tourism destination image model. *Tourism*, *52*, 165–182.

James, J., & Von Wald, D. (2006). The development of the. *Tourism Culture & Communication*, *6*(3), 191–203.https://doi.org/10.3727/109830406778134135

Kastenholz, E. (2012). *The role and marketing implications of destination images on tourist behavior: The case of Northern Portugal.* https://ria.ua.pt/bitstream/10773/1838/1/2005001493.pdf

Li, C., Cao, M., Wen, X., Zhu, H., Liu, S., Zhang, X., & Zhu, M. (2022). MDIVis: Visual analytics of multiple destination images on tourism user generated content. *Visual Informatics*. https://doi.org/10.1016/j.visinf.2022.06.001

Liu, J., & Jo, W. (2020). Value co-creation behaviors and hotel loyalty program member satisfaction based on engagement and involvement: Moderating effect of company support. *Journal of Hospitality and Tourism Management*, *43*, 23–31. https://doi.org/10.1016/j.jhtm.2020.02.002

Luque-Martínez, T., Del Barrio-García, S., Ibáñez-Zapata, J. Á., & Rodríguez Molina, M. Á. (2007). Modeling a city's image: The case of Granada. *Cities*, *24*(5), 335–352. https://doi.org/10.1016/j.cities.2007.01.010

Lv, X., Li, C., (Spring), & McCabe, S. (2020). Expanding theory of tourists' destination loyalty: The role of sensory impressions. *Tourism Management*, *77*, 104026. https://doi.org/10.1016/j.tourman.2019.104026

MacKay, K. J., & Fesenmaier, D. R. (1997). Pictorial element of destination in image formation. *Annals of Tourism Research*, *24*(3), 537–565. https://doi.org/10.1016/S0160-7383(97)00011-X

McCartney, G., Butler, R., & Bennett, M. (2008). A strategic use of the communication mix in the destination imageformationprocess. *Journal of Travel Research*, *47*(2), 183–196. https://doi.org/10.1177/0047287508321201

Martins, H.; Silva, C., Pinheiro, A., & Gonçalves, E. (2021). A importância da marca no turismo: o caso da entidade regional Turismo do Porto e Norte de Portugal. *PASOS Revista de Turismo y Patrimonio Cultural*, *19*(4 SE-), 753–762. https://doi.org/10.25145/j.pasos.2021.19.049

Martins, H. (2022a). Tourism in protected areas: the example of Peneda-Gerês National Park (Portugal). *PASOS Revista de Turismo y Patrimonio Cultural*, *20*(5), 1113–1128. https://doi.org/10.25145/J.PASOS.2022.20.075

Martins, H. (2022b). Os impactos económicos da Covid-19 em eventos. *Revista Turismo & Desenvolvimento*, *38*, 265–280. https://doi.org/10.34624/rtd.v38i0.25863

Martins, H, Carvalho, P., & Almeida, N. (2021). Destination brand experience: A study case in touristic context of the Peneda-Gerês National Park. *Sustainability*, *13*(21). https://doi.org/10.3390/su132111569

Martins, H., Carvalho, P., & Almeida, N. (2022). O turismo em Áreas Protegidas: uma análise ao perfil do turista no Parque Nacional da Peneda-Gerês (Portugal). *Cadernos de Geografia*, *46*, 77–91. https://doi.org/10.14195/0871-1623_46_6

Martins, H.; Carvalho, P., & Almeida, N. (2023). Destination brand experience and place attachment: A study at the Peneda-Gerês

National Park. *Tourism: An International Interdisciplinary Journal*, *71*(1), 106–120. https://doi.org/10.37741/t.71.1.7

McDaniel, C., & Gates, R. (2004). *Pesquisa de marketing* (Thomson Le).

Milman, A., & Pizam, A. (1995). The role of awareness and familiarity with a destination: The central Florida Case. *Journal of Travel Research*, *33*(3), 21–27. https://doi.org/10.1177/004728759503300304

Neves, J. M. de O. (2012). Imagem de destino turístico : contributo para uma sistematização da leitura científica. *Cogitur: Journal of Tourism Studies*, *5*, 39–68. http://hdl.handle.net/10437/5237

Pike, S. (2002). Destination image analysis—A review of 142 papers from 1973 to 2000. *Tourism Management*, *23*(5), 541–549. https://doi.org/10.1016/S0261-5177(02)00005-5

Prayag, G. (2010). Images as pull factors of a tourist destination: A factor-cluster segmentation analysis. *TourismAnalysis*, *15*(2), 213–226. https://doi.org/10.3727/108354210X12724863327768

Prebežac, D., & Mikulić, J. (2008). Destination image and key drivers of perceived destination attractiveness.*Market-Tržište*, *20*(2), 163–178. https://hrcak.srce.hr/53067

Romero, I., Fernández-Serrano, J., & Cáceres-Carrasco, F. R. (2020). Tour operators and performance of SME hotels: Differences between hotels in coastal and inland areas. *International Journal of Hospitality Management*, *85*, 102348. https://doi.org/10.1016/j.ijhm.2019.102348

Shani, A., Wang, Y., Hudson, S., & Gil, S. M. (2009). Impacts of a historical film on the destination image of South America. *Journal of Vacation Marketing*, *15*(3), 229–242. https://doi.org/10.1177/1356766709104269

Silva, R., & Correia, A. (2017). Places and tourists: Ties that reinforce behavioural intentions. *Anatolia*, *28*(1), 14–30. https://doi.org/10.1080/13032917.2016.1240093

Tan, W. K., & Wu, C. E. (2016). An investigation of the relationships among destination familiarity, destination image and future visit intention. *Journal of Destination Marketing and Management*, *5*(3), 214–226. https://doi.org/10.1016/j.jdmm.2015.12.008

Tapachai, N., & Waryszak, R. (2000). An examination of the role of beneficial image in tourist destination selection. *Journal of Travel Research*, *39*(1), 37–44. https://doi.org/10.1177/004728750003900105

Tavitiyaman, P., Qu, H., Tsang, W. L., & Lam, C. R. (2021). The influence of smart tourism applications on perceived destination image and behavioral intention: The moderating role of information search behavior. *Journal of Hospitality and Tourism Management*, *46*, 476–487. https://doi.org/10.1016/j.jhtm.2021.02.003

Turism of Portugal. (2017). *Estratégia Turismo 2027*.

Vengesayi, S., Mavondo, F. T., & Reisinger, Y. (2009). Tourism destination attractiveness: Attractions, facilities,and people as predictors. *Tourism Analysis*, *14*(5), 621–636. https://doi.org/10.3727/108354209X12597959359211

White, C. J. (2004). Destination image: To see or not to see? *International Journal of Contemporary Hospitality Management*, *16*(5), 309–314. https://doi.org/10.1108/09596110410540285

Zhang, H., Wu, Y., & Buhalis, D. (2018). A model of perceived image, memorable tourism experiences and revisit intention. *Journal of Destination Marketing & Management*, *8*, 326–336. https://doi.org/10.1016/j.jdmm.2017.06.004

Cultural Heritage, Tourism and Sustainable Development. The Model of the Cultural Heritage Digital Media Lab

Fernando Faria Paulino and Tiago Cruz

Abstract

The protection of cultural heritage must, at present, be seen as a determining factor, taking an active role with local communities, resident populations and the environment, for local development, in what is now known as sustainable tourism, articulating all stakeholders. Institutions, organisations, private industries and local populations should play, in close coordination with each other, an active role in the socio-cultural development of the region (s) either as a place of construction and representation of intangible cultural heritage, or as a place of valorisation and revaluation of this same heritage, transforming it into a factor of self-esteem and cultural resource of the populations, considering the potential development of local communities. The Cultural Heritage Digital Media Lab (CHDML) is an ongoing project and aims to contribute to the participatory governance process, to the rescue and enhancement of cultural heritage and memory and to the creation of opportunities for economic development and appreciation of local products, bearing in mind the market potential. New modalities of appropriation and reappropriation of the popular and the traditional emerge, now shaped under the designation of "heritage". Cultural industries are born and developed around the concept of culture, now considered fundamental in the creative economy, tourism and sustainable development. The main goal of this paper is to present and introduce the CHDML as a platform that supports the development of projects and activities related, mainly, with the preservation of intangible cultural heritage of Portuguese-speaking countries. In this context, the authors present some examples of projects that follow the methodological approach presented in this article and underline the three main characteristics that are essential to the CHDML: (i) it is focused just on Portuguese-speaking countries; (ii) the educational aspect of working with students; (iii) the importance of working on heritage reinterpretations.

Keywords

Cultural heritage • Tourism • Development • Sustainability

1 Through Heritage and Tourism

The notion of collective imaginary has, in its essence, a strong connection to what is currently called intangible heritage, a concept that in itself raises problems within the academy, both on an epistemological, conceptual and methodological level. It will be pertinent to move forward from now on with an elementary definition of intangible heritage, broad and comprehensive, without, however, gathering the unanimity of the researchers. The Intangible Cultural Heritage "manifests itself among other domains in traditions and oral expressions, in social practices, rituals and festive events, in knowledge and practices related to both nature and the universe" (UNESCO, 2003). It is thus presented as a "vast set of manifestations and expressions of an intangible character that have memory as a means of preservation and orality as a means of transmission" (UNESCO, 2003). "Legends, myths, folk tales, rituals and festivities, as well as the entire universe of knowledge and experiences of popular cosmogony", are then encompassed within this immateriality (UNESCO, 2003).

Separating tangible from intangible, within the scope of issues related to cultural heritage, will obviously be a complex task with immense conceptual uncertainties. Ultimately, heritage objects or experiences will never be just tangible or intangible. According to Jorge (in Ramos, 2005:11), "If there is something that defines heritage as a cultural heritage

F. F. Paulino (✉) · T. Cruz
University of Maia, Maia, Portugal
e-mail: fpaulino@umaia.pt

T. Cruz
e-mail: tcruz@umaia.pt

and that is common to all its contemporary meanings, that something is immateriality". Defining something as tangible heritage implies the existence of an underlying immateriality and, on the other hand, the heritage designated as intangible is constantly associated with materialities such as spaces, utensils and various instruments, clothing, gestures, etc.

Despite the problems surrounding the "tangible/intangible" dichotomy, we assume that, "contrary to the so-called tangible cultural heritage, the intangible differs from the first in terms of its supports, which are highly fragile and, consequently, easily perishable" (Paulino, 2010a: 311). In other words, this fragility related to immateriality presents particular challenges with regard to its preservation and dissemination. It is in this sense that "the urgent need to recover, collect and preserve them", within the scope of projects such as the CHDML, "turns out to be decisive, having as objectives their inventory, their treatment —we refer specifically to interpretive studies—as well as their dissemination, determinant for the collective memory and identity of a group or society" (Paulino, 2010a: 311). Lorentç Prats refers in this regard that heritage activations, namely those promoted by tourism, are identity representations (1997: 46).

This present and growing interest in intangible cultural heritage was essentially born in 2008, when the Portuguese State joined the Convention for the Safeguarding of Intangible Cultural Heritage, drawn up by UNESCO in 2003. That is, one gets the idea that the discovery of the existence of the so-called intangible heritage seems to have emerged only today. However, this whole set of concerns, underlying the Convention, was born with Anthropology itself. Intangible heritage has always been a field of interest for anthropological science. Already in the nineteenth century, debates arose on how to preserve traditions and what their interest was in the identification of particular groups of people. If the notion of culture appeared from the 1920s onwards, anthropology dealt until the same time with the notion of tradition (Fabietti, 2001; Kilani, 2002).

It is also worth noting that this "re-invention" of intangible heritage arises in opposition to the UNESCO Convention for the Protection of Cultural and Natural Heritage of 1972, before which, a year later, Bolivia warned of the need for a convention that would also defend the most intangible issues of heritage. The notion of "intangible heritage thus appears as an expression of reaction, as an expression of concerns within an area not covered by those who work with tangible heritage (Paulino, 2010a: 312). The distinction between tangible and intangible "turns out to be a merely political distinction, first debated and worked on within UNESCO, at a supranational level, becomes part of the list of concerns of States, to finally reach the academy" (Paulino, 2010a: 312). Societies, groups and individuals holding the heritage remained far from the debate.

It is in this way that we understand that the notions of tangible and intangible are culturally determined notions, and that it will belong to each culture, society or group to establish a boundary between what is tangible and intangible. The Convention for the Safeguarding of Intangible Cultural Heritage thus appears to be a legal-political instrument that defines a clear separation between tangible and intangible cultural heritage, which above all contradicts the notion of cultural diversity, if understood as an instrument of cultural homogenisation. An instrument of the States, of central power, of local authorities, which is why it seems pertinent to emphasise the lack of protagonism given to local communities in taking the initiative to preserve and safeguard their local heritage.

The relationship between intangible cultural heritage and tourism has, particularly over the last decade, sparked debates and discussions between professionals and academics in the areas of tourism, culture, heritage, development, the arts and creative industries. The interdisciplinarity of approaches has revealed challenges with little consensus in the attempt to protect the intangible cultural heritage through tourism, bringing to the debate the promotion of sustainable tourism as well as the recovery of communities through development policies.

Since the signing of the UNESCO Convention, in 2003, for the Safeguarding of Intangible Cultural Heritage, publications (available online on UNESCO websites and partners working directly with the Organisation) have multiplied with the dominant emphasis of a paradigm for safeguarding cultural heritage. Cultural practices around the so-called intangible cultural heritage, especially in processes in which communities, groups and individuals play central roles in safeguarding and protecting their intangible cultural heritage.

Thus, shortly after the signing of the aforementioned Convention, there was a rapid growth of scientific literature in the field of Intangible Cultural Heritage, with authors defending the UNESCO paradigm, and others rejecting the Organisation's position (Hafstein, 2018; Meskell, 2015; Ramos, 2003; Stefano et al., 2014; Smith & Akagawa, 2009). Questions were thus raised regarding the respective policies of UNESCO's national lists and inventories, the processes of patrimonialisation of cultural practices, as well as community participation as a guiding principle of the Convention.

There should be two premises that must necessarily be present when working in the field of intangible heritage. "Since the notion of intangible cultural heritage is almost an anthropological definition of culture, it should therefore be recognised as something dynamic" (Paulino, 2010b: 210), something that is in constant and permanent evolution. A festival, a ritual or an agricultural practice must be "recognised as processes, reflections of the way of life of communities, groups, but which will only exist as long as they make sense for individuals or communities" (Paulino, 2010b: 210). This festival, this ritual or this agricultural practice, "can never be fixed, crystallised, in the sense of imposition" (Paulino, 2010b: 210). They will have to be seen as dynamic processes, subject to re-significations, subject to modifications due to changes in the socio-cultural context—and which socio-semiotics scientifically explores and deconstructs—in a process very similar to myths and their respective characteristics (evolutionary and not-universals). Carlos Fortuna refers, in relation to what he called the "creative destruction of identities [...] on the other hand, with the space, immediate or represented, in which they interact" (1999: 29).

The second question, which also emanates from the quote by Carlos Fortuna, has to do with the need for an integrated approach between intangible heritage and material culture. The basic question is to ignore the holistic character of anthropology and, in particular, the anthropological concept of culture. Indeed, separating tangible from intangible culture is something that shakes Mauss's theory (1988) by which each cultural manifestation is considered, and we do, as a "total social fact". If, on the one hand, "intangible heritage appears dependent on a tangible space—a territory —that gives it meaning, or a landscape that evokes it, intangible heritage always shows a subject/space relationship, even if both are implied" (Paulino, 2010a: 313). In other cases, this dependence results from the connection of an object, a utensil, with a traditional practice. Seeing the intangible heritage as an element in itself will result in a "decontextualisation and, consequently, in a loss of meaning" (Paulino, 2010a: 313). It will therefore be impossible to work with a myth or legend by decontextualising it geographically, it is the territory itself, the space that also gives its meaning (see Edensor, 1998).

Likewise, this second question, translated into a tangible/intangible separation, raises yet another problem. Intangible heritage, by implying a collection (carried out through a text, audio, video, photographic record, among others) "leads to a process of materialisation of the intangible, that is, the record sacralises the intangible heritage, giving it a status that will allow it to appear in a museum, as a tangible object" (Paulino, 2010a: 313). This process of secularisation is in line with the way in which the Convention operates, whose objectives also pass, in addition to its already mentioned political interest, by a logic of action in the economic sphere, essentially for tourism purposes, an issue raised by João Ramos referring to the idea by which a heritage-based tradition places the place where it takes place on the map of international tourism (Ramos, 2005: 67 76).

Therefore, the Convention ends up having a strong impact on the tourist industry, whether at a national, regional or local level. In other words, intangible heritage is important as a tourist resource, firstly for the States, followed by local authorities and finally for the holders and their respective communities. This need for preservation arises when the tradition begins to disappear, and then it is only from that moment onwards an object of concern. In some cases, even when tradition no longer exists as such, the process of preservation ends up transforming itself into a process of patrimonialisation, of musealisation, practically transforming it into an object of "authenticity" (staged authenticity) with the objective of tourist exploitation.

The definition of intangible cultural heritage raises a problem in its connection with tourism. The preservation of authenticity for the tourism development of a place must be experienced by the tourist (Cohen, 1979, 1988; MacCannell, 1973, 1976; Paulino, 2007a, 2007b; Urry, 1991; Wang, 1999). Defining characteristic of cultural tourism, in which the product is centred on the representation of the other and the discovery of other ways of life in different communities. The staging of the cultural traditions of a territory and the visited community emerge as necessary elements for the production of a memorable tourist experience, in which preservation is a determining factor. The tourist will experience not the everyday life of the community, but a practice based on events produced for consumption, which emphasises the exotic, folkloric side of the visited community, a "staged authenticity" (Paulino, 2007a, 2007b), eventually preserved.

2 The CHDML

The Cultural Heritage Digital Media Laboratory (CHDML) project is centred on the collective imaginary of Portugal, Cape Verde, Brazil, Guinea-Bissau, São Tomé e Príncipe and Macau—Portuguese-speaking countries—through the intangible cultural heritage and its role in the identity construction of the territory. The CHDML is focused on Portuguese-speaking countries due to two aspects. The most important one is the fact that there are connections between the various social practices of the various countries and, consequently, it becomes pertinent to work not only on the intangible heritage of these countries but also to explore these interrelationships between the different practices, contexts and activities. The second aspect relates to the fact that, through this delimitation, the CHDML defines a clear frontier of action, differentiating itself from other similar projects.

The projects carried out that determined the birth of the CHDML proved to be decisive in the experience and field practices accumulated in different situations and contexts. Between 2012 and 2019, the Visual Anthropology and New Media Workshop was held annually in partnership with the University of Cape Verde and under the scientific coordination of Fernando Faria Paulino, a project that put students in direct contact with audiovisual technologies and a methodology for collecting images and respective dissemination of Cape Verde's intangible heritage. Also under the scientific coordination of Fernando Faria Paulino, the project to collect the Intangible Heritage of the Porto Metropolitan Area involved Portuguese students in the use of audiovisual media, culminating in the production of 17 documentaries related to each of the municipalities that make up the Metropolitan Area of Porto. The concepts of "participation" and the role of "communities", "groups" and "individuals" (Sousa, 2018), key concepts for UNESCO and currently still under debate and discussion, were applied in the participatory methodologies used, derived from the research methods in anthropology, both in Portugal and Cape Verde.

Based on the Cooperation Protocol between the University of Maia and the University of Cape Verde, signed in 2012, the 1st Visual Anthropology and New Media Workshop was held in 2013 at the University of Cape Verde, in Cidade da Praia (see Fig. 1), intended for students of the Degree in Technologies, Multimedia and Communication. It was the first contact of Cape Verdean students with the audiovisual area and the respective production of content at the level of video documentary, in this case, specifically in the area of heritage. Observing, producing and sharing were the specific objectives of the workshop carried out with the following structure: (i) presentation and theoretical module of approach and methodologies; (ii) fieldwork and production and shooting phases; (iii) analysis of the collected

images, montage and edition and respective public presentation of the developed works.

As an example, one of the documentaries made was about the Mercado da Assomada (city of Assomada, island of Santiago), as a space for selling products, but also as a meeting place for life stories and experiences, and its importance as a cultural space of the Municipality of Assomada (see Figs. 2, 3 and 4). The issues carried into the documentary, the result of the relationship work between students and social actors, were centred on the patrimonial value of the space, on the identity of the space, on the passing of history from generation to generation, on the preservation of the memory of the market by users themselves and the importance of space as a tourist attraction. The results obtained placed the entire process of observation and collection of elements by Cape Verdean students in a process of relationship with social actors, thus activating the (re)discovery of the local heritage by the populations that inhabit that space. Equally interesting was the awareness of the existence of products (understood here in its broadest sense, cultural goods, tangible and intangible heritage, among many other aspects) within their region, until then unknown to the subjects. This fact implied enjoying the region itself directly, becoming aware of the unnoticed because of its everyday nature and therefore considered banal. It is in this way that local development policies should have repercussions on increasing the collective awareness of populations as members of a territory and respective integration within it.

Fig. 1 University of Cape Verde, Campus Palmarejo, Praia, May 2013

Fig. 2 Assomada Market, Assomada, May 2013

Fig. 3 Assomada Market, Assomada, May 2013

Likewise, the Intangible Heritage Collection Project of the Porto Metropolitan Area, carried out in 2013 (see Fig. 5), served as the basis for audiovisual experimentation in the collection of intangible heritage, in a collaborative and participatory process and the respective construction of a digital platform (Barbosa & Paulino, 2018: 397–409). The project was carried out by the University of Maia and the Área Metropolitana do Porto (a region comprising 17 municipalities in the north of Portugal), with the objective of producing 17 documentaries on the collection of intangible

Fig. 4 Assomada Market, Assomada, May 2013

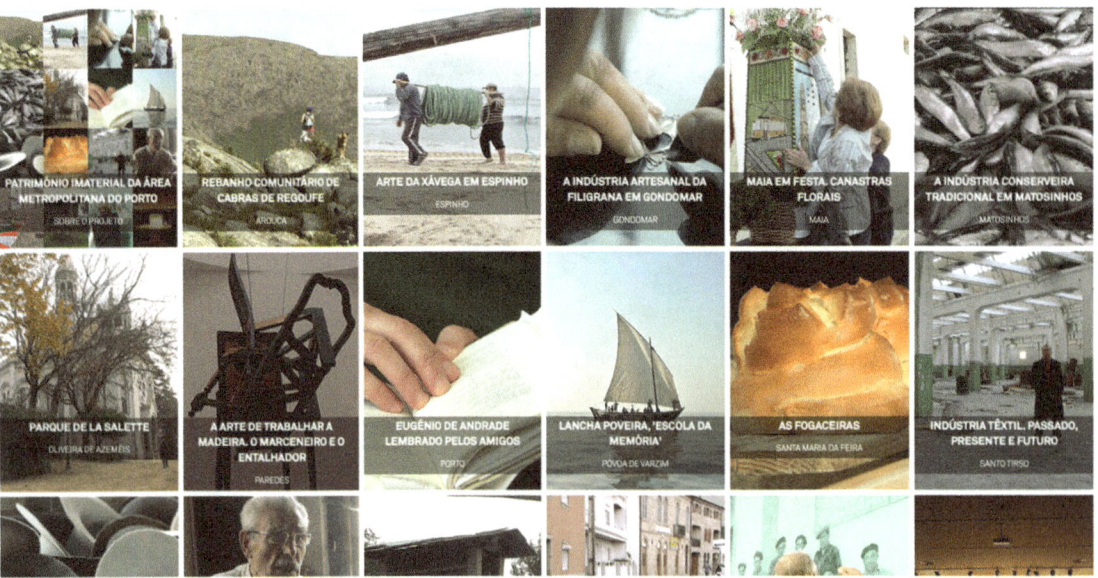

Fig. 5 PIAMP platform (http://piamp.amp.pt)

heritage, addressing cultural practices, rituals and events festivals placing emphasis on the identity construction of the territory through the collective imagination, memory and identity, determining axes of the project itself.

The entire process of pre-production, production, shooting and editing was carried out by undergraduate and master's students from the scientific area of Multimedia Communication at the University of Maia (see Figs. 6, 7 and 8).

Fig. 6 Shooting of the
documentary, Regoufe, 2013

Fig. 7 Shooting of the documentary, Trofa, 2013

3   Goals and Actions

There are three axes of action on which the entire project is based: (i) collection and dissemination of intangible heritage; (ii) educational activity related to the collection and dissemination of intangible heritage; (iii) artistic/documentary production based on intangible heritage.

The recovery, collection and preservation of this heritage involves inventorying, processing the information and its dissemination. These activities are crucial for the collective memory and identity of groups and societies.

The second point, related to the educational component, is essentially related with two interests. On one hand, the CHDML's participatory and collaborative aspect. It is important to involve the local communities in the process of

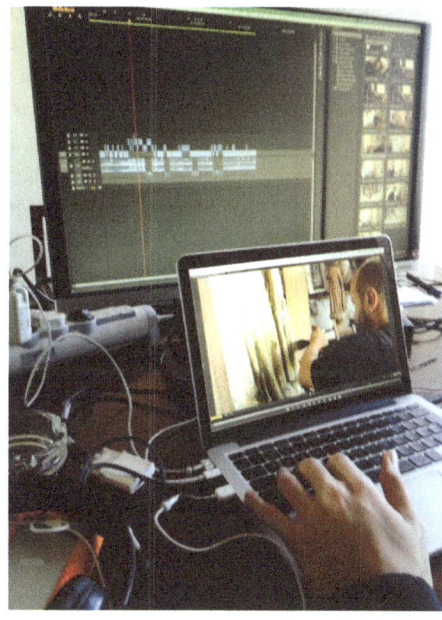

Fig. 8 Editing process

collecting and disseminating heritage and, in this sense, it is intended that the educational component of the project supports this interest through the provision of audiovisual material, training on its use, workshops, among others. On the other hand, student involvement is equally important in this context. It is intended that students from different universities play an active role not only in collecting and disseminating heritage, but also in heritage reinterpretation processes.

Finally, the third aspect is closely related not only to the collection and dissemination of heritage, but also with the reinterpretation process. It is intended that the contents produced may have a documentary and/or artistic character and, in this sense, different genres may be explored. Not only in the documentary field but also in the artistic field, such as installations, sound art, interactive art, among others. To make this possible, CHDML intends to make all the collected resources available so that anyone can use the information and, from it, create communicative artefacts of different nature.

From these three main axes, focusing on the collective imaginary, memory and identity, of Portugal, Cape Verde, Brazil, Guinea-Bissau, São Tomé e Príncipe and Macau, the CHDML has the following general objectives: (i) the collection and treatment of intangible heritage; (ii) the creation of a digital platform (hosted in Portugal) that brings together bibliography, audio recordings, audiovisual records, photography, drawing, among others; (iii) contribute to the dissemination of intangible cultural heritage through documentaries, exhibitions, dramatisations and other forms of artistic and documentary expression; (iv) create an educational service aimed at raising awareness of the issue and at providing conceptual, artistic and technical tools that support the practice of preserving and disseminating intangible heritage; (v) play an active role in the socio-cultural development of the region(s) either as a place of construction and representation of intangible cultural heritage, or as a place for the revaluation of this heritage, transforming it into a factor of self-esteem and cultural resource of the populations; (vi) use the materials collected and produced within the scope of the digital platform for educational purposes with the school-age populations of the respective countries/regions involved; (vii) contribute to pedagogical training in different languages (sound, visual and audiovisual) in the collection of heritage, as well as training activities in the use of new media (media and digital literacy).

The CHDML should aim not only to increase awareness of heritage among local populations, but also to raise awareness of the active participation of populations in the construction of the aforementioned platform, that is, in the collection of heritage, memory, identity and its collective imaginary.

Therefore, CHDML has the following specific objectives: (i) to increase the value of heritage, as a way of promoting the integrated and sustainable development of the region(s) and countries; (ii) creation of a digital platform containing the contents collected and produced; (iii) collection of heritage (historical, cultural and traditional) and its promotion; (iv) inventory of intangible heritage; (v) documentary collection of historical/cultural information on intangible heritage; (vi) construction and animation of environment houses, culture and heritage interpretation; (vii) promotion of literary works through heritage collection and inventory activities.

An active participation of the population is sought and defended in the collection of the intangible heritage, of memory, of their collective imaginary. Populations, groups and individuals will have to review themselves in this process, which is, above all, their belongings. The tasks of collecting, selecting and valuing the heritage is something in which they should participate, assuming a (pro)active role. CHDML will have the primary task of providing training to populations, in areas as diverse as the use of audiovisual, new technologies and ethnographic inquiry, because training and pedagogical action are also fundamental tasks in the new museology today (Carvalho, 2011: 155–157). This entire process of collecting, disseminating and (re)interpreting the intangible cultural heritage will therefore imply an active participation of the subjects, that is, the full exercise of citizenship, which is crucial for sustainable development.

In order to achieve these goals, the CHDML assumes certain lines of action. Thus, the construction—which must be carried out in a collaborative and participatory way—of the digital platform is of paramount importance. This is intended to protect, defend, value and share the intangible heritage, memory and identity. Following the previous points, it is intended to implement cultural policies that lead to the strengthening of heritage and boost related creative industries. It is also intended to promote training in expressive languages such as photography, video, audio, drawing, performance and installation, among others, in digital literature and the use of new media, with the aim of being used not only in the collection of heritage but also in its dissemination.

The project aims to contribute to the participatory governance process, to the rescue and enhancement of cultural heritage and memory and to the creation of opportunities for economic development and enhancement of local products, bearing in mind the market potential. New modalities of appropriation and reappropriation of the popular and the traditional were thus configured, now shaped under the designation of "heritage". Cultural industries were born and developed around the concept of culture, now considered fundamental in the creative economy.

Culture, in all its variety of forms, expressions, practices and knowledge, has gained international recognition as a fundamental element for development. The Convention for the Safeguarding of Intangible Cultural Heritage prepared by UNESCO in 2003 turns out to be a reflection not only of this recognition but also of the appreciation of heritage. CHDML intends to be an active agent in this process through an openly interpretive approach, either together with local actors, with local communities, or with elements outside the communities that intend to assume the intangible cultural heritage as a resource in the creation of new discourses. Thus, the CHDML assumes the interpretative character of the heritage representation process, also seeking to promote the use of information in processes of reinterpretation of collected and disseminated materials. Heritage is thus something dynamic, in constant transformation and involved in constant processes of reinterpretation through artistic and documentary practices.

In the context of these potential reinterpretations of heritage, and assuming heritage re-creations as an important element in the process of its preservation/dissemination, the CHDML assumes that heritage can be an important semiotic resource available for the construction of new discourses. Thus, we open the project to the possibility of exploring the semiotic potential of the collected materials, thus contributing to the construction of new artefacts and communicative experiences.

Finally, in a close relationship with the previous point, the CHDML seeks to encourage not only the use of photography, video and text, among others, in isolation, but also forms of communication and expression that unfold, multiply and hybridise in different semiotic genres (Cruz et al., 2019; Giannetti, 2012; Santaella, 2003). The aim is to create hybrid communicative artefacts and experiences that are part of the universe of installation, performance, sound art, video art, conceptual art and media art, among others.

4 Visual Anthropology and Methodology

This whole set of issues leads us to the methodological issue in collecting intangible heritage. It is difficult to define a proper methodology, however, as mentioned, the process of identification and collecting is of paramount importance. A process of stimulation in which the researcher—the ethnographer—and the person who owns the heritage—the informant or social actor participate. In this process, which involves the "evocation of memory, it is up to the anthropologist, the ethnographer, to bring the social actor to the present, and in some cases, to rescue him from the past. Likewise, it implies that the ethnographer is aware that collective memory is always conditioned by belonging to a community or group, as well as by the physical transformations of the territory" (Paulino, 2010b: 212). When it is transformed, there are no longer spatial references on which the memory is supported and it dissipates.

Before moving on to the methodological question regarding collecting intangible heritage, it is worth to emphasise the fact already mentioned, of assuming anthropology as an interpreted science (Geertz, 1993; Rabinow, 1977). The anthropologist constantly has access not to facts but interpretations, which give rise to new interpretations, to end up in an interpretation by the reader. As Bourdieu states, the proper dimension of fieldwork is based on the work of constructing social reality (in Rabinow, 1977: 163). It is the set of interpretations that leads us to the construction of social and cultural representations. All the facts that the anthropologist collects on the ground are, as mentioned,

interpretations. "The facts we interpret" turn out to be "constructed and reconstructed facts" (Rabinow, 1977: 150). Anthropology thus leaves the sphere of objective science to assume itself as an interpreting science (Geertz, 1993, 1999).

Thus, within the scope of collecting the so-called intangible heritage, the methodology applied in the scope of visual anthropology seems appropriate. Sarah Pink refers to Applied Visual Anthropology in a practice that implies the use of Visual Anthropology theory, methodology and practice without academic goals that generally serve to solve problems and are at the service of "cultural mediation" (Pink, 2007: 6). The collection of a myth, a legend or a tale does not only mean the collection of content, but also of a form, the way in which they are verbalised. The participation of "gestures, the emphasis placed on words, facial expressions, among others, constitute a whole performance that is also part of the narrative" (Paulino, 2010a: 314). In this way, we have access not to the text, but to a wide set of elements that participate in the story and in the interpretation of the same, bringing the social actor to the scene, and giving him an active role.

In this way, by the production and representation of knowledge in anthropology through the use of images, a reflective approach becomes necessary. This reflective approach to images in anthropology means, among other things, first of all that researchers are aware of the theories that guide their own practice of image production and the theories that guide the approaches of subjects/social actors in relation to those same images. Thus, reflexivity, in the ethnographic phase, which goes beyond the mere collection of data, implies, first of all, the development of an awareness of the way in which ethnographers play their role as image producers in particular cultural contexts, the way in which they produce a certain image and why they choose specific subjects. Second, it also implies a consideration of how these choices are related to both academic expectations and local visual cultures (informing subjects/actors). Finally, it also implies an awareness of the theories of representation that guide their images (Paulino, 2007a, 2011).

New knowledge, new social and cultural processes are constantly transported to the interior of the images. A complex system in semiotic terms, a web of tension lines, polarised by different generations, genres, markedly marked social characteristics, different experiences, rural and urban, local and global. Thus, the images, by integrating this infinity of parallel knowledge from the context of production and this complexity of dimensions that intersect when they are read, at an epistemological level, should be considered as a product of a relationship, as a result of a construction, a reflective look. The images, by proposing a shared way of seeing, result in a high significant density, assuming the status of a reflective (significant) whole.

The anthropologist therefore has to assume the image as the result of a process of construction between himself and the social actors, as well as being aware of the different readings that the images will have depending on the different contexts of reading. To observe does not just mean to see, but leads to an enormous amount of information, in a process that is itself a relationship. The reflective paradigm on which anthropologists rely favours the participatory tendency, revealing the ethnographic encounter that took place (Paulino, 2011).

More than a data collection method, as Sarah Pink says, ethnography is a process of creating and representing knowledge based on the ethnographer's personal experiences. The ethnographer is not responsible for producing objective and true testimonies of reality, but should aim to make known his experiences of reality, which should be as faithful as possible to the context, negotiations and intersubjective relationships through which knowledge was produced. This fact implies participatory, reflective and collaborative methods (Pink, 2001: 18). Thus, the practice of reflexivity in terms of anthropological work, therefore, implies that anthropologists reveal themselves, and reveal their methodology as instruments that also generate data, conceiving the production of communication as a coherent whole: producer—process—product (Paulino, 2011:144).

In moving from the image to the production of knowledge, we will necessarily have to address the "classic" uses of images as a research methodology: (i) images as a method of collecting information; (ii) the use of images in interviews or conversations with informants/social actors; (iii) the concept of "shared" image; (iv) the images as a process of exposing the research results.

1. The use of images in gathering information is crucial both in terms of the physical environment, space, as well as objects, actions or events in which subjects/social actors participate. However, these images should be treated as representations of certain aspects of culture and not objective records of the totality of a culture or as symbols with complete and accurate meanings.
2. Regarding the role of images in conducting interviews and conversations, images are not a simple visual record of reality, but representations that are interpreted in terms of different understandings of reality. When an informant/social actor sees the images produced by the anthropologist, his interpretation of how that same anthropologist sees reality immediately begins. In an interview or conversation, in which the images are present, the anthropologist and informant/social actor will have the opportunity to discuss their different interpretations of the images, in a process of confrontation between two ways of seeing, two different experiences about a reality. It is an ongoing negotiation process.
3. The concept of "shared" image combines the intentions of both the anthropologist and the informants or social

actors, representing the outcome of their negotiations. In most cases subjects very quickly inform the ethnographer of the type of images they would like to see, taking into account their personal and cultural uses in relation to the practice of image production (photography/video). "What do they want me to film/photograph?", "How do they want me to film/photograph them?".
4. Finally, the images as a process of exposing the research results. The descriptive power of the images, polysemic in their essence, also shows us their enormous power of synthesis. Images and sounds emerge as a means capable of synthesising in themselves, much more than what is possible to convey through the written word. They thus become an epistemological tool, which makes it possible to synthesise aspects of a cultural and social reality, obtained through interviews and participant observation, constructed through a constant encounter between anthropologist and subjects.

Thus, the methodological steps explored in the projects referenced above and which have been applied in the CHDML project itself are:

Displacement to the field;
Observation without a camera (observation of the real);
Setting a research objective (idea);
Visual and documentary research on the theme to be worked on;
Construction of the "subjectivity" of the look. Customisation of look and style. Reflexivity;
Return to the field. Adaptation to the field. Fieldwork;
Moving on to images (production of meaning);
Analysis of the images produced;
Construction of a visual discourse;
Elaboration of the research report.

5 Conclusions

CHDML is a project that seeks not only the preservation of intangible heritage, through its collection and dissemination, but also assumes all this "preserved" information as something dynamic in constant mutation. The authors are aware that the collection and dissemination of heritage is an activity directly involved with issues of representation and, in that sense, it is always at stake a point of view of someone in the context of production, dissemination and consumption of information. Therefore, CHDML makes available the collected, untreated information, so that others can use this data as a resource for other activities such as reinterpretations, production of other communicative artefacts, educational purposes, among others.

There is a working method associated with CHDML as previously exposed. This approach is important but is by no means imposed on project contributors. In fact, the approach is quite experimental and, as such, the method is not something static but flexible to adjust to the particular contexts of each community, country, context, etc. The methodology has a strong conceptual basis related to how we see the problem and how we see the issue of preservation. However, it invites the introduction of new techniques and processes that respond to the needs and objectives of the moment.

Due to the involvement of several universities, and taking into account that the project was born with an academic base, it is important to involve students in the production of content. Therefore, CHDML has an educational component that seeks not only to develop educational activity with local communities, but also to involve students in the task. CHDML thus serves as a platform that also supports the teaching of specific subjects related to communication, arts and design.

Finally, as already mentioned, the CHDML focuses on Portuguese-speaking countries and, also highlighted above, there are two aspects that justify this delimitation in terms of action. On the one hand, there are links between activities in different countries and activities that exist in more than one country. CHDML is interested in exploring these crossings, how they influence each other, what brings them together and what separates them, common roots, etc. On the other hand, it was important to define a more circumscribed, more limited field of action, so that CHDML is not just a platform related to the preservation of intangible heritage in general/global terms. The working field would be too vast. In this way, there is a geographic and conceptual coherence that justifies the pertinence of CHDML's existence.

Acknowledgements We would also like to highlight the various protocols and partnerships made so far. Among them: University of Maia (Portugal), UniCV-University of Cape Verde (Cape Verde), Amílcar Cabral University (Guinea-Bissau), University of São José (Macau), Federal University of Minas Gerais (Brazil), Federal University of Maranhão (Brazil), INEP-National Institute of Studies and Research (Guinea-Bissau), CIAC-Center for Research in Arts and Communication/University of Algarve (Portugal), IPC-Institute of Cultural Heritage of Cape Verde (Cape Verde), Porto Digital Association (Portugal).

References

Barbosa, R., & Paulino, F. (2018). Recording the intangible heritage of the city in the Metropolitan Area of Porto. In *Cities' identity through architecture and arts*. Taylor & Francis.

Carvalho, A. (2011), Os Museus e o Património Cultural Imaterial. Estratégias para o desenvolvimento de boas práticas, Lisboa: Edições Colibri.

Cruz, T., Paulino, F., & Tavares, M. (2019). CulturalNature Arga #2. In I. Management Association (Ed.), *Geospatial Intelligence: Concepts, methodologies, tools, and applications* (pp. 804–812). IGI Global.

Edensor, T. (1998). *Tourists at the Taj*. Routledge.

Fabietti, U. (2001). *Storia dell'Antropologia*. Zanichelli editore.

Fortuna, C. (1999). *Identidades, Percursos, Paisagens Culturais*. Celta Editora.

Geertz, C. (1993). *The interpretation of cultures*. Fontana Press.

Geertz, C. (1999). O Saber Local, Novos ensaios em antropologia interpretativa. Editora Vozes.

Giannetti, C. (2012). Estética Digital: Sintopia da arte, a ciência e a tecnologia. C/Arte.

Hafstein, V. Tr. (2018). *Making intangible heritage. El Condor Pasa and Other Stories from UNESCO*. Indiana University Press.

Kilani, M. (2002). *Antropologia. Una introduzione*. Dedalo.

Mauss, M. (1988). *Ensaio sobre a dádiva*. Edições 70.

Meskell, L. (Ed.). (2015). *Global heritage: A reader*. Wiley Blackwell.

Paulino, F. (2007a). Dos documentos de terreno ao hipermédia. In *Antropologia Visual e Hipermedia*. Ed. Afrontamento.

Paulino, F. (2007b). Turismo e Autenticidade. O papel da publicidade na representação dos lugares turísticos. In *Imágenes de la cultura/Cultura de las imágenes*. EditUM.

Paulino, F. (2010a). Recuperação do Imaginário Coletivo da Serra d'Arga. Pela projeção turística dos lugares simbólicos. In *Dinâmicas de Rede no Turismo Cultural e Religioso*. Ed. ISMAI.

Paulino, F. (2010b). *Cultura Visual e Turismo. Natureza e Cultura no Alto Douro Vinhateiro*. Universidade Aberta, tese doutoramento.

Paulino, F. (2011). "L'utilisation de l'hypermédia dans le discours anthropologique" in Inter Media. *Littérature, cinema et intertextualité*. L'Harmattan.

Pinheiro, A., & Paulino, F. (2022). Urban tourism and World Heritage: Relations and effects of the classification. *Pasos Revista De Turismo y Patrimonio Cultural, 20*(5), 1234–1254.

Pink, S. (2001). *Doing visual ethnography*. Sage Publications.

Pink, S. (2007). Applied visual anthropology. Social intervention and visual methodologies. In S. Pink (Ed.), *Visual interventions. Applied visual anthropology, Nova Iorque*. Bergahn Books

Prats, L. (1997). *Antropología y patrimonio*. Editorial Ariel.

Rabinow, P. (1977). *Reflections on fieldwork in Morocco*. University of California Press.

Ramos, M. (coord.). (2003). *A Matéria do Património—Memórias e Identidades*. Edições Colibri.

Ramos, M. (2005). Breve nota crítica sobre a introdução da expressão "património intangível" em Portugal. In V. O. Jorge (Org.), *Conservar para quê?* (pp. 67–76). DCTP-FLUP – CEAUCP-FCT

Santaella, L. (2003), Culturas e Artes do Pós-Humano, São Paulo: PAULUS

Smith, L., & Akagawa, N. (Eds.). (2009). *Intangible heritage*. Routledge.

Sousa, F. (2018). *A Participação na Salvaguarda do Património Cultural Imaterial. O papel das Comunidades, Grupos e Indivíduos*. Memória Imaterial CRL.

Stefano, M., Davis, P., & Corsane, G. (Eds.). (2014). *Safeguarding intangible cultural heritage*. The Boydell Press.

Unesco. (2003). *Convenção para a salvaguarda do património cultural imaterial*. UNESCO.

The World Heritage Classification in Urban Tourism Destinations: Perspectives for the City of Porto, Portugal

António José Pinheiro and Hugo Martins

Abstract

Cities are notable tourist destinations, responsible for important tourist flows. These are territories that present different dynamics, evolutions and functional transformations, visible, especially in their historic centres. The material and immaterial evidence of these dynamics configure important representations of cultural heritage that are the basis for the development of urban tourist attractions. The recognition of heritage value results from cultural perceptions and visions of different types of actors, which can be varied according to locations, aims and relationships with culture in general. However, there are relatively more consensual validations of the notion of cultural heritage, being, eventually, the most consensual on a planetary scale, the World Heritage classification by UNESCO. In terms of tourist attraction, the World Heritage seal is an expressive communication tool. As such, several researchers dedicate efforts to understanding the importance that the World Heritage classification has for the development of tourist destinations. The literature review allows finding case studies that show a high importance of classification for the growth of destinations, but others do not show that classification is a determining factor. Therefore, this study intends to contribute to this discussion with a dual purpose; to carry out a collection of case studies that demonstrate the different types of effects and to develop a case study applied to the city of Porto (Portugal). Understanding the importance of World Heritage classification for the success of this tourist destination led to the methodological option of a mixed study that includes tourism supply and demand. In terms of supply, through a sample of 29 tourist accommodations, interviews were carried out with managers and analysis on websites to understand the importance that is given and expressed in terms of communication about the World Heritage. On the demand side, surveys were carried out with 415 guests of tourist accommodations to assess how the World Heritage influenced their decision to visit the city.

Keywords

Cultural tourism • Urban tourism • World Heritage • Historic centre • Porto

1 Introduction

Cities are territories where a large part of the world's population is concentrated. They are spaces for issuing and receiving flows of people, goods and services and also are privileged places for cultural interactions. In different proportions and weights, urban centres are multifunctional where functions such as housing, industry, commerce, services and leisure are created, adapted and evolved. It is with a focus on this last feature that it can be said that cities are, in the present millennium, the main tourist destinations, both in terms of international and domestic tourism (Heeley, 2011). Despite the multiplicity of users and functions existing in cities, it seems reasonable to establish that urban tourism essentially brings together leisure and cultural attraction factors. As such, cities with greater historical weight and quality of vestiges of the past have the potential to become true cultural tourism destinations. As Richards (2022) mentions, cities have a central role in the increase of cultural tourism, due to changes in the focus of cultural consumption, now more focused on everyday culture, intangible events and experiences.

Recognition of the heritage value of historic urban centres is a different process depending on the territories and entities

A. J. Pinheiro (✉) · H. Martins
Centre of Studies in Geography and Spatial Planning (CEGOT), University of Maia, Maia, Portugal
e-mail: ajpinheiro@umaia.pt

H. Martins
e-mail: hugomartins@umaia.pt

involved. It depends on the definition and valuation of concepts such as aesthetics, authenticity or inheritance. Despite this heterogeneity, some organizations seek to consolidate, standardize and globalize the valorization of heritage. An essential milestone in this action is the Convention concerning the Protection of the World Cultural and Natural Heritage, issued by the United Nations Educational, Scientific and Cultural Organization (UNESCO) in 1972. Its global character, in terms of membership by countries and public visibility, make the World Heritage List, resulting from the convention, a powerful tool for communicating cultural heritage. The list includes cities, monuments and cultural and natural landscapes, a significant part of which are high-performance tourist destinations. The relationship between the existence of the World Heritage (WH) classification and the tourist performance of a place is a matter of interest in several studies that seek to confirm a cause-and-effect relationship, which in theoretical terms seems logical. It is expected that the attribution of the classification will be an important contribution to the notoriety of a place that can thus develop as a tourist destination. However, given the variety of classified sites and the heterogeneity of the various countries where they are located, with very different tourist development stages, it is cautious to consider that the relationship between classification and tourism is not necessarily the same in all cases.

The city of Porto is a relevant case study, as its historic centre was inscribed on the UNESCO WH List in 1996 and is an important European urban tourist destination with more than 2 million tourists per year. It is, therefore, a consolidated tourist destination with critical mass, both in terms of supply and demand to provide a meaningful analysis. It is from this perspective of analysis, of the demand and supply components, that the study was envisaged, as the combined reading of the perceptions of tourists and tourism managers will eventually be the most complete way to achieve the purpose of understanding the importance of the WH classification of the historic centre for the positive tourist performance of the city. In operational terms of application of the investigation, tourist accommodation was chosen. In a tourist destination, accommodation has a very important weight in the measurement of the performance of the destination, in addition to the fact that, in the specific case of cities, it is an impacting agent of urban refunctionalization. In practical terms, it is also an opportunity to get in touch with guests, the overwhelming majority of whom are tourists. Indeed, the final purpose of the study will be ensured by achieving two essential aims: (i) understand if the WH classification is a determining factor for the decision to visit the historic centre; (ii) perceive whether tourist accommodations actively use classification in the strategy and operationalization of their communication.

2 Theoretical Framework

2.1 Urban Tourism

Cities are considered culturally-based tourist attractions, at least since the seventeenth century, with trips that have been defined in history as the Grand Tour, covering mainly the monumental cities of France and Italy (Davidson, 1998; Gunn, 1997). However, the modern vision of cities as tourist destinations is much later, starting to take shape in the second half of the twentieth century, where other forms of tourism, such as cultural and nature, appear as an alternative to mass tourism of sun and sea. It is especially in the last 30 years that it has become evident that tourism has become a central part of the management of cities as urban culture assumes itself as a consumer good (Henriques, 2003; Richards, 2022). The consumption of urban culture within the tourist activity is particularly intense in Europe due to the existence of an important number of cities with strong and varied cultural attractions, boosted by the relative proximity between them and facilitated by the mobility brought by low-cost airlines that connect destinations with outbound markets. These facilitating factors make it possible to understand the affirmation of the city break concept as a type of short-term trip that encourages high volumes of urban tourists. Cultural consumption in cities is carried out by both intentional and accidental users, and in both cases, they can be tourists or residents in the cities' area of influence (Ashworth & Tunbridge, 2000; Henriques, 2003). Especially for the intentional users of cities, the existence of cultural activities is crucial for the decision to visit the city. The affirmation of historic centres as leisure areas is the result of a refocusing of the importance of these areas, in the context of urban rehabilitation processes (Boavida-Portugal & Kastenholz, 2017; Salgueiro, 1999).

2.2 WH Classification and Urban Tourist Destinations

In a historical approach, the interest in the vestiges of the past can be in the nineteenth century. The conservation and preservation movements that sought to raise awareness and pressure public authorities to take actions to safeguard ruins and monuments date from this period (Ashworth & Tunbridge, 2000), as well as the first laws on heritage (Choay, 2001). Progressively, the value attributed to the monuments spread, at the same time that different trends were affirmed as to the way to preserve them. An essential milestone is found in the first international charter on heritage conservation: the Charter of Athens (ICOMOS, 2004). Another important milestone occurred in 1964 with the publication of the

Venice Charter, which expands the notion of monument, as it is now seen as an element surrounded by its own setting, so the areas surrounding monuments must also be preserved (ICOMOS, 2004). Therefore, the 1960s and 1970s were a period of widespread awareness of the importance of safeguarding heritage, as evidenced by the high number of official international initiatives to recognize heritage and the need for measures for its conservation (Ashworth & Tunbridge, 2000). Amongst these, the Convention concerning the Protection of the World Cultural and Natural Heritage, issued by UNESCO in 1972, stands out, which starts to attribute the distinction of outstanding universal value to certain heritage elements.

Prior to 1972, the purposes were fundamentally of protection of cultural property, particularly in wartime. With the convention, a paradigm shift begins, in the words of Di Giovine (2017), from politics to ethical action, instilling a sense of collective responsibility towards cultural diversity through governmentality. To this end, the World Heritage convention has 5 strategic objectives, known as the 5Cs: Credibility, Conservation, Capacity Building, Communication and Communities (UNESCO World Heritage Centre, 2023). The integrated fulfilment of the 5 objectives supersedes an initial purpose of safeguarding the heritage. In fact, the 1972 convention is part of a deeper perspective of intervention in society, as mentioned by Di Giovine (2017, p. 83), is "a fundamentally ethical framework aimed at slowly cultivating a new, and ostensibly more peaceful, world system by appealing to communities at a grassroots level to responsibly embrace and act on a particular conception of heritage".

From a tourism perspective, as Drost (1996) refers, the recognition of special qualities at the historical, scientific or aesthetic level that the classification of WH confers, allowed this places to constitute themselves as tourist attractions of high visibility. The coexistence between WH and internationally renowned tourist destinations mentioned by several other authors, such as Jones et al. (2017) or Boavida-Portugal and Kastenholz (2017), can be supported by a simple statistical analysis. Considering the 5 countries with the highest number of international tourists, 4 of them are the countries with the highest number of classified properties: Italy, China, France and Spain (Table 1).

Since the last inclusion in the WH List, in 2021, there have been 1154 sites, of which 28.9% are located in cities or are part of urban areas (Pinheiro & Paulino, 2022). Considering again Table 1, it can be seen that in the 3 European countries, about two thirds of classified properties are located in urban areas. This finding reinforces the relationship between cities, WH and tourism triangle. For cities with these characteristics, the WH can be an important differentiating element from other destinations and fundamental argument of tourist attraction.

The relation between tourism and WH is ambiguous, complex and cannot be generalized. The use of WH by the tourism phenomenon raises ethical questions, being what Vecco and Caust (2020) define as a paradox. The dissemination of heritage brought by the WH List encourages tourist demand and an inherent commoditization of heritage. This represents the raising of funds that can help conserve the heritage but, on the other hand, an excessively high frequency of sites represents risks of physical degradation. As a result, tourism, political conflicts and climate change are considered the main risk factors for the destruction of heritage sites (Vecco & Caust, 2020). It is at the turn of the current millennium that UNESCO recognizes the impacts of the interaction of tourism with heritage (Di Giovine, 2017). In summary, it is accepted that the recognition of the unique character brought by the WH nomination guarantees international visibility, and is therefore a marketing tool (Frey & Steiner, 2011; Vecco & Caust, 2020). The possible increase in the number of tourists will mean economic growth, especially in terms of income and jobs (du Cros & McKercher, 2015). Associated effects are raising funds for the physical conservation of the sites and raising awareness of the need for safeguarding and sustainability issues in general (Frey & Steiner, 2011; Gao & Su, 2019). From a perspective of negative impact, it is understood that a possible economic dependence on tourism may represent a progressive loss of intrinsic cultural values in order to adapt

Table 1 Location of WH sites in the countries with the most international tourists

Country	WTO ranking	WH sites ranking	Sites in urban areas	Total number of sites	% Sites in urban areas (%)
France	1st	4th	33	49	67.3
Spain	2nd	4th	34	49	69.4
United States	3rd	12th	7	24	29.2
China	4th	2nd	23	56	41.1
Italy	5th	1st	41	58	70.7
Total	–		138	236	58.5

Source Data from the World Tourism Organization (2020) and UNESCO World Heritage Centre (2021)

the destination to the preferences and needs of tourists (Vecco & Caust, 2020). In this way, the identity of the local community can be affected by the adulteration or loss of cultural values (Urry, 1990) visible in the denigration of social customs, the alienation of residents and the creation of place homogeneity (Arthur & Mensah, 2006).

Considering the focus of this investigation on urban tourism, it is intended to limit the analysis to WH properties located in urban areas. Most studies rely on an econometric approach to look for relationships and trends in the volume of tourist demand before and after classification. Several studies demonstrate a positive relationship between WH sites and growth in the number of tourists, as in a study covering 66 countries, made by Su and Lin (2014) or in a case study in Kaiping Diaolou and Villages site (China), where in the 8 years following the classification, there was an average annual increase of 12.8% in the number of tourists (Han et al., 2019). However, a number of other cases fail to establish this relationship. In a study for several regions of Italy, covering several classified historic centres, the authors claim that there is no statistical evidence that the classification accelerated the growth rates that Italian destinations were already registering (Ribaudo & Figini, 2017). In the case of several Chinese WH cities, Gao and Su (2019) state based on statistical analysis supported with robustness measures that the classification did not promote an increase in tourists and revenue. In the particular case of Macao, there was also no significant long-term effect, apart from some short-term impact (Huang et al., 2012). In Sintra (Portugal), the researchers found no significant changes in the search pattern in the next 9 years after classification (Soares et al., 2007). In an approach to a set of this type of studies, (Patuelli et al., 2013) interpret that most sites register annual increases in the number of tourists between 1 and 5%, but the relationship between the UNESCO distinction and these increases is difficult to prove, especially in mature tourist destinations, that is, which were already attracting tourists with notoriety before classification.

Another class of studies is based on the profile of demand and their attitudes and behaviour towards the WH inscription. There are several that demonstrate that most tourists know the WH classification before the trip, as in Tarragona (Spain), where, after dividing tourists into clusters, it appears that between 65 and 76% of tourists knew (Bové-Sans & Ramírez, 2013). But as for the level of influence on the decision to visit, there are different indicators. In Portugal, in the village of Sintra, a study (Soares et al., 2007) concluded that 65.8% of respondents were aware of the classification and around 50% acknowledged that this influenced the decision to visit. Also, in the city of Évora, another study reached a similar conclusion (Marujo et al., 2012), when stating that the classification influences the decision to visit

in clusters of tourists whose motivations are leisure and culture. In Asia, other studies demonstrate a lower expressiveness of the influence of WH on the decision to visit. In Lijiang (China), most international tourists knew about the classification (63%) but only 36.7% admit that this was one of the reasons for making the visit (Mahadevan & Zhang, 2022). In Melaka (Malaysia), in a study aimed at the Chinese market, 62% of respondents consider the status of WH as important, with 36% of respondents considering visiting classified sites as mandatory in their holiday options (Moy & Phongpanichanan, 2014). In Nazareth (Israel), it was shown that tourists are moderately aware of the existence and importance of classification, but that it has little effect on their visiting behaviour (Poria et al., 2011). At this level, mention is made of a generalized study covering 105 WH cities worldwide, (Koufodontis & Gaki, 2022) concluded that in larger cities, with a greater number and variety of tourist attractions, the importance of classification in tourist markets is less, but in smaller cities with less attractions, the UNESCO distinction can be important for the communication strategy in order to attract flows of tourists. It is also worth mentioning a study with participants from Israel (Poria et al., 2013) which suggests that the classification may have no impact on the decision to visit, or even have a negative impact for tourists who consider that the notoriety brought by UNESCO is too much, which will result in a mass destination and a less authentic tourist experience.

The different conclusions of these studies lead authors such as du Cros and McKercher (2015) to conclude that the positive, neutral or negative relationship of the classification with the number of tourists depends on the particularities of tourist destinations, their history, previous notoriety, types of issuing markets and their communication strategies. It is from this perspective, of the particularities of each case that a study in the city of Porto is envisaged, contributing yet another example to this theme. Previously, there was a study of tourist motivations in Porto, which concluded that 59.21% of respondents consider the classification as something important or very important, and the main reason for visiting is leisure, at a considerable distance from culture and heritage (Ramires et al., 2018). The present investigation seeks to deepen the tourist's perception, questioning their attitude towards classification and influence on the decision to travel, in addition to the perspective of the tourist offer.

3   The Historic Centre of Porto as a Tourist Destination

Porto is the second most populous city in Portugal. The city's historic centre is a repository of medieval base streets combined with others from the Renaissance period and buildings from later historical phases of the city's development and expansion. It has important elements of civil and religious architecture that represent different styles and stages of evolution of the history of the city and the country, with the main emphasis on the riverside area, where the Douro River is the focus of tourist attraction. The historical and cultural value of the city led to its historic centre being classified as a WH Site in 1996 (Fig. 1). Of the six existing criteria for the classification of cultural assets, criterion iv was recognized, which considers that the historic centre is an exceptional example of a type of building or architectural or technological ensemble, or landscape that illustrates one or more significant periods of human history (UNESCO, 2017).

It can be said that cultural tourism, in conceptual terms, is what best fits most of the experiences of visitors to the city of Porto. It is considered a broad definition that cultural tourism provides tourists with experiences that imply an expansion of their cultural knowledge (Calabuig & Ministral, 1999) and that the reasons for tourist attraction are based on artistic and monumental values that are expressions of culture and way of life of local communities (Almeida & Araújo, 2012). As Richards (2018) argues, cultural tourism today is a set of niches, such as heritage, gastronomy, creative, arts and film, which in the case of a city tend to be complementary and connected. In fact, this is what happens in Porto: (i) buildings representing various architectural

styles and monuments made by notable architects and engineers such as Nasoni and Eiffel, (ii) varied gastronomy, where the famous Port wine stands out, (iii) craft activities and workshops, (iv) the art gallery quarter and (v) the city's connection with the author of the literary work and the associated films of the Harry Potter Saga.

The performance of tourism in the city has registered a remarkable growth in the last decade. Considering some statistical data (Table 2), there is a joint growth in demand and supply of accommodation. Accommodation capacity, measured in beds, grew by 35.4% in the period 1999–2009, but this growth was 123.6% in the following 10 years. In terms of guests, there is an increase of 52.5% in 10 years (from 1999 to 2009), but in the most recent period from 2009 to 2019, the growth was 183%. This evolution also results in an improvement in occupancy rates, which increased from 36.2% in 1999 to 55.1% in 2019.

The reception of 2,245,291 guests in legal accommodation, in 2019, is the highest number recorded in the history of the city and which consolidates it as the second with the highest number of tourists in Portugal (Instituto Nacional de Estatístca I.P., 2020). The type of tourism is essentially leisure with cultural motivations, with the average stay being 2 nights. It is added that it is a destination with some seasonality, visible in the concentration of 31.7% of overnight stays in 3 months of the year (July till September).

Fig. 1 Plan of Porto historical city centre with the classified area and the protection area. *Source* UNESCO (2017)

Table 2 Porto accommodation statistics (1999–2019)

Statistics	1999	2009	2019
Accommodations	92	92	373
Capacity (beds)	7687	10,405	23,270
Guests	520,192	793,315	2,245,291
Average Stay	1.8	1.8	2.0
Occupancy Rate (%)	36.2	38.9	55.1
% of overnight stays (July–September)	–	32.5%	31.7%

Source Data from the Anuários Estatísticos da Região Norte (Instituto Nacional de Estatístca I.P., 2001, 2010, 2020)

4 Methodology

To obtain a better understanding of the phenomenon, a mixed approach was chosen, focused on the vision of the offer and also of the tourist demand in the historic centre of Porto. It is a descriptive investigation, as it is a type of investigation that seeks to describe phenomena and look for relationships between variables (Fortin, 2003). Three research instruments were developed: two to collect data on supply (guide for interviewing managers and analysis grid for documental research on the communication practices of accommodation on their websites) and one for collecting data on demand (survey by questionnaire to guests). The area under study is the historic centre, by the perimeter delimited when it was classified as WH. The research units considered were all tourist accommodations within that area with a front-desk service for guests, this being a criterion established by the fact that accommodations with reception are larger and this increases the volume of potential respondents. From the contact with all existing tourist accommodation—a total of 48—29 agreed to collaborate in this study, that is, 55.4% of the total. Based on this voluntary collaboration, it was possible to take 24 interviews with managers of tourist accommodation and 415 questionnaire surveys to guests who stayed in these same accommodations. Data were collected from May to August of 2017.

5 Findings

The results are organized by the two analysed poles: tourist offer and demand.

5.1 Supply Insights

Of the 24 interviews carried out with managers, there is a significant majority who still work for a short time in the accommodation sector (Table 3): 66.6% have been working for 5 years or less. It appears that all the interviewees, except for one, entered tourism years after the WH classification, so they are unable to compare the current situation to a pre-classification situation. As for the length of stay in the accommodations, where they were interviewed, it is even shorter, as 83.3% have worked in the respective place for 5 years or less. These situations are probably because a large number of accommodations are recent and it is a sector that attracts young labour.

In the interview, it was directly asked: How important is it for the accommodation to be in an area classified by UNESCO as a WH Site? The answers were analysed and categorized into two domains: positive aspects that the managers consider to be direct effects of the classification and aspects that secondary to the importance of classification compared to other aspects considered more influential to justify the tourist situation in their accommodations and the city (Table 4). Clearly, only 2 managers consider that the WH classification is important for the city's tourism performance. Most managers have a shared view that classification may have been important in the more distant past but that it is currently of little relevance. They claim that the various international events that the city has hosted (such as the Primavera Sound music festival or the Red Bull Air Race) and the prizes and mentions given to the city (such as the European Best Destination that Porto won 3 times) guarantee a greater notoriety than the WH. Based on these arguments, 14 managers consider that the location of accommodation in the historic centre is important in itself, 9 managers say that

Table 3 Managers' experience

Experience by years	Career time		Time at the accommodation	
	f	%	*f*	%
< 1 year	2	8.3	2	8.3
1–5 years	14	58.3	18	75.0
6–10 years	5	20.8	2	8.3
11–20 years	2	8.3	1	4.2
> 20 years	1	4.2	1	4.2

Source Interviews with managers

Table 4 Aspects of the WH classification

Category	Aspects	f
Positivity	Classification is decisive	1
	More visibility	1
Secondaryization	Location in the historic centre itself	14
	The city's appeal alone	9
	Positive situation of tourism in Portugal	1
	No importance	5

Source Interviews with managers

the city is appealing by nature and there were 5 who do not see any importance in the classification.

The analysis of the interviews with the managers can be complemented with the observation of the references to the WH on the official websites of each accommodation. This diagnosis supports the managers' perception of devaluation of the value of the WH classification (Table 5). It appears that 37.9% of the accommodations have no mention of WH. Almost half of the accommodations have references to the WH—48.3%—although only 34.5% do so on the homepage, although mostly in a discreet and barely visible way.

Table 5 Reference to WH on the website

References	f	%
None	11	37.9
In text on the home page	10	34.5
In image on home page	0	0.0
In text on inner page	4	13.8
WH Logo	0	0.0
No website	4	13.8
Total	29	100.0

Source Documentary research on the accommodations' websites

5.2 Demand Insights

The questionnaire survey of 415 guests comprised a series of questions aimed at obtaining a socio-demographic characterization (Table 6). There was a higher proportion of female tourists (57.1%), an average age of 43.6 years, with 46.5% of tourists aged between 31 and 50 years. The vast majority live outside Portugal (94.9%), with 74% of tourists having European nationalities, where France, Spain and the United Kingdom are the three main outbound markets. In terms of scholarity, 65.8% of respondents have a higher education level.

The aspects collected on the characterization of the trip (Table 7) allow placing the average stay in the city at 3 nights, with most guests (59.5%) staying between 2 and 3 nights. Most respondents (79%) visited the city for the first time. It can also be seen that most tourists travel with a partner (87.2%), with the trip as a couple being the most common (49.6% of the responses).

The main reason for travelling to the city of Porto (Table 8) was clearly the search for leisure (83.6% of respondents), followed by culture (49.6%), so that these two reasons overlap in a very significant number of tourists.

Regarding the classification of the historic centre as a WH site, tourists were asked if they already knew about this distinction before making the trip (Table 9). Two groups are relatively close, as 54.9% knew about the classification before the trip and 45.1% only knew at the destination itself.

Considering only tourists who already knew about the classification before making the trip, they were asked if *the classification of the historic centre as WH site influenced your decision to visit the city?* (Table 10). The results show that 59.6% consider that the classification had little or no influence on their decision to visit Porto. On the opposite

Table 6 Socio-demographic characterization of guests

Gender	f	%	Nationality	f	%	Residence	f	%
Female	237	57.1	Europe	307	74.0	Portugal	21	5.1
Male	178	42.9	France	60	14.5	Abroad	394	94.9
Total	415	100	Spain	47	11.3	Total	415	100
			U.K	40	9.6			
Age group	f	%	Germany	31	7.5	Scholarity	f	%
16–20	5	1.2	Portugal	25	6.0	Elementary school	6	1.4
21–30	83	20.0	America	78	18.8	Middle school	11	2.7
31–40	97	23.4	Brazil	28	6.7	High school	79	19.0
41–50	96	23.1	U.S.A.	24	5.8	College degree	273	65.8
51–60	82	19.8	Oceania	21	5.0	Other	23	5.5
> 60	52	12.5	Asia	9	2.2	Did not answer	23	5.5
Total	415	100	Total	415	100	Total	415	100

Source Questionnaire surveys carried out to guests

Table 7 Characterization of the trip

Nights of stay	f	%	Previous visits to the city	f	%	Travelling company*	f	% of cases
1	50	12.0	0	328	79.0	Couple	206	49.6
2	122	29.4	1	44	10.6	Family	108	26.0
3	125	30.1	2	22	5.3	Friends	58	14.0
4	67	16.1	3	10	2.4	Alone	53	12.8
5	26	6.3	> 3	11	2.7	With kids	17	4.1
> 5	25	6.0	Total	415	100	Total	442	n = 415
Total	415	100				*Multiple answer question		

Source Questionnaire surveys carried out to guests

Table 8 Reason for travel

Reason	Answers*		% of cases
	f	%	
Leisure	347	51.0	83.6
Culture	206	30.2	49.6
Food	67	9.8	16.1
Work	8	1.2	1.9
Sport	12	1.8	2.9
Specific event	14	2.1	3.4
Other reason	27	4.0	6.5
Total	681	100	n = 415

*Multiple answer question

Source Questionnaire surveys carried out to guests

Table 9 Knowledge of the WH Classification

Knowledge	f	%
Knew before trip	228	54.9
Knew in the destination	187	45.1
Total	415	100

Source Questionnaire surveys carried out to guests

Table 10 Influence of WH classification on travel decision

Influence	Answers		% of cases
	f	%	
None	71	31.1	17.1
Little	65	28.5	15.7
Rather	49	21.5	11.8
Much	27	11.8	6.5
Totally	16	7.0	3.9
Total	228	100	n = 415

Source Questionnaire surveys carried out to guests

pole, 18.8% consider that the classification had a great or even total influence on the decision to visit, with the remaining 21.5% recognizing an average influence.

6 Conclusion

The WH is a very visible and global facet of heritage processes. Its expansion and correspondence with many well-known tourist destinations makes it, in practical terms, an argument for attracting visitors and, in academic terms, an interesting growing object of study, both in terms of the number of case studies and different methodological approaches. The case study of Porto is another contribution to the theme, having as its most innovative aspect the approach to the perspectives of tourist demand and supply.

Regarding the specific aim of understanding whether the classification of WH is a decisive element in the decision to visit the historic centre of Porto, it verified the notoriety of the classification amongst most tourists: 54.9% of those surveyed in the study knew about the distinction before the trip. However, within this group, there is a majority that did not value the WH classification as important in the decision-making of the visit. To better understand this situation, the respondents were divided into two groups: those with a positive influence, that is, tourists who knew about the existence of the classification and who admitted that it influenced them rather, much or totally in the decision to visit Porto, and those with neutral influence, that is, tourists

Table 11 Positive and neutral influence of the WH classification

Influence	f	%
Neutral	323	77.8
Positive	92	22.2
Total	415	100

Source Questionnaire surveys carried out to guests

who answered that the classification had little or no influence on the decision to visit, plus those who were unaware that the historic centre of Porto was a WH site. With this division (Table 11), it is clear that the group of tourists in which the WH classification exerted a positive influence is clearly in the minority (22.2%).

To deepen this question, the two groups—with neutral and positive influence—were analysed in terms of their socio-demographic characteristics to identify possible trends (Table 12). In terms of gender or residence, there are no significant differences, but aspects can be pointed out in other variables. Regarding age, in the group of tourists with a neutral influence of the classification, there is a more homogeneous distribution, whilst in the tourists with a positive influence, the greater presence of older tourists is clear, as 69.6% are older than 40 years old. As for scholarity, the fact that WH is a cultural classification would lead to the theoretical expectation of finding a greater

representation of tourists with more qualifications in the group of positive influence. However, the opposite happens, that is, the proportion of individuals with college degree is even higher in the group of tourists with a neutral influence. Regarding the reason for travel, a similar expectation could be expected, in which the weight of tourists with cultural reasons for visiting would be greater in the group of tourists with a positive influence, but the opposite happens.

Regarding the specific aim of evaluating the use of classification as an argument for communication in the tourist offer, the interviews carried out with accommodation managers showed that the classification is secondary to other appeals to visit Porto, mainly due to the intrinsic nature of the historic centre regardless the classification. In operational terms, the analysis of tourist accommodation websites confirmed that less than half of the accommodations have a website with references or information about the WH, which reinforces the perception of the secondary role of the classification. To make this point even clearer, guests who did not know that the historic centre was a WH site were asked when they became aware of this (Table 13). It was found that only 4.3% obtained this information at the accommodation during their stay, and the majority (56.1%) only knew because of this study.

The analysed data lead to the general conclusion that Porto is a tourist destination with some maturity and international notoriety, especially in Europe. The classification of

Table 12 Crossing with the variables of socio-demographic characterization

Variables		Neutral		Positive	
		f	%	f	%
Gender	Female	185	57.3	52	56.5
	Male	138	42.7	40	43.5
Group age	16–20	5	1.6	0	0.0
	21–30	73	22.6	10	10.9
	31–40	79	24.5	18	19.6
	41–50	70	21.7	26	28.3
	51–60	61	18.9	21	22.8
	> 60	35	10.8	17	18.5
Residence	Portugal	20	6.2	5	5.4
	Abroad	303	93.8	87	94.6
Scholarity	Elementary school	6	2.0	0	0.0
	Middle school	11	3.6	0	0.0
	High school	52	17.2	22	24.7
	College degree	221	72.9	59	66.3
	Other	13	4.3	8	9.0
Reason for travel (multiple answer question)	Leisure	270	83.6	77	83.7
	Culture	165	51.1	41	44.6
	Food	56	17.3	11	12.0

Source Questionnaire surveys carried out to guests

Table 13 Knowledge of the WH classification in the destination

Knowledge in the destination	f	%
Found out at the accommodation	8	4.3
Found out in the city	74	39.6
Found out in the survey	105	56.1
Total	187	100

Source Questionnaire surveys carried out to guests

its historic centre is a recognized fact, both in terms of tourism supply and demand. However, it is seen as an accessory factor as a communication tool by the managers of tourist accommodation, whilst most tourists consider the WH classification irrelevant for their decision to visit. Basically, supply and demand converge in the perception that the historic centre has intrinsic factors of tourist attraction as well as well-known events and distinctions that dilute the weight of WH.

It is important to put into context that the development of tourism in Porto, reflected in the data shown in Table 2, takes place in a context in which Portugal has also experienced historical growth in terms of tourist flows, revenues and notoriety. Intrinsic and extrinsic factors contribute to this performance. Since the 1990s, Portugal began to enter the route of major international events, which contributed to the greater tourist visibility of the country: the European Capitals of Culture in Lisbon (1994) and Porto (2001), the Universal Exhibition of 1998 in Lisbon or the European Football Championship in 2004. The European growth of the low-cost aviation model and the city break segment also greatly benefited the cities of Porto and Lisbon. In extrinsic terms, Portuguese competitiveness also benefited from the turmoil in terms of security that affected competing destinations, such as the Arab Spring, in North Africa from 2010 onwards and the various terrorist attacks in European cities. The attractive variety of the territory, the quality of the service providers, the climate, the hospitality and the security of the country have kept Portugal's tourist notoriety at a high level, visible in successive distinctions in the tourism and travel sector. These aspects, which result in a good tourist image of Porto and of Portugal in general terms, are motivators, in themselves, for most tourist visits. It is, thus, clear that the WH List's power of attraction is secondary to a large part of the tourist demand. With regard to the tourist offer, it is natural that, given the quantitative success of the tourist performance of Porto and Portugal, the efforts by accommodation managers to attract tourists are minimal, as they do not experience difficulties in capturing tourist demand and therefore do not have to make efforts to appeal. It should also be noted, according to the research, that they are young managers in businesses that emerge during the

growth process of tourist demand, so they cannot see any added value in the use of the UNESCO seal.

In academic terms, it is a case study that is part of the group of studies that seek to understand the perceptions of tourist markets, proving a significant awareness of the WH. But, it also shows a limited ability to influence both tourists in the decision to visit the city and managers in terms of the destination image projected through tourist accommodation. In practical terms, the conclusions of the study can serve to recall the potential of communicating the WH distinction in a more evident and qualified way, as there is still a significant part of the demand that is not aware of it, and the benefits of its dissemination are not disrespectful. Greater promotion of Porto based on the WH List is not related to increasing tourist flows to the city, but rather contributing to its qualification and awareness. According to the purposes of the WH convention, it is essential to promote the sustainable use of heritage, favouring social and cultural sustainability over economic sustainability, so the inclusion of Porto in the WH List should deserve greater reflection and use in the promotion of the tourist destination based on these principles.

References

Almeida, P., & Araújo, S. (2012). *Introdução à Gestão de Animação Turística*. Lidel.

Arthur, S. N. A., & Mensah, J. V. (2006). Urban management and heritage tourism for sustainable development. *Management of Environmental Quality: An International Journal, 17*(3), 299–312. https://doi.org/10.1108/14777830610658719

Ashworth, G. J., & Tunbridge, J. E. (2000). *The tourist-historic city*. Taylor & Francis.

Boavida-Portugal, L., & Kastenholz, E. (2017). Paradigmas Territoriais dos Destinos Turísticos em Portugal: O Caso das Áreas Costeiras e Áreas Urbanas Históricas. In F. Silva & J. Umbelino (Eds.), *Planeamento e Desenvolvimento Turístico* (pp. 393–408). Lidel.

Bové-Sans, M. À., & Ramírez, R. L. (2013). Destination image analysis for Tarragona cultural heritage. *Review of Economic Analysis, 5*(1), 103–126. Retrieved from https://ojs.uwaterloo.ca/index.php/rofea/article/view/1404

Calabuig, J., & Ministral, M. (1999). *Manual de Geografía Turística de España (2.ª edición revisada)*. Editorial Sintesis.

Choay, F. (2001). *A alegoria do património*. Estação Liberdade—Editora UNESP.

Davidson, R. (1998). *Travel and tourism in Europe* (2nd ed.). Longman.

Di Giovine, M. (2017). UNESCO's world heritage program: The challenges and ethics of community participation. *Between Imagined Communities of Practice, 2015*, 83–108. https://doi.org/10.4000/books.gup.213

Drost, A. (1996). Developing sustainable tourism for world heritage sites. *Annals of Tourism Research, 23*(2), 479–492.

du Cros, H., & McKercher, B. (2015). *Cultural tourism* (2nd ed.). Routledge.

Fortin, M.-F. (2003). *O Processo de Investigação: da concepção à realização* (3.ª). Lusociência.

Frey, B. S., & Steiner, L. (2011). World heritage list: Does it make sense? *International Journal of Cultural Policy, 17*(5), 555–573. https://doi.org/10.1080/10286632.2010.541906

Gao, Y., & Su, W. (2019). Is the world heritage just a title for tourism? *Annals of Tourism Research, 78*(May), 102748. https://doi.org/10.1016/j.annals.2019.102748

Gunn, C. (1997). *Vactionscape: Developing tourism areas* (3rd edn.). Routledge.

Han, W., Cai, J., Wei, Y., Zhang, Y., & Han, Y. (2019). Impacts of the world heritage list inscription: A case study of kaiping diaolou and villages in China. *International Journal of Strategic Property Management, 24*(1), 51–69. https://doi.org/10.3846/ijspm.2019.10854

Heeley, J. (2011). *Inside city tourism—A European perspective*. Channel View Publications.

Henriques, C. (2003). *Turismo, Cidade e Cultura—Planeamento e Gestão Sustentável*. Edições Sílabo.

Huang, C. H., Tsaur, J. R., & Yang, C. H. (2012). Does world heritage list really induce more tourists? Evidence from Macau. *Tourism Management, 33*(6), 1450–1457. https://doi.org/10.1016/j.tourman.2012.01.014

ICOMOS. (2004). *International charters for conservation and restoration* (2nd ed.). ICOMOS.

Instituto Nacional de Estatístca I. P. (2001). *Anuário Estatístico da Região Norte 2000*. Instituto Nacional de Estatística I.P.

Instituto Nacional de Estatístca I. P. (2010). *Anuário Estatístico da Região Norte 2009*. Instituto Nacional de Estatística I.P.

Instituto Nacional de Estatístca I. P. (2020). *Anuário Estatístico Regional 2020*. Retrieved from https://www.ine.pt/xportal/xmain?xpid=INE&xpgid=ine_doc_municip_2020

Jones, T. E., Yang, Y., & Yamamoto, K. (2017). Assessing the recreational value of world heritage site inscription: A longitudinal travel cost analysis of Mount Fuji climbers. *Tourism Management, 60*, 67–78. https://doi.org/10.1016/j.tourman.2016.11.009

Koufodontis, N. I., & Gaki, E. (2022). UNESCO urban world heritage sites: Tourists' awareness in the era of social media. *Cities, 127* (May), 103744. https://doi.org/10.1016/j.cities.2022.103744

Mahadevan, R., & Zhang, J. (2022). Tourism in UNESCO world heritage site: Divergent visitor views to Lijiang on experiences, satisfaction and future intentions. *Journal of China Tourism Research, 18*(3), 670–688. https://doi.org/10.1080/19388160.2021.1965061

Marujo, N., Serra, J. M., & Do Rosário Borges, M. (2012). Visitors to the city of Évora: Who are they? *European Journal of Tourism, Hospitality and Recreation, 3*(2), 91–108. Retrieved from www.ejthr.com

Moy, L. Y. Y., & Phongpanichanan, C. (2014). Does the status of a UNESCO world heritage city make a destination more attractive to mainland Chinese tourists? A preliminary study of Melaka. *Procedia—Social and Behavioral Sciences, 144*, 280–289. https://doi.org/10.1016/j.sbspro.2014.07.297

Patuelli, R., Mussoni, M., & Candela, G. (2013). The effects of world heritage sites on domestic tourism: A spatial interaction model for Italy. *Journal of Geographical Systems, 15*(3), 369–402. https://doi.org/10.1007/s10109-013-0184-5

Pinheiro, A. J., & Paulino, F. (2022). Urban tourism and world heritage: Relations and effects of the classification. *PASOS Revista De Turismo Y Patrimonio Cultural, 20*(5), 1243–1254. https://doi.org/10.25145/j.pasos.2022.20.084

Poria, Y., Reichel, A., & Cohen, R. (2011). World heritage site-is it an effective brand name? A case study of a religious heritage site. *Journal of Travel Research, 50*(5), 482–495. https://doi.org/10.1177/0047287510379158

Poria, Y., Reichel, A., & Cohen, R. (2013). Tourists perceptions of world heritage site and its designation. *Tourism Management, 35*, 272–274. https://doi.org/10.1016/j.tourman.2012.02.011

Ramires, A., Brandão, F., & Sousa, A. C. (2018). Motivation-based cluster analysis of international tourists visiting a world heritage city: The case of Porto, Portugal. *Journal of Destination Marketing and Management, 8*, 49–60. https://doi.org/10.1016/j.jdmm.2016.12.001

Ribaudo, G., & Figini, P. (2017). The puzzle of tourism demand at destinations hosting UNESCO world heritage sites. *Journal of Travel Research, 56*(4), 521–542. https://doi.org/10.1177/0047287516643413

Richards, G. (2018). Cultural tourism: A review of recent research and trends. *Journal of Hospitality and Tourism Management, 36*, 12–21. https://doi.org/10.1016/j.jhtm.2018.03.005

Richards, G. (2022). Urban tourism as a special type of cultural tourism. In J. van der Borg (Ed.), *A research agenda for urban tourism* (pp. 31–50). Edward Elgar. https://doi.org/10.4337/9781789907407.00009

Salgueiro, T. B. (1999). *A cidade em Portugal* (3.ª). Edições Afrontamento.

Soares, J. O., Neves, J. O., & Fernandes, F. (2007). The impact of world heritage classification on the development of tourist destinations: The Sintra case study. In *Advances in tourism economics*. Vila Nova de Santo André.

Su, Y. W., & Lin, H. L. (2014). Analysis of international tourist arrivals worldwide: The role of world heritage sites. *Tourism Management, 40*, 46–58. https://doi.org/10.1016/j.tourman.2013.04.005

UNESCO. (2017). Historic centre of Oporto, Luiz I Bridge and Monastery of Serra do Pilar.

UNESCO World Heritage Centre. (2021). World Heritage List. Retrieved September 30, 2022, from https://whc.unesco.org/en/list/

UNESCO World Heritage Centre. (2023). World heritage policy compendium. Retrieved February 15, 2023, from https://whc.unesco.org/en/compendium

Urry, J. (1990). *The tourist gaze*. Sage.

Vecco, M., & Caust, J. (2020). UNESCO, cultural heritage sites and tourism: A paradoxical relationship. In H. Pechlaner, E. Innerhofer & G. Erschbamer (Eds.), *Overtourism: Tourism management and solutions*. Routledge.

World Tourism Organization. (2020). World tourism barometer. *UNWTO World Tourism Barometer, 18*(7).

The Role of Community-Led Initiatives in the Circularity-Based Heritage Revitalization

Yasmine Tira and Handan Türkoğlu

Abstract

Community-led revitalization is a catalyst for promoting and valuing heritage centers. It enhances their conservation while empowering the community to take ownership of their cultural heritage and recognize its potential to drive local development. Several historic cities that had previously been significant for their heritage value for local identities, became under threat of decay or have become derelict. Some historic cities are suffering due to the legacy of centralized planning systems, limited capacity, and resources at the local level, which constrains them to deal with the growing responsibilities of decentralization and economic transition. In many countries going through economic and political transitional periods, innovative ways of addressing the challenges faced by heritage revitalization became needed for activating innovative powers in communities, i.e. circularity-based initiatives. One such example of inner old city witnessing a transitional political period and a gradual move from the linear economy to the circular economy in its governance strategy is the Medina of Tunis UNESCO World Heritage City. Several circular approach-based civil society initiatives are reconnecting the community with their tangible and intangible heritage based on local revitalization programs. This study presents key points on the community-led revitalization policy under the arch of the circular city concept. It provides food for thought on both opportunities and issues community-led heritage revitalization initiatives can present when meeting new trends in the economy.

Y. Tira (✉) · H. Türkoğlu
Department of Urban and Regional Planning,
Faculty of Architecture, Istanbul Technical University,
Taşkışla Campus, 34367 Şişli, Istanbul, Turkey
e-mail: tira18@itu.edu.tr

H. Türkoğlu
e-mail: turkoglu@itu.edu.tr

As a methodology, this study uses empirical research followed by a case study exploration. It opens up an examination of the existing relationship between circularity-based heritage revitalization and community-led initiatives. It concludes that the Medina of Tunis community-led revitalization initiatives are a witness to the sharing economy and circularity.

Keywords

Cultural governance • Circular approach • Community-led revitalization • Heritage revitalization

1 Introduction

The advent of globalization resulted in tensions in modern urban conservation. A dichotomy appeared between conservation principles based on the Western experience, and the processes of change accelerated by social, economic, and political changes. The increase of heritage-based urban transformations and tourism uses in many historic cities have caused threats to the idealized image of heritage (Bandarin & Van Oers, 2012). Consequently, the possibility to re-integrate urban conservation principles and practices into urban development through the Historic Landscape Approach (HUL), has been initiated. The UHL has indeed the objective to define effective principles able to redefine urban heritage as the center of the spatial development process. This is what Bandarin and Van Oers describe as the fact "*to recognize and position the historic city as a resource for the future*" (Bandarin & Van Oers, 2012, p. 19). It follows from the aforementioned assumptions that, heritage is a notion not limited to what is inherited from the past, it is an entry point for reaching the sustainability goals in shaping the future (Pappalardo, 2020). Cultural heritage is recognized as an influential indicator for urban sustainable development (Nocca, 2017, p. 3; Nocca et al., 2021, p. 107).

© The Author(s), under exclusive license to Springer Nature Switzerland AG 2024
J. Chica-Olmo et al. (eds.), *Sustainable Tourism, Culture and Heritage Promotion*,
Advances in Science, Technology & Innovation, https://doi.org/10.1007/978-3-031-49536-6_17

In the paradigm shift to more sustainable urban development policies, the circular economy can play a pivotal role (Nocca, 2017, p. 107). Each city has its unique characteristics and its own social and economic structure. Thus, to implement the circularity approach to its different assets, each city needs to assess the appropriate starting point. In accordance with the SDGs and the EU's 2050 target of 0% emission, the EU launched the Green Deal strategy based on circular economy principles. Covering all sectors of the economy including transport, energy, buildings, agriculture, and industries, the initiative showed that the circularity concept can go beyond the economic sphere and that cities also can adopt it to attend a long-lasting sustainable future (EC, December 11, 2019).

According to the European Green Deal, transiting to a holistic integrative circular approach in cities is among the European key priorities. Putting culture at the heart of the Green Deal means adopting cultural heritage as an entry point to reach Europe's aim of 2050 (Culture Action Europe, European Cultural Foundation & Europa Nostra, 2020).

Historic inner cities regeneration and renewal approaches are seen as an effective entry point for the implementation of a circularity-based sustainable future (Nocca et al., 2021, p. 111). This means the relation between cultural and economic values of heritage is not only between *"intrinsic"* cultural considerations and *"extrinsic"* economic considerations, but it is rather a multidimensional sphere of friction and potential interaction (Stoffelen, 2022, p. 18). According to the New Urban Agenda (NUA), culture plays a crucial role in community empowerment and inclusion in development initiatives. In this regard, Pappalardo (2020) talks about *"a humanization of cities and human settlements"* by culture. Culture contributes to the rehabilitation, and revitalization of urban areas, and emphasizes public participation (NUA, point 38). Urban heritage, in particular, plays a crucial role in strengthening local identities, empowering communities, and boosting economic revitalization. Based on the circular economy implementation, current trends in heritage revitalization focus on the share economy, making it possible to use short-term investments for many services, including communities of place, communities of interests, and communities of practice in the process, instead of saving massively to make major investments for services.

The present study discusses innovative community-led heritage revitalization initiatives, questioning how they highlight the transition to the circular economy and the circular city principles. In this perspective, the Medina of Tunis UNESCO World Heritage City has been taken as an example.

2   Research Aim

The present study aims to reveal the interaction between heritage revitalization and community-based initiatives in the circular city debate. It aims to put forward the changing role of cultural governance in the globalized economic transition led by the sustainability debate. It also draws upon the opportunities and issues community-led heritage revitalization initiatives can present when meeting new trends in the economy. The present research tries to put forward the link between circularity-based heritage revitalization and community-led initiatives through one case study exploration. It tries to narrow the aforementioned relationship down to a case that illustrates a group of initiatives that put forward distinctive characteristics of the chosen site—the Medina of Tunis historic city. In line with this purpose and in the light of the existing literature review, an interview has been conducted with one investor in the Medina of Tunis who is involved in several community-led heritage initiatives.

3   Theoretical Background

In light of the prerequisites defined to respond to the sustainable development goals drawn for 2050 (EC, 11 December 2019), the present research illustrates the active role of cultural heritage governance and puts forward the shift it presented from the "governing for culture", to "the governing through culture". Governing cultural heritage assets entered the debate around more sustainable futures and the shift from the linear economy to the circular economy in the path toward sustainability achieving. Revitalization of historic inner cities, in particular, has a role to play in the circularity-based sustainable futures. In doing so, the community-led heritage revitalization presents challenges and opportunities that will be briefly explained in this research.

3.1   Governing Cultural Heritage Assets

Schmitt (2009) believes that the acceleration of globalization and modernization phenomena in the past decades made the local assets at risk of being replaced by *"globally standardized architecture and consumer cultures"* (Augé, 1992). This resulted in an increased interest in heritage. What stimulated this increase is also the adoption in 2003 of the UNESCO Convention for the safeguarding of the Intangible Cultural Heritage (a sister convention of the World Heritage Convention of 1972). For the general public, the UNESCO World Heritage list is a reference for *"what is worth preserving for future generations*" (Schmitt, 2009, p. 2). Cultural governance is not a new concept. Yet, the current effects and content of culture have changed. The stress on cultural governance has progressively changed from stimulating nationalism, and artistic appreciation, to market-oriented commercial revenue gathering, characterized by economically competitive creativity and aesthetics. According to Wang (2013), cultural governance is a notion associating the global, national, and local tendencies of political and economic developments with culture: "*Cultural governance means governing through culture [...] and the regulation of the economy through culture*" (Wang, 2013, p. 11). However, as it has been reclaimed by Barile and Saviano (2015), the governance of heritage assets should opt for equilibrated approaches for action able to preserve the structure and maintain the systems' viability of cultural heritage. This calls for the need for a holistic comprehensive vision including the different perspectives, and where cultural heritage governance becomes based on conservation, protection, and enhancement (Barile and Saviano, 2015, p. 85).

Several European contexts illustrated the shift from governing heritage to the holistic governance of heritage assets through the circular city debate. Pioneers of adopting the circularity approach in heritage-led projects are the Regeneration and Optimization of Cultural Heritage in Creative and Knowledge cities (ROCK) and the Circular models Leveraging Investments in Cultural heritage adaptive reuse (CLIC). Therefore, they assume that cultural heritage is used as a tool to experiment with collaborative models of urban development (Garzillo et al., 2018; ROCK, 2019). Both of these examples applied the circular economy principles to cultural heritage revitalization achieving environmental, social, cultural, and economic sustainable urban development. The circular paradigm doesn't concern economic growth only but also promotes human development (ICLEI, 2020). Following a heritage revitalization based on the circularity approach, historic sites became laboratories to prove how cultural heritage can be a powerful engine of regeneration and revitalization, sustainable development, and economic growth for the whole city (ROCK, 2019). Among the main principles and lessons learned from these two projects is that sustainability and future plans of the model cities made local communities, and stakeholders take ownership and have a better understanding of their cultural heritage.

3.2 Revitalization of Heritage Sites: A Long-Lasting Sustainability Catalyst

Heritage development is an economic revitalization approach that augments the local life quality through preservation, conservation, and community partnership. Heritage revitalization consists of reorganizing existing city structures following improvement concepts of the urban environment in line with the economic and social changes (Lam et al., 2022, p. 4). Historic areas revitalization initiatives employ or associate one or more approaches, i.e. (1) the adaptive reuse and cost recovery based on stimulating private investments and raising funds through commercially viable uses that respect the historic value; (2) the integrated area development based on addressing an integrated revitalization of the entire historic city center; (3) the full commercialization of historic city centers allowing plots to be sold to commercial entities for rehabilitation; (4) the transfer of development rights offering owners alternative plots of land suitable for development; (5) the modernization of commercial activities and the prevention of traditional economic activities from being lost; (6) feedback between increasing land use values and public revenues permitting to afford more funds from the increased value of revitalized urban fabric; (7) levying a heritage tax on touristic activities and assigning it to further revitalization efforts; and (8) the conservation of historic monuments through charging entrance fees for tourists and exempting locals (Steinberg, 2011, pp. 13–16).

Based on the aforementioned approaches, revitalization processes of the historic urban fabric involve the renewal of the physical fabric, economic enhancement, and community empowerment and inclusion. This makes heritage revitalization a process that can be achieved following three phases, i.e. physical regeneration, economic renewal, and social rebirth (Danisworo, 1996 cited in Jauhar et al., 2021). Therefore, social rebirth means creating opportunities for social inclusion for the whole community. In this regard, heritage revitalization can be considered an engine for responding to sustainable development prerequisites. As reclaimed by Musyawaroh et al. (2018) "*revitalization covers the management of the physical aspects as well as the social and economic impacts of the region concerned*". However, it faces two conflicting encounters; on the one hand, the historical potential should be maintained, and on the other hand, technological requirements of modern-day urban life should be considered. Thus, the solution for this is "*sustainable revitalization*" embedding a non-linear governance approach and calling for community inclusion (Musyawaroh et al., 2018, pp. 2–3).

It follows from these assumptions that heritage revitalization can be achieved only through a sustainable strategy involving all stakeholders under the arch of a holistic approach. The problem at stake in heritage revitalization is that not all initiatives can consider the potential of community inclusion and not every revitalization attempt gives the right to heritage to everyone. So, what can make community-driven heritage revitalization work?

3.3 What Makes Community-Driven Heritage Revitalization Work?

Four fundamental operational factors for making differences in the heritage revitalization outcomes have been defined by Han (2003), i.e. (1) small-scale activities by "place entrepreneurs", or "individuals with active financial, social, and ethnic commitments to place"; (2) "the creation and promotion of heritage narratives by heritage institutions (such as museums)"; (3) "the public institutions and policies through which heritage is developed"; and (4) the symbolic memory and place evocation through "ephemeral urbanism", comprising daily patterns of street life and "invented traditions" like festivals (Han, 2003).

According to Steinberg (2011), besides refurbishment or redevelopment, one main strategy for heritage revitalization is the "neighborhood approach" of Haberer et al. (1980), which deals with urban renewal as a social, financial, and organizational matter. He advances community participation as a pre-condition to the success of any revitalization program. In relocation projects, for example, the participation of residents should go hand in hand with the project design phase rather than during the implementation (Haberer et al. 1980 cited in Steinberg, 2011, p. 12).

> Regardless of the policy stance in force, once the decision to revitalize a historic urban area has been taken, soliciting the participation of current residents is of utmost importance (Steinberg, 2011, p. 12).

Sustainable tourism principles are also seen as a pre-condition for better heritage revitalization practices. If based on cultural tourism development, principles of heritage revitalization can increase the attractiveness of the heritage in question, while preserving its physical, economic, and socio-cultural sustainability (Jelinčić and Mansfeld, 2019).

It is also believed that ephemeral urbanism has a role to play in the success of community-led urban planning initiatives. It contributes to accepting "non-permanence and adaptability" as a planning and place-making feature. Andres et al. (2019) reclaim that ephemeral urbanism, or what they also call "temporary urbanism" is a form of place-making based on a more holistic understanding of the valorization of space. It is the form of space valuing that transcends price or monetarization. It is a concept mostly related to social and cultural constructs (Andres et al., 2019, p. 36).

Among ephemeral urban planning principles, the example of fairs, festivals, and urban art events for heritage revitalization purposes can be given. Known as "*the creative economy-based urban development*", ephemeral urbanism represents a concentration of creative professionals to make lively, colorful, and thrilling cultural spaces able to regenerate neighborhoods culturally and economically (Kumar, 2020, p. 400). Ephemeral creative activities are capable to bring life back to the sites (Ferreri, 2015, p. 182).

3.4 Issues of Community-Led Heritage Revitalization

There are several community-based bad practices worldwide threatening the values of cultural heritage and driving some of the world's unique heritage assets to risk.

An oscillation between concentrating on economic gain and a will to respond to the conservation prerequisites generally emerges while trying to adopt a community-led heritage revitalization. This bipolar trait of cultural heritage governance creates tensions between revitalization and conservation. The regrouping of conservation, protection, and enhancement in a harmonious way is apparent on account of intangible cultural heritage. Enhancement itself creates protection needs. This leads to an existing virtuous circle between conservation, protection, and enhancement. A circle acting in a systemic way putting forward the cultural value. However, protection does not herald enhancement, since the recognition of value behind the expressed protection need represents the first step in enhancement (Barile and Saviano, 2015, p. 80). Thus, the choice between protection and enhancement in cultural heritage conservation is proved to be a false dilemma.

Besides the conflict between revitalization and conservation constraining heritage-led revitalization attempts, the loss of ownership awareness and local community connection to cultural heritage should also be considered. Loss of awareness stimulates the vandalism of cultural heritage assets (Jelinčić and Glivetić, 2021, p. 17). Graffiti paintings are, for example, one of the most severe threats to cultural heritage revitalization.

Kumar (2020) talks about "*heritage-based*" and "*creative economy-based*" urban redevelopment practices and policies as a catalyzer for making the indentities of historic sites a container for upscale "*branding*" (Kumar, 2020). The problem with upscale city branding through heritage-based urban development is that low-income groups in the local community run the risk of being estranged by heritage representations that underscore attracting global tourists (Nash 1989; Urry, 1990 cited in Kumar 2020).

It is true that city branding through heritage revitalization is a way to attract tourists and a way to afford services for novel mixed-use developments. However, it, at the same time, amplifies the fact of making "*downtowns like stages*". This is what Grodach calls "*staging the city for consumption*" (Grodach, 2017, p. 68). Aestheticizing downtowns with the purpose to attract tourists drive heritage sites into the trap of becoming "*cultural ghettos*" (Mouhli, 2016).

4 Materials and Methods

This study is a qualitative research based on the authors' observation of the community-led revitalization initiatives led at the Medina of Tunis. As previously mentioned in the literature review, in a European context, there have been examples where the circular economy principles are applied to cultural heritage revitalization. In developing countries, and regardless of the aforementioned European context, the Medina of Tunis is chosen as a case study based on its being a UNESCO world heritage city that is slowly living the shift toward the circular economy application in its governance strategy.

In order to support the observation and investigations, and to answer the research questions and attend the objectives defined, Expert Interviews methodology has been adopted as a complementary research strategy. As it is the case in the present study, the non-formalized experts' interview method, suitable for cultural studies where field experience is noteworthy (Libakova & Sertakova, 2015, p. 117), has been adopted.

Discussions with local community members, government representatives, and other stakeholders involved in the Medina of Tunis conservation strategy were also considered.

4.1 Case Study Exploration: Description of the Medina of Tunis Historic City Center and Its Significance

The Medina of Tunis is a Mediterranean historic city lying at the very heart of Tunisia's capital. Similar to other Mediterranean historic cities, the Medina of Tunis results from the coexistence of different ethnic and religious groups reflecting the different heritage of architectural styles, and buildings techniques (Micara et al., 2006, p. 11).

The Medina of Tunis is a UNESCO-protected heritage site since 1979. The central Medina (Area A in Fig. 1) went through different changes spanning a history of several civilizations, i.e. Roman, Byzantine, *Aghlabids*, *Almohads*, *Ottomans,* and lastly *Husseynids.* From its very beginning until the sixteenth century, the Medina of Tunis had a structure of an Arabic city. Its history dates back to the seventh century and it refers to the establishment of *Al-Zaytuna* Mosque in 695. As reported by Santelli, the Medina of Tunis was built over the derelict Roman city morphology (Santelli, 1995).

From the beginning of the seventh century until the beginning of the ninth century, the Medina of Tunis passed through noticeable evolutions, i.e. city walls reconstruction,

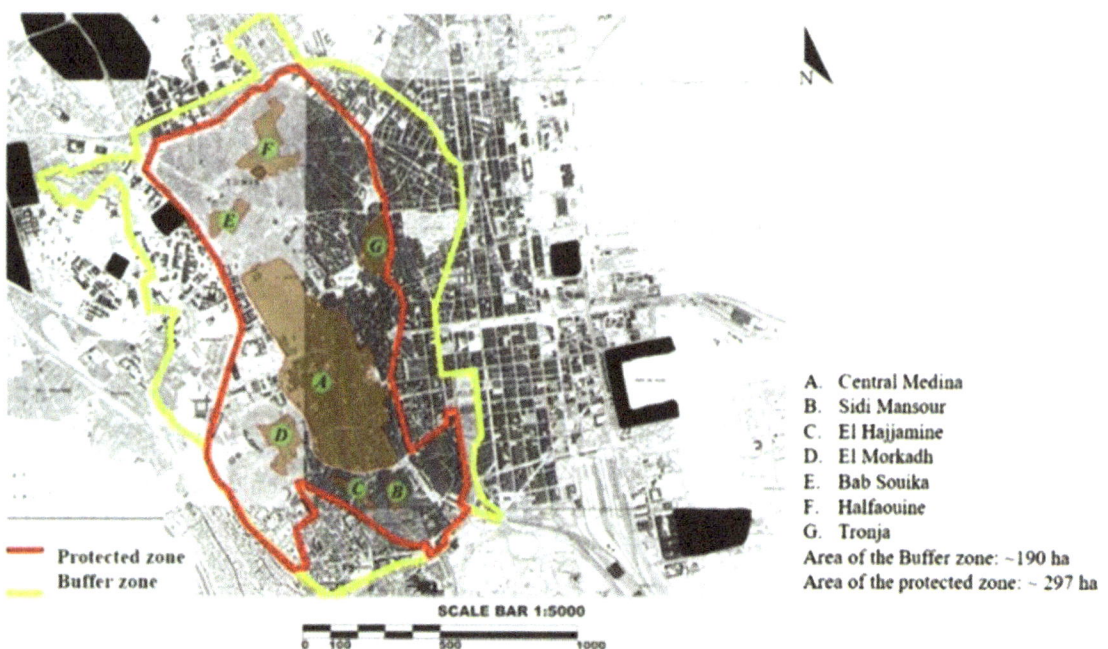

A. Central Medina
B. Sidi Mansour
C. El Hajjamine
D. El Morkadh
E. Bab Souika
F. Halfaouine
G. Tronja
Area of the Buffer zone: ~190 ha
Area of the protected zone: ~ 297 ha

Protected zone
Buffer zone

SCALE BAR 1:5000

Fig. 1 Map showing the boundaries of the Medina of Tunis Protected and buffer zone, WWW.WHC.UNESCO.ORG)

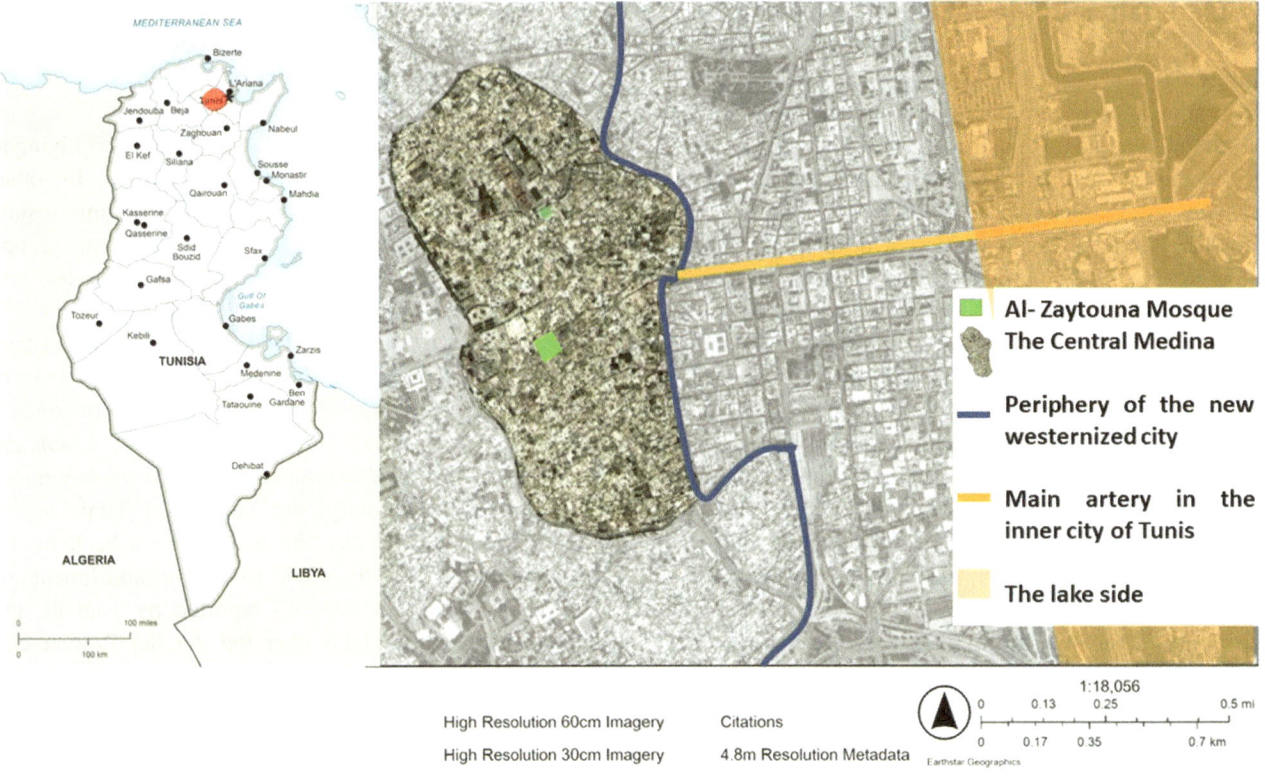

Fig. 2 Urban context of the Medina of Tunis, Authors elaboration after Google Earth

and the redefinition of the center and principle Souks around *Al-Zaytuna* mosque by the *Aghlabids* (Santelli, 1992).

At the beginning of the twelfth century, under *Almohads* regency, the Medina's urban structure again lived remarkable changes. After 1228, the *Hafsids* regency started and lasted three and a half centuries. Their regency is considered a cornerstone in the development of the urban fabric, and the economic and social sectors in the Medina of Tunis (Daoulatli, 2009).

Following these changes, the most noticeable modifications that the Medina of Tunis saw have been made mainly by *Ottomans* and *Husseynids* from the beginning of the sixteenth century until the end of the nineteenth century. By September 1574, *Ottomans* took control of Tunis's coastal side. Until the end of the sixteenth century, the Medina experienced conflicts between the Spanish and *Ottomans* over the seizure of the region. It was ultimately taken over by the *Ottomans* who launched the reparations of city walls and renovated the main central governance building; the Diwan (Saadaoui, 2010).

From its early beginning until the middle of the nineteenth century, the Medina of Tunis saw political, social,

economic, and urban history marked by changes. The Medina of Tunis's overall development got influenced mostly by the French Protectorship of Tunisia (Fig. 2). After the 1850s, the urban form of the Medina of Tunis started to change (Ammar, 2010, p. 13). A new westernized city occupied by Europeans has been developed neighboring the Central Medina (Mouhli, 2016). The French city initiated a French lifestyle and heightened the Tunisians' attachment to the central Medina which became a symbol of resistance against the colonizers (Tira, 2018, p. 22).

The Medina of Tunis was a gated community having defined city limits and gates. Connection axes between these gates are following cardinal directions. They intersect at the central point of the central Medina; *Al-Zaytuna* mosque. The Central Medina lies at the very heart of the capital Tunis, and it occupies an area of around 2.97 km^2 (Tira, 2018, p. 81).

Being the old governance and trade center, the Medina of Tunis still embraces some of these functions within its urban scenery. Its northern side is boarded by the sea roads leading to *La Goulette*, *Carthage* archaeological site neighboring the presidency area. From the extreme north, the Medina neighbors Carthage International Airport (Fig. 3).

Fig. 3 Transportation map around the Medina of Tunis, Authors elaboration after OpenStreetMap

4.2 Coupling Heritage Conservation and Economic Development in the Medina of Tunis

The Medina of Tunis has a management structure specified by the National Institute of Heritage INP, and the ASM, Association for the Safeguarding of the Medina, which is under the Municipal Authority of Tunis. Heritage conservation strategy in Tunisia has always been related to the country's political history and land ownership practices (Halleb, 2018, p. 10). However, for more than two centuries, Tunis reinforced its role as a capital city, against a background of major urban socio-cultural, and economic changes (Akrout-Yaïche, 2002).

As a way to fit the urban conservation agenda with modernization ideals of Bourguiba,[1] in 1974, the *Rapport Sauvegarde et mise en valeur de la Médina de Tunis* suggested to link cultural heritage conservation and economic development through tourism (Nardella & Cidre, 2016, p. 9). Thereby, as reported by Dhaher et al. (2020), Tunisia was one of the first Muslim countries that attempted to explore the potential of tourism where tourism and heritage become entwined (Dhaher et al. 2020).

In the late 1990s, the World Bank, a lender in historic cities' revitalization, produced the "*Cultural heritage and development, a framework for action in the Middle East and North Africa*" as a strategic paper to position itself in the culture and sustainable development arena and boost its credentials in linking cultural heritage conservation and economic development (Nardella and Mallinson, 2014 cited in Nardella & Cidre, 2016). The document generated a new

line of projects, such as the Tunisia Cultural Heritage Project (TCHP) (Nardella and Cidre, 2016, p. 10).

In partnership with the Association for the Safeguarding and Conservation of the Medina of Tunis (ASM), the municipal government implemented a policy that is a part of the revitalization of the historic city. The adopted policy is based on a comprehensive view of sustainable development which made Tunis, be the first to consider urban conservation from an overall perspective in the Arab region (Mouhli, 2016).

4.3 Community-Led Heritage Revitalization Initiatives in the Medina of Tunis: How Culture and Economic Transition Meet?

The Medina of Tunis revitalization attempts followed an integrated area development strategy. However, the contribution of the private sector remains modest (Steinberg, 2011, p. 13). Observations revealed that several community-led heritage revitalization initiatives have been achieved or are still under process.

On this point, the activity of the *Collectif Créatif*[2] association should be considered. It is a group of cultural actors interested in various forms of creative activities, such as cultural heritage, contemporary design, contemporary art, artistic education, artistic mediation, audience development, and media tools.

To date, the *Collectif Créatif* has developed nearly a dozen of projects. It contributed to making fruitful community meetings in the context of socio-cultural effervescence in post-revolutionary Tunisia. It supported the

---

[1] The first president after the French colonization regime (1957–1987).   [2] www.collectifcreatif.org/propos/.

Fig. 4 *EL-WARCHA* initiative in *Al Hafsia* district in the Medina of Tunis. *Source* www.elwarcha.org/

Medina's fertile ground for public participation in heritage enhancement and valuation by promoting the development of alternative cultures and by bringing existing initiatives together.

Among these attempts, the following initiatives can be given:

– *EL WARCHA*[3] (From 2016): Located in *Al-Hafsia* since 2016. It is a collaborative design studio monitored by the French designer Benjamin Perrot. *EL WARCHA* is a participative public design association that encourages children and teenagers' participation in refurbishing public spaces of the Medina of Tunis. It allows them to appropriate the streets and explore their urban space like adults. It tends to promote alternative forms of education and civic actions through the making of urban furniture and art installations. The main principle of this design studiois "*learning by making*", where all the activities revolve around live projects. In partnerships with other organizations, the studio works on public art, public participation, and private commissions (Fig. 4).

– *INTERFERENCE* (From 2016 biennial): "*Interference is the first international light art biennial on the African continent*" (Tira, 2021, p. 181). It is a modern-day urban art initiative hosted in the Medina. It is based on a series of improvisations combining video footage, sound, and texts. It is a community-based digital urban art festival launched in 2015 by two curators, i.e. Bettina Pelz (German), and Aymen Gharbi (Tunisian).

– *INTERFERENCE* is considered a novel exhibition of art in public space launched and influenced by the Arab Spring. It contributes to highlighting the role "*digital urban art and the city museum concept*" play in historic sites. *INTERFERENCE* plays a medium in unpacking unchallenged social relations in the context of social and political influences, through the changing night-time experience of the heritage space (Tira, 2021, p. 186). Based on care for artistic potentials and freedom of

expression, *INTERFERENCE* denotes both, the freedom to generate artistic expressions and also the citizen's right to be active in the cultural circle (Tira, 2021, p. 181).

– *JOURNAL DE LA MEDINA*[4] (From 2016): It is a participative newspaper that posts news about the Medina of Tunis including cultural events.

– *El HOUMA KHIR* (From 2017): It is a field-based think-tank initiative targeting the public space of the Medina. *El-Houma Khir* initiative took place in the Medina of Tunis in July 2017, in cooperation with, Oecumene Studio, Collectif Créatif, Le Journal de La Medina de Tunis, ICOMOS, *Dar El Harka*, Association Enauvateur supported by the Arab Council for Social Sciences under the auspices of SIDA (Swedish International Development Cooperation Agency) (Fig. 5).

– *FANTEK* (2018): Exchange platform around digital arts.

– *DOORA FEL HOUMA* (From 2018): It is an interactive guided tour.

– *DOOR WAHDEK* (From 2019): *DOOR WAHDEK* is a mobile app offering an interactive urban experience in the Medina of Tunis. Based on the city digitization, the idea of reinventing the historic city and its streets completes several initiatives that highlight the Medina of Tunis and its cultural heritage. *DOOR WAHDEK* is initiated by the organization *Collectif Créatif*, developed by *DIGINOV*, and supported by *TFANNEN*.[5]

– *J'NINA FEL M'DINA* (From 2019): Urban vegetable garden in the Medina of Tunis. It is an urban agriculture community-based movement. It has an educational and entrepreneurial focus. It mainly aims to combine environmental commitment and economic development opportunities by strengthening social ties in the Medina (Fig. 6).

---

[3] www.elwarcha.org/about.

[4] www.issuu.com/journaldelamedina.
[5] *TFANEN* (2016–2021) is a pioneer bilateral cultural project financed by the European Union (EU). It is implemented by the British Council on behalf of the EU National Institutes for Culture (EUNIC) and in collaboration with the Tunisian Ministry of Cultural Affairs (MAC). The project falls under the Programme to Support Cultural Sector in Tunisia (PACT) of the MAC. (UN Tunisia).

Fig. 5 El HOUMA KHIR initiative in the Medina of Tunis, November 30th, 2019. *Source* Facebook page of Al-Houma Khir

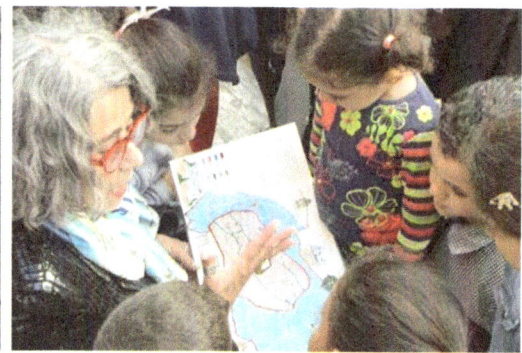

Fig. 6 J'NINA FEL M'DINA initiative in the roofs of Dar El-Harka, April 2020, the Medina of Tunis. *Source* Facebook page of Collectif Créatif

- *Carismed* (Capitalization for Resetting Innovation and Sustainability in MED-Cities): *Carismed* is a 24-month-long project co-funded by the European Union under the ENI CBC MED Program. It aims to develop an integrated urban policy to improve the sustainability of old buildings in Mediterranean city centers. It does so by the capitalization of the circular economy and creative approaches for achieving a better quality of life in urban areas (*Collectif Créatif*).

To support the agreement regarding the aforementioned community-based initiatives, an in-depth interview has been conducted with one investor in tourism retail in the Medina of Tunis. During the interview, she talked about the GIE (*Groupe d'Intêret Économique*) Group of Economic Interest. It consists of an economic lobby formed from 16 bodies having small businesses in the Medina of Tunis. The idea of making this group dates back to 2016. It was reported in the interview that they have common issues; among them is that people avoid visiting the Medina for several reasons including security, lack of hygiene, lack of parking spaces, lack of signage, and the risk of getting lost in the alleys. Although these problems are common, none of the investors could solve them alone. Since several small businesses are grouped for the same purpose, a force having more power emerged. Consequently, a legal background able to negotiate with the government has been formed. According to the

interviewee's explanation, the GIE led many community-based revitalization attempts in cooperation with several activists and associations:

> The most important ongoing project under the arch of the GIE is *M'dinti* project. It consists of revitalizing the Medina of Tunis during the weekends because generally, people avoid the Medina during that time. This means that on weekends the Medina stagnates, there is no movement, and the shops are closed. What we are actually trying to do is to make meetings before weekends to decide about what can be organized. Cultural events, festivals, workshops with craftsmen, and chess competitions are few examples of what we planned. This kind of movement makes the Medina a priority destination for the community on weekends. This is not limited to residents of the Medina, this concerns all those who wish to visit the Medina. It is in a way a circular logic, we revive the Medina via the cultural circle, we establish certain security and this contributes to the economic gain and the continuity of craftsmanship. Someone who chooses the Medina of Tunis as a destination on the weekend will come to consume. I, myself, will be able to maintain the two guesthouses I already have. I may be able to open other guest houses. A certain economic dynamic is created. With new visitors, a movement is created. It is this economic dynamic that contributes to maintaining the Medina of Tunis's cultural heritage. (Interview with Laila Ben Gacem, inverstor in retail tourism in the Medina of Tunis and member of the GIE, February 16, 2022).

The other initiative described by the interviewee is *La Médina Experience* (Experience of the Medina); it is an initiative that tends to give visitors the experience of craftsmen workshops. A calligrapher, for example, can make the visitor experience the ink making, and the use of traditional calligraphy in an authentic atmosphere.

5 Findings and Discussion

It follows from the foregoing empirical research and case study exploration that challenges can be spotted in the economic, socio-cultural, environmental, and political considerations concerning the community-led heritage revitalization initiatives. They are usually denoted as the lack of funding, resulting in using inadequate materials or equipment (physical revitalization); the lack of heritage revitalization specialists; the poor control of cultural heritage assets, and the weak management of cultural tourism activity (cultural and economic revitalization); and the inappropriate community inclusion in the revitalization process (social revitalization).

In the case of the Medina of Tunis, community-led heritage revitalization initiatives are proved to be engines for attracting tourists and also "*heritage local communities*" (Table 1). Socio-cultural revitalization initiatives based on art events and ephemeral urbanism are stimulating the Medina of Tunis's

branding and contributing to the creation of a closed-loop process; economic enhancement, creation of jobs, justice in appropriating heritage assets, security providing, and also in some cases, environmental sustainability (the example of *J'NINA FEL M'DINA* urban agriculture initiative).

However, some limitations can be drawn upon with the recognition of community-based heritage revitalization initiatives as an engine for the economic transition and circular economy implementation. The existence of community-based initiatives as a promoter/coordinator of many revitalization initiatives does not automatically guarantee trustworthy community empowerment. Questions remain open in terms of defining the international cooperation-based revitalization projects as a community-based: who are the participants in each initiative?; how permeable is it to the wider contribution for the community of the Medina of Tunis ?; and what is the effort in terms of community empowerment?; does the community recognize the role of the existing activist associations as an engine for the economic transition and the share economy principle?

The practices for enhancing democratic participation are mostly absent, with some exceptions. This lack puts forward

Table 1 Challenges and opportunities of the Medina of Tunis's community-led revitalization initiatives	Revitalization category	Revitalization initiative	Challenges and opportunities
	Socio-physical revitalization	*J'NINA FEL M'DINA*	The Medina of Tunis's organic morphology, and being a UNESCO World Heritage site, do not allow the creation of green public spaces. This initiative made it possible to collaboratively create spaces for sustainable urban gardens Besides its being a circularity-based initiative, it represented an opportunity for socialization
		El HOUMA KHIR	All citizens including marginalized minorities, children, elderly participated in the design process of public places in the Medina
		Carismed	Improving the sustainability of old buildings Several dilapidated buildings could not be upgraded due to ownership clearance problems
	Socio-cultural revitalization	*EL WARCHA community-based co-design studio*	Even inclusion of the citizen in art and design-based activities of the Medina
		INTERFERENCE	It represents a space for public voice expression through an innovative display of art in urban heritage Giving a new nightlife to the Medina of Tunis's heritage making it securer
		El HOUMA KHIR	It represented an innovative way for a "heritage community" development through cultural heritage It gave the community the possibility to better understand the importance of their history and identity
		FANTEK	It presented augmented reality technologies to enhance people's perception of the heritage site
		DOOR WAHDEK	An opportunity to overcome the digital divide that can be found in the low-income population of inner cities
		DOORA FEL HOUMA	A way to attract visitors and stimulate small businesses in the Medina

other questions, i.e.: which other tools have been used in order to guarantee wide community participation in the revitalization initiatives? Then, in terms of stakeholders' cooperation, the knowledge of experts is not always easy to trace with the local knowledge, but in some illustrated initiatives, a record of public meetings and workshops have been intended as a form of exchange between residents and experts (the example of *EL HOUMA KHIR* and Mdinti initiative). The involvement of craftspeople, artists, and experts is distinguished; underpinning the voice of local communities does not always apply. In some initiatives, it is rather a "heritage community" voice underlying.

In light of these findings, it can be assumed that such knowledge about community-led heritage revitalization challenges and opportunities is essential to enhance the making discourse and develop new policies which can further support the global shift toward the circular economy.

6 Conclusion

The foregoing research offers insights into the changing role, heritage revitalization plays in the current economic transition debate. There is an attention shift from the cultural heritage in itself, toward communities and their active participation in their heritage revitalization process and the recognition of the values held in it (Fabbricatti et al., 2018). In order to raise the awarness of local citizens about the importance of their local, but at the same time significant world heritage, and to strengthen their sense of pride and identity, the community-led revitalization attempts in the Medina of Tunis:

1. Provided local citizens free access to the local culture and heritage when making them aware that free access does not make heritage a lower value and this is achieved by making a balance between conservation needs and economic enhancement demands.
2. Proved that investment in preserving and restoring old buildings creates jobs, and stimulates business and tourism. This, in itself, contributes to supporting small businesses and generating local economic activity. Such circular processes underline the shift from a linear economy to a closed-loop circular economy principle in the Medina's heritage revitalization attempts (case of *Dar Ben Gacem* guest house).
3. Exposed a balanced staging and branding of the historic city, right in parallel to the contribution to giving citizens a touchstone to the past.

The forgoing work is not free from methodological limitations that can be a basis for future research. Further statistical studies and survey-based methods are required to better evaluate the true extent of the community-led revitalization initiatives of the Medina of Tunis in the current global economic transition. Statistical studies concern the number of participants in the revitalization attempts, supported by analytical studies on variations according to the initiative categories. A collaboration with the *Collectif Creatif* association to assess the number of participants, and community-based donations can be implemented. This can quantitatively evaluate the weight given to the community-driven revitalization attempts in the circular city debate from the perspective of the Medina of Tunis.

Acknowledgements The authors wish to thank Mme Faïka Bejaoui, Tunisian Architect and Urban planner active in the ASM, and Mme Laila Ben Gacem owner of Dar Ben Gacem Pasha and Dar Ben Gacem Kahia guest houses, for sharing their knowledge about the Medina of Tunis revitalization initiative.

References

Akrout-Yaïche, S. (2002). Local involvement in urban management: The experience of the city of Tunis. *International Social Science Journal, 54*(172), 247–252. https://doi.org/10.1111/1468-2451.00376

Ammar, L. (2010). *Tunis d'une ville à l'autre: cartographie et histoire urbaine*, 1860–1935. Nirvana.

Andres, L., Bakare, H., Bryson, J. R., Khaemba, W., Melgaço, L., & Mwaniki, G. R. (2019). Planning, temporary urbanism, and citizen-led alternative-substitute place-making in the Global South. *Regional Studies.* https://doi.org/10.1080/00343404.2019.1665645

Augé, M. (1992). *Non-places: An introduction to anthropology of super modernity* (p. 79). Le Seuil.

Bandarin, F., & van Oers, R. (2012). The historic urban landscape. *In the historic urban landscape.* https://doi.org/10.1002/9781119968115

Barile, S., & Saviano, M. (2015). From the management of cultural heritage to the governance of the cultural heritage system. In: Golinelli, G. (eds) *Cultural Heritage and Value Creation.* https://doi.org/10.1007/978-3-319-08527-2_3

Culture Action Europe, European Cultural Foundation, and Europa Nostra. (2020). "A Cultural Deal for Europe," Accessed August 20, 2021. From: https://cultureactioneurope.org/files/2020/11/statement-Final-1.pdf.

Daoulatli, A. (2009). *Tunis capitale des Hafsides* (pp. 19–24), ALIF les éditions de la méditerranée.

Dhaher, N., Seyfi, S., & Michael Hall, C. (2020). Cultural heritage and tourism in Tunisia: Evolution, challenges, and perspectives. In *Cultural and heritage tourism in the Middle East and North Africa: Complexities, management, and practices* (pp. 87–101). Taylor and Francis. https://doi.org/10.4324/9780429279065-5

EC. (2015). Circular Economy Action Plan. European Union. Accessed September 14, 2021. From: https://ec.europa.eu/environment/pdf/circular-economy/new_circular_economy_action_plan.pdf

EC. (2019, December 11). *The European green deal.* Accessed September 16, 2021. From: http://eur-lex.europa.eu/resource.html?uri=cellar:208111e4-414e-4da5-94c1-852f1c74f351.0004.02/DOC_1&format=PDF

Fabbricatti, K., et al. (2018). Community-led practices for triggering long-term processes and sustainable resilience strategies. The case of the eastern Irpinia, inner periphery of southern Italy. 5965.https://doi.org/10.3390/ifou2018-05965

Ferreri, M. (2015). The seductions of temporary urbanism. In Bialski, P., et al. (Eds.), 'Saving' the city: Collective low budget organizing and urban practice, *Ephemera, 15(1)*, 181.

Garzillo, C., Gravagnuolo, A., & Ragozino, S. (2018). Circular governance models for cultural heritage adaptive reuse: the experimentation of Heritage Innovation Partnerships. *Urbanistica Informazioni, 278*, 17–22. https://doi.org/10.4324/9781315672168.Nobre

Grodach, C. (2017). Urban cultural policy and creative city making. *Cities, 68*, 82–91. https://doi.org/10.1016/j.cities.2017.05.015

Haberer, P., G. De Kleyn, and W. De Wit. (1980). The neighbourhood approach: improvement of old neighbourhoods by and on behalf of their inhabitants. Netherlands: Ministry of Housing and Physical Planning.

Halleb, Y. (2018). The Politics of Cultural Heritage Management in Tunisia. A Thesis Submitted to the Graduate Faculty of The University of Georgia in Partial Fulfilment of the Requirements for the Degree Master of Historic Preservation. Athens, Georgia.

Han, C. (2003). *Re-constructing place and community: Urban heritage and the symbolic politics of neighborhood revitalization*. Massachusetts Institute of Technology.

ICLEI. (2020). "Adaptive Reuse of Cultural Heritage an Examination of Circular Governance Models from 16 International Case Studies Synthesis Report," no. March. www.clicproject.eu

Interview with Ben Gacem, Laila, February 16, 2022.

Jauhar, J., Setijanti, P., & Hayati, A. (2021). Revitalization of cultural heritage area with sustainable tourism approach, case study: Tindoi Fort, Kabupaten Wakatobi. *Journal of Architecture & Environment, 20(2)*, 75. https://doi.org/10.12962/j2355262x.v20i2.a10719

Jelinčić A, D., & Glivetić, D. (2021). Cultural heritage and displacement. In *Language issues in comparative education II*. https://doi.org/10.1163/9789004449671_008. Decentralizationand economic transition

Jelinčić, D. A., & Mansfeld, Y. (2019). Applying cultural tourism in the revitalisation and enhancement of cultural heritage: An integrative approach. *Cultural Urban Heritage*, 35–43.https://doi.org/10.1007/978-3-030-10612-6_3

Kumar, V. (2020). When heritage meets creativity: A tale of two urban development strategies in Kampong Glam, Singapore. *City and Community, 19(2)*, 398–420. https://doi.org/10.1111/cico.12427

Labadi, S., Giliberto, F., Rosetti, I., Shetabi, L., & Yildirim, E. (2021). Heritage and the sustainable development goals. https://openarchive.icomos.org/id/eprint/2453/1/ICOMOS_SDGs_Policy_Guidance_2021.pdf

Lam, E. W. M., Zhang, F., & J. K. C. H. (2022). *Effectiveness and advancements of heritage revitalizations on community planning: Case studies in Hong Kong* (pp. 1–20).

Libakova, N. M., & Sertakova, E. A. (2015). The method of expert interview as an effective research procedure of studying the indigenous peoples of the North. *Journal of Siberian Federal University. Humanities & Social Sciences, 1*(2015 8), 114–129. https://doi.org/10.17516/1997-1370-2015-8-1-114-129

Micara, L., Petruccioli, A., & Vadini, E. (2006). The Mediterranean Medina. *International Seminar, 17–19 June 2004, School of Architecture, Pescara, 570*.

Mouhli, Z. (2016). Cultural policy: Valuing heritage for more sustainable development. *The second International Award "UCLG—Mexico City—Culture 21* (January–May of 2016).

Musyawaroh, M., Pitana, T. S., Masykuri, M., & Nandariyah. (2018). Sustainable revitalization in cultural heritage Kampong Kauman Surakarta supported by spatial analysis. *IOP Conference Series: Earth and Environmental Science, 123*(1). https://doi.org/10.1088/1755-1315/123/1/012043

Nardella, B. M., & Cidre, E. (2016). Interrogating the 'implementation' of international policies of urban conservation in the medina of Tunis. In S. Labadi & W. Logan (Eds.), *Urban Heritage* (pp. 57–79). Routledge.

Nardella, B. M., & Mallinson, M. (2014). Only foreigners can do it?: Technical assistance, advocacy and brokerage at Aksum, Ethiopia. *Museums, Heritage and International Development*, 188–210. https://doi.org/10.4324/9780203069035-15

Nash, D. (1989). 2. Tourism as a form of imperialism. In V. Smith (Ed.), *Hosts and Guests: The Anthropology of Tourism*, 37–52. Philadelphia: University of Pennsylvania Press. https://doi.org/10.9783/9780812208016.37

Nocca, F. (2017). The role of cultural heritage in sustainable development: Multidimensional indicators as a decision-making tool. *Sustainability, 9*(10). https://doi.org/10.3390/su9101882

Nocca, F., De Toro, P., & Voysekhovska, V. (2021). Circular economy and cultural heritage conservation: A proposal for integrating level (s) evaluation tool. *Aestimum, 78*, 105–143. https://doi.org/10.36253/aestim-10119

Pappalardo, G. (2020). Community-based processes for revitalizing heritage: Questioning justice in the experimental practice of ecomuseums. *Sustainability, 12*(21), 1–18. https://doi.org/10.3390/su12219270

Regeneration and Optimization of Cultural heritage in creative and Knowledge cities (ROCK). (2019). "New Governance Models for Creative, Sustainable and Circular Cities" no. 730280. https://bologna.rockproject.eu/

Saadaoui, A. (2010). *Tunis ville Ottomane trois siècles d'urbanisme et d'architecture*. Centre de publication universitaire.

Santelli, S. (1992). *Medinas: Traditional architecture of Tunisia* (1st edn.). Dar Ashraf Publications.

Santelli, S. (1995). *La ville le creuset méditerranéen Tunis*. CNRS.

Schmitt, T. M. (2009). Global cultural governance. Decision-making concerning World Heritage between politics and science. *Erdkunde, 63*(2), 103–121. https://doi.org/10.3112/erdkunde.2009.02.01

Steinberg, F. (2011). Revitalization of historic inner-city areas in Asia. In *ICOMOS 17th General Assembly (Issue August)*.

Stoffelen, A. (2022). Revitalising place-based commercial heritage: A Cultural Political Economy approach to the renaissance of lambic beers in Belgium. *International Journal of Heritage Studies, 28*(1), 16–29. https://doi.org/10.1080/13527258.2020.1862275

Tira, Y. (2018). *Cultural continuity of traditional Bazaars in the globalized world: The Medina of Tunis and Istanbul Grand Bazaar*. LAP Lambert Academic Publishing.

Tira, Y. (2021). Digital urban art in historic city centers in times of democratic transition. In *Transforming urban nightlife and the development of smart public spaces* (pp. 172–190). IGI Global.

Urry, J. (1990). *The tourist gaze: leisure and travel in contemporary societies*. London; Newbury Park: Sage Publications

UNESCO. (1972). Convention concerning the protection of the world cultural and natural heritage. Adopted by the General Conference at its seventeenth session. Paris, 16 November 1972.

UN. Department of Economic and Social Affairs Sustainable Development (n.d.). Tfanen-Tunisie Créative/Strengthening the cultural sector in Tunisia British Council, Tunisian Ministry of Cultural Affairs (Government). Available from: https://sdgs.un.org/partnerships/tfanen-tunisie-creative-strengthening-cultural-sector-tunisia. Visited 08.06. 2022.

Wang, C. H. (2013). Heritage formation and cultural governance: The production of Bopiliao Historic District, Taipei. *International Journal of Heritage Studies, 19*(7), 676–691. https://doi.org/10.1080/13527258.2012.687696

Addressing Connectivity Issues Between the Historical and Natural Touristic Heritage Sites of Egypt

Mohamed Elkaftangui

Abstract

Tourism in Egypt is a significant contributor to the country's economy, with both natural and historical heritage sites playing a crucial role. However, tourists face several challenges navigating the country, particularly when moving between the natural heritage sites along the green corridor of the Red Sea and the historical sites in Luxor, Aswan, and Cairo. This study aims to address the connectivity issues faced by tourists in Egypt, to improve the tourist experience and boost the tourism industry. The results of the study showed that a lack of proper infrastructure, limited transportation options, and insufficient signage were some of the major challenges faced by tourists in navigating Egypt. Furthermore, tourists also expressed frustration with the availability of information, particularly in terms of route planning and navigating the country's landscape. The results suggest that addressing these connectivity issues would improve the tourist experience and enhance the country's competitiveness in the global tourism market. Based on the findings, the study concludes by recommending the development of a comprehensive tourism strategy that addresses the connectivity issues between the historical and natural heritage sites of Egypt. This could include the creation of better transportation options, the installation of proper signage, and the provision of accurate and up-to-date information to tourists. By improving connectivity between the different tourist sites, the study argues that Egypt can enhance its position as a leading destination for both natural and historical heritage tourism.

Keywords

Green Corridors • Cultural tourism • Heritage • Economic development • Egypt

1 Introduction

Egypt is one of the world's most popular tourist destinations, attracting millions of visitors each year with its rich cultural and natural heritage. However, travel within the country can be a challenge for tourists, who face several difficulties navigating the landscape, including poor infrastructure, limited transportation options, and a lack of information about the many historical and natural heritage sites scattered throughout the country. Despite its many attractions, Egypt has yet to fully realize its potential as a top destination for tourists due to these connectivity issues.

Despite these difficulties, tourists are motivated to visit Egypt for several reasons. Some come to see the famous historical sites and learn about the rich cultural heritage of the ancient civilization (El-Khadrawy et al., 2020), while others come to enjoy the natural beauty of the Red Sea coast and engage in outdoor activities, such as snorkeling, diving, and beach-based recreation. Many also come to experience the unique culture and hospitality of the Egyptian people and explore the country's vibrant cities like Cairo, Luxor, and Aswan.

Given the importance of tourism to Egypt's economy, it is critical to address the challenges of connectivity and access to information to enhance the overall tourist experience and maximize the potential of this sector. This research paper aims to do just that, by examining the connectivity issues between the historical and natural heritage sites of Egypt, and exploring ways to improve the overall travel experience for visitors. Through the examination of case studies, such as the green corridor along the Red Sea coast and the historic sites of Luxor, Aswan, and Cairo, the study will provide

M. Elkaftangui (✉)
College of Engineering, Faculty of Architecture,
Abu Dhabi University, Abu Dhabi, United Arab Emirates
e-mail: mohamed.elkaftangui@adu.ac.ae

insights into the challenges tourists face and identify potential solutions to improve their travel experience. By exploring the potential for enhanced connectivity and greater access to information, the study will contribute to efforts to further enhance Egypt's position as a leading destination for tourists from around the world.

This research paper aims to address the challenges of navigating Egypt by examining the connectivity issues between the historical and natural heritage sites of the country. Through the examination of case studies, such as the green corridor along the coast of the Red Sea, which includes destinations like Sharm el Sheikh, Hurghada, all the way to Marsa Alam, and the historic sites of Luxor, Aswan, and Cairo, the study will analyze the challenges tourists face when trying to visit these sites and explore ways to improve the overall tourist experience. By exploring the potential for improved connectivity and greater access to information, the study will help to shed light on how Egypt can further enhance its position as a leading destination for tourists from around the world.

2 Problem Statement

One of the main challenges facing tourists in Egypt is the lack of adequate infrastructure. The country's roads, bridges, and highways are often in poor condition, making travel between destinations slow and difficult. This can be especially challenging for tourists traveling to remote or rural areas, where access to basic services, such as water and electricity, is limited. In addition, the lack of investment in public transportation systems, such as trains and buses, makes it difficult for tourists to move between destinations in a timely and affordable manner.

The political instability in Egypt has also had a significant impact on the country's tourism industry. The ongoing conflict and unrest in the country have created safety and security concerns for tourists, which can deter them from visiting. Furthermore, the political situation can also lead to sudden changes in travel restrictions and safety measures, making it difficult for tourists to plan and book their trips.

Social issues, such as poverty, inequality, and limited access to education and healthcare, can also impact the connectivity of touristic areas in Egypt. For example, in some areas, poverty and unemployment levels are high, leading to a lack of investment in basic infrastructure and services. Additionally, limited access to education and healthcare can result in a lack of local support for tourism development, making it more difficult for destinations to grow and attract visitors.

Urbanization and rapid development in some areas of Egypt can also create connectivity issues for tourists. For example, overcrowding and rapid development can lead to a

lack of open space, green areas, and public spaces, which can reduce the quality of life for residents and visitors alike. In addition, rapid development can also result in the loss of cultural heritage sites and natural landscapes, making it more difficult for tourists to experience the unique character of these destinations.

3 Problems in the Historical Heritage Sector

One of the major problems faced by the historical heritage sector in Egypt is a lack of investment in preservation and maintenance. Despite the cultural and economic benefits that these sites bring to the country, many are in a state of disrepair and are at risk of further damage and degradation. The inadequate funding and resources available for maintenance and preservation work are a major concern, as these sites are crucial to the country's cultural heritage and historical identity.

Another major challenge faced by the historical heritage sector in Egypt is the lack of infrastructure and access to information. Many of the sites are located in remote areas with limited transportation options and limited access to basic amenities, such as toilets, water, and food. In addition, the lack of information about these sites, and the surrounding areas, makes it difficult for tourists to plan and make the most of their visits.

The issue of security is also a significant concern for the historical heritage sector in Egypt. Political instability and the threat of terrorism have had a negative impact on the tourism industry and the security of the sites themselves. This has led to a decline in the number of visitors and a decrease in the revenue generated from the tourism sector, which is a major contributor to the country's economy.

4 Problems in the Natural Heritage Sector

Egypt's Natural Heritage Sector is facing several challenges that are impacting the preservation and promotion of its unique natural attractions. The mountainous areas in Sinai and Dahab, for example, are facing numerous threats, including degradation of their natural habitats, illegal hunting, overgrazing, and encroachment of human settlements.

One of the major problems is the degradation of the natural habitats in the mountainous areas, which is causing a decline in the biodiversity of the region. This degradation is largely a result of unsustainable tourism practices, such as the construction of tourist facilities without proper planning and environmental assessment, and the increased human presence in the area. In addition, the illegal hunting of wildlife, particularly in the reserves and protected areas, is also a significant problem.

5 Literature Review

Tourism is an essential aspect of the global economy, with millions of people traveling every year to experience different cultures, historical sites, and natural wonders. In Egypt, the tourism industry plays a significant role in the country's economy, attracting millions of visitors each year to its ancient ruins, rich cultural heritage, and stunning beaches. However, despite its popularity, tourists face a range of challenges navigating the landscape of Egypt, including difficulties in traveling between the country's historical and natural heritage sites.

This lack of connectivity between these heritage sites is a major challenge, particularly when it comes to attracting tourists who are interested in experiencing the country's cultural and natural offerings in their entirety. In the past, there have been several studies conducted to examine the difficulties of navigating the landscape of Egypt, with a focus on its cultural heritage sites (Abdel Wahed Ahmed & Abd El Monem, 2020; Manière et al., 2021). However, there has been limited research conducted on the challenges of connecting its historical and natural heritage sites, which form an integral part of the country's tourist offering.

One of the main difficulties in connecting these heritage sites is the lack of comprehensive transport infrastructure in the country. The limited connectivity between the cities and towns of Egypt, combined with the vast distances between the different heritage sites, makes it difficult for tourists to travel between them, limiting their experience of the country's cultural and natural offerings. Additionally, the lack of clear information and signage, as well as a shortage of guides, has made it difficult for tourists to navigate the country's heritage sites, leading to frustration and disappointment.

Another challenge in connecting these heritage sites is the lack of integration between the different tourism offerings. Despite the availability of both historical and natural heritage sites in Egypt, there has been limited effort to integrate these offerings, making it difficult for tourists to experience the country's cultural and natural offerings in a single trip. This lack of integration also makes it difficult for the tourism industry to attract tourists who are interested in experiencing both the country's historical and natural offerings.

To address these challenges, it is essential to conduct research into the connectivity issues between the historical and natural heritage sites of Egypt, with a focus on exploring ways to improve the transport infrastructure, information and signage, and overall experience for tourists. This research paper will examine the challenges of connecting Egypt's historical and natural heritage sites, drawing on existing research on the difficulties of navigating the landscape of Egypt, as well as the importance of integrating its cultural and natural offerings. The paper will also provide recommendations for addressing these challenges, to improve the overall experience for tourists visiting Egypt.

6 Tourist Motivations

Tourist motivations play a crucial role in understanding why individuals choose to travel to certain destinations and participate in specific activities during their vacations. Motivation theories in tourism are rooted in consumer behavior literature and argue that motivations represent the internal forces that drive individuals' actions and behaviors. Travel motivations are related to human needs and reflect why people travel to selected destinations and engage in ecotourism activities.

The most widely referenced motivation theory with psychological roots is the theory of escaping and seeking (Gnoth, 1997), which includes psychological motives, such as relaxation and exploration, and cultural motives, such as education and novelty. However, few studies have explored tourists' motivations in the context of ecotourism in mountainous areas (Bokhari, 2021). A study by the Countryside Commission (1995) identified key visitor motivations including relaxation, health, and peace. Similarly, studies by Devesa et al. (2010) and Yi et al. (2011) recognized different groups of rural tourists based on their interests, such as tranquility and interaction with nature, cultural experiences, gastronomy, visiting family and friends, or relaxation.

Categorizing tourists based on their values, activities, or socioeconomic characteristics has a crucial role in shaping urban policies that balance local community needs, economic growth, and sustainable development. Therefore, understanding tourists' motivations in mountain regions is essential for developing clear urban and regional policies that enhance ecotourism.

7 Connectivity Issues in Egypt from a Tourist Perspective

Connectivity issues are a major hindrance for tourists visiting Egypt and hinder their ability to fully experience the country's rich natural and historical heritage. These connectivity issues stem from a variety of factors, including poor infrastructure, limited transportation options, safety and security concerns, and a lack of information and signage (Faajir & Zidan, 2016).

Poor infrastructure can make it difficult for tourists to reach certain destinations within the country. This is particularly true for Egypt's mountainous areas, such as Sinai and Dahab, which are often only accessible by rough roads and limited public transportation options. This can make it difficult for tourists to explore these areas and experience the unique natural beauty they have to offer. As is evident in Figs. 1 and 2, Egypt attains a variety of historical and natural heritage sites, some of which reside along the Nile (nodes 1, 7, 8, and 9 in Fig. 2), in Sinai (3 and 2 in Fig. 2), in nature preserves in the Sahara (nodes 10, 13, and 14 in Fig. 2), along the Mediterranean sea (nodes 11 and 12, Fig. 2), and the coast of the Red Sea (nodes 4, 5, 6, and 15 in Fig. 2).

Safety and security concerns are another major factor that affects tourist connectivity in Egypt. Despite the country's efforts to improve safety and security, incidents such as terrorism and crime remain a concern for tourists and can limit their ability to explore certain destinations (Zakarriya, 2021). This is particularly true for tourists visiting remote areas, such as Egypt's mountainous regions, which may not have a strong security presence.

Finally, a lack of information and signage can make it difficult for tourists to navigate Egypt's heritage sites and

Fig. 2 Depicts the nodes of the proposed connectivity network

understand the cultural and historical significance of what they are experiencing. This can lead to missed opportunities and a lack of appreciation for the country's rich heritage.

8 Results

Improving the infrastructure and access to information, increasing investment in preservation and maintenance, and enhancing the security of the sites are all crucial to ensuring the long-term viability of this important sector. Addressing these challenges will help to enhance Egypt's reputation as a top tourist destination and ensure that its rich cultural heritage is preserved for future generations.

In the context of tourism, the concept of circular economies can be applied by promoting locally sourced and sustainable products, reducing the carbon footprint of the industry, and improving the local economy (Brown, 2018).

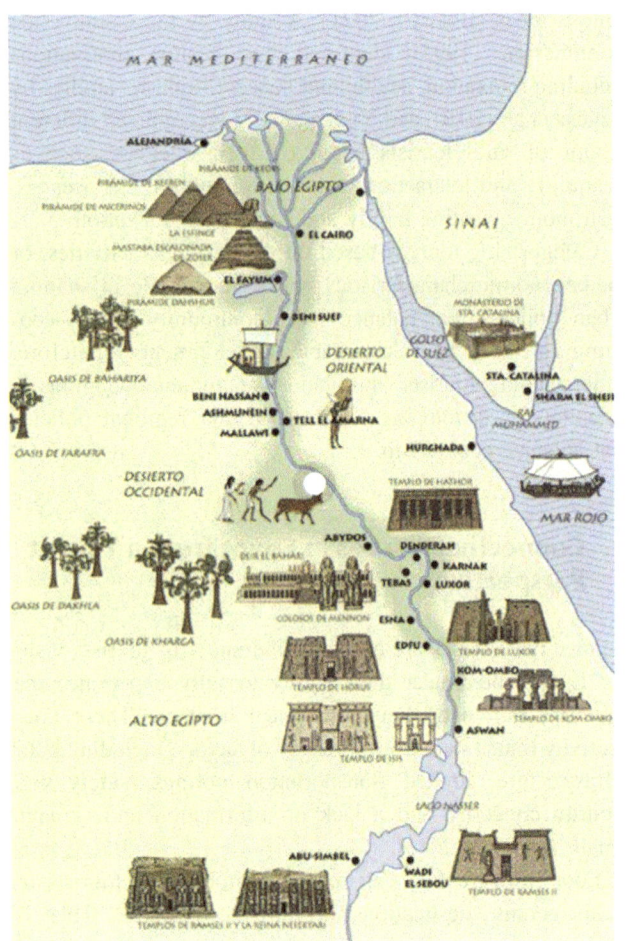

Fig. 1 Depicts the most famous heritage and touristic sites in Egypt

For instance, in Egypt, local communities can participate in the production and distribution of souvenirs and other tourism-related products, which can contribute to the development of the local economy. This approach can also help to reduce the negative impact of tourism on the environment and preserve the cultural heritage of the country.

Community participation is another key aspect that can contribute to the sustainability of tourism in Egypt. By engaging local communities and including important factors in development programs such as program benefits, objectives, and leadership in the planning and management of tourism, the industry can be aligned with the needs and interests of the local population (Naku et al., 2021). This can help to reduce the negative impact of tourism on the environment and enhance the cultural heritage of the country. Additionally, community participation can help to ensure that the benefits of tourism are shared equitably among the local population.

Sustainable cultural tourism is another approach that can help to solve the problems facing tourism in Egypt. By promoting cultural heritage and preserving the unique character of the country, sustainable cultural tourism can help to attract more visitors and improve the local economy. This approach can also help to preserve the cultural heritage of the country and reduce the negative impact of tourism on the environment. In addition, sustainable cultural tourism can help to promote cultural exchange and mutual understanding between different cultures.

9 Recommendations

Based on the findings of this study, several recommendations can be made to address the connectivity issues facing tourists in Egypt. Firstly, there should be a promotion of direct flights from different parts of the world to various Egyptian airports, to increase accessibility and ease of travel in its cities. In addition, there is a need for improving the infrastructure and connectivity in the country by renovating existing airports and railway stations and investing in new transportation options.

A prioritization toward the development of circular economies in tourist areas is direly needed. This includes promoting the use of renewable energy, reducing waste through recycling and reusing resources, and supporting local businesses to stimulate the local economy. By adopting a circular economy approach, touristic areas in Egypt can become more sustainable, resilient, and attractive to visitors.

Communities living in and around touristic areas should be encouraged to participate in the planning and management of tourism activities. This can be achieved through capacity-building programs, awareness-raising campaigns, and the provision of opportunities for local people to engage in tourism-related businesses. Community participation can help ensure that tourist areas are developed in a way that benefits both visitors and local people, and that the negative impacts of tourism are minimized.

There is also a need to improve accessibility to tourist sites and provide necessary information to tourists. This can be achieved by providing informative maps and brochures at airports and tourist sites and by offering multilingual guides to help tourists navigate the country. Lastly, the government should prioritize security measures and improve the safety and security of tourists in the country.

Overall, additional urban planning studies are needed to improve the connectivity of tourist sites from heritage, historical, and natural contexts. The main objective of which is to ease movement within Egypt's cities, and provide a fulfilling and rewarding tourist experience.

References

Abdel Wahed Ahmed, A. W., & Abd El Monem, N. (2020). Sustainable and green transportation for a better quality of life case study greater Cairo-Egypt. *HBRC Journal, 16*(1), 17–37.

Bokhari, A. (2021). Understanding tourists' motivations: The case of Al Baha mountainous region in Saudi Arabia. *International Journal of Environmental Science & Sustainable Development, 6*(1), 75–87. https://doi.org/10.21625/essd.v6i1.793

Brown, H. (2018). Resiliency and regeneration in the Pannonian Region of Hungary: Towards a circular economy. *International Journal of Environmental Science & Sustainable Development, 3*(2), 2–13. https://doi.org/10.21625/essd.v3iss2.372

Countryside Commission. (1995). Public attitudes to the countryside. Countryside Commission.

Devesa, M., Laguna, M., & Palacios, A. (2010). The role of motivation in visitor satisfaction: Empirical evidence in rural tourism. *Tourism Management, 31*(4), 547–552.

Eli Avraham. (2016). Destination marketing and image repair during tourism crises: The case of Egypt, Journal of Hospitality and Tourism Management.https://doi.org/10.1016/j.jhtm.2016.04.004

El-Khadrawy, R. K., Attia, A. A., Khalifa, M., & Rashed, R. (2020). Sustainable tourism and culture: A symbiotic relationship. *International Journal of Environmental Science & Sustainable Development, 5*(1), 54–67. https://doi.org/10.21625/essd.v5i1.717

Faajir, A., & Zidan, Z. H. (2016). An analysis of the issues and challenges of transportation in Nigeria and Egypt. *The Business & Management Review, 7*(2), 18.

Gnoth, J. (1997). Tourism motivation and expectation formation. *Annals of Tourism Research, 24*(2), 283–304.

Manière, L., Crépy, M., & Redon, B. (2021). Building a model to reconstruct the Hellenistic and roman road networks of the eastern desert of Egypt, a semi-empirical approach based on modern travelers' itineraries. *Journal of Computer Applications in Archaeology, 4*(1), 20–46.

Naku, D. W. C., Kihila, J., & Mwageni, E. (2021). Towards programs sustainability: Assessment of institutional determinants for effective community participation in development programs in Tanzania.

International Journal of Environmental Science & Sustainable Development, 6(2), 60–72. https://doi.org/10.21625/essd.v6i2.838

Yi, S., Day, J., Cai, L., & Trends. (2011). Rural tourism demand: Duration modeling for drive tourists length of stay in rural areas of the United States. *Journal of Tourism Challenges*, 4(1), 147–168.

Zakarriya, J. (2021). Security, dividedness and green activism in Egypt. *Journal of International Women's Studies*, 22(9), 174–189.

Elbe—A Tourist Line Cycling Through Europe

Michaela Štěbetáková, Petr Červinka, Michal Hořejš, Šárka Lukešová, and Leona Vaculovičová

Abstract

The Lower Elbe region, the last part of the Elbe River on Czech territory, has been a major tourist destination for several hundred years, an important development element in the past and today. It is not only the river that is important, but also its hinterland. The Elbe has influenced the geological development of this landscape, giving life and livelihood. It motivates tourists to visit the area and discover the unique local landscape. Today, the Lower Elbe area is protected by several special protection areas with a relatively high degree of protection—Bohemian Switzerland NP, NPR Elbe Canyon, number of Natura 2000 sites. The Elbe is also followed by one of the most important cycle paths in Europe, which runs from the source of the Elbe up to the estuary of the Elbe. Not only with the development of the cycle path, a concentration of tourist destinations, activities, and services, as well as visitors, is emerging in the Lower Elbe area. However, tourism is not an isolated activity, but also affects its surroundings, residents, and landscape, and influences local life, economy, and culture. The main question of this thesis is how residents perceive tourism development in the region and whether they perceive a conflict between tourism, development, and conservation in the area. This study is based on a case study of tourism in the Lower Elbe and it used content analysis of media, development documents, interviews with key actors, and field research. The research shows that the cycle path is important for the municipalities, however, its potential is much higher than the municipalities are currently able to exploit.

Keywords

Tourist lower Elbe • Cycle path • Tourism in protected areas • Tourist resident interaction • Tourism development

1 Introduction

The Elbe Cycle Route (LCS) is one of the most important cycling routes in Europe. It is one of the longest European routes, crosses several countries, and attracts a large number of tourists. From a tourism perspective, the cycle path can be described as a development line that links several tourism services (Hose, 2018). The route passes not only through cities but also through smaller municipalities that respond differently to the existence of the LSC (Skrinar et al., 2019). How municipalities approach the development of tourism potential in their strategic documents should be explained in this article.

Cycling routes in general can be an interesting topic for tourism geography. The development of tourism around cycle routes has been addressed by many authors (Adam et al., 2020; Pavluković et al., 2020; Schneider, 2000; Skrinar et al., 2019). A large part is devoted to the use of cycling as a sustainable mode of transport (Pavluković et al., 2020). Others, on the other hand, point out that cycle paths can be a cornerstone for the emergence of overtourism or a territory of conflicts between tourists and residents (Schneider, 2000). Weight is also given to cycle paths as educational routes for approaching the functioning of landscape–human relations (Adam et al., 2020). In all these cases, the environment through which the cycle path passes

M. Štěbetáková (✉) · P. Červinka · M. Hořejš · Š. Lukešová · L. Vaculovičová
Department of Geography, Faculty of Science, University of Jan Evangelist Purkyně in Ústí Nad Labem, Ústí Nad Labem, Czechia
e-mail: michaela.stebetakova@ujep.cz

Š. Lukešová
e-mail: sarka.lukesova@ujep.cz

M. Štěbetáková
Faculty of Social and Economic Studies, University of Jan Evangelist Purkyně in Ústí Nad Labem, Ústí Nad Labem, Czechia

plays a role. The environment is largely managed by municipalities, regions, or authorities involved in the management of protected areas, etc. These authorities influence the development of the background of the cycle path. This article aims to identify the attitude of local government units toward tourism development in the context of the Elbe cycle path. How the municipalities approach this role should be explained by the research questions:

1. *How do municipalities implement tourism in their strategic documents?*
2. *Does the area offer all the services needed to realize this form and type of tourism?*
3. *If the supply of services is not sufficient, are there other possibilities for the development of tourism services?*

The approach in this study is constructivist, which is mainly based on the use of qualitative research. The research procedure will be based on deductive foundations and a study of literature from the field of cycling and cycling tourism development. The research will also include an analysis of the strategic plans of each municipality in the study area. The study is also complemented by field research in the study area. The field research was mainly focused on identifying the infrastructure adjacent to the LSC.

2 Literature Review

Tourism, like other industries dependent on the transformation of society and the growing trend of leisure sufficiency, is changing (Oliviera et al., 2020). Traditional tourism requires undiscovered landscapes, which is an essential element for adventure in the form of attractive and inaccessible landscapes (Saarinen et al., 2017). A traditional approach to tourism considers strong prerequisites of good transport accessibility, such as multi-modal transport, a connection of the cities to the motorway, a connection to the Prague-Dresden railway, tourist connections, and the Elbe Cycle Route (LCS). Additionally, river transport, which dates back to 950 (Klír, 1908), also operates here. However, in recent years, interest in river transport has been declining (DÚK, 2020). The decline is also related to climate change and lower water levels during which navigation is not possible (Povodí Labe, 2021). Tourist cruises can be found, for example, in Řivnáč's guidebook (Řivnáč, 1882), which describes a steamboat journey from Litoměřice to Hřensko and from there to Dresden. "This voyage touches, especially from Lovosice downwards, perhaps the most beautiful landscapes of Bohemia, which in many places are equal to the famous Rhineland landscapes. Whoever, therefore, has not much time to spare, let him use the steamer, by which the journey, though more tedious, is much more pleasant and delightful" (Řivnáč, 1882, pp. 194–195).[1] The high density of marked hiking trails is also favorable.

Good transport accessibility is not only used by tourists, a large part is used by freight carriers. Smaller villages are burdened by the increased negative effects of road and rail traffic (Aletta et al., 2018). In this respect, the re-development of river transport would help, but under current navigation conditions, an increase in capacity is not possible.

Modern tourism transcends this inaccessibility by bringing tourism closer to all, across age groups, nationalities, time zones, and transport options. Cycle paths, which could be considered as traditional infrastructure that primarily serves residents to spend their leisure time, are also becoming a development element of tourism nowadays. If we compare the similar area of the ViaRhona cycle path, here, the creation of the cycle path has led to an increase in tourist and recreational activity (Adam et al., 2020). The experience it offers contributes to the increased value that users place on the river. This is due to the change in the image of the river following the rediscovery of its natural environment. However, linking cycle paths between cities, regions, or countries (long-distance cycle routes) allows a new mode of transport that is less demanding and can be managed by almost everyone. Cycling is also seen as a means of reducing emissions when transporting tourists to their destination instead of using fossil fuel energy vehicles (Pavluković et al., 2020).

Just as the demands and needs of tourists change concerning the types of tourists according to the absorption of the number of perceptions associated with other tourists or residents, the demands and needs of different types of cyclists are changing. Tourists are divided into psychocentrics, mesocentrics, and allocentrics according to their need for service use during their trips (Huang & Hsu, 2009). The space for tourists' movement is almost unlimited, or only the space of allocentrics is unlimited. This type of tourist does not have exaggerated demands on tourist infrastructure and, instead, seeks out places not marked by tourism. The allocentric prefers meeting locals more than meeting other tourists. Increasing demands on tourist services define as mesocentrics or psychocentrics. The latter can no longer manage without tourist infrastructure and require care and adequate tourist services. They move mainly to tourist destinations and meet many other tourists. The space of cyclists is limited in a similar way to that of mesocentrics or psychocentrics. Mostly, they cannot just

---

[1] *"Tato plavba dotýká se, obzvláště od Lovosic dolů, snad nejkrásnějších krajin českých, které na mnohých místech rovnají se vyhlášeným krajinám porýnským. Kdo tedy nemusí příliš časem spořit, nechť užije parníku, kterýmž jest cesta sice zdlouhavější, ale mnohem příjemnější a rozkošnější."* (Řivnáč, 1882, s. 194-195).

move freely in nature, they need some infrastructure to move around. Like ordinary tourists, we can divide them according to the purpose of spending time on a bicycle and the duration of this time into day-trippers, sport cyclists, and wandering cyclists (Adam et al., 2020). As the length of time spent on the road increases, the cyclist/tourist perceives more of their surroundings. Adam's research shows that sport cyclists and day-trippers do not perceive the environment of the cycle path as much compared to itinerant cyclists who purposefully choose the route they take and want to explore the environment more. Just as non-local cyclists/tourists want to get to know the environment more than cyclists/tourists who live closer.

Cycling is a long-term emerging trend that is close to almost all age groups and popular not only among families with children but also due to the development of e-bikes in productive to post-productive groups (Hose, 2018). A large part of European cycling routes are lined by river landscapes, which are friendly for most cyclists due to their mostly gentle relief. Many European cities also give an important component to watercourses and areas within their territory. One example is Bratislava, which wants to make use of the Danube, its arms, and riparian areas to create a high-quality environment for its residents and visitors/tourists (Skrinar et al., 2019). Adam et al. (2020) further argue that cycle paths create important connections between physical activity and the landscape, but mostly fail to connect deeper connections to ecological functioning. In many cases, there is a lack of active educational tools to understand the environmental interconnectedness of these places. It also states that the use of cycle paths can lead to overtourism especially as awareness of a particular cycle path increases with a marketing strategy in the world/region. Increasing visitation may exclude residents from the users of the cycle path and may encourage conflicts between residents and visitors/tourists (Adam et al., 2020).

Tourism development is currently considered a developmental economic activity (Paulauskienė, 2013). Cycling tourism can be considered a sustainable mode of transport that contributes to tourism development and leaves behind a minimum of negative impacts (Adam et al., 2020). However, the implementation of cycling tourism requires facilities, which are mainly the responsibility of the municipalities on whose territory the cycle paths are located (Andergassen et al., 2017).

3 Methodology and Study Area

Study area: The Elbe Cycle Route (LCS) can be described as an important international cycling route in the Czech Republic (Labská cyklostezka, 2022). As its name suggests, it follows the route of the Elbe River from source to estuary.

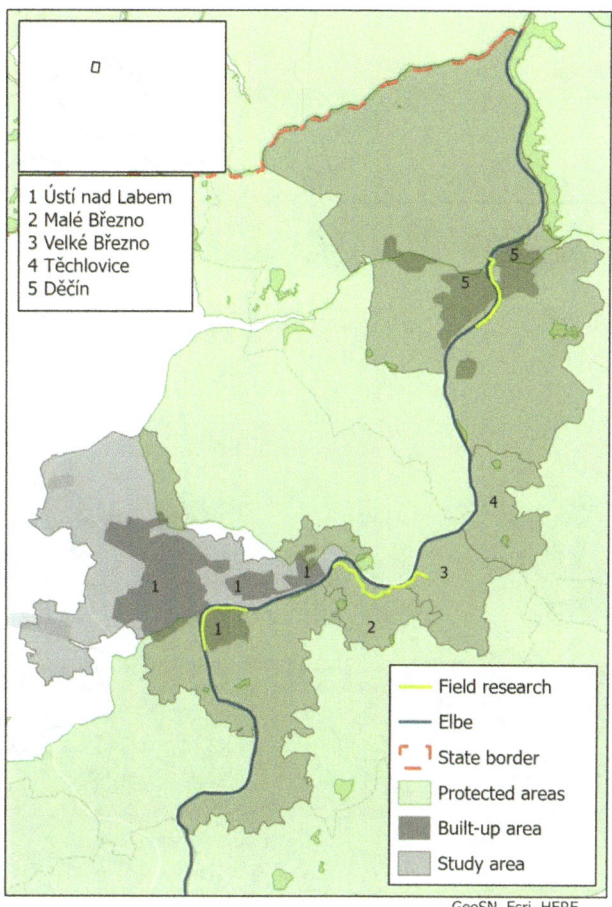

1 Ústí nad Labem
2 Malé Březno
3 Velké Březno
4 Těchlovice
5 Děčín

Field research
Elbe
State border
Protected areas
Built-up area
Study area

GeoSN, Esri, HERE,

Fig. 1 Study area of lower Elbe; by author

The beginning of the cycle path is in Vrchlabí, the end is at the shores of the North Sea. Important cities along the route, especially because of their size (population), are Hradec Králové, Pardubice, Kolín, Nymburk, Mělník, Litoměřice, Ústí nad Labem, Děčín, Dresden, Leipzig, Magdeburg, and Hamburg. On Czech territory, the cycle path is 358.6 km long. The total length is 1300 km.

The study area extends between the state border with Germany and the city of Ústí nad Labem. The Area is mainly formed by the towns of Ústí nad Labem and Děčín (Fig. 2) and then by smaller villages between the towns (Malé Březno, Velké Březno, Těchlovice, Povrly, Malšovice). The main line of the study is the Elbe River as a development factor of the area (Fig. 1).

The relief is formed by the volcanic part of the Bohemian Central Highlands[2] (Fig. 3) and then by the Elbe Sandstones[3] (AOPK, 2022). The Elbe River, which cuts into a deep valley, is an important element of the relief. Specific to this landscape is also the large number of specially protected

---

[2] CHKO České Středohoří.
[3] CHKO Labské Pískovce.

Fig. 2 Lower Elbe landscape, LSC in Děčín, by author

areas, which are almost the entirety of this region in several levels of protection (Table 1). Tourism in this area is focused on nature tourism, cycling, and marginal also water tourism (Štěbetáková et al., 2022). The area also offers several cultural and historical attractions—castles—Děčín, Velké Březno, Střekov, Blansko, museums, theatres, galleries—Ústí nad Labem, Děčín, etc. The main preconditions for the development of tourism in this area include a large number of natural tourist attractions. Most of them are located in the Protected Landscape Area of the Elbe Sandstone (or NP České Švýcarsko)—Pravčická brána, gorges of the Kamenice River, viewpoints to the Elbe Canyon, Jetřichovice Rocks, Via Ferrata in Děčín, or the Protected Landscape Area of the Czech Central Highlands.

Methods: The research was divided into three phases. The first was an analysis of the theoretical background, i.e. literature and media reports/papers and websites. A search of academic articles was conducted using publicly available online search engines for scholarly literature—Web of Science EBSCO and Google Scholar. The keywords used for the search were "tourism", "river", "cycling", and "geotourism".

Furthermore, an analysis of strategic plans, land-use plans, and media analysis related to five selected municipalities of Děčín, Malé Březno, Těchlovice, Velké Březno, and Ústí nad Labem was carried out. The analysis of the strategic plans focused on the attitudes of the individual municipalities toward tourism in general and the improvement of conditions for cycling. Based on the spatial plans, it was determined whether there are other cycling routes than the Elbe Trail in the examined municipalities and whether there is space for their connection to the Elbe Trail or the completion of missing sections of the Elbe Trail. The analysis of the media associated with the municipalities aimed to find mentions of the Elbe Trail and some forms of its promotion.

The research described above was also complemented by personal visits to three sites as part of field exercises (Fig. 1). The first of these took place in Ústí nad Labem and involved walking a section along the right bank of the Elbe from the former shipyard to the Masaryk Weir in Střekov. The second visit included the section of the Elbe Trail between Malé Březno and Valtířov. The last trip was to Děčín. The purpose was to gain personal knowledge from the field and to compare the claims described in the analyzed documents to the real situation on at least some parts of the Elbe Trail within the area of interest. The exact dates of the fieldwork were 4, 11, and 18 May. It was always on Wednesdays in the morning (approximately 3 h). This time was chosen because of the start of the tourist season and the more pleasant morning temperatures that may encourage visitors to spend time on the bike path.

Table 1 Specially protected areas with the year of establishment and subject of protection

NP[4]	České Švýcarsko Bohemian Switzerland	2000	Ecosystems of meadows, heaths, forests, the vegetation of watercourses, caves
CHKO (see Footnote 4)	Labské pískovce Elbe Sandstone Mountains	1972	Geomorphological diversity
CHKO (see Footnote 4)	České Středohoří Central Bohemian Highlands	1976	Varied geological development, rich fauna, and flora
PO[5]	Labské pískovce Elbe Sandstone Mountains	2004	Significant bird species
EVL (see Footnote 5)	České Švýcarsko Bohemian Switzerland	2005	Favorable condition of habitat types and plant and animal species
EVL (see Footnote 5)	Labské údolí Elbe Valley	2016	Geological landscape, water ecosystems, forest vegetation
EVL (see Footnote 5)	Dolní Ploučnice Lower Ploučnice	2008	Riparian vegetation, aquatic animals
NPR[6]	Kaňon Labe Canyon Elbe	2010	Natural forest cover
NPP (see Footnote 6)	Březinské tisy	1969	Yew forests, palaeontological site, a geological feature
NPP (see Footnote 6)	Pravčická brána Prebischtor	1963	Geological phenomenon
PP[7]	Nebočadský luh	1994	River channel of the Elbe, an important nesting and wintering site for bird

Source AOPK (2022)

The work is complemented by interviews from the project *"Děčín Navigation Stage, a step towards sustainability or loss? The potential of shipping from the perspective of the tourism stakeholders concerned."*[8] The research was conducted through semi-structured interviews among municipal government actors. The topics covered included the current state of tourism, the future development of tourism in municipalities, and transport issues. The interviews were realized from September to November 2021. The interviewees were the municipal leadership or persons responsible for tourism development in the municipality. The respondents were 3 men and 2 women of working age.

4 Results and Discussion

Tourism according to strategic documents: According to the development strategy of the city of Ústí nad Labem, the Elbe cycle path does not represent a key part of the city's economy for the time being, and its current importance speaks for its overall attractiveness. International cooperation is crucial for tourism in Ústí nad Labem, especially with Germany, but the level of cooperation at the local (city) level is still insufficient. The strengths of the municipality are the quality of natural conditions, both in the town and especially in its surroundings. This represents the greatest potential for the development of hiking, cycling, and other forms of leisure tourism. The Elbe cycle path is considered a key development potential. The downside of the trail in the neighborhood of Ústí nad Labem is that it lacks quality additional tourism infrastructure and its unsafe connection to the city, which reduces its leisure and tourist use (Magistrát města Ústí nad Labem, 2020).

The document of the concept of tourism development in the city of Ústí nad Labem shows that cycling is one of the important forms of tourism in the city, which contributes to the overall attractiveness of the destination for visitors. The document shows that the main problem of the cycle path is perceived as its insufficient connection to the city and other regional and long-distance cycle routes. The real usability of the Elbe River for tourist purposes is limited due to the undeveloped infrastructure and water conditions (KPMG ČR, 2020).

Within the framework of the action plan for the development of cycling in Ústí nad Labem, the Ústí nad Labem City Council aims to support the completion (design, construction) of the Elbe Trail in the city (i.e. in Církvice). In some sections, the cycle path does not have its infrastructure but is routed along the local road II/261. Furthermore, it also

---

[4] Large protected areas according to national legislation.
[5] European Protected Areas.
[6] Minor protected areas according to state legislation.
[7] Nature park.
[8] Plavební stupeň Děčín, krok k udržitelnosti nebo ke ztrátám? Potenciál lodní dopravy z perspektivy dotčených aktérů v oblasti cestovního ruchu.

Fig. 3 Landscape of lower Elbe, protected areas, and municipalities, by author

aims at the gradual completion of other sections, e.g. the elevation in Sebuzín and the stairs by the dam (Magistrát města Ústí nad Labem, 2012).

The strategic plan of the municipality of Malé Březno shows that the municipality supports tourism in the municipality and has plans for the future to further develop tourism. There are several objectives, namely to improve and develop the accompanying tourism infrastructure, in the form of route marking, thematic and educational trails, sanitary facilities, furniture, rest areas, bicycle parking, bicycle storage, and horse mooring (Malé Březno, 2022a, 2022b). In the municipal urban plan, the Elbe Trail and its surroundings are marked as structural green space. The development of the cycle path is not indicated in the plan.

The strategic plan of the municipality of Velké Březno shows that the municipality promotes tourism in its territory, but the SWOT analysis shows that the municipality perceives the lack of tourism promotion as a weakness. The Elbe Trail is perceived by the municipality as an unambiguous sports attraction and in the SWOT analysis, the trail

is included among the strengths of the municipality. In addition to the strengths of the municipality, the Elbe Trail also appears its weaknesses concerning the low offer of additional services for cyclists coming to the municipality along the Elbe Trail (e.g. accommodation or bike storage). In the future, there are plans to reconstruct the first floor of the FK Jiskra Velké Březno building (football club building) into accommodation for tourists and to build a public small harbor (Velké Březno, 2022). In the spatial plan, we can see that there is no indication in the spatial plan that anything could change in terms of the possible future development of the Elbe Trail. The area around the cycle path is designated as green space (Velké Březno, 2022).

The strategic plan of the municipality of Těchlovice shows that the municipality does not show much interest in tourism, like the other municipalities in our research. This statement is supported by the opinion we can read in the strategic plan, namely "the village will continue to serve mainly a residential function while maintaining a quality living and living environment". At present, the municipality of Těchlovice does not exploit the potential of the area in terms of tourism, although many cyclists pass through the municipality from spring to autumn along the Elbe Trail. Nevertheless, development plans linked to the Elbe Trail can be found in the strategic plan. The aim is to provide or cooperate in the creation of several resting places for tourists and cyclists. At the same time, the Elbe Trail is seen as an opportunity for possible tourism development and promotion of the municipality (Těchlovice, 2022). In the spatial plan, the Elbe Trail is designated as a stabilized cycle path along the Elbe River. No new cycle paths are proposed in the municipality. The area around the cycle path is designated as green space (Těchlovice, 2022).

According to the strategic plan, Děčín has significant potential for tourism because of the Elbe River, its location, and its natural surroundings. Tourism is heavily concentrated in the summer season and visitor stays tend to be shorter. The SWOT analysis shows that the strengths of the city include the Elbe Trail and its important position for tourism. At the same time, the absence of cycle paths and cycle lanes for daily commuting and the lack of direct connectivity between the left and right banks of the city center for pedestrian and cycle traffic is perceived as weaknesses. Another negative of the Elbe Trail is that the cycle path is not sufficiently widened for daily use and there is a lack of nearby capacity for parking or bicycle storage. As the municipality aims to increase the use of walking and cycling by residents, expand barrier-free accessibility, develop supporting infrastructure for cycling and walking in the city, and use cycling for daily commuting and recreation, it plans to make changes by 2027 to meet these goals. The construction of a footbridge over the Elbe River, the construction of significant cycle paths, the modernization of existing cycle paths (surface repairs, signage), and the creation of accompanying infrastructure (racks, rest areas) and services for the cycle path (bike rental, storage, service, bike sharing, charging points for e-bikes, etc.) should be implemented (Statutární město Děčín).

The spatial plan shows that in terms of the possible future development of the Elbe Trail, there is no indication that anything will change in the future. The Elbe Trail is drawn in the land-use plan as a publicly beneficial transport structure, specifically classified as a cycle route (H1—cycle route along the Elbe valley Nebočady—Prostřední Žleb). The areas around the cycle path are designated as green areas (Statutární město Děčín, 2022).

Media analysis: References about the Elbe Route in connection with Ústí nad Labem are very frequent on various websites, as Ústí is one of the larger and more important centers located on this cycle route. From the leisure portal of the city of Ústí nad Labem, the official cycling guide Labská stezka 2022 can be downloaded. The city also promotes the Labská stezka on its Facebook page and Instagram profile (Statutární město Ústí nad Labem, 2022a, 2022b). Mentions about the cycle path can also be found on travel portals such as Kudyznudy.cz, labskastezka.stredohori.cz or autovylet.cz. The Labská cyklostezka has its official website, which provides information about the route itself, links to accommodation, restaurants, and tips for trips (Labská cyklostezka, 2022).

The official website of the municipality of Malé Březno has a tourism tab, but there are no tips for trips and no mention of the Elbe Cycle Path (Malé Březno, 2022). A search for the Elbe Trail in connection with Malé Březno found a total of five relevant results in the search results.

On the official website of the municipality of Velké Březno, a list of tourist destinations and sights can be found (Velké Březno, 2022). From this page, it is then possible to click through to the official website of the Elbe Trail. When searching for the Elbe Trail in connection with Velké Březno, a total of thirteen relevant results were found in the search results.

Most of the articles about the municipality in connection with the Elbe Trail only mention that the cycle path passes through the municipality but no other connection between the cycle path and the municipality can be found. On the official website of the municipality, one can find a tab that is dedicated to sports and tourism in the vicinity of the municipality, but no mention of the Elbe Trail is included (Těchlovice, 2022).

On the official website of the city of Děčín, a link to the tourist portal (Děčín je boží, 2022) of the municipality of Děčín can be found in the "links" tab, where information about the Elbe Trail can be found. References to the Elbe Trail in connection with Děčín are very frequent on various websites, since, like Ústí, Děčín is one of the larger and more important centers along this cycle route (Děčín je boží, 2022).

Field research: **Ústí nad Labem** The Elbe Trail in Ústí nad Labem is visually in quite a well-preserved condition, but there are still negative features that limit or make it uncomfortable for visitors to use. During the field exercise, we observed that in the part of the former shipyard over the Mariánský Bridge to the Dr. Edvard Beneš Bridge, the environment on the cycle path is disturbed by noise from road and rail traffic on the left bank of the Elbe. Specifically, these are the busy roads, which are the I/62 road and the Prague-Děčín-Drážďany railway corridor. The section from the Railway Bridge to the dam was significantly quieter in terms of noise. There is a residential area in the vicinity of this section, specifically around the Střekovské nádraží railway station, and there is a vegetation cover on the side away from the station, which means that there is not as much noise (Aletta et al., 2018) as in the case of the parts located opposite the city center.

In terms of amenities and facilities for cyclists, the cycle path has only one facility, which is located immediately next to the Dr. Edvard Beneš Bridge on Střekovské nábřeží. Here, according to Mapy.cz (2021), there is a resting place, information boards, a place for bike repair, a bike rack, and a refreshment place called KOLOcafé. In addition to these facilities, rest areas in the form of benches were located along the entire route of the cycle path. The area around the cycle path could be further developed, there is space to improve the river banks and access to the river. Of course, this depends on the possibilities given by the water level, which is not stable due to the increasingly frequent fluctuations in rainfall and dry periods. Possible modifications do not only concern the right bank of the Elbe but also the left bank, which is between the Střekov weir and the Railway Bridge, e.g. burdened by the brownfields of the former sugar factory, shipyard, and others.

Malé Březno The Elbe Trail in Malé Březno is visually quite attractive. Near the trail in the direction of Velké Březno, the trail is surrounded by greenery and several bunkers (remnants of the Second World War), which are, however, in an unfavorable condition, can be an interesting element for visitors. Their repair and change of use could increase the attractiveness of the trail/region. This part of the cycle path looked quieter than the part in Ústí nad Labem, however, this part is also bordered by a main road and a busy railway.

The equipment of the route in this part was somewhat worse than in the Ústí nad Labem part. There were very few rest areas along the route. In the part of the trail closest to the town, there is a large gazebo with an information board and a playground. A little further on, there is the only refreshment stand U Tůně, which was closed at the time of our visit.

Velké Březno The LCS is visually quite attractive in the area of Velké Březno as it is in the previous part of the research area. The cycle path is partly through the green area and partly through the residential area. This part of the Elbe Trail was the calmest of all the parts we visited, especially in the section that runs through the residential part. Here too, the main road and the railway line are close to the cycle path.

The trail facilities in the part of Velký Březno were somewhat better compared to the part in the surroundings of Malé Březno, but even so, the area around the trail could be equipped with more rest areas. For variety, it is possible to use the ferry to get to the other side of the Elbe River, where you can use the cycle path that leads from Povrly to other villages. There is also the Museum of Czechoslovak Fortification, which can also be a pleasant diversion for tourists. It is also possible to have a snack in the Tivoli restaurant, which is probably the only place that offers this possibility to tourists in the village.

Děčín The last part of the research area was the Elbe Trail in the territory of the city of Děčín in the section from the Křešice district to the Tyrš Bridge (Tyršův most) in the city center. This part of the cycle path was very attractive for the most part (Fig. 2), but one part of it ran along a road in a residential area, which, in our opinion, is quite an unfortunate solution, as tourists have to avoid cars, even though this road is not very busy. The first three kilometers or so were along a busy road, which makes it difficult to enjoy the trail peacefully. The second half of our journey was in the center of the town, where the great surprise was the abundance of greenery and picturesque views of, for example, the Děčín Castle or the Shepherd's Wall.

The attendance at the cycle path was quite high considering the fact that we visited the cycle path in the morning hours on a weekday. According to our information, this section of the cycle path is very busy, both on weekdays and especially on weekends.

In terms of amenities and facilities for cyclists, this section has the most facilities for visitors to the cycle path. In the part that we had the opportunity to walk through, there were three refreshment facilities and one campsite. One refreshment stand at the cycle path is located in the Křešice district. The other two are located right next to each other, namely the restaurant Kocanda, which also serves as a hotel and has a playground and a renovated bunker, and the second object is the Kozinec buffet, which is famous for selling fish. Camp Děčín is located under the road bridge of road 13 and is, thus, the only camp we saw during our field exercises. However, we are unable to assess the occupancy of the above sites as none of the facilities were open at the time of our visit. These listed facilities are located outside the main town center, however, as the cycle path also passes through the town center, there is the possibility of eating there as well, for example, there is a newly renovated library right

next to the cycle path, which includes an information center and café. Bicycle parking is available at the building. There were three bicycle parking options on the cycle path, two of which were located right next to the library building, where the bicycle repair shop was also located, and the second location was at the Kocanda restaurant. In terms of rest areas, the section did not have sufficient rest areas. We counted about three benches in total from the beginning of our journey, but there were more resting places in the center. Near the Kocanda restaurant, there is a plot of land that could be further worked with, as it is a large concrete area, which, according to the cadastral office, is owned by the city of Děčín. Due to the size of this area, one could come up with countless ideas on how this area could be used, as the place is currently not used in any way.

Interviews: Interviews of municipal representatives were used to supplement the research. The interviews serve to complement the attitude of the municipality toward tourism and cycling development in the municipalities.

> …if I put it in real common sense, yes, there will be a resurgence and I see it mainly in the freight traffic, that it will relieve and reduce the trucks a little bit.[9]
>
> I see a lot of rail traffic because lately the frequency and speed of the trains have been increasing, they're always doing some repairs, so I guess they're preparing for the increase in speed. I've lived here since I was a kid and I'm used to the noise.[10]
>
> That would make the road a lot easier, what can we say, and it's not as much of a polluter as the trucks are. And it's nice when the boats come by, you see it sometimes and it's nice. There used to be orchards here and the fruit was delivered by water. So the traffic on the Elbe was …[11]

Controversial among the realization preconditions is the capacity of the tourist infrastructure represented by accommodation facilities. The cities of Ústí nad Labem, Děčín, and also Hřensko have a large capacity. Smaller municipalities have zero or very low accommodation capacity.

Although this area is exposed to tourism, already from the information described above, not all municipalities focus on tourism.

> Tourists, yes, so that maybe there are some of those businessmen, but when I see how these people behave. It's going to be a burden, that kind of organized tourism, fifty or sixty people come in a little steamboat and you know, they're not going to

---

[9] "…když to vemu vopravdu selskym rozumem tak ano, bude to voživení a já to vidim hlavně v tý nákladní dopravě, že se uleví a trochu se omezí ty kamiony."

[10] "Železniční dopravu vnímám hodně, protože poslední dobou se zvyšuje frekvence a rychlost vlaků, pořád tu dělají nějaký opravy, tak to asi připravují na navýšení rychlosti. Já tu bydlím odmala a na tenhluk už jsem si zvyknul.."

[11] "To by se ulevilo silnici, co si budeme povídat a není to takový znečišťovatel jako jsou kamiony. A je to hezký, když tady jezděj lodě, občas to vidíte a je to pěkný. Dřív tady byly sady a ovoce se rozváželo po vodě. Takže ta doprava na Labi byla ..."

walk on the trail, it's messy everywhere...I'm not surprised that maybe the conservation group is a little bit worried about it.[12]

The fault of the lack of use of cruise transport is due to a lack of presentation and perhaps promotion by the city and the tourism industry as a destination.[13]

We as a municipality don't have that many attractive places, we're more of a connector ... It would be nice (to develop tourism) but I don't have anything here to attract those people. If there was a dam here, it might come back to having shops and pubs, but now it's not worth it. Now we've got one lady who rents a pub here and she's been dragging it out for nine years.[14]

Some of the villages see the cycleway, they also see that it could help community development, and tourism and improve services for residents. The cycleway offers new opportunities and acts as a pull factor for businesses. The cycle path can also be seen as a tool for reducing emissions and an option for use in normal daily transport.

> ...there was a gentleman here who wanted to do a campground, that the bike path was here, with the pond - he wanted to clean that up.[15]

Discussion: Depending on the fact that bicycle paths can increase tourist and recreational visits to the area, it is necessary to prepare other infrastructure for tourist services. At the same time, the needs of visitors change depending on the type of visitor. The relatively sparse range of services—the possibility of accommodation in Ústí nad Labem and then only in Děčín—might suit only long-distance cyclists, but these will not help the development of smaller communities. Some municipal representatives admit that they would like to see tourism development and acknowledge that tourism could help the communities economically. However, their planning documents do not quite show this (Table 2). According to the traditional concept (Klír, 1908; Saarinen et al., 2017), the development of cycling tourism could be helped by providing multi-modal transport. An example could be the connection of boat transport, which also from the perspective of modern tourism will ensure attractiveness by offering a higher range of services (Oliviera et al., 2020) and bringing the destination closer to a higher range of

visitors, including those who are less mobile (Pavluković et al., 2020).

From an e-cycling perspective, the concentration of services is sufficient (Hose, 2018), as it is possible to cover a greater distance on an e-bike. On the other hand, there are no facilities for bike maintenance or repair in the whole area.

When comparing personal knowledge with the information from the strategic plan of Ústí nad Labem, it can be concluded that the condition of the cycle path in Ústí nad Labem corresponds to the information in the document. We can read that the Elbe Trail is not a key part of the economy for the city of Ústí, so the city does not have that much effort to improve the cycle path. In addition, the document says that the Elbe cycle path is considered a key development potential, but according to its condition, there is no indication that the potential has been used. This is evidenced by the fact that the objectives set out in the Action Plan for the Development of Cycling in Ústí nad Labem have not been met to date, and this is a ten-year-old document. If the document of the concept of tourism development of the city of Ústí nad Labem states that cycling is one of the important forms of tourism in the city, which contributes to the overall attractiveness of the destination for visitors, the city should focus at least on the above-mentioned objectives within the action plan for the development of cycling in Ústí nad Labem.

When comparing personal knowledge to the information from the strategic plan of Malé Březno, it can be noticed that the municipality is aware of the deficiencies that are found on the Elbe Trail in this part and they are responding to this with plans written in the strategic plan. From our knowledge, the cycle path is mostly lacking rest areas and other accompanying infrastructure and it is these deficiencies that should be completed by 2027.

If we compare our observations to the information from the strategic plan of Velké Březno, we can notice that the municipality of Malé Březno perceives the low offer of additional services for cyclists as a weakness of the cycle path, which we had the opportunity to see ourselves. This is generally a problem of this cycle path across all the municipalities we had the opportunity to visit. Almost nowhere was there a possibility of bike storage or some kind of parking outside of Děčín, where this possibility existed, but only in the city center. But besides that, the strategic plan does not mention the pros, cons, or plans with the Elbe cycle path, so there is not much to compare. The municipality perceives the Elbe Trail as an unambiguous sports attraction, which would be appropriate, as it is a small municipality that does not offer many sporting activities apart from the cycle path.

Field research in Těchlovice has not been carried out in this village, however, from the available sources (Mapy.cz, 2021), it can be concluded that the amenities of the village

---

[12] "Turisti ano, aby tu třeba byli nějaký ty podnikatele, ale když vidim jak se ty lidi chovaj. Bude to zátěž, taková organizovaná turistika, přijede padesát šedesát lidí na parníčku a znáte to, oni po chodníčku nepůjdou, všude nepořádek... já se nedivim že třeba to chákáočko z toho má trošičku obavy."

[13] "Chyba nedostatečného využití osobní lodní dopravy je nedostatečná prezentace a možná i osvěta ze strany města a cestovního ruchu jako destinace."

[14] "My jako obec nemáme zase až tak atraktivní místa, my jsme spíš taková spojnice ... Bylo by to hezký (rozvíjet CR) ale nemam tady nic, co by ty lidi přitáhlo. Kdyby tady bylo vodní dílo, tak by se sem zase mohlo vrátit to, že by tu byly krámečky a hospody, ale teď to nemá cenu. Teď tu máme jednu paní co si pronajímá hospodu a ta to chudák nějak devět let už táhne."

[15] "...byl tady pán, kterej tu chtěl udělat kemp, že je tu ta cyklostezka, s rybníkem - ten chtěl vyčistit.."

Table 2 Overview of statements from the Strategic Plans of Research Municipalities

Ústí nad Labem (1)	Tourism is not a key part of the economy	LCS has a key potential for development	Intermittent, noisy
Malé Březno (2)	Support for tourism development	By 2027 planned development, equipment	Attractive, quiet
Velké Březno (3)	Insufficient promotion of tourism Insufficient services	LCS is a strength of the municipality	Attractive, quiet, tourist attractions
Těchlovice (4)	Municipality focuses on residents, tourism comes second	LCS is an opportunity for possible tourism development	Attractive, quiet, tourist attractions, outside the center
Děčín (5)	The sparse density of cycle paths, insufficient bank connections	LCS is a strength of the municipality	Refreshments, accommodation

Source Magistrát města Ústí nad Labem (2012), Malé Březno (2022), Velké Březno (2022), Těchlovice (2022), Statutární město Děčín (2022)

are similar to those of Malé and Velké Březno. However, this municipality takes the position of targeting residents rather than tourists. According to the strategic plans, although the LCS is important for the village, it runs outside the center and therefore does not directly develop the village.

When comparing personal observations to the information from the strategic plan of Děčín, it can be concluded that the condition of the cycle path corresponds to the information in the document, as it shows the shortcomings that we saw on the cycle path. According to the strategic plan, the LCS is very important for the city of Děčín, as evidenced by the fact that a relatively large number of tourists were on the trail at the time of our visit. The lack of a direct connection between the left and right bank of the city center for pedestrian and cycling traffic is perceived as a minus, which we saw ourselves. It is only possible to cross the Tyrš Bridge, which concentrates both pedestrian and bicycle traffic, as well as other types of road traffic. The city is planning to build a footbridge to allow better movement between the banks of the Elbe River. In addition, the city plans to add accompanying infrastructure in the future, which we believe is also lacking on the cycle path, especially outside the main city center. The same is the case with services for cyclists, which are more in Děčín compared to other visited municipalities, but the question is whether there are enough of them. We observed only one service, and no storage or rental facilities in the vicinity of the cycle path. As a result, although, in our opinion, Děčín was the best in terms of facilities and attractiveness, there is still plenty of space for improvement, because, if the number of visitors is high in the summer season, these facilities may not be sufficient.

The results of the paper point to the fact that despite the ability of cycling tourism to develop the territory, and efforts to implement more tourism in strategic plans, the implementation and the actual state of services of the study area do not correspond to the set requirements and needs of cyclists.

5 Conclusion

The main objective was to identify the attitude of local government units toward tourism development concerning the existence of one of the major European cycle routes. The results of the work indicate that despite the ability of cycling to develop the territory and the efforts to implement it more in strategic plans, the implementation and the actual state of services of the study area do not correspond to the established requirements and needs of cyclists. For the completeness of the thesis, field research was implemented as a verification of the theoretical knowledge obtained from professional articles in the field of tourism and tourism development to the presence of bicycle paths, the study of the strategic plans of each municipality, as well as media research related to the Elbe cycle path. Three research questions were intended to clarify the objective:

1. *How do municipalities implement tourism in their strategic documents?*
2. *Does the area offer all the services needed to realize this form and type of tourism?*
3. *If the supply of services is not sufficient, are there other possibilities for the development of tourism services?*

The attitude of most municipalities toward tourism development is identified in strategic documents (1). At the same time, the municipalities perceive the potential for its development in their territory and in the surrounding areas, which are surrounded by protected landscape areas. Even though municipalities perceive the potential, they do not consider it to be crucial for their development, or point to the lack of tourism infrastructure. During the field survey, it was found that the service offered in the studied section is rather limited (2), offering only basic equipment (correction of bicycle defects), and only in the part near Děčín. As far as services such as accommodation facilities or restaurants are concerned, they are mainly offered only by the cities of Děčín and Ústí nad Labem. Some of the smaller municipalities offer only bistros or refreshment bars and no accommodation services. There were no sanitary facilities (e.g. toilets) in any of the places in the studied area. The strategic plans state that there is a lot of scope for tourism development and this is confirmed by the statements of the municipality representatives (3). Some strategic documents point to the use of unused space for restaurants or new areas for campsites and other low-cost accommodation facilities. Municipalities mostly want to develop tourism, but some of them are more focused on improving services for residents. This is evident from the strategic documents and the statements of mayors. Although the Elbe Cycle Route (LCS) is one of the most important cycling routes in Europe and ranks among the long-distance European routes, development in its surroundings is not ideal. Communities would welcome the development of services, but some are aware of the threats that tourism development brings. The seasonality of the cycle route is also a problem, as although it offers a wide potential in the summer, its business potential would be untapped in the winter.

References

Adam, M., Cottet, M., Morardet, S., Vaudor, L., Coussout, L., & Riviere-Honegger, A. (2020). Cycling along a river: New access, new values? *Sustainability, 12*(22), 9311.

Aletta, F., Van Renterghem, T., Botteldooren, D. (2018). Influence of personal factors on sound perception and overall experience in urban green areas. A case study of a cycling path highly exposed to road traffic noise. *International Journal of Environmental Research and Public Health, 15*(6), 1118.

Andergassen, R., Candela, G., & Figini, P. (2017). The management of tourism destinations: A policy game. *Tourism Economics, 23*(1), 49–65.

AOPK. (2022). Agentura ochrany přírody a krajiny České republiky [online]. [cit. 2022-08-14]. Available: https://www.nature.cz/homepage

Děčín je boží. (2022). Turistický portál statutárního města Děčín [online]. [cit. 2022-08-14]. Available: https://www.idecin.cz/

DÚK. (2020). Doprava Ústeckého kraje [online]. [cit. 2021-06-20]. Available: https://www.dukapka.cz.ustecky.cz/gs/

Hose, T. A. (2018). A wheel along Europe's rivers: Geoarchaeological trails for cycling geotourists. *Open Geosciences, 10*(1), 413–440.

Huang, S., & Hsu, C. H. (2009). Travel motivation: Linking theory to practice. *International Journal of Culture, Tourism and Hospitality Research, 3*(4), 287–295.

Klír, A. (1908). Stavby komise pro kanalisování řek Vltavy a Labe v Čechách. V Praze: Ant. Klír., 1908. s. [45]. Dostupné také z: https://kramerius5.nkp.cz/uuid/uuid:dac45ef0-7994-11e3-ae4b-001018b5eb5c

KPMG. (2020). Koncepce rozvoje cestovního ruchu města Ústí nad Labem do roku 2030. [cit. 2022-08-14]. Available: https://www.usti nad labem.cz/files/cz/uredni-portal/o-meste/rozvoj-mesta/koncepce-cestovniho-ruchu-unl_20200806.pdf

Labská cyklostezka. (2022). [online]. [cit. 2022-08-14]. Available: https://www.labska-stezka.cz/

Magistrát Města Ústí Nad Labem. (2012). Akční plán č. 1 – Rozvoj cyklistické dopravy v Ústí nad Labem. [cit. 2022-08-14]. Available: https://www.usti-nad-labem.cz/files/civitas/magul_akcni-plan-1.pdf

Magistrát Města Ústí Nad Labem. (2020). Strategie rozvoje města Ústí nad Labem 2021–2030. [cit. 2022-08-14]. Available: https://www.usti-nad-labem.cz/files/cz/uredni-portal/o-meste/rozvoj-mesta/strategie-rozvoje-mesta-unl_20210401.pdf

Malé Březno. (2022). Vše o obci [online]. [cit. 2022-05-08]. Available: http://www.malebrezno.cz/vse-o-obci/d-18813/p1=51

Mapy.CZ. (2021). [online]. [cit. 2021-09-15]. Available: <https://mapy.cz/turisticka?x=13.9797996&y=50.6584705&z=11

Oliviera, F., Costa, D. G., Duran-Faundez, C., & Dias, A. (2020). Bikeway: A multi-sensory fuzzy- based quality metric for bike paths and tracks in urban areas. *IEEE Access, 8*, 227313–227326.

Paulauskienė, L. (2013). Tourism governance principles and functions at the local level. *Management Theory and Studies for Rural Business and Infrastructure Development, 35*(1), 101–112.

Pavluković, V., Kovačić, S., & Stankov, U. (2020). Cycling tourism on the Danube cycle route in Serbia: Residents' perspective. *Eastern European Countryside, 26*(1), 259–285.

Povodí Labe. (2021). Stavy a průtoky na vodních tocích [online]. [cit. 2022-08-14]. Available: http://www.pla.cz/portal/sap/cz/PC/Mereni.aspx?id=1042&oid=3

Řivnáč, F. (1882). Řivnáčův průvodce po království Českém. s. [I]. Dostupné také z: https://kramerius5.nkp.cz/uuid/uuid:0bb6a822-7774-4b64-8e75-53c8e5e3a39e

Saarinen, J., Rogerson, C. M., & Hall, C. M. (2017). Geographies of tourism development and planning. *Tourism Geographies, 19*(3), 307–317.

Schneider, T. (2000). Bike path Phobia: Selling skeptics on urban greenway bike path safety. *Parks and Recreation-West Virginia, 35* (8), 62–69.

Skrinar, A., Misik, M., & Janota, M. (2019, September). River restoration as an element in sustainable urban development. In *IOP Conference Series: Materials Science and Engineering* (Vol. 603, No. 3, p. 032031). IOP Publishing.

Statutární Město Děčín. (2022). Strategický plán rozvoje města Děčín 2021–2027. Available: https://mmdecin.cz/ostatni-dokumenty/rozvoj-strategie-mesta/strategicky-plan-rozvoje-mesta-decin-2021-2027

Statutární Město Ústí Nad Labem. (2011). Územní plán Ústí nad Labem. Available: https://www.usti-nad-labem.cz/cz/uredni-portal/seznamy-zprav/dalsi-informace-z-odboru/odbor-uzemniho-planovani-stavebniho-radu/uzemni-plan-usti-nad-labem.html

Statutární Město Ústí Nad Labem. (2022a). Územní plány obcí v ORP Ústí nad Labem [online]. [cit. 2022-05-08]. Available: https://mapy.usti-nad-labem.cz/apps/up_obce/?run=upd&kod_obce=568350&nazev=VelkeBrezno_1zmena

Statutární Město Ústí Nad Labem. (2022b). Volnočasový portál—Ústí nad Labem [online]. Available: https://www.usti-nad-labem.cz/cz/volny-cas/

Štěbetáková, M., Hruška, V., & Raška, P. (2022). Bohemian Switzerland: Long-term spatiotemporal transformations of tourism facilities in rural peripheries between the regulations and access for all. *European Countryside, 14*(2), 328–345.

Těchlovice. (2022). Dokumenty obce [online]. Available: https://docplayer.cz/13104168-Strategicky-plan-rozvoje-obce-techlovice.html

Velké Březno. (2022). Dokumenty obce [online]. [cit. 2022-05-08]. Available: https://www.velke-brezno.cz/dokumenty-obce-s21CZ

Residents' Perceptions of the Socio-economic Benefits of Restaurants in the Township

Zimkitha Bavuma

Abstract

With the emerging change and rapid development within the tourism industry, we have seen the rise of a number of restaurants in the townships of South Africa. Residents are now moving away from franchise restaurants or restaurants located in the city and more towards authentic restaurants within the township which provide a local experience. This has been largely due to a change of perception in the township regarding township-based tourism establishments and the understanding of socio-economic benefits associated with the existence of these establishments. Therefore, the aim of this research was to investigate the residents' perception of the socio-economic benefits of restaurants in the township. This study focused on a qualitative approach. A non-probability technique was selected to ensure convenience sampling of residents within the Langa township, in Cape Town, South Africa. The semi-structured face-to-face interviews were conducted October 2019, and the data collected has been analysed through descriptive analysis which outlined the socio-economic benefits that emerged in the research. The results from this study outlined the importance of restaurants in townships and the significance in tourism development within township. The results further outline the socio-economic benefits of the restaurants in townships, thus contributing to the knowledge of research in tourism, and especially within the South African context.

Keywords

Perceptions • Qualitative method • Restaurants • Socio-economic benefits • South Africa • Sustainable tourism • Township tourism

Z. Bavuma (✉)
Cape Peninsula University of Technology, Cape Town, South Africa
e-mail: bavumaz@cput.ac.za

1 Introduction

The tourism industry still remains a key driver of South Africa's (SA) national economy. It adds to job creation within the country and is a noteworthy supporter of the work of locals and residents despite the devasting effect that the COVID-19 pandemic had on the industry (South African Government, 2022). According to World Travel and Tourism Council (2022) and the South African Government (2022), the tourism industry contributed about 3.7% to the nation's Gross Domestic Product (GDP) in 2019 and dropped to 1.3% in 2020. There was also a decline of 32.4% in employment and 72.6% reduction in international arrivals in 2020 (South African Government, 2022). In comparison with global statistics, there was a 5% decline in the contribution of travel and tourism to the global GDP; as represented in Fig. 1, statistics signalled 10.3% in 2019 and 5.3% in 2020 (World Travel and Tourism Council, 2022). Although the impact of COVID-19 brought about a sudden and immediate halt to the industry, we still relied on the industry as a valuable contributor to the economy, especially since domestic travel become the "positive spin" during the pandemic. Whilst it is improbable that the domestic tourism was able to balance for the losses of the international arrivals (Helble & Fink, 2020), it should be noted that during the pandemic, the focus on the domestic market helped SA and other countries (Organization for Economic Cooperation and Development, 2020). According to South African Government (2022), SA was able to build a better understanding of the domestic market and through exposure drove the market to provide diversity within its offerings.

Some of the unique offerings that continue to be produced in the townships across SA include events, festivals, township tours, and restaurants. With food being largely linked to culture and being the glue to connecting people, food-related entrepreneurship spirit has increasingly become vigorous within townships, especially in the recent years (Salie, 2022). Specifically, the emergence of restaurants in

Fig. 1 Tourism statistics 2019 versus 2020. *Source* Figure created by author

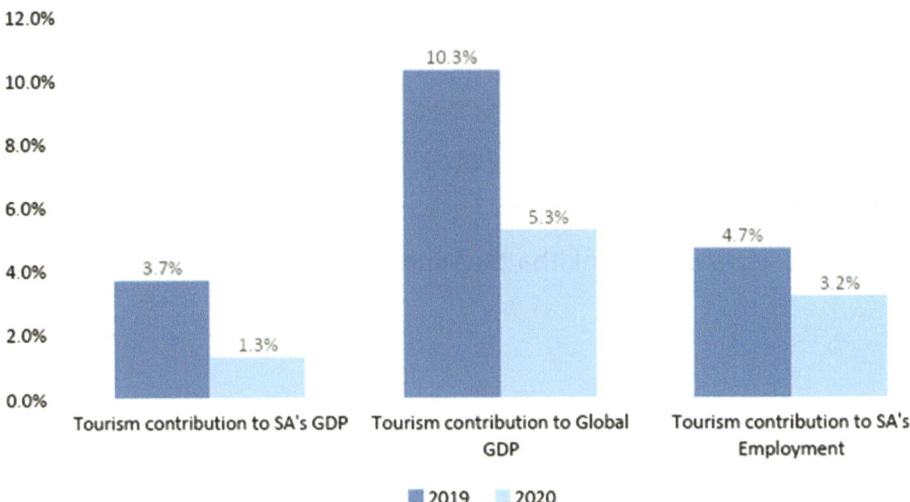

townships has inevitably become a popular tourism attraction for surrounding locals and residents of various townships in SA (SME SA, 2021). This has been as a result of the change of perception in the communities where people are now starting to believe in the concept of supporting local businesses and trusting in the brands that they see emerging in front of them. In local South African communities, residents have become more trusting in brands whose content is well shared and publicised within the heart of their community (Pineda, 2013). Significantly, according to findings of the 2022 SA Township CX Report, there is "a leaning towards building and leveraging Kasi brands, where township residents are looking more to *their own* for services and products" (Rogerwilco, 2022: 25).

Research conducted by Kuzay (2018) shows that majority of South Africans prefer comfort and convenience in terms of restaurant location and are cost conscious when it comes to service reception in restaurants. This is how it became clear that there was a gradual change in perception towards restaurants within townships. The change in the experience and perception towards township restaurants plays an instrumental role in enabling development in the sector as well as the community (Richards, 2011). This has also cultivated a change from the pre-conceived notion that township restaurants are not able to be compared to franchised restaurants or are not seen as adequate enough to compete with the franchises. The change in perception allows for township restaurants to succeed in the market by making enough money to invest towards developing the business and giving back to the community in order to encourage residents to create tourism establishments and developed businesses within the township (Kozak & Kozak, 2015). According to Cárdenas et al. (2015), this is not only for their [restaurant owners'] personal gain, but it is also for residents to understand socio-economic benefits that come from

having establishments like these in their communities. Tourism can help reduce poverty; therefore, it is important to equip residents with ways in which they can make informed decisions that will result in them knowing how tourism can empower them sustainably (Auala et al., 2019).

The importance of having restaurants in townships is that international visitors get to engage with the locals, learn more about the residents, and hopefully provide their perception of how South African townships have changed (Auala et al., 2019; Rogerwilco, 2022). Although, the importance lies in visitors' arrivals, residents are the heart of the community and are the primary concern when it comes to considering the impacts of various establishments in their township (Auala et al., 2019). Therefore, the main aim of the study rested in its investigation of the *Residents' perceptions of the socio-economic benefits of restaurants in the township*, specific to the residents of Langa township. The novelty of this research is that was important to probe this subject in order to (1) understand the opinions of residents in townships, (2) obtain vital information for planning further restaurants in townships in the future, and (3) increase research that covers township tourism as well as restaurants within the township, which scarcely exists.

Included in this paper is a literature review which discusses the following aspects: pro-poor tourism, township tourism, a comparison between pro-poor tourism and township tourism as well as perceptions on socio-economic impacts and sustainability in tourism. Furthermore, the methodology is addressed to provide clarity on which methods were selected and how the research was conducted. The results provided in this paper provide a connection between three key aspects, township tourism, sustainability, and socio-economic benefits in relation to residents' perceptions of restaurants in the township of Langa. The paper then concluded with recommendations for future research.

2 Literature Review

2.1 Pro-poor Tourism and Township Tourism

2.1.1 Pro-poor Tourism

Over the recent years, tourism has been increasingly recognised for its economic potential and its contribution to poverty alleviation (Odhiambo, 2021). This has been especially noted as a great tool for the poorer communities and countries (Ayoo, 2022). However, Sharpley and Telfer (2002) point out that tourism benefits often do not filter down to poor communities in the way in which they aim to. Therefore, it is critically important to investigate approaches to township tourism that will afford local people (residents) real opportunities for economic empowerment (Booysens, 2010). Pro-poor tourism (PPT) as defined by Amoako et al. (2022) speaks to establishing a creative tourism market aimed at targeting the lower class. This can be done by ensuring the introduction of a viable market which creates jobs for residents within a community and a chance at building businesses which will establish generational wealth that will build a safe and reliable source of living. Furthermore, PPT does not have to be a specific product or sector of tourism; however, it needs to be a viable and working approach to the industry for a specific community's needs (Roe, 2018).

Small restaurants within the townships are now the emerging leaders of the food industry compared to commercial franchise restaurants which were the most trusted brands before township residents identified with local restaurant owners, before residents changed their perceptions on how these businesses are ran and before residents got to understand the quality of services they provided (Kuzay, 2018). These small restaurants in the township have become one of key drivers of PPT, ensuring that they are aligned with specific community needs. According to Turok et al.

(2021), this is why tourism is considered as work concentrated and is therefore seen as a device that can be used to employ sizeable segments of the neighbourhood population (residents). Tourism enables the poor to use characteristic assets, which at times are the essential resources that they have in order to create movement within their local economy and the tourism economy (Spenceley & Seif, 2003). This shows how tourism is an effective tool in fighting poverty (Manzoor et al., 2019). The below diagram shows the linkage between tourism and poverty alleviation.

Figure 2 represents all the key aspects that need to be employed to ensure that residents are exposed to all the social and economic impacts that will benefit them long term (Gantait et al., 2021; Valentin, 2015). Restaurants are developing businesses in the township, and it is significant that residents gain something from that. The restaurants can employ its staff from the township in order to play a role in decreasing unemployment and the crime rate caused by unemployment in the township.

2.1.2 Township Tourism

Township tourism was birthed post-Apartheid elections in 1994 (Spencer et al., 2021) and has decades later become widely popular as international tourists have come to SA to gain a more cultural and "authentic" experience which is different from the other breath-taking attractions (Joseph, 2013). Township tourism according to Booysens (2010) and Frenzel (2014) is any tourism activity and tours which are undertaken within the informal settlements or underdeveloped parts of the city also known as slums. In recent years, township tourism has evolved to being creatively utilised through the use of several avenues for capitalising on the economic opportunities and the upgrading of the township's physical spaces that will ensure that locals benefit directly from the development of a particular establishment (Sithole, 2017). Furthermore, Bavuma (2021: 502) states that

Fig. 2 Linkages between tourism and poverty alleviation. *Source* Valentin (2015: 40)

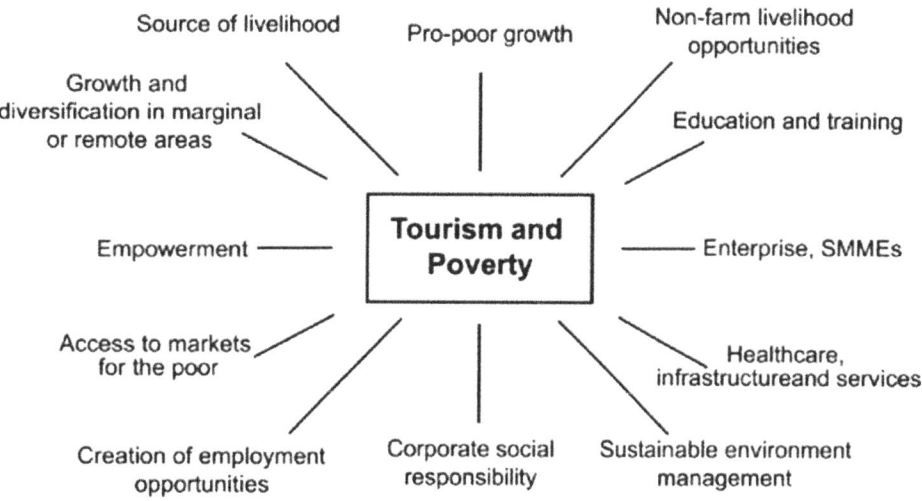

Table 1 Comparison between pro-poor tourism and township tourism

Comparison categories	Pro-poor tourism	Township tourism
Business/economic focus	Pro-poor tourism is about the upliftment of tourism businesses in SA regardless of the community and the societal construct of the area.	Township tourism concentrates predominantly on the upliftment and development of township businesses to the level which they can compete economically with mainstream tourist establishments
Development area focus	Pro-poor tourism addresses the development which will improve the living conditions of the community and ensure the success of the industry. It incorporates both township and rural areas in the development process, so as to ensure equal success in rural and township areas	Township tourism speaks to creating an enabling market where township establishments (restaurants in the case of this study) can compete with urban/city-based establishments
Community focus	Pro-poor tourism is more inclusive compared to township tourism. It looks beyond the community and moves towards building a more inclusive system	Township tourism prioritises businesses in township areas only and leaves out rural communities in the process

Source Table created by author

"township tourism has a positive impact on the township economy through entrepreneurship and township revitalisation programmes as well as policy facilitated by government to help grow enterprises within the community".

Township tourism has created the development of establishments, such as restaurants in townships, which has in turn created opportunities for unemployed youth to gain skills (Koens, 2014). The acquired skills have positioned them better in the pursuit of furthering their career aspects. The establishment of these restaurants has also forced public safety to be channelled to township areas, and residents also have gained an opportunity to socialise while feeling safe in their township (Mavundza, 2019). Township tourism also ensures that the tourism industry is not only focused on international tourists, but also ensures the inclusion of residents who would like to access the tourist experience within their township (Koens, 2014; Auala et al., 2019). On a greater scale, this is also an expression of how township tourism promotes the sharing of business within townships, thus creating equal opportunity for each business to operate. These are all qualities of creating a viable economy, which township tourism advocates for; the benefit of economic elevation of residents (Rolfes et al., 2009).

2.1.3 Comparison Between Pro-poor Tourism and Township Tourism

It is important to understand that PPT and township tourism are tools that can be used to alleviate poverty and can be used as a means of marketing townships as attractions (Sloan et al., 2015). However, it should be noted that they each have a vital role in demonstrating their purpose accurately (Rolfes et al., 2009). Whilst PPT and township tourism are joint collectives, it is evident that each is equally important in their own spheres. Included in Table 1 is a comparison of the two concepts according to business or economic focus, development area focus, and community focus (National Department of Tourism, 2012; Van Wyk, 2005).

From the two concepts and in cohesion with the purpose of this research project, township tourism is at the centre of the long-term impact that restaurants in townships aim to create (Booysens & Rogerson, 2019). Whilst this is a positive step in the right direction, Leonard and Dladla (2020: 900) state that there is still a "lack of tourism skills and education, there is still poor tourism business support, a lack of gender equality and disability support, and high crime" [within townships]. This is the gap which government as well as tourism industry stakeholders need to fill. This can be done by providing support in order to assist township residents to grow their small businesses and to support township residents adequately, in order to harness and grow their tourism skills (Leonard and Dladla, 2020; Sekhaolelo, 2019).

2.2 Perception on Socio-economic Impacts and Sustainability in Tourism

Over the last decade, the tourism industry underwent significant development worldwide, with important consequences being at top levels of decision-making due to the importance of the attitudes and perceptions of the local communities for sustainable tourism planning strategies (Harun et al., 2018). Since SA hosted the 2010 World Cup, there has been huge development in societal perception towards rural and township tourism. Communities started seeing business opportunities in selling their culture and environment with tourists for monetary value and in a way, which benefits the whole community (Rasool et al., 2021; Rutherford, 2006). Thus, the emergence of local businesses such as restaurants in townships. Society now believes that the tourism industry has huge potential to establish a viable income stream even for the poorest (Manzoor et al., 2019). A lot of job opportunities are created through township tourism establishments which benefit the whole community, and societal perception has changed over the years in a positive light when it comes to tourism (Booysens & Rogerson, 2019; Nkemngu, 2015). The importance is to note that communities differ significantly in terms of physical and infrastructural attributes as well as socio-economic profiles (Bob & Majola, 2011). Historically, disadvantaged SA; rural and township communities, were marginalised from tourism initiatives and benefits (Auala et al., 2019; Sloan et al, 2015). These communities were at a disadvantage because of the lack of infrastructure development, accessibility problems, and lack of investor interest to fully develop rural and township tourism (Bvuma & Marnewick, 2020). However, new kinds of tourism such as eco-tourism and cultural tourism offer rural communities an opportunity to attract more visitors and investors to their areas, thereby helping to diversify rural and township economies (Bob & Majola, 2011; Leonard and Dladla, 2020; Viljoen & Tlabela, 2007).

This means that tourism is taking huge strides in township communities, and the hosting of events and festivals attracts people to the township and rural communities. Furthermore, restaurants which never existed before in the townships only came to succeed due to change in societal perception which brought about strategic thinking for local residents who are now owners of the restaurants and contributors to township tourism (Bavuma, 2021). Therefore, sustainability of the industry is dependent on the change of societal perception which will determine the way residents, foreign tourists, and investors see the industry (Zhuang et al., 2019). In order for the businesses to succeed, the community needs to develop so that it can create an enabling environment for the business to attract locals, residents, and international tourists. Figure 3 outlines the four interconnected elements which speak to Global Sustainable Tourism Council (GSTC)'s criteria on the sustainability of the tourism. It indicates the cycle of impacts that are connected in creating a long-term benefit for society, taking into account not only socio-economic impacts, but environmental and cultural impacts as well (Global Sustainable Tourism Council, 2022).

In relation to Fig. 3, the GSTC criteria perspectives that are used for sustainable tourism can be considered determining the following questions in relation to restaurants in the township, especially Langa township in the case of this research.

Fig. 3 Sustainable tourism criteria. *Source* Figure created by author

- Socio-economic: How can Langa implement different programs to boost economic activity, job creation, and sustainable development is in motion? How much economical flow and money can restaurants bring into the community of Langa?
- Cultural: How can Langa showcase their culture, values, and community to tourists/locals who want to experience their history and local customs?
- Environmental: How can the environment be protected and improved? How can safety of locals and residents of Langa be guaranteed?
- Sustainable management: How can locals get to be involved in applying tourism strategies and practices in the community of Langa? How can the conservation of materials happen whilst building tourism in Langa? How can the community be educated in terms of tourism, its benefits, and its suitability within the community of Langa?

To concur with the above mentioned, the United Nations Environment Programme (UNEP) and World Trade Organization (WTO) (2005: 12) state that sustainable tourism takes place when "full account of its current and future economic, social and environmental impacts, addressing the needs of visitors, the industry, the environment, and host communities". Therefore, all of these key criteria need to be considered, planned, and managed accurately in order to ensure sustainable tourism is practiced in the township of Langa and not for the restaurants only (Mowforth & Munt, 2009; Global Sustainable Tourism Council, 2022).

3   Research Methodology

The research project was undertaken under ethical considerations, which firstly included the application of ethical clearance for the research (Research ethics number: 2019FOBREC737). Secondly, followed a sequence of ensuring that permission was obtained from the targeted participants of the study, whilst also ensuring that their responses are kept anonymous and confidential. The sub-sections below further discuss how the research methods were applied for the study in order to obtain the data presented further in this paper.

3.1   Research Area and Research Sample

The study took place in the township of Langa, in Cape Town, SA. Langa township is one of the oldest townships in SA and was established in 1927 as one of the areas which enforced the segregation of Africans after the 1923 Urban Areas Act had been passed (South African History Online, n. d.). Langa township has a rich history and heritage which is celebrated and showcased through its vibrant tourism activities in the post-apartheid era (Spencer et al., 2021). As reported by Statistics SA (n.d.), Langa has a population estimated at 52,401 residents. Therefore, the targeted sample size of this study was to set at obtaining at least 12 interviews, because Boddy (2016: 1353) states that this is where "data saturation occurs among a relatively homogeneous population". However, because the sample is of a large population, the researcher aimed to obtain a maximum of about 50–60 interviews in order for the sample to be represented more widely (Vasileiou et al., 2018). The participants of this research included the residents of Langa township, which were the suitable research sample in order to provide insights on the research topic.

3.2   Research Method and Data Analysis

This study utilised a qualitative research approach, which according to Mohajan (2018) aims interpret the meaning of data by assisting the researcher in understanding the social life and experiences through the target population of the study. The researcher utilised the non-probability technique to collect the necessary data and employed convenience sampling. According to Stratton (2021), this technique is suitable because it allows the researcher to select participants that are available around a particular location. Therefore, this research method aligned to the purpose of this research, which was to investigate the residents' perception of socio-economic benefits of restaurants in the township. Semi-structured interviews were conducted in the

month of October 2019. The interview questionnaire consisted of three sections: (1) a resident's profile, (2) questions on restaurants in Langa as well as a (3) perceptions section where residents responded to particular statements on restaurants in Langa. The researcher designed these questions in order to fulfil the objectives of the research and to align them to the criteria for sustainability in tourism (Fig. 3). Furthermore, the advantage of the researcher having designed the questions and carried out interviews is that it gave strength to the data collection process. In terms of the data analysis of the semi-structured face-to-face interviews, the researcher utilised descriptive analysis to be able provide an in-depth interpretation of each result provided of the residents. According to Gutcheck (2020), descriptive analysis is utilised in order to explain or describe the research objective, based on the responses received from the research sample. As further described by Rawat (2021), this method of analysis assists the researcher in describing as well as summarising the data in a constructive manner that will reveal patterns and perceptions that fulfil the objective of the research.

4   Results and Discussion

This section discussing the findings of the research conducted with the selected participants of the study, which make up of the residents in Langa township. The sub-sections included in the results pertain to the profile of the respondents that participated in the research as well as their perceptions socio-economic benefits of restaurants in Langa.

4.1   Profile of Residents

The profile of the residents that participated in this study is presented in Table 2. The results show that a high majority of the respondents were between the ages of 21–30 years old, followed by fourteen of the respondents being between the ages of 31–40 years old. Showing that there is a high number of young individuals who were willing to participate in the research. Majority of the targeted population that responded to this questionnaire were males residing in Langa township. The reason for a higher male participation could be that, during the week of data collection, the male gender was dominating in presence within the township or were more willing to participate in the research. According to Maila and Ross (2018), most township residents may be considered to not have completed secondary education due to various reasons which are dominated by financial difficulties. However, results show that most of the respondents had completed High School and pursued education beyond

Table 2 Profile of residents

Demographic variables		n	$\%$
Age	21–30 years old	34	58
	31–40 years old	14	24
	41–50 years old	4	7
	51–60 years old	5	9
	61 years old and above	1	2
Gender	Male	30	52
	Female	28	48
Level of education	Senior certificate/matriculation	14	24
	Certificate/diploma	19	33
	Undergraduate degree	14	24
	Postgraduate degree	7	12
	Did not complete high school	4	7
Employment status	Employed	21	36
	Unemployed	14	24
	Student	11	19
	Self employed	9	16
	Retired	3	5
Length of residency	Less than one year	3	5
	1–5 years	18	31
	6–10 years	16	28
	11–15 years	21	36
Frequency of visitation	Once a month	23	39
	Every two weeks	16	27
	Spontaneously	11	20
	Once a week	8	14

that. Only four of the respondents had not completed secondary education. This is significant for this research in that the research participants possessed the sufficient and considerate knowledge on tourism in order to respond to the research questions.

Although majority of the respondents indicated that they were employed, a quarter of the respondents indicated that they were unemployed. It should be noted that some of the restaurants in Langa have provided job opportunities around the area; however, this cannot alleviate the unemployment rate of the entire township. While some of the respondents were students, some of the respondents had positively found a way in making their own income by starting their own businesses and making use their own skills to develop themselves. A great majority of the respondents had lived in the area for a long time, with more than half of the respondents having lived in Langa for more than 5–15 years. This shows that they were well aware of the changes that have occurred during the recent decade, which was significant for the results of the study. The respondents provided great enthusiasm in terms of frequency of visiting the

restaurants in Langa. Majority of the respondents indicated that they visited the restaurants in Langa at least once a month and the reasons could be that individuals were occupied with school and work or attended different spaces on certain ends of the week. Some of the respondents indicated visitations of every second week to restaurants within Langa. Whilst other respondents visited the restaurants in Langa once a week or visited them spontaneously. These results indicated high level of frequency which meant that these residents enjoy the food and atmosphere at the restaurants in their township.

4.2 Perceptions About Socio-economic Benefits of Restaurants in Langa

In recent decades, there has been a great effort in order to equip and develop townships from within, through skills development and through entrepreneurship (Sekhaolelo, 2019). The research conducted included a questionnaire with a perceptions section where residents responded to particular

Table 3 Statements on socio-economic perceptions (in %, n = 58)

Socio-economic perceptions	SD	D	N	A	SA
Restaurants in townships created more business opportunities for residents	–	8	19	28	45
Restaurants in townships are good for linked residents' businesses	–	10	19	31	40
Restaurants in Langa created jobs for residents in Langa	–	12	22	28	38
The existence of restaurants in townships has improved the relationship between locals, residents of Langa, and tourists	–	5	19	–	76
The success of restaurants in the township has boosted the image of Langa	–	5	19	–	76

statements on restaurants in Langa. These perceptions which are linked to the criteria for sustainability in tourism (as discussed in Sect. 2.2). A five-point Likert scale (strongly agree—SA, agree—A, neutral—N, disagree—D, and strongly disagree—SD) was used to each of the various statements. However, for the purpose of this paper, there are three key themes that were identified and utilised for perceptions in relation to the socio-economic benefits of restaurants in the township. The sub-sections (Table 3) discuss the findings from the responses provided by the participants of this research.

4.2.1 Business Opportunities

In terms of business opportunities, two key statements were posed to the participants of this study. Majority of the respondents strongly agreed or agreed to the fact that *restaurants in townships created more business opportunities for resident*. This could be due to the fact that some of the residents were service providers to the restaurants that are within Langa township. For example, one respondent stated that his business entailed providing Deejay services to venues, and he had been able to secure a residency appointment by Deejaying every Sunday at one of the restaurants in Langa. This not only provided him with a source of income, however, with a platform to showcase and grow his business as diners at the restaurant were his potential clients or future business associates. Some of the respondents felt indifferent about this statement because to them there was not much of an indication that the restaurants in townships *specifically* created more business opportunities for residents. In relation to the second statement, majority of the respondents strongly agreed and agreed that *restaurants in townships are good for linked residents' businesses*, whilst some were neutral to this statement. The reason for such a great agreement to this statement would be because some of the respondents were business owners who were benefiting from the existence of the restaurants in Langa. For example, one of the respondents stated that he owned a tour operating company, and

the existence of these restaurants provided a good link for him to show the tourists the attractions of Langa township whilst also visiting these restaurants as part of that the tour package he offered to his clients. Again, it can be noted that those respondents that provided a neutral response may have been indifferent due to the fact that this link was not necessarily for the majority of the businesses that are within the Langa township.

4.2.2 Job Creation

According to the Western Cape Economic Development Partnership and Human Science Research Council (2019), investing in promising projects within the township as well as skills development of township residents are the key strategies towards alleviating poverty and strengthening township economies. This research reveals that a large number of the respondents strongly agreed and agreed with the statement *restaurants in Langa created jobs for residents in Langa*. This can be largely contributed to the fact that the respondents are only aware of the individuals within their township that have benefitted from job opportunities offered by these restaurants. It should also be noted that some of the respondents were either neutral to the aforementioned statement or totally disagreed. This could be due to the fact that there are residents that have benefited from jobs created by these restaurants; however, some of the other employees at these restaurants are not all from Langa township, as there are employees who reside in other townships or areas within Cape Town. This shows that a majority of the township community has gained skills and employment from these restaurants, and as shared by one of the respondents, one of the restaurants in Langa provides a hospitality skills and training program which has helped some community members gain certification and find employment in five-star restaurants and hospitality establishments. One of the restaurant owners in the township of Langa is a product of this program, the owner used all his skills, training, and travel experience to open a restaurant within the community of Langa.

4.2.3 Improve Image and Living Standards

Nkemngu (2015) suggests that in relation to tourism, residents usually support the development of establishments and attractions if they understand how it works, and how they will benefit from it. Manzoor et al. (2019) further postulate that this needs to be represented mainly in the form of individual or community quality of life improvement. In this study, about three quarters of the respondents expressed that *the existence of restaurants in townships has improved the relationship between locals, residents of Langa, and tourists*. The change and improvement in relationships are as a result of social exchange occurring between residents and tourists, as well as the continued relationships that are built between the residents and the owners of the establishments. The restaurants which are situated in the heart of the community are surviving as a result of the community support and have improved living standards that contribute to the community. Some respondents were indifferent to the statement, with three of them disagreeing to the statement. Their responses are linked to the fact that the image of the township remains as an impoverished area which the respondents feel that the restaurants do not change that. However, there is acknowledgement that these restaurants do bring something different within the township environment. The responses collected from the research population remained the same for their perceptions of the following statement: *the success of restaurants in the township has boosted the image of Langa*. This shows that restaurants in the township have brought a significant change in how the residents view the idea of these establishments. These restaurants are not only regarded as privately owned businesses, but as tourism establishments and landmarks which the township is known for. They are establishments that residents can proudly boast about as successful and positive contributors to the image of Langa township.

4.3 Residents' Statements of Socio-economic Benefits of Restaurants in Langa Township

The development of restaurants in the township have become popular, and it has been noted that the township's food, the lively atmosphere, and culture is a big drawcard for these restaurants being noticed by residents and other visitors (Business Insider, 2019). When asked about what benefits they experienced from the existence of restaurants in their township, more than half of the respondents expressed that the restaurants in Langa have not benefited them at all. However, the other respondents provided a number of socio-economic benefits they have experienced through the existence of these restaurants in their community. Some of these socio-economic benefits include: *the close proximity of these restaurants*, which does not require residents to travel far in order for them to *experience professional restaurant standards*. Some of the respondents expressed that they were able to *socialise and meet new people*. The existence of these restaurants also created *a safe space* where they can *relax with their family and friends*, whilst enjoying *good entertainment* and *quality food* at an *affordable price*. One of the critical statements provided by one of the respondents is that: *Restaurants in Langa township change people around me and bring sense of sanity in my township*.

The above-mentioned socio-economic benefits can be linked to answering the key questions posed in relation to sustainability of tourism (Sect. 2.2) including the key perceptions that emerged in the research (Sect. 4.2). Figure 4 demonstrates the interconnection of the literature provided in this research and the results.

The results gathered from the residents' perceptions of the socio-economic benefits of restaurants in Langa show that there are clear aims or attempts in providing a means of **poverty alleviation** through the existence of these restaurants in Langa.

Fig. 4 Concept map of the research. *Source* Figure created by author

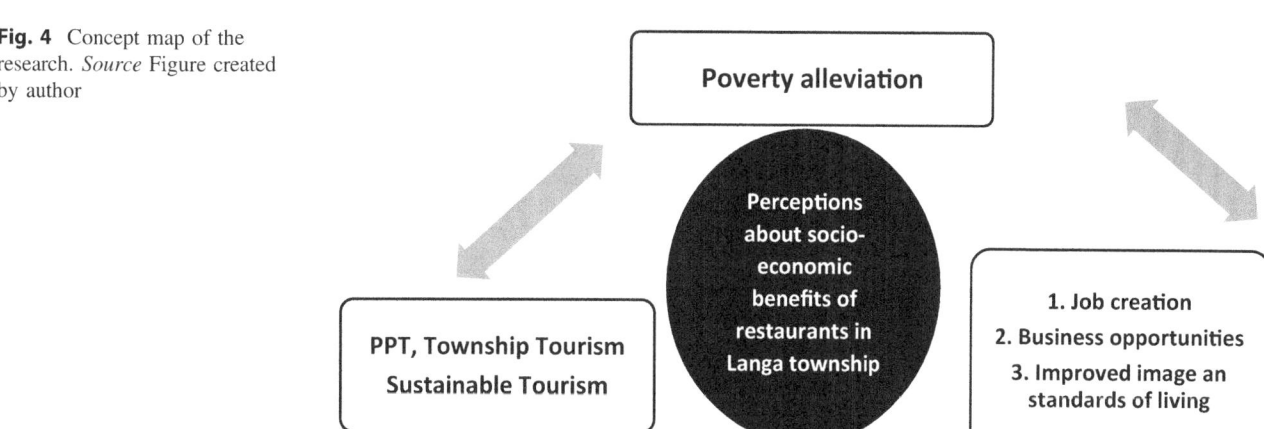

5 Conclusion and Recommendations

To conclude the findings of this research, it is vital to note that residents positively embrace the establishment of restaurants within townships, as they bring about many social benefits including the importance of safety and convenience. Residents also acknowledged that even though the restaurants are privately owned businesses, they do create jobs for residents within the township. The grave factor that comes through in the study is that the restaurants do not primarily create money for the community; however, through their existence and each restaurant owner's affiliations, restaurants in townships have been able to encourage residents in establishing their own businesses. This shows that the restaurant owners in townships have given back to the community by assisting in job opportunities and projects that benefit the community, both in economic prospects and by elevating the residents' standards of living. Thus, encouraging sustainability and growth towards the image and legacy of Langa township. The first recommendation would be to conduct the study again (within 3–5 years) in order to compare the results of the perceived socio-economic benefits of restaurants in the Langa township, especially in the context of a post COVID-19 era. The second recommendation would be to conduct the study of a longer period of time in a quantitative measure to obtain more participants that can contribute to the research. Lastly, the third recommendation for the study is to further continue the study in other townships, not only in the Western Cape (Cape Town), however, on a national scale in order to obtain a broader basis of how to identify and manage socio-economic benefits of restaurants (and other food establishments) within SA townships.

References

Amoako, G. K., Obuobisa, T., & Marfo, S. O. (2022). Stakeholder role in tourism sustainability: The case of Kwame Nkrumah Mausoleum and centre for art and culture in Ghana. *International Hospitality Review, 36*(1), 25–44. https://doi.org/10.1108/IHR-09-2020-0057

Auala, L. S. N., van Zyl, S. R., & Ferreira, I. W. (2019). Township tourism as an agent for the socio-economic well-being of residents. *African Journal of Hospitality, Tourism and Leisure, 8*(2), 1–11. https://www.ajhtl.com/uploads/7/1/6/3/7163688/article_3_vol_8_2_2019.pdf

Ayoo, C. (2022). Poverty reduction strategies in developing countries. *IntechOpen.* https://www.intechopen.com/chapters/79838

Bavuma, Z. (2021). Restaurant owners' perceptions of restaurant developments in the Township of Langa, Cape Town, South Africa. In *Conference proceedings of the 5th international conference on events*, November 2021. https://makingwavesinevents.org/legacies#aad04639-9b7a-4aa3-9858-fd20dbd07ccd

Bob, U., & Majola, M. (2011). Rural community perceptions of the 2010 FIFA World Cup: The Makhowe community in KwaZulu-Natal, Makhowe. *Development Southern Africa, 28*(3), 387–399. https://doi.org/10.1080/0376835X.2011.595999

Boddy, C. R. (2016). Sample size for qualitative research. *Qualitative Market Research*, 1352–2752. https://www.emerald.com/insight/content/doi/10.1108/QMR-06-2016-0053/full/html

Booysens, I. (2010). Rethinking township tourism: Towards responsible tourism development in South African townships. *Development Southern Africa, 27*(2), 273–287. https://doi.org/10.1080/0376835 1003740795

Booysens, I. & Rogerson, C. M. (2019). Re-creating slum tourism: Perspectives from South Africa. *Urbani Izziv, 30*, 52–63. https://urbaniizziv.uirs.si/Summary_s/id/72/id_k/s/idc/5

Business Insider. (2019). *Township homes are being turned into pop-up restaurants—Thanks to this award-winning start-up.* https://www.businessinsider.co.za/down-with-invites-tourists-to-townships-2019-2. September 20, 2022.

Bvuma, S., & Marnewick, C. (2020). Sustainable livelihoods of township small, medium and micro enterprises towards growth and development. *Sustainability, 12*(8), 3149. https://doi.org/10.3390/su12083149

Cárdenas, D. A., Byrd, E. T., & Duffy, L. N. (2015). An exploratory study of community awareness of impacts and agreement to sustainable tourism development principles. *Tourism and Hospitality Research, 15*(4), 254–266. https://doi.org/10.1177/1467358415580359

Frenzel, F. (2014). Slum tourism and urban regeneration: Touring inner Johannesburg. *Urban Forum, 25*, 431–447. https://doi.org/10.1007/s12132-014-9236-2

Gantait, A., Mohanty, P., Singh, K., & Singh, R. (2021). Pro-poor tourism in India—Reality of hyperbole. *Psychology and Education, 58*(2), 9672–9682. https://doi.org/10.5281/zenodo.4608946

Global Sustainable Tourism Council. (2022). What is sustainable tourism. https://www.gstcouncil.org/what-is-sustainable-tourism/. March 28, 2023.

Gutcheck. (2020). *What is descriptive research.* https://www.gutcheckit.com/blog/what-is-descriptive-research/. February 25, 2023.

Harun, R., Chiciudean, G. O, Sirwan, K., Arion, F. H., Muresan, I. C. (2018). Attitudes and perceptions of the local community towards sustainable tourism development in Kurdistan Regional Government, Iraq. *Sustainability, 10*(9), 2991. https://www.mdpi.com/2071-1050/10/9/2991

Helble, M., & Fink, A. (2020). *Reviving tourism amid the COVID-19 pandemic.* Asian Development Bank.

Joseph, M. (2013). *Does township tourism contribute to Government's strategic goals for the tourism sector?* University of Cape Town.

Koens, K. (2014). *Small businesses and township tourism around Cape Town.* Leeds Metropolitan University.

Kozak, M., & Kozak, N. (2015). *Tourism development.* Cambridge Scholar Publishing.

Kuzay, M. K. (2018). *Consumers of local food systems in Cape Town—Perceptions and preferences, using the example of harvest of hope.* Humboldt University.

Leonard, L., & Dladla, A. (2020). Obstacles to and suggestions for successful township tourism in Alexandra Township, South Africa. *e-Review of Tourism Research, 17*(6), 900–920. https://ertr-ojs-tamu.tdl.org/ertr/index.php/ertr/article/view/531

Maila, P., & Ross, E. (2018). Perceptions of disadvantaged rural matriculants regarding factors facilitating and constraining their transition to tertiary education. *South African Journal of Education, 38*(1). https://doi.org/10.15700/saje.v38n1a1360

Manzoor, F., Wei, L., Asif, M., Haq, M. Z., & Rehman, H. (2019). The contribution of sustainable tourism to economic growth and employment in Pakistan. *International Journal of Environmental Research and Public Health, 16*(19), 3785. https://www.mdpi.com/1660-4601/16/19/3785

Mavundza, B. (2019). *Township homes are being turned into pop-up restaurants—Thanks to this award-winning start-up.* https://www.

news24.com/news24/bi-archive/down-with-invites-tourists-to-townships-2019-2. March 10, 2023.

Mohajan, H. K. (2018). Qualitative research methodology in social sciences and related subjects. *Journal of Economic Development, Environment and People, 7*(1), 23–48. https://mpra.ub.uni-muenchen.de/85654/1/MPRA_paper_85654.pdf

Mowforth, M., & Munt, I. (2009). *Tourism and sustainability. Development, globalisation and new tourism in the third world.* Routledge.

National Department of Tourism. (2012). *National rural tourism strategy.* https://www.tourism.gov.za/AboutNDT/Branches1/domestic/Documents/National%20Rural%20Tourism%20Strategy.pdf. March 28, 2023.

Nkemngu, A. P. (2015). Quality of life and tourism impacts: a community perspective. *African Journal of Hospitality, Tourism and Leisure, 4*(1). http://www.ajhtl.com/uploads/7/1/6/3/7163688/article_11_vol_4_1_2015_jan-june.pdf

Odhiambo, N. M. (2021). Tourism development and poverty alleviation in sub-Saharan African countries: An empirical investigation. *Development Studies Research, 8*(1), 396–406. https://doi.org/10.1080/21665095.2021.2007782

Organization for Economic Cooperation and Development. (2020). *Tourism policy responses to the coronavirus (COVID-19).* https://www.oecd.org/coronavirus/policy-responses/tourism-policy-responses-to-the-coronavirus-covid-19-6466aa20/. September 10, 2022.

Pineda, F. M. (2013). *Tourism and environment.* Complutense University.

Rasool, H., Maqbool, S., & Tarique, M. (2021). The relationship between tourism and economic growth among BRICS countries: A panel cointegration analysis. *Future Business Journal, 7*(1). https://doi.org/10.1186/s43093-020-00048-3

Rawat, A. S. (2021). *An overview of descriptive analysis.* https://www.analyticssteps.com/blogs/overview-descriptive-analysis#:~:text=Descriptive%20Analysis%20is%20the%20type,for%20conducting%20statistical%20data%20analysis. September 10, 2022.

Richards, G. (2011). *Cultural tourism: Global and local perspectives.* Routledge.

Roe, D. (2018). *Pro-poor tourism: Harnessing the world's.* International Institute for Environmental Development.

Rogerwilco. (2022). *The 2022 South Africa township CX report.* https://www.rogerwilco.co.za/2022-south-africa-township-cx-report. March 25, 2023.

Rolfes, M., Steinbrink, M., & Uhl, C. (2009). *Townships as attractions: an empirical study of township tourism in Cape Town.* Universitätsverlag Potsdam.

Rutherford, D. L. (2006). *Towards a development strategy for small businesses in the tourism industry of the Southern Cape.* University of Pretoria.

Salie, I. (2022). *#StartupStory: Delivery Ka speed—A township fast-food delivery service.* https://www.bizcommunity.com/Article/196/842/229778.html. September 18, 2022.

Sekhaolelo, E. (2019). *Four ways to give township entrepreneurs a much-needed boost.* https://www.gsb.uct.ac.za/ideas-exchange/entrepreneurship-and-innovation/fo-four-ways-to-give-township-entrepreneurs-a-much-needed-boost. September 28, 2022.

Sharpley, R., & Telfer, D. J. (2002). *Tourism and development: Concepts and issues.* Channel View Publications.

Sithole, N. (2017). *The contribution of tourism to local community development: the case of Shakaland Zulu Cultural Village.* Durban University of Technology.

Sloan, P., Simons-Kaufmann, C., Legrand, W., & Perlick, N. (2015). Township tourism in South Africa—A successful tool for poverty alleviation? *Advances in Hospitality and Leisure, 11*, 153–168. https://doi.org/10.1108/S1745-354220150000011009

SME South Africa. (2021). *Township business trends opportunities.* https://smesouthafrica.co.za/township-business-trends-opportunities/. September 15, 2022.

South African Government. (2022). *Deputy minister Lindiwe Sisulu: Tourism dept budget vote 2022/23.* https://www.gov.za/speeches/deputy-minister-lindiwe-sisulu-tourism-dept-budget-vote-202223-19-may-2022-0000. September 10, 2022.

South African History Online. (n.d.). *Langa township.* https://www.sahistory.org.za/place/langa-township-cape-town. February 25, 2023.

Spenceley, A., & Seif, J. (2003). *Strategies, impacts and costs of pro-poor tourism approaches in South Africa.* Department for International Development.

Spencer, J. P., Ndzumo, P., Muresherwa, G., & Dube, C. N. (2021). Cape Town's township tourism: A case study of Langa. In *Conference Proceedings of the International Conference for Tourism Research*, May 2021. https://doi.org/10.34190/IRT.21.004

Statistics SA. (n.d.). *Langa.* https://www.statssa.gov.za/?page_id=4286&id=318. September 15, 2022.

Stratton, S. J. (2021). Population research: Convenience sampling strategies. *Prehospital and Disaster Medicine Journal, 36*(4), 373–374. https://doi.org/10.1017/S1049023X21000649

Turok, I., Visagie, J., & Scheba, A. (2021). *Social inequality and spatial segregation in Cape Town.* In M. van Ham, T. Tammaru, R. Ubarevičienė & H. Janssen (Eds.), *Urban socio-economic segregation and income inequality. The urban book series.* Springer.

United Nations Environment Programme (UNEP) & World Trade Organization (UNWTO). (2005). *Making tourism more sustainable—A guide for policy makers.* http://www.unep.fr/shared/publications/pdf/DTIx0592xPA-TourismPolicyEN.pdf

Valentin, B. (2015). *Determining the contribution of tourism to poverty alleviation in Mozambique: Case studies of Praia Bilene and Macaneta.* North West University. https://doi.org/10.13140/RG.2.2.33309.51688

Van Wyk, S. (2005). *Doing business differently.* https://mg.co.za/article/2005-07-01-doing-business-differently/. March 26, 2023.

Vasileiou, K., Bernett, J., Thorpe, S., & Young, T. (2018). Characterising and justifying sample size sufficiency in interview-based studies: systematic analysis of qualitative health research over a 15-year period. *BMC Medical Research Methodology, 18*, 148. https://doi.org/10.1186/s12874-018-0594-7

Viljoen, J., & Tlabela, K. (2007). *Rural tourism development in South Africa—Trends and challenges.* HSRC Press.

Western Cape Economic Development Partnership & Human Science Research Council. (2019). *Strengthening township economies.* https://www.ukesa.info/download/u7Kj6DRcliJNtFdBvVGQPnEaHrAk8fmq/STRENGTHENING-TOWNSHIP-ECONOMIES_2019.pdf. September 29, 2022.

World Travel and Tourism Council. (2022). *Economic impact.* https://wttc.org/research/economic-impact. September 10, 2022.

Zhuang, X., Yao, Y., & Li, J. (2019). Sociocultural impacts of tourism on residents of world cultural heritage sites in China. *Sustainability, 11*(3), 840. https://doi.org/10.3390/su11030840

Architecture of Historical Mosques: A Typological Study on the Archaeological Site of Barobazar

Humayra Alam, Nomrota Sarkar, and Md. Khalid Hossain

Abstract

The focus of this study is on the architectural organization and typological characteristics of ancient mosques in the archaeological remains of Barobazar, a region with historically rich infrastructure and vast natural resources. While a small number of historic structures in Bangladesh have received minor conservation measures, the conservative decisions rarely reflect the built environment's period-centric architectural traits. Barobazar's excavated site has a variety of mosque archetypes, such as the Sultanate, as well as ancient mounds and tanks, all of which are indicative of Bangladesh's historical architectural scene. Despite efforts by the Department of Archaeology in Bangladesh to locate and safeguard the old mosques at Barobazar as antiquity, some of the unearthed building forms have been modified in contrast to their prior state, while others have remained deteriorated and unrestored. The study of these mosques provides opportunities to learn about existing instances of ancient religious spaces as well as the decisions that must be made to preserve the physical structures' history and "sense of space". The study focuses on the historical, socio-cultural, and architectural effects on the composition of traditional Islamic religious spaces in the Barobazar area. Based on the results of the archaeological excavation by the Department of Archaeology, Bangladesh, historical mosques were studied, and eight cases were researched to create a typological comparison based on architectural and structural traits. As part of the methodology, field surveys, existing literature reviews on each mosque's social and architectural history, interviews with locals, and data collection based on prior and current structural and architectural conditions were compiled.

The research presented in this paper can aid in the design considerations needed to preserve Barobazar's government-protected sites.

Keywords

Historical mosque • Barobazar • Mosque typology • Religious architecture • Conservation

1 Introduction

Religious practice can be a powerful indicator of a community's cultural identity, manifesting itself in the built environment as reflected through the longtime material and symbolic ties of cities to numerous religions (Goh & van der Veer, 2016). Religious architecture can be defined as manifestations of a religion's dependable fundamental ethical, historical, scriptural, and doctrinal principles—what it professes and accepted standards, and how the architecture works to symbolize and broadcast these (Barrie, 2020). Religious structures can serve as symbols, media, sites of cultural memory, and hubs of social identification (De Wildt et al., 2019). Over time, religious architecture has formed the tangible heritage of cities and nations throughout the Indian subcontinent and the yearning for cultural and religious recognition has grown in importance over the past few decades and is still relevant today (Arab, 2013). As a result, religious structures are being repaired as heritage monuments universally (Verkaaik, 2013).

In the Islamic religion, it is followed that an architectural philosophy must "surrender" to the law and will of Allah SWT since Islam does not submit to the demands of any other, and this is accomplished by incorporating Islamic religion into the construction and decoration of mosques (Longhurst, 2012). This is supplemented with the fact that traditional architecture in Islamic contexts has evolved in response to several characteristics that are unique to each

H. Alam (✉) · N. Sarkar · Md. KhalidHossain
Khulna University of Engineering & Technology,
Khulna, Bangladesh
e-mail: humayra@arch.kuet.ac.bd

context, such as the climate, the available building materials, the sophistication of construction technology, the level of social prosperity, and the local architectural traditions and practices that existed in that context before the introduction of Islam (Sidawi, 2013). Like many other nations throughout the world, Bengal's geographic situation has a significant impact on its history and political boundaries. Due to the silt deposition of great rivers like the Ganga, Brahmaputra, and Meghna, the biggest deltaic region was endowed with extremely rich soil and life-sustaining supplies. Bengal has long drawn new immigrants, traders, and conquerors due to its abundant natural resources, pleasant temperature, and legendary wealth (The Physical Geography of Bangladesh| The University Press Limited, n.d.). In Bengal, remains of ancient urban centers or sizable cities that date back more than 2000 years have been discovered. Even though the economy was mostly agrarian, both domestic and foreign trade (particularly cotton and muslin, a very fine fabric for garments) developed. Cities primarily began to form as centers of trade, production, and expansion. However, there have also been administrative, military, and even cities for centers of learning and information collecting (Polin et al., 2019).

Between the fourteenth and mid-sixteenth centuries, a large number of people in Bengal converted to Islam which gave rise to the construction of mosques during that period. As a result, a new form of building was developed which would influence the architecture of future mosques, and this development was a combination of both tradition and adaption of global trends in mosque architecture (Hasan, 1989). History has shown that Ikhtiyar Uddin Mohammed bin Bakhtiyar Khilji was drawn to the north after conquering Nadia, oblivious to the south or southeast. His kingdom remained vast in the north as a result. Later, some of Jashore and Khulna were brought under the dominion of Nasiruddin Mahmud Shah, the son of Shamsuddin Ilyas Shah. Finally, Khan Jahan Ali was crowned victorious in the southeast (বাংলাদেশ, n.d.). Historically, the construction of mosques in the context of Bengal was aided by rulers and soldiers to assist in the growth of Islam (Tariq & Jinia, 2013). Khan Jahan traveled a long way to Bagerhat, via Nadia (currently a district in West Bengal, India), along the river Bhairab, and by crossing the mighty Ganges River. He then made an appearance at Barobazar in the district of Jhenaidah. As a result, of the four townships he founded in Bengal, Barobazar was the first. The others were Khalifatabad, Murali

Qashba, Poyogram Qashba, and Murali Qashba (Mitra, 1914). The Department of Archaeology conducted an investigation and found 14 mounds in this region. But in 1989, only two mosques, Satgachia Gayebana Mosque and Gorar Mosque were targeted for protection. The latter was being used for prayer and was partially uncovered by the locals. The remnants of a 35-domed mosque were discovered in 1989 as a result of excavation work done by the Department of Archaeology in the site's surviving area, Satgachya. The government designated these locations as protected monuments, and a program of routine archaeological excavation was launched to uncover the region's cultural value and reconstruct its past. Jorbangla, Galakata, and Kharer Dighi were placed under spade operation that year. The sites' structural remnants had been made visible as a result. These include mosques, cemeteries, ports, and nonreligious structures (Barobazar—Banglapedia, n.d.). The small township of Barobazar spans an area of approximately 6.44 km and continues to be significant due to the presence of these ancient mosques.

Due to inadequate publicity, initiatives from local authorities, and a lack of surrounding accommodations, travelers are unaware of the former splendors of the Sultanate era in Baro Bazar (Independent, n.d.). Thus, the study focuses on the spatial organization of these excavated mosques to explain the sense of space as developed throughout history in these religious architectures, not only as a result of tradition but also as a typological interconnection between local and historical influence.

2 Methodology

The study was conducted in multiple phases (Fig. 1). First of all, for the secondary data collection, relevant pieces of literature, articles, and online portals were researched and studied based on each mosque's social and architectural history. As part of the primary data collection, field surveys of each case were conducted and these surveys included recording the architectural features, the existing condition of the built forms, and the surrounding land use along with the activity. Interactive and open-ended interviews with the local residents were performed to collect intangible data like the existing functions, local legends, myths, and current socio-cultural aspects.

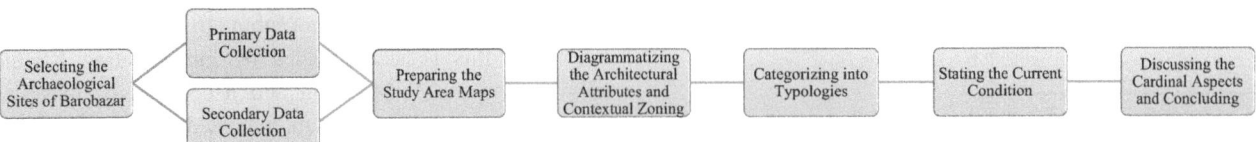

Fig. 1 Methodology diagram

In the analytical phase, during the first step, the current and initial mappings of the study area are done as per the references from literature and survey locating the mosques. Then, the collected data were compiled according to the architectural features and contextual zoning and categorized as per the fundamental typologies obtained from the plans, sections, elevations, and images.

After stating the current condition of each structure, the article discusses the cardinal attributes and reaches a conclusion that has a scope for further relevant research and interventions.

3 Study Area

Jhenaidah district is situated in southwest of Bangladesh. It is a subdivision of the Khulna Division. It is bounded to the north by Kushtia district, to the south by Jessore district and West Bengal, India, to the east by Rajbari district and Magura district, and to the west by Chuadanga district and West Bengal, India. Barobazar is a town located on the Bhairab River's dead north bank (Fig. 2). Its name comes from the twelve number of Aulia, Baro. It is a union parishad located in the Kaliganj Upazila, Jhenaidah. It is located 12 km south of Kaliganj Bazar and 16 km north of the Jessore district center. Through Barobazar market, the Jessore-Jhinaidah highway and the Khulna-Parvatipur railway run parallel. Right now, Barobazar is a thriving commercial hub.

3.1 Mohammadabad

At Barobazar in the Kaliganj Upazila of Jhenaidah, an old city called Mohammadabad (Fig. 3) occupies a space of around 3 km^2. There are 19 mosques from the Sultanate period that are underground. Seven mosques that date back more than 700 years are still present underground. Barobazar is the present name for historic Shah Mohammadabad. For visitors, the government has revealed a map of the historic city of Mohammadabad.

3.2 Location of the Mosques

Barobazar mosques are located close to one another (Fig. 4). Majority of the mosques in this area can be found on the Hakimpur-Barobazar road.

Pathagar Mosque: Pathagar Mosque is a single-domed mosque which is located in the Belat Daulatpur area of Barobazar. This mosque is surrounded by a community. The adjacent land use type is residential (Fig: Location of the Mosques). The Department of Archaeology excavated the mosque from the ground to repair it in 2007. During the Sultanate period, the mosque was supposed to be library-centered. In front of the mosque, there is a big dighi (larger version of a pond) which is known as Pithagora Pond.

Fig. 2 Map of Barobazar (produced by authors)

Fig. 3 Map of Mohammadabad (Department of Archaeology, Bangladesh: reproduced by authors)

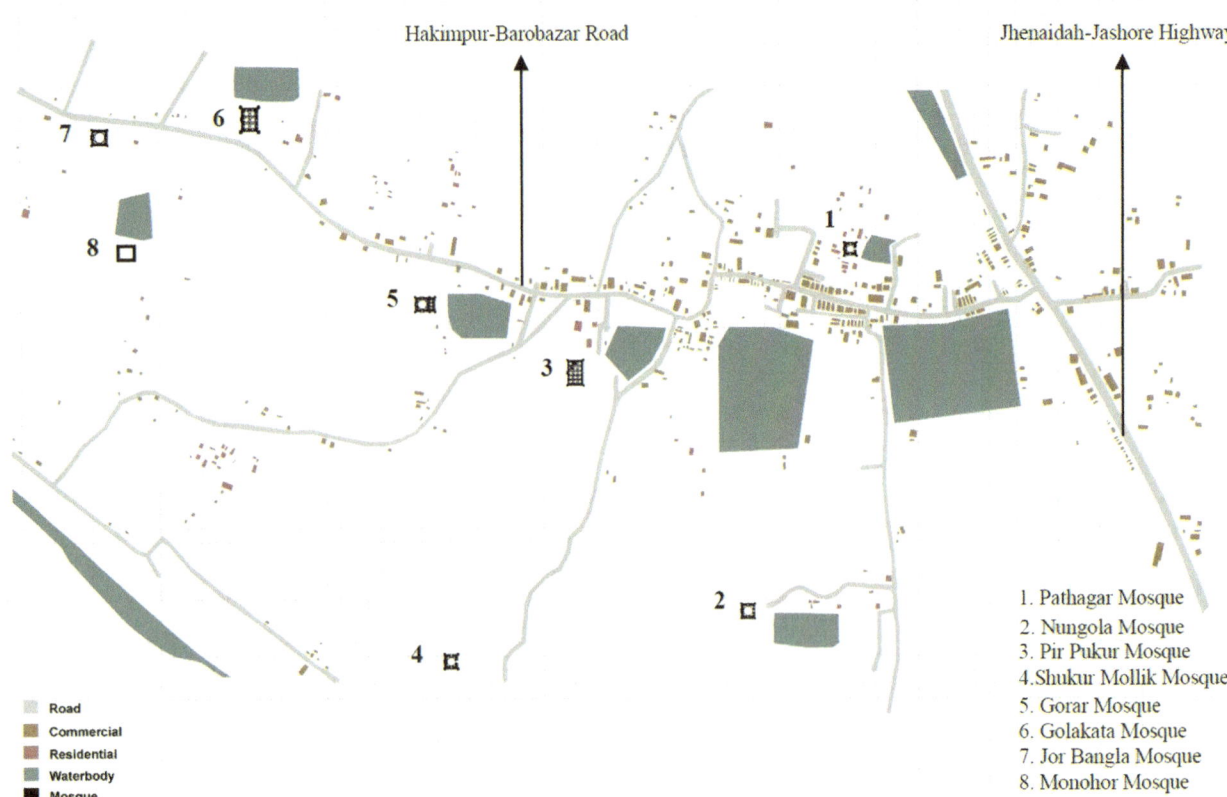

Fig. 4 Location of the mosques (produced by authors)

Nungola Mosque: Single-domed Nungola Mosque is located in the village of Mitha Pukur of Barobazar. This mosque is sometimes referred to as Mitha Pukur mosque, the same as the village name. It is the biggest mosque in the area, despite having a dome. There are three semicircular Mihrabs located inside the mosque. There is a large pond that runs north to south on the mosque's east side. The mosque has three entrances: one on the east side, one on each of the north and south sides, and a sturdy arch at each of the three side entrances. On the southern side of the mosque, there are two paved graves (Home|BaroBazar, n.d.).

Pir Pukur Mosque: Pir Pukur Mosque is also located in the Belat Daulatpur area of Barobazar. It is a 16-domed mosque situated to the west of the Pathaagar Mosque (Fig: Land Use Map). The mosque was discovered in 1994 while digging from an earth cemetery. The mosque is constructed of red brick and has no roof, only walls.

Shukur Mollik Mosque: Shukkur Mollik Mosque is located in the Hasilbag area of Barobazar. To the locals, it is also known as Hasilbag Mosque. It is the smallest mosque in the Barobazar neighborhood, with a single domed. The mosque has only one entrance on the east side. As this mosque is adjacent to the Bhairab River, there were formerly Damdam Jahajghata, Cantonments, and River Ports.

Gorar Mosque: Gorar Mosque is also located in the Belat Daulatpur area of Barobazar and situated just beside the Hakimpur-Barobazar road. Among all the mosques from that historic city, it has the most popularity. The mosque was named after Gorai, a dorbesh (a Sufi or a religious mendicant). On the mosque's eastern side, there is a huge pond. This Gorar Mosque attracts many visitors, many of whom are not tourists. They only go there to make or fulfill a desire. This is a form of blind obsession. From that passion, people occasionally sacrifice hens, contribute money, and so on (বাংলাদেশ, n.d.).

Golakata Mosque: Golakata Mosque is a square-shaped mosque located on the Hakimpur-Barobazar road's northern side (Fig: Land Use Map). It is said that at Barobazar, and there was a tyrannical ruler who chopped off people's necks and tossed them into the dighi in front of the mosque. That is why it is called Golakata. It is unknown who built this mosque and when it was erected exactly. Bangladesh's Archaeology Department recently renovated and shaped the mosque. Every day, people gather here to pray as well. There are several trees all around the mosque (বাংলাদেশ, n.d.).

Jor Bangla Mosque: Jor Bangla Mosque is a single-domed mosque on a high pedestal located on the Hakimpur-Barobazar road's southern side (Fig: Land Use Map). During the excavation of the mosque in 1993, a brick with the Arabic inscription written by `Shah Sultan Mahmood ibn Hussein, Eight Hundred Hijri' was discovered. Because of the twin hut that existed next to the mosque, it was formerly known as Jor Bangla Mosque. Every day, people gather here to pray as well. A medium-sized water tank, known locally as Ondho Pukur, is situated next to the mosque and is utilized by the people to wash before prayers.

Monohar Mosque: There aren't enough transit options near the Monohar Mosque because of its location (Fig: Land Use Map). A medium-sized pond is situated next to the mosque which is used by the people to wash before prayers.

3.3 Data Collection

Historical Data:

Mohammadabad, a historic city, is located around three square kilometers away at Barobazar in Jhenaidah's Kaliganj Upazila. Barobazar is the present name of the historic Shah Mohammadabad city, the capital of the fifteenth century. It is surrounded by hundreds of ponds and natural beauties on both sides of the Jhenaidah-Jashore Highway. Barobazar in Jhenidah is believed to have derived its name from a certain twelve (Baro) obscure Muslim saints (Aulias) who settled here before Khan Jahan Ali (The Buildings of Khan Jahan in and Around Bagerhat|The University Press Limited, n.d.). The Muslim saints (Aulias) are Enayet Khan, Abdal Khan, Daulat Khan, Rahmat Khan, Shamsher Khan, Murad Khan, Haibot Khan, Niamat Khan, Syed Khan, Belayet Khan, and Shahadat Khan. Mosque architecture in the Barobazar area has generally flourished through the efforts of these twelve Muslim saints (Aulias).

The 19 mosques, 11 lakes, and 4 cemeteries from the Sultanate period are only a few of the gems that make Barobazar noteworthy. These mosques were constructed around the year 1500 CE and were founded in 1987. Due to a lack of sufficient care and monitoring, they have lost most of their attractive creative elements. The design and structure of the red terracotta plaques on the 500-year-old mosques nevertheless serve as a testament to the extensive heritage of the archaeological sites, even though time and nature have mostly ruined them.

Rather than the Khan Jahan Ali style that emerged in adjacent Bagerhat, Barobazar's architecture is more comparable to the Illyas Shahi architecture in Gaur and Pandua. Neither structure has an inscription, but based on stylistic similarities to the Gaur and Adina architecture, including a square room with a foreroom, an octagonal turret, a vertical offset, and projected shallow entrance panels, it may be assumed that this structure was constructed in the Illyas Shahi period, which preceded the Khan Jahan architecture, in the early fifteenth or late fourteenth century (Ghorar Mosque at Barobazar, Jhenaida, 2016).

History has shown that Ikhtiyar Uddin Mohammed bin Bakhtiyar Khilji was drawn to the north after conquering Nadia, oblivious to the south or southeast. Later, during Nasiruddin Mahmud Shah's authority, he placed some of

Jhenaidah, Jashore, and Khulna under his control. Finally, Khan Jahan Ali was crowned victorious in the southeast. After observing the difficulties in finding water to drink along the trip, he created several dighi and ponds in the region. The reservoirs were allegedly excavated the same night, according to rumors. As a result, the Barobazar region still has 84 acres of ponds and dighi (বাংলাদেশ, n.d.).

Locally Conducted Survey:

The locally conducted survey has been a very essential part of this study as most of the data collection has been performed either from observation, visual survey, or direct interviews with the local residents. The local authority or community approaches toward these mosques along with the locally contained myths and facts have been an imperative resource here.

Due to proper publicity and a lack of surrounding accommodations, travelers are unaware of the former splendors of the Sultanate era in Jhenaidah. Because of this, the next generation will not have the opportunity to learn about these historic riches. Locals have long wished that the government would take appropriate action to beautify and provide amenities around the mosques so that visitors from home and abroad might appreciate the ancient structures. The government and local residents will both profit financially from tourism. The abundance of ancient structures at the archaeological sites will provide a limitless opportunity for visitors to learn about the history and customs of the area.

The soil and water of these ancient mosques are highly revered by the people. Every Friday, people travel great distances to the mosques to collect soil and water, believing them to be the panacea, a cure-all. Some devoted people touch the mosque's wall in the hope of healing their fatal illnesses.

There goes a myth about Jor Bangla mosque. There were supposedly twin huts next to the mosque. The mosque may have been given the name Jor bangla as a result. This mosque's single dome is said to have been constructed under Sultan Gias Uddin Mahmud Shah's rule. Terracotta flower art is attractively displayed on the walls of the modest mosque. On the east side, there is an entrance with three pointed arches. Now, most of the mosque's renovations are complete.

There allegedly was a tyrannical king in Barabazar. In front of the mosque, he killed his people and dumped their heads in the adjacent pond, or dighi. Golakata mosque is the reason for its name. But, it has no historical validity on this myth.

Another myth says, that once, there was found a beheaded body or head inside the mosque. From then on, the mosque has been named Golakata. A myth about Pathagar and Pir Pukur mosque says, the mosques were buried or had

been tried to demolish to suppress the local Muslim community in the British Colonial Period.

A resident of that area who is also a heritage enthusiast and activist, working actively to promote this particular heritage belt of Barobazar, reported that they have been set directional boards at places with maps and information regarding these historical assets to let people know about them, thus, stimulate cultural tourism in the historical sites of Barobazar. He further added that sometimes there are some difficulties encountered regarding the balance of tourism, architectural conservation, and the daily lives of the local stakeholders who live in those neighborhoods. They expect the Department of Archaeology, Bangladesh to take this issue into sincere account.

The local authority, the Barobazar Union Parishad under Kaliganj Pourasava is also informed and acknowledges these amazing heritage belts very well. They wish to develop this historical city in a prominent tourist site like the mosque city of Bagerhat and seek attention from the top brass.

Apart from these intangible data, the Mosques' prominent architectural features and materials have been observed and noted in Table 1.

The historical mosques of Barobazar can be grouped according to their contextual locations. These current contextual locations do not resemble or justify the initial or original context or period when the mosques were established. Most of the mosques of Barobazar formerly known as Mohammadabad were established to preach and spread the light of Islam around that region. According to historians, it is believed that this particular area of Barobazar hosted a well-flourished civilization in the then period. Though after many years of continuous evolution and transition, several timelines went by, thus, leaving several different historical layers.

This made these substantial historical entities to be concealed underneath for years. However, they have been discovered gradually through excavations performed by local authorities and the Department of Archaeology, Bangladesh. A part of them was partially dilapidated. All of them are preserved and protected antiquities by the Department of Archaeology. Some of them have already been restored as per the original design as much as possible though several details in decorative motifs might have been misinterpreted. Nevertheless, after getting some recognition, a very recent neighborhood grew surrounding the mosques. That is why the current contextual zoning shows most of them located in either residential areas or agricultural areas. There is another attribute behind these current zoning as none of them are situated exactly in the commercial vicinity. Being located slightly diverted from the main access route and commercial hubs, none of the mosques stands exactly at central commercial zones.

Table 1 Architectural features and materials

SL	Title	Architectural features	Materials
1	Pathagar mosque	I. Single dome, square II. Four octagonal turrets at the four corners III. One pointed arched entrance on the east IV. Two arched openings—one each on the north and south sides V. Geometric bands on the turrets and around openings VI. Four octagonal corner turrets VII. A blind mihrab on either side of the central mihrab VIII. Decorative motifs on the mihrabs IX. Adjacent tank on the east side	Exposed red brick
2	Nungola mosque	I. Single dome II. The largest mosque in this region III. Three openings on the east facade, the central one is larger (entrance door) IV. Pointed arches along openings V. Eight pilasters in the interior, two each on the four walls VI. Three semicircular mihrabs in the qibla wall in the interior VII. Two small niches one on each of the northern and southern walls in the interior VIII. Two small outlets for water on the north and south sides of the mosque IX. Four octagonal corner turrets X. Screens on the north and south wall XI. Rammed concrete floor on the east XII. Adjacent tank on the east side XIII. Geometric bands on the turrets and around openings	Lime plaster, sand, stone, rammed concrete, exposed red brick
3	Pir Pukur Mosque	I. Located on the southwest side of the Pathagar Mosque II. 16 domes but currently dilapidated III. Four octagonal corner turrets and six interior columns IV. Pointed arches along all openings V. Two interior stairs VI. Adjacent tank on the east side VII. Underground stair connected directly to the tank VIII. Geometric bands on the turrets and around openings	Exposed red brick
4	Shukur Mollik mosque	I. Single dome, square II. Four octagonal corner turrets III. Three openings on the east facade, the central one is larger (entrance door) IV. Pointed arches along openings V. Eight pilasters in the interior, two each on the four walls VI. Semicircular mihrab on the west with one blind mihrab on either side of it with terracotta bell and chain motifs VII. Side mihrabs have multi-cusped arches and rosettes at the top VIII. Geometric bands on the turrets and around openings	Exposed red brick
5	Gorar mosque	I. Four domes, square central chamber II. The central chamber with a dome at the top is original and larger III. The foreroom collapsed, but it has been reconstructed with three domes IV. Four octagonal corner turrets are attached to the main building and two other corner turrets are attached to the Foreroom V. Three pointed arched openings on the east facade, the central one is larger(Entrance Door) VI. Northern and southern entrances are bordered by rectangular frames of geometric designs VII. Walls, turrets, fenestration, and recesses are decorated with geometric and leafage terracotta motifs VIII. Adjacent tank on the east side	Exposed red brick
6	Monohar Mosque	I. Multiple turrets II. Adjacent waterbody	Exposed red brick
7	Golakata mosque	I. Six domes II. Three openings on the east façade, the central one is larger (entrance door) III. Pointed arches along openings IV. The walls are about five feet wide and in the middle, there are two long black stones V. 18 arched blind recesses on four facades (4 on the East, North, and South, 6 on the West) VI. Adjacent tank on the North side VII. Four octagonal corner turrets I. viii. Geometric bands on the turrets and around openings	Stone, exposed red brick

(continued)

Table 1 (continued)

SL	Title	Architectural features	Materials
8	Jor Bangla mosque	I. Single dome, square II. Placed on a raised platform III. Four octagonal corner turrets IV. Three pointed arched openings on the east facade, the central one is larger (entrance door) V. Two blind pointed arched recesses on the north and south sides VI. Three semicircular mihrabs VII. Four small niches in the wall of the mosque; two each on the northern and southern walls in the interior VIII. Four octagonal turrets at the four corners IX. Geometric bands on the turrets and around openings X. Adjacent tank on the north side that has been separated by a later constructed road	Exposed red brick, lime plaster

3.4 Neighborhoods Surrounding Mosques

Based on the field survey, it can be determined that the mosques are mainly located in residential, agricultural, and commercial neighborhoods (Table 2; Fig. 5, and Appendix 2).

Mosques in Residential Neighborhoods:

Pathagar mosque is located in a very vibrant residential neighborhood surrounded by several households adjacent to a large waterbody that has been in existence for years. The location is a bit oblique to the major accessway which is the Barobazar-Hakimpur road. An alleyway leads the visitors from the main road to this ancient mosque situated at a very domestic locale.

Pir Pukur Mosque stands in a partially dilapidated condition amid a residential locality. The neighborhood grew along as the area was excavated and a prominent historical asset was found. The adjacent homesteads have gardens, orchards, and cattle sheds which also complement this neighborhood to be noted as a particular context.

The existing context of Gorar Mosque is a bit different from the other mosques that are situated in residential neighborhoods. This mosque is extended as a complex including a general store, toilet, and ablution zone just adjacent to the existing pond. Some accessory service areas serve to arrange free meals on Fridays. There is a familial graveyard beside.

Table 2 Neighborhood type

Sr.	Title	Agricultural	Residential
1	Pathagar Mosque		✓
2	Nungola Mosque	✓	
3	Pir Pukur Mosque		✓
4	Shukur Mollik Mosque	✓	
5	Gorar Mosque		✓
6	Monohar Mosque	✓	
7	Golakata Mosque		✓
8	Jor Bangla Mosque		✓

Golakata Mosque is situated adjacent to the Barobazar-Hakimpur road but a bit away from the commercial zone. This historical multi-domed mosque also stands amid a residential vicinity that is some feet depressed from the actual access road but has direct visual connectivity from there.

Jor Bangla mosque stands just adjacent to the Barobazar-Hakimpur road which makes it visually and physically accessible from there directly. The surrounding neighborhood is a residential one with tin-shed homesteads with adjoining bamboo and banana trees.

Mosques in Agricultural Neighborhoods:

A vast chunk of agricultural land surrounds this antiquity. A residential neighborhood is located within walking distance from there. Being located among cultivable fields the mosque is not at all to be discovered from the cardinal accessway. A narrow alleyway within the residential neighborhood leads to the fields, thus, this particular historical asset. Beside the southern wall, there is a graveyard just contiguous to this structure.

Shukur Mollik Mosque holds onto a huge surrounding of agricultural fields. It has a direct visual connection from the access road. A slender earthen way leads visitors to this historical mosque. The closest commercial and residential zones are comparatively away from this locality.

Monohar Mosque is still in a dilapidated condition and is located amid an agricultural neighborhood though there is an adjacent residential locale nearby. There is an absence of a proper accessway that could lead the visitors to this ancient mosque.

3.5 Typological Classification

Islamic structures with architectural characteristics such as domes, mihrabs, and minars hold their location-wise uniqueness and possess individuality influenced by climate, geographical context, construction material availability, and local artisans. The mosque architecture of Bangladesh is somewhat adapted to the local environment. Their design and building techniques are ingenious and distinctive.

Pathagar Mosque   Nungola Mosque   Pirpukur Mosque   Shukur Mollik Mosque

Gorar Mosque   Manohar Mosque   Golakata Mosque   Jor Bangla Mosque

Agricultural Land
Commercial
Residential
Waterbody
Mosque

Fig. 5 Contextual zoning of the historical mosques of Barobazar (produced by authors)

Mosques in Bangladesh can be classified into two major types: (1) Courtyard mosques and (2) Enclosed mosques. The mosques of Barobazar are certainly of the Enclosed type. The mosque plans are further categorized as rectangular and square type and considering the number of domes, single-domed and multi-domed type of mosque architecture was adapted during the growth of mosque design in Bangladesh. The Minaret or tower has also been an integral part of traditional mosques and instead, as a similar feature to some extent, turrets were constructed either to reinforce the corners or were purely aesthetic additions that always formed a significant element of the mosques (Chowdhury, 2021).

The fundamental typological classification can be identified and formulated on the number of domes and the basic shape of the plan (Table 3). These can be categorized as square, single-domed mosques and rectangular, multi-domed mosques.

Typological Characteristics of Single-Domed Square Mosques

1. The Jor Bangla mosque, a single-domed square mosque, is entered by a flight of stairs via an arched brick entryway. The mosque is situated on a raised mound of soil. The elevated open shan or courtyard was originally enclosed by a low border wall, which is now at the level of the plinth. Although having a square shape, the mosque seems to be used on Fridays because of its wider frontal shan or courtyard.
2. The prayer halls specifically have four octagonal turrets in four corners. The moldings split each turret into four sections: a group at the base, two in the center, and one at the top.
3. The tiny brick pendentives construct the circular platform upon which the hall's dome is supported. The terracotta friezes with continuous merlons and many moldings are visible on the dome's shoulder when viewed from within.
4. Double or triple mihrab niches are noted on the qibla walls.

Typological Characteristics of Multi-domed Rectangular Mosques

1. Foreroom is a very notable characteristic of one of the Barobazar's multi-domed rectangular mosques. The foreroom in the Gorar mosque features three pointed-arch entrances, one on each side and one in the center. From the foreroom, three arched doors lead to the prayer chamber. Using pendentives at each corner, three equal hemispherical domes are used to cover the foreroom.
2. The Pir Pukur mosque has a hypostyle interior prayer hall that is prominently significant among all the mosques in Barobazar.

Table 3 Typological classification

SL.	Title	Single domed	Multi-domed	Rectangle shaped	Square-shaped	Tetra-turrets	More than four turrets
1	Pathagar mosque	✓			✓	✓	
2	Nungola mosque	✓			✓	✓	
3	Pir Pukur mosque		✓	✓		✓	
4	Shukur Mollik mosque	✓			✓	✓	
5	Gorar mosque		✓	✓			✓
6	Monohar mosque		✓	✓			✓
7	Golakata mosque		✓	✓		✓	
8	Jor Bangla mosque	✓			✓	✓	

3. The qibla wall particularly has multiple, even more than three mihrab niches that correspond to the front entry.

4. The Golakata mosque, another multi-domed rectangular mosque, is notable for its freestanding stone pillar with a chain and bell motif on the round shaped shaft and a petal capital, multi-unit prayer hall, octagonal tower, vertical offset, and recessed niche with moldings.

The pillar's base is square, but the capital has fluted petals and is smaller than the base. This provides the image of a shaky construction for the dome's supporting arch. Two freestanding stone pillars and eight partly disguised pilasters support the ceiling of six equal hemispherical domes.

5. There are just four octagonal corner towers in every multi-domed mosque too. But interestingly, the Gorar mosque contains six corner turrets—one on either end of the foreroom and one at each corner of the main hall. Four of the frontal turrets are octagonal in design, while the other two are square in shape.

Common Attributes

1. Islamic architecture is known for its ornamentation, which is both a surface craft and a structural component. Terracotta tile decoration with floral and geometric abstract patterns was widely employed in both single-domed square mosques and multi-domed rectangular mosques.

2. Each entry aperture of these mosques has a recessed rectangular frame around it that is topped with high brick moldings and parallel terracotta accents.

3. In the interior of the northern and southern walls, there are multi-cusped arched niches.

4. The center mihrab is taller and broader than the side ones, and it projects from the exterior.

5. The squinches at each corner of the prayer hall are supported by brick pilasters and enclosed walls.

6. The main prayer hall's northern and southern walls either have two arched openings or blind recesses.

7. All the mosques have eastern entry apertures that face the qibla on the parallel surface (Chowdhury, 2021).

3.6 Current State

Within a quite close radius of Barobazar, there are a large number of historic tanks and cultural mounds that are dispersed around the area. They are mostly found to the west of the current roadway and on the north side of the river. Only the Gorar Mosque, located in the community of Belat Daulatpur, was left standing among the rubble and was in a terrible condition of preservation. Others suffered severe damage. The Department of Archaeology conducted an investigation and found 14 mounds in this region. But only some of them were targeted for protection (Table 4). They were being used for prayer and made partially accessible by the residents of that locality. Right now, the mosques that have been restored, renovated, or partially reconstructed are functionally active, used by the local people, and visited by tourists. The government designated these locations as protected monuments, and a program of routine archaeological excavation was launched to uncover the region's cultural value and recreate its past. Most of the mosques' structural remnants have been made visible as a result, however, there are more, yet to be worked on (Barobazar - Banglapedia, n.d.).

The government designated these locations as protected monuments, and a program of routine archaeological excavation was launched to uncover the region's cultural value and recreate its past. However, most of the mosques' structural remnants had been made visible as a result but there are more, yet to be worked on (Barobazar—Banglapedia, n.d.).

4 Discussion and Conclusion

The mosques at Barobazar reflect the traditional form of architecture as a reflection of spirituality, society, and culture in sacred space, as well as the implementation of local materials and construction systems. The development of these mosques in the past has been significant in influencing

Table 4 Current state of mosques

SL.	Title	Restored	Preserved	Partially Reconstructed	Mostly Dilapidated	Protected Antiquity
1	Pathagar Mosque	✓	✓			✓
2	Nungola Mosque	✓	✓			✓
3	Pir Pukur Mosque		✓		✓	✓
4	Shukur Mollik Mosque	✓	✓			✓
5	Gorar mosque	✓	✓	✓		✓
6	Monohar Mosque		✓		✓	✓
7	Golakata Mosque	✓	✓			✓
8	Jor Bangla Mosque	✓	✓			✓

the spatial and formal organization of religious spaces in contemporary times. In this study, a spatial investigation into the mosques of Barobazar produced four significant insights into the architectural elements of these mosques.

The difference in dome numbers is quite formulated on the shape of the plan. All the square mosques are single domed whereas all the rectangular ones are multi-domed. Because the dome structure doesn't require internal support, a broad range of floor plan configurations is possible. This enables us to make the most of the mosques' spacious interiors. Due to the structure's tightness, they preserve a significant quantity of energy, which lowers the cost of operating and heating them.

Instead of building larger mosques and expecting people to travel great distances during the prolonged rainy season, it was more feasible to build smaller mosques to accommodate the people living in the small clusters of huts that make up the Bangladeshi village. That is why square-shaped, single-domed mosques were built in large numbers.

The central axis of the eastern openings lines up with the position of the mihrab niches. The concept of positioning a niche right across from an entrance is reminiscent of Buddhist temple design (Chowdhury, 2021).

Most of the mosques are of tetra-turrets which are dominantly octagonal. Only some of the rectangular mosques that might have sections in interior or semi-interior functions have turrets of more than four. For example, in the Gorar Mosque, there is a verandah-type semi-interior space before accessing the cardinal interior monospace.

The foreroom was integrated into the mosque's single square unit that made one of the mosques rectangular in shape. It was more prudent and reasonable to add a foreroom to increase the available area rather than construct a larger single dome with all of its related structural difficulties (Chowdhury, 2021).

Most of the mosques are of tetra-turrets which are dominantly octagonal. Only some of the rectangular mosques that might have sections in interior or semi-interior functions

have turrets of more than four. For example, in the Gorar Mosque, there is a verandah-type semi-interior space before accessing the cardinal interior monospace.

All of the mosques are particularly mono-spatial which leads the interior space to appear gigantic and monumental. They have vaulted ceilings to hold the eminent interior scale and break the monotony of extensive slabs. Arched openings and recesses are other obvious features to mention for all. The common material used for all of them is Red brick (Appendix 1—Figs. a1–g4) which has been a local and traditional material in Bangladesh because of its location in the Deltaic region. Adjacent waterbody is another feature that was designed to provide ablution facilities as well as mitigate the demand for fresh drinking water in that region.

The mosques that are comparatively larger and could have not been supported by single domes are parted into smaller multiple domes to accommodate more built area. The necessity might have arisen because of the then neighborhood context or community demand or initial functional need.

These findings from the study show that, even though the Department of Archaeology has identified the sites as being protected, there is still much work to be done before these buildings can be said to be truly preserved. There is an urgent need for protection planning in Bangladesh, particularly for religious spaces, which differs from general cultural heritage protection. In the case of Barobazar, we propose that the study into the archaeological sites of the mosques be taken into consideration where the protection of heritage is concerned, to ensure the retention of original characteristics while also achieving structural protection. Additionally, the attempt to the protection of mosques in Barobazar, and the preservation and restoration of their traditional characteristics can also serve as an impetus for the boost in national recognition and visitors that the sites currently lack. Further study into this can be critical to the conservation strategies required in the unprotected and dilapidated structures at Barobazar.

Appendix 1

Plan, Section, Elevation, Images of Mosques (Photographs Are Taken By Authors)

Title	Plan	Section	Front Elevation	Image
Pathagar Mosque	Appendix 21.A Figure a₁	Appendix 21.A Figure a₂	Appendix 21.A Figure a₃	Appendix 21.A Figure a₄
Nungola Mosque	Appendix 21.A Figure b₁	Appendix 21.A Figure b₂	Appendix 21.A Figure b₃	Appendix 21.A Figure b₄
Shukur Mollik Mosque	Appendix 21.A Figure c₁	Appendix 21.A Figure c₂	Appendix 21.A Figure c₃	Appendix 21.A Figure c₄
Jor Bangla Mosque	Appendix 21.A Figure d₁	Appendix 21.A Figure d₂	Appendix 21.A Figure d₃	Appendix 21.A Figure d₄
Pir Pukur Mosque	Appendix 21.A Figure e₁	Appendix 21.A Figure e₂	Appendix 21.A Figure e₃	Appendix 21.A Figure e₄
Gorar Mosque	Appendix 21.A Figure f₁	Appendix 21.A Figure f₂	Appendix 21.A Figure f₃	Appendix 21.A Figure f₄
Golakata Mosque	Appendix 21.A Figure g₁	Appendix 21.A Figure g₂	Appendix 21.A Figure g₃	Appendix 21.A Figure g₄

Appendix 2

Data Collected on Field Survey

No.	1	2	3	4	5	6	7	8
Name of mosque	Pathagar mosque	Nurtgola mosque	Shukur Mollik mosque	Pir Pukur mosque	Gorar mosque	Monohar mosque	Golakata mosque	Jor Bangla mosque
Year	Sixteenth century	Late Sultanate	Sixteenth century	Sixteenth century	Sixteenth century	Sixteenth century	Sixteenth century	Sixteenth century
Area/location	Mithapukur. Barobazar. Jhenaidah	Mithapukur Area, Hasilbagh, Barobazar, Jhenaidah	Hasilbagh, Barobazar, Jhenaidah	Hakimpur-Barobazar Road, Barobazar, Jhenaidah	Belat, Daulatpur, Barobazar, Jhenaidah	Jessore-Jhenaidah Highway, Barobazar, Jhenaidah	Hakimpur-Barobazar Road, Barobazar, Jhenaidah	Hakimpur-Barobazar Road
Area type	Residential	Agricultural land	Agricultural land	Residential	Residential	Agricultural	Residential	Residential
Approach road type (ex. alley/ highway)	Alley	Alley	No approach road	Alley	Alley	No approach road	Secondary road	Secondary road
Adjacent built form function/ adjacent plot function	Residence	Agricultural land	Agricultural land	Residence	Family graveyard	Agricultural land	Grove of miscellaneous tress	Residence
Form	Square	Square	Square	Square	Rectangular	Rectangular	Rectangular	Square
No. of entry and exit	1	1	1	1	1		1	1
Location of entry (e.g.: front, rear etc.)	Front	Front	Front	Front	Front	Front	Front	Front
Functions accommodated inside the mosque	Prayer hall	Prayer hall	Prayer hall	Prayer hall	Prayer hall	Prayer hall	Prayer hall	Prayer hall
Additional facilities	–	Toilet	Toilet, tin-shed semi-outdoor extension	Ablution zona, toilet	General store, ablution zone, toilet, tube well	–	Ablution zone	–
Roof type	Domed	Domed	Domed	Domed	Domed	Domed	Domed	Domed
Multi domed/ single domed	Single	Single	Single	Multi (16)	Multi (4)	–	Multi (6)	Single
Use of jali screens	Yes	Yes	Yes	Yes	No	–	Yes	No
Floor type	Rammed concrete floor	Rammed concrete floor	Rammed concrete floor	Rammed concrete floor	Rammed concrete floor	–	Rammed concrete floor	Rammed concrete floor
Construction materials	Exposed red brick	Lime plaster, sand, stone, rammed concrete, exposed red brick	Exposed red brick	Exposed red brick	Exposed red brick	Exposed red brick	Exposed red brick	Exposed red brick. Lime plaster

(continued)

No.	1	2	3	4	5	6	7	8
Minaret (none/no.)	4	4	4	4	6	–	4	4
Changes made (small, moderate; major)	Small	Small	Major	Moderate	Small	–	Small	Small
Current state	Restored, preserved, protected antiquity	Restored, preserved, protected antiquity	Restored, preserved, protected antiquity	Mostly dilapidated, preserved, protected antiquity	Restored, reconstructed, preserved, protected antiquity	Mostly dilapidated, protected antiquity	Restored, preserved, protected antiquity	Restored. preserved, protected antiquity

References

Arab, P. T. (2013). The biggest mosque in Europe!: A symmetrical anthropology of Islamic architecture in Rotterdam. In O. Verkaaik (Ed.), *Religious architecture* (pp. 47–62). Amsterdam University Press. https://www.jstor.org/stable/j.ctt6wp6sx.5

Barobazar—Banglapedia. (n.d.). Retrieved October 8, 2022, from https://en.banglapedia.org/index.php/Barobazar

Barrie, T. (2020). *Architecture of the world's major religions: An essay on themes, differences, and similarities.* BRILL. http://ebookcentral.proquest.com/lib/tamucs/detail.action?docID=6319544

Chowdhury, T. (2021, January 21). *Mosque architecture of Bangladesh.* RTF|Rethinking The Future. https://www.rethinkingthefuture.com/architectural-styles/a2976-mosque-architecture-of-bangladesh/

De Wildt, K., Radermacher, M., Krech, V., Löffler, B., & Sonne, W. (2019). Transformations of 'sacredness in stone': Religious architecture in urban space in 21st century Germany—New perspectives in the study of religious architecture. *Religions, 10*(11), 602.

Ghorar mosque at Barobazar, Jhenaida. (2016, June 18). Context BD. https://contextbd.com/ghorar-mosque-at-barobazar-jhenaida/

Goh, D. P. S., & van der Veer, P. (2016). Introduction: The sacred and the urban in Asia. *International Sociology, 31*(4), 367–374. https://doi.org/10.1177/0268580916643088

Hasan, P. (1989). Sultanate mosques and continuity in Bengal architecture. *Muqarnas*, 58–74.

Home|BaroBazar. (n.d.). Retrieved October 10, 2022, from http://visitbarobazar.net/

Independent, T. (n.d.). *Barobazar: Thorp of mosques. Barobazar: Thorp of Mosques*|Theindependentbd.Com. Retrieved October 8, 2022, from https://www.theindependentbd.com/magazine/details/49855/Barabazar:-Thorp-of-Mosques

Longhurst, C. E. (2012). Theology of a mosque: The sacred inspiring form, function and design in Islamic architecture. *Lonaard Journal, 2*(8), 3–13. https://search.ebscohost.com/login.aspx?direct=true&db=asu&AN=88842177&site=eds-live&scope=site&authtype=shib&custid=s8516548

Mitra, S. (1914). *Jashohar- Khulnar Itihas* (Vol. 1). http://archive.org/details/in.ernet.dli.2015.289452

Polin, F., Mahboob, F., & Alam, D. (2019, August 26). Trails of Khan Jahan Ali. *The Daily Star.* https://online.thedailystar.net/in-focus/news/trails-khan-jahan-ali-1790449

Sidawi, B. (2013). Understanding the vocabulary of the Islamic architectural heritage. *Global Built Environment Review, 8*(2).

Tariq, S. H., & Jinia, M. A. (2013). The contextual issues in the Islamic architecture of Bengal mosques. *Global Journal Al-Thaqafah, 3*(1), 41–48.

The Buildings of Khan Jahan in and Around Bagerhat|The University Press Limited. (n.d.). Retrieved October 8, 2022, from http://www.uplbooks.com/book/buildings-khan-jahan-and-around-bagerhat

The Physical Geography of Bangladesh|The University Press Limited. (n.d.). Retrieved October 8, 2022, from http://www.uplbooks.com/book/physical-geography-bangladesh

Verkaaik, O. (2013). Religious architecture: Anthropological perspectives. In O. Verkaaik (Ed.), *Religious architecture* (pp. 7–24). Amsterdam University Press. https://www.jstor.org/stable/j.ctt6wp6sx.3

বাংলাদেশ D. B. :: ডেইলি. (n.d.). *Barobajar: An ancient city of mosque.* Daily Bangladesh. Retrieved October 8, 2022, from https://www.daily-bangladesh.com/english/feature/30506

Tourism as a Driver of Soft Power: The Case of South Korea

Jessica L. Quijano Herrera and Gema Pérez-Tapia

Abstract

In recent years, Korea has become more well known due to the expansion of its culture and entertainment industries. According to the four regions proposed by Ooi (Encyclopedia of Tourism. Springer, Cham, Switzerland, pp. 1–2, 2016), (i) Tourism has increased knowledge about Korea, fostering empathetic understanding and familiarity with the destination. (ii) The principal goal of the tourism authority is to enhance Korea's reputation as a tourist destination. (iii) Korea has organized and invested in several events, exhibitions, and games that serve as a major tourist draw, thereby enhancing its geopolitical position. (iv) Through their behavior in the destination country, foreign tourists reflect the image of their home country, so altering preconceived notions of the other countries. This paper examines how tourism has become a driver of the country's soft power, taking into account the shift in the rebranding of the country toward the cultural and entertainment industry "Hallyu" and the country's ability to design public policies that promote attractive and accessible tourism for international visitors. In addition to their ability to organize large-scale events, the peninsula has demonstrated its political, social, and economic development to the world. Additionally, the manner in which tourism has been approached during and after COVID has made it possible to maintain the confidence of tourists and capitalize on the popularity of Hallyu stars as a clear strategy to attract tourists and position the country as a tourist destination, thereby boosting tourism and, consequently, enhancing soft power.

Keywords

Soft power • Korea • Tourism • Hallyu • Creative industries

1 Introduction

Many countries choose diplomacy to act on the international stage, avoiding armed conflicts and economic pressures. Since the end of the Cold War, international relations have played a crucial role in world stability. These international relations are sometimes unstable and changing, and there is a risk of deployment of traditional military forms of power or also called "Hard Power." To avoid the latter, soft or intangible power resources such as culture, ideology, and institutions have been resorted to.

Thus, arises what Professor Nye (1990) establishes as "soft power" and which he defines as the ability of some countries to persuade others, through their cultural manifestations, values, social model or international institutions, thus avoiding the use of force or coercion. It is a form of power that enhances the attractiveness of a country, an attraction that will be crucial to attract other states to institutional alliances, coalitions, international cooperation, and development aid.

Nowadays, we find a very changing context as far as geopolitical relations are concerned, and it is in this scenario where the use of hard power is giving way to a softer approach or soft power to attract investment, economic development, and wealth to their countries (ElMassah, 2015;

The original version of the chapter has been revised: The author Jessica L. Quijano Herrera affiliation has been updated. A correction to this chapter can be found at https://doi.org/10.1007/978-3-031-49536-6_27

J. L. Quijano Herrera (✉)
University of Málaga, Málaga, Spain
e-mail: jessica.quijano@uma.es

Evangelical University of El Salvador, San Salvador, El Salvador

G. Pérez-Tapia
Department of Economy and Business Administration,
University of Málaga, Málaga, Spain
e-mail: gema.perez@uma.es

Nye, 1990). This transformative nature is now reflected in the dynamism, ease, and solidity of international relations (Rawnsley, 2016). Chitty (2017) defined soft power as "a cultural artifact that represents a body of thought that is associated with resources invested in attraction power as well as with strategies for using such resources to further actors' interests." It is unwise to underestimate soft power, since it is intimately related to tourism as the widespread idea that tourism plays a crucial role in international diplomacy (Xu et al., 2020).

However, and despite the fact that some countries still use hard power to achieve their objectives, its use is decreasing and this is giving way to a much more active role of diplomacy, involving universities, sports organizations, and NGOs. These strategies also called "soft power packages" are used by states to calm their internal audiences and strengthen external positions (Grix et al., 2021).

For a long period of time, South Korea has been considered a manufacturing powerhouse, and the name "Made in Korea" conveyed an image of quality and value of its products. However, and due to the growth and boom of other countries in terms of production and manufacturing such as China and other emerging countries, Korea has lost weight on the international stage. For this reason, the Korean government has seen the need to change its approach and therefore its country image, and has based this on a cultural image or brand, that is, it has based its strategy on Korean popular culture or Hallyu (Holt, 2004).

The term Hallyu or Korean wave is used to refer to Korean entertainment products and their popularity abroad. These entertainment products (movies, TV dramas (K-drama), pop music (K-pop), and online gaming) have undergone exponential growth since the mid-2000s (Oh, 2018).

This new approach or cultural image of Korea based on the entertainment industry has generated impacts in other areas such as the export of products from the music industry (Chae, 2014) and the formation of perceived image abroad (Lee, 2011; Lee & Workman, 2015). Since it has spread across all industries, it is causing the effect of the Korean wave.

It seems that tourism has not been left out of this effect. Tourism is cause and effect of soft power. As a cause, it is considered as an activity through which countries show their culture and values (Hollinshead & Hou, 2012). As economic effect, it could be reflected in exports, export effect in other industries, and the economic ripple effect (Lee, 2011) This fact is the starting point of this research, since it mainly tries to analyze, using the four main areas framed by Ooi's (2016), the soft power of South Korea and its relationship with tourism, that is, if the four interrelated areas are suitable to evaluate and understand the relationship between both variables in a country whose notoriety is growing day by day. The reason of choosing South Korea is the progress and status it has achieved in recent years in terms of soft power.

Ko (2012) focuses on the Korean wave as the cause of the increase in foreign tourists. One of the motivations that tourists have when visiting Korea is its cultural industry, that is, foreigners visit Korea and visit Hallyu tourist attractions due to the influence of this Korean wave (Lee, 2011).

Because of this, and based on previous research that relates tourism with the entertainment industry and the Korean wave, this study is justified, which raises the need to analyze whether South Korea is conveniently using its creative and entertainment industry to attract tourists and thus strengthen its soft power. In addition, another fact that justifies this research is that most of the existing literature has focused on China and Japan (Otmazgin, 2008; Shambaugh, 2015), and South Korea has not been examined in the framework of the soft power of tourism.

2 Methodology

The methodology is based on the development of a case study from secondary sources. Likewise, and starting from the four main areas proposed by Ooi's (2016), the research will try to analyze the actions in the field of tourism that South Korea is carrying out and that will have an impact on its soft power.

Ooi (2016) has proposed four elements that links soft power and tourism:

(i) Tourism has generated a greater knowledge of Korea, generating empathy and greater familiarity with the destination.
(ii) The primary objective of the tourism authorities is to improve the image of Korea as a tourist destination.
(iii) Korea has organized and invested in numerous events, exhibitions, and games which constitute an important attraction for tourists and which in turn improves their geopolitical position.
(iv) Foreign tourists show the image of their own country through their behaviors in the destination country and thus change the stereotypes established against those other countries (Ooi, 2016).

Based on these four elements or frames of reference, this study will carry out an exhaustive review of the existing literature to analyze whether Korea is enhancing its tourist resources in an ideal way to improve the soft power.

Findings

(i) *Tourists' Understanding and Empathy*

To achieve nationally significant global trade, investment, and tourism goals, more governments are becoming increasingly aware of the importance of building, enhancing and promoting their nation brands in order to create a crucial competitive advantage in the global market (Dinnie, 2008; Moilanen & Rainisto, 2009). The majority of studies analyzing the image of Korea as a tourist destination emphasize "Hallyu" or the Korean wave that is creating a surge in its popularity. According to these surveys, tourist satisfaction has increased as a result of this (Chen, 2005; Kim, 2007a, 2007b; Suh & Suh, 2004).

Tourism related to popular culture, in which individuals visit a location associated with movies or TV programs (Iwashita, 2008), has garnered a growing amount of interest over a relatively brief time span (Connell, 2012). The impact of pop culture/film-induced tourism has been studied considering the tourists' experiences, destination image, intention to visit, future attitudes, destination brand development, as well as residents' responses (Busby & Klug, 2001; Busby, Brunt, & Lund, 2003; O'Connor & Bolan, 2008; Roesch, 2009; Ward & O'Regan, 2009).

Thus, the market of cultural goods has been dominated by Japan and Hong Kong. However, it is spreading to new markets as the Korean, where Korean cultural goods are increasing its popularity in Asia (Seabrook, 2012).

According to Lee (2015), the "Korean Wave 4.0" has introduced a brand-new trans-border portrayal of Korea in the global market. It is a blend of social media and musical skills facilitated by Korea's digital technology investment. Korea has utilized its K-pop artists, dramas, and movie business to grab the attention of people around the world.

As illustration, if we think about what could inspire individuals all over the globe to look information related to Korea, Psy's music video Gangnam Style made it with its 3.4 billion views in YouTube as of September 2019 is one of the music videos most seen for international community attracted by Korea (Nye & Kim, 2013).

The dramas and cinema industries also contribute to spread Korea interest and gained a global crowd. The melodrama "Descendants of the Sun" has attracted approximately one billion watchers in 32 countries, and the movie Parasite, directed by Bong Joon-Ho, was the first language other than English movie in Oscar's history to win for best picture (Trisni et al., 2019).

The definitive success is related to the entry into the US market of the BTS. Its consolidation of success was when they reached the top ten US Billboard chart in early summer 2018 (Băjenaru, 2022). BTS contending with Justin Bieber and Ariana Grande was seen as a hit of Korean K-pop industry added to its invitation to the United Nations Organization to make a speech during a session of the General Assembly in 2018 and again in 2021.

Korea national public diplomacy strategy has been well developed with the image and support of Bangtan Boys (BTS). Its members: RM, Jin, Suga, J-hope, Jimin, V and Jungkook enthusiastically recognized their cultural and diplomatic impact. Their partnership with UNICEF with the successful campaign "Love yourself," which seeks to end violence and abandonment and promote self-esteem, has generated about five million tweets and over fifty million engagements, including likes, retweets, replies, and comments (UNICEF, 2021).

According to Mukarromah et al. (2022), innovations and technical advancements make entertainment items such as movies, series, and dramas more accessible to the general public. As a result, several streaming platforms such as Netflix, Amazon Prime, HBO Max, Star Plus, and Disney Plus have emerged. Since Netflix is the greatest streaming platform in the entire world, it has been a popular medium for reaching consumers of Hallyu items, as proven by the fact that its total members reached 203.7 million from its inception till 2020 (Mukarromah et al., 2022). Squid Game is watched by millions of people throughout the world, and while being a thriller drama, it shows part of the Korean traditional culture games.

All of the aforementioned demonstrates a vast array of cultural manifestations that are core component of the Korean cultural offer, i.e., they are constructed as motifs or features of the destination that draw the attention of visitors, creating and widening the level of understanding of the venue, as well as compassion and understanding and appreciation toward it.

Based in tourist perceptions and image about Korea from its rebranding strategy, according to the work carried out by Choi et al. (2011), most Russian tourists have a favorable impression of Korea. They highlight the beauty of its landscapes and its historical and cultural attractions. They also highlight the variety of restaurants, quality of their accommodation, and transportation services.

Pop culture has a wide spread and positive impact in the likeness of destination image and in the leading effect over indirect experiences of culture such as its food, language, and sense of fashion in their homelands of visitors from Chinese, Japanese, Malaysian, and Mexicans according to Lee and Bai (2016) study. This has prompted them to wish to visit South Korea in the future and strengthen the positive impression through experiences of K-pop stars events and trips which have a strong influence on two aspects of future actions: intention to return and word of mouth.

(ii) *Korean Tourism Policy*

In 2019, the National Tourism Strategy was embodied in the so-called "Korea's Tourism Innovation Strategy." Although the growth of recent years has been impressive both in number of visits and in expenses, the sector faces several challenges. These challenges include visa facilitation strengthening the competitiveness of small and medium-sized enterprises in response to future tourists tendencies, marketing less-visited regions, and improving the allure and quality of tourism-related products and services (OECD, 2020).

From the initiative "Tourism Strategy 2020," Korea is seeking several policy responses to address these challenges.

South Korea has five objectives for its National Tourism Innovation Strategy include the creation of a smart tourist environment. The Smart Tourist Strategy of Korea began with the development of smart tourist facilities. This all-encompassing strategy includes a digital visitor information network to enhance the traveler experience, free Wi-Fi at main tourist attractions, and big data analytics for reference in the future and enhancement.

In contrast, the state will relax visa requirements to attract more tourists (multiple entry), four local localities will be established into tourist centers, and tour developers will start embracing more cultural content, such as K-pop with much more heavy advertising, to attract the foreign fans of Korean pop culture. In addition, the state would boost tourist subsidies to the same level as those granted to the industrial industry (OECD, 2020).

In addition, life cycle leisure programs are developed for all age ranges, from youngsters to elderly. Independent travelers will benefit from a dedicated Internet platform and enhanced digital advertising strategies. Developing suggested tour itineraries with the option to buy things (OECD, 2020).

The developing of new tourism areas related to Hallyu has required some efforts to develop tour programs with cultural content. For example, Lee (2021, October 20) in the Korea Herald stated that as a result of the global success of the Netflix-streamed series "Squid Game," s local governments attempt to market their communities as cultural tourism destinations. Locations including Wolmido's theme park My Land, Kyodong Elementary School, Seongapdo, Dokdo, Deokjeokdo, and Guleopdo islands have been promoted as tourism attractions out from Seoul, turning into new tourist sites.

One tourism strategy that was done by the Korea Government during the Coronavirus pandemic with the BTS phenomenon was the creation of tourism advertisement on YouTube featuring BTS named SEOUL X BTS SEE YOU IN SEOUL in an attempt to increase the number of global visitors who recognize BTS as K-pop stars (Oktaviani et al., 2022).

In 2018, the Hyundai Research Institute (HRI) estimated that BTS contributed approximately $3.54 billion USD annually to the nation's economy. HRI also said that as many as 800,000 international visitors visit Korea annually due to BTS's promotion of the tourism sector (Suntikul, 2019).

The case of the Korean tourism commercial version of Seoul x BTS has demonstrated that the rebrand carried out by the government, and BTS as core resource in the strategy have done a financial rippling impact (Lee, 2011) reaching also the K-Beauty, K-Food, and K-Fashion industries.

(iii) *Investing in Big Events*

Globally, the use of sporting events or mega events to achieve certain national and external policy objectives has become a widespread practice (Ikenberry, 2018). Mega events, in general, refer to "cultural events, both commercial and sporting, on a large scale that have a dramatic character, a great popular appeal and international significance" (Roche, 1994).

China, Japan, and South Korea are considered great powers in terms of mega sporting events (Lee & Tan, 2019). South Korea hosted the Olympic Games in 1988, the FIFA World Cup in 2002, and the PyeongChang Winter Games in 2018. Specifically, Korea has a clear sporting approach to improving its soft power; this being in a double sense: they invest in the athletes themselves and invest in the organization of sporting events (Grix et al., 2021).

The holding of the 1988 Olympic Games allowed Korea to show itself to the international community as a democracy based on traditional loyalty and patriotism as well as to promote Taekwondo and emit positive images abroad (Kang & Houlihan, 2021).

In relation to the FIFA World Cup in 2002, Kim (2007a, 2007b) argued that the main reason that prompted them to organize it was to improve relations with Japan, show their technological advances, and promote Hallyu, a concept well known at that time in East Asia. Other authors corroborate this motivation and this achievement, such as Horne and Manzenreiter (2004) who assured that the World Cup was a perfect scenario to show technological advances and make Hallyu something beyond a regional phenomenon. The Hallyu was currently considered mainstream regionally and internationally, signaling its strengths as a new regional contender and raising the nation's global positioning (Jeong, 2021).

Finally, regarding the celebration of the PyeongChang Winter Olympic Games (2018), South Korea had the opportunity to show and promote peace on the Korean peninsula, boost and develop infrastructure in the province of Gangwon, and also attract tourism to the area where the Asian Winter Games had been held in 1999 in Yongpyong (Merkel, 2008). The country considered that through this event, it could improve the perception of its technological products (Kang & Kim, 2019) and improve its status of pop culture at the global level (Choi, 2019).

Grix et al. (2021) identifies mega events with soft power, and this is the main reason that leads Korea to hold global events, as they allow it to develop its soft power strategy and shape its national identity and country image regionally and internationally. However, this policy of attraction seems to work especially when the resource of soft power is organic and refers to the cultural sector, which is increasingly influential (K-Pop, cinema, and television hits).

With festivals for everything from fireflies to pine mushrooms and swimming in icy water, Koreans get hold of numerous traditional celebrations. These regional festivals show how varied and rich Korean culture is.

In addition, a variety of cultural activities sponsored by the Ministry of Foreign Affairs (MOFA) and others related to the Korean language and its culture are organized by Korea Foundation (Korea Foundation, 2013).

Another example that illustrates this is the collaboration between MOFA, the Korean Broadcasting System (KBS), and the Ministry of Culture, Sports, and Tourism for the organization of the World K-Pop Festival, a music and dance competition in which groups from all over the world participate. MOFA also worked together with the World Taekwondo Federation and the World Taekwondo Peace Corps, to send teachers abroad and thus promote the knowledge and practice of martial arts, contributing to their globalization (Băjenaru, 2022).

(iv) Change of Stereotypes Against the Other Countries

The policies applied in relation to tourism in Korea show that there are no restrictions based on origin, ethnicity, or culture. This facilitates social and cultural exchange that avoids the creation of stereotypes.

Tourist arrivals to South Korea have grown across the board since 2000. However, as in the rest of the world and due to the Coronavirus pandemic, this trend was abruptly modified, reducing by more than 85%. Overall, with the exception of the pandemic years, each year almost 47% of all arrivals in South Korea are based on mainland China. It is followed by Japan, Taiwan, and the United States. In 2019, South Korea received a total of 17,503,000 tourists in 2019 (World Bank, 2020).

South Korea makes great efforts and resources to boost its tourism sector both domestically and internationally. Proof of this is the importance of tourism in relation to other economic activities, representing 5.1% of the Gross Domestic Product and employing 1.5 million citizens, that is, 5.8% of total employment (Trends, 2018).

Given the importance that the tourism sector is acquiring in South Korea, there is no evidence to justify that tourism flows have contributed to the elimination of stereotypes. To be able to affirm this, that is, that tourists entering and leaving South Korea have changed the preconceived image of this country and its citizens, needs a deep,

multidimensional, and longitudinal analysis (Hussin, 2018). There are no previous studies that have specifically measured how tourism has contributed to eliminating previous stereotypes.

One might consider the idea that there are certain stereotypes related to the South Korea–North Korea duality. It is clear that there is a latent threat, and this can provoke a certain sense of insecurity. Aware of this reality, the Ministry of Foreign Affairs and Trade has as its mission to seek a Peace Treaty and participate in the six-party negotiations.

In recent years, certain events have occurred that have been able to modify or strengthen certain pre-existing stereotypes. The health crisis caused by COVID-19 has led all governments to make extreme decisions and protection measures that have influenced the image they project abroad.

It is also evident that the tourism sector has been very sensitive to this fact, since tourists show a greater aversion to risk in times of economic and health crises (Kim et al., 2020).

In this scenario, "contactless" tourism, or also called "Untact" has been widely used in Korea during the pandemic (Bae & Chang, 2021). Since it began to be used in 2017, it has represented a trend of avoiding human contact, along with the development of technology-based services that avoid physical contact (Lee & Lee, 2020). Bae and Chang (2021) showed that this phenomenon improved the perception of COVID-19 risk, positively influencing the attitude toward contactless tourism in Korea, which increased the intention and predisposition to do this type of tourism.

3 Conclusion

This research tried to analyze if Korea is enhancing its tourism resources in an ideal way to improve its soft power, based on the study of the four areas proposed by Ooi (2016).

In view of the analysis carried out, it can be said that the four areas are being conveniently used. Research suggests that South Korea has efficiently used its resources to rebrand the country through the creative industries and its traditions and as effect, promote tourism. It has been a strategy well used and made possible to be the latter being the main claim for potential tourists. In this way, and through the promotion of these four areas, South Korea is improving the country's notoriety in the international context, boosting tourism and, therefore, enhancing the soft power.

After the analysis, it is noted that the government rebranding toward to make cultural industries more materialized in the so-called Hallyu constitute the main element on which they base their tourism strategies. Also, tourism innovation must accompany and enhance a sector that is becoming more relevant every day. The K-Pop stars and its success in Asia, USA, Latin-American, and all over the world are a key, in the government strategy for tourism, diplomacy, and the soft power spread.

This confirms what other authors have claimed and is the emergence of a new market segment driven by pop culture, TV shows, k-dramas, movies, music, and celebrities (Miller, 2006). The choosing process of who will be in the marketing strategy and the combination with the diplomacy with BTS made possible to show a different image of Korea to a new market—far away for its manufacturing positioning after its economic miracle—moving forward to a country with a rich culture and what it seems, a very good recipe in music and entertainment groups production, to all over the world. The impact of the boy bands is also a positive image for the country brand, tourism, and has an effect in other industries as the medical, fashion, food, etc. as Lee (2011) suggested.

In addition, recent research shows that a country's pop culture, such as television drama series, positively influences perceptions of the country (Kim et al., 2008; Lee et al., 2008) and a great influence on image formation of Korea to foreigners (Lee, 2011; Lee & Workman, 2015).

Korea have been working on the scenario to jump to the world and get a position of preference for tourism. Since the efforts done to make possible to host big sport events—and the cultural promotion with the support of the Ministry of Foreign Affairs and the Korea Foundation—show to the world Korea advances in democracy, technology, and its culture to the Hallyu, these mega events helped place South Korea on the world tourism market.

Based on the four areas proposed by Ooi (2016), tourism is a driver of soft power in Korea. During the pandemic, the policies adopted, and the use of technology at the foreign tourist view, it was a trust. Its strategy at the domestic market with untact or "contactless" tourism, the smart tourism, and its online strategies made possible to keep on mind the country for traveling.

References

Bae, S. Y., & Chang, P. J. (2021). The effect of coronavirus disease-19 (covid-19) risk perception on behavioural intention towards 'Untact' tourism in South Korea during the first wave of the pandemic (March 2020). *Current Issues in Tourism, 24*(7), 1017–1035.

Bǎjenaru, I. R. (2022). Bangtan boys (Bts)–part of South Korea's cultural diplomacy and soft power strategy. *Romanian Review of Political Sciences & International Relations, 19*(1).

Busby, G., Brunt, P., & Lund, J. (2003). In Agatha Christie Country: Resident perceptions of special interest tourism. *Tourism, 51*(3), 287e300.

Busby, G., & Klug, J. (2001). Movie-induced tourism: The challenge of measurement and other issues. *Journal of Vacation Marketing, 7*(4), 316e332.

Chae, J. Y. (2014). The need and composition of establishing Korean wave theme infrastructure. Korea Culture & Tourism Institute.

Chen, Y. B. (2005). A study on the impact of Korean Wave on Korea's tourism image. *The Journal of Tourism Policy, 12*(1), 63–78.

Chitty, N. (2017), "Introduction". In N. Chitty, J. Li, G. D. Rawnsley & C. Hayden (Eds.), *The Routledge handbook of soft power* (pp. 1–6). Routledge, London.

Choi, J. G., Tkachenko, T., & Sil, S. (2011). On the destination image of Korea by Russian tourists. *Tourism Management, 32*(1), 193–194.

Choi, K. J. (2019). *The Republic of Korea's public diplomacy strategy: History and current status.* Usc Center On Public Diplomacy.

Connell, J. (2012). Film Tourism—Evolution, Progress and Prospects. *Tourism Management, 33*, 1007–1029.

Dinnie, K. (2008). *Nation branding: Concepts, issues, practice.* Butterworth-Heinemann.

Elmassah, S. (2015). Islamic economy option: Swot case study analysis. *Advances in Management & Applied Economics, 5*(3), 63–84.

Grix, J., Jeong, J. B., & Kim, H. (2021). Understanding South Korea's use of sports mega-events for domestic, regional and international soft power. *Societies, 11*(4), 144.

Hollinshead, K., & Hou, C. X. (2012). The seductions of "soft power": The call for multifronted research into the articulative reach of tourism in China: '软实力'的诱惑—以中国旅游为例. *Journal of China Tourism Research, 8*(3), 227–247.

Holt, D. B. (2004). *How brands becomes icons.* Harvard Business School Press.

Horne, J. D., & Manzenreiter, W. (2004). Accounting for mega-events: Forecast and actual impacts of the 2002 football world cup finals on the host countries Japan/Korea. *International Review for the Sociology of Sport, 39*(2), 187–203.

Hussin, H. (2018). Gastronomy, tourism, and the soft power of Malaysia. *SAGE Open, 8*(4), 2158244018809211.

Ikenberry, G. J. (2018). The end of liberal international order? *International Affairs, 94*(1), 7–23.

Iwashita, C. (2008). Role Offilms and television dramasininternational tourism: The case of Japanese Tourists to the UK. *Journal of Travel & Tourism Marketing, 24*, 139e151.

Jeong, J. (2021). *How nations use sport mega-events to leverage soft power: A New rise in East Asia* (Doctoral Dissertation, Manchester Metropolitan University).

Kang, M., & Kim, H. (2019). Global and local intersection of the 2018 Pyeongchang winter olympics. *International Journal of Japanese Sociology, 28*(1), 110–127.

Kang, Y., & Houlihan, B. (2021). Sport as a diplomatic resource: The case of South Korea, 1970–2017. *International Journal of Sport Policy and Politics, 13*(1), 45–63.

Kim, J. (2007b). Why does Hallyu matter? The significance of the Korean wave In South Korea. *Critical Studies in Television, 2*(2), 47–59.

Kim, J., Kim, J., Lee, S. K., & Tang, L. R. (2020). Effects of epidemic disease outbreaks on financial performance of restaurants: Event study method approach. *Journal of Hospitality and Tourism Management, 43*, 32–41.

Kim, S. S., Agrusa, J., Chon, K., & Cho, Y. (2008). The effects of Korean pop culture on Hong Kong residents' perceptions of Korea as a potential tourist destination. *Journal of Travel & Tourism Marketing, 24*(2–3), 163–183.

Kim, Y. H. (2007a). The relations among Hanryu attitudes, tourism destination image, satisfaction of Chinese in visit to Korea. *Convention Research, 7*(1), 143–159.

Ko, J. M. (2012). The Korean wave culture and tourism. *Korea Tourism Policy, 49*, 33–40.

Lee, J., & Lee, C. (2020). Over the half of the adults in Korea have experienced Corona Blue. April 14, 2020.

Lee, J. W., & Tan, T. C. (2019). The rise of sport in the Asia-Pacific region and a social scientific journey through Asian-Pacific sport. *Sport in Society, 22*(8), 1319–1325.

Lee, S. (2015). Introduction: A decade of Hallyu scholarship: Toward a new direction in Hallyu 2.0. *Hallyu, 2*, 1–27.

Lee, S., & Bai, B. (2016). Influence of popular culture on special interest tourists' destination image. *Tourism Management, 52*, 161–169.

Lee, S., & Nornes, A. M. (Eds.). (2015). *Hallyu 2.0: The Korean wave in the age of social media*. University Of Michigan Press.

Lee, S., Scott, D., & Kim, H. (2008). Celebrity fan involvement and destination perceptions. *Annals of Tourism Research, 35*(3), 809–832.

Lee, S. H., & Workman, J. E. (2015). Compulsive buying and branding phenomena. *Joitmc, 1*, 3.

Lee, S.-J. (2021, October 20). Local goverments seek to promote tourism with "Squid Game" contents. The Korea Herald. https://www.koreaherald.com/view.php?ud=20211020000824

Lee, W. H. (2011). *A research on policy direction for voluntourism*. Korea Culture & Tourism Institute.

Merkel, U. (2008). The politics of sport diplomacy and reunification in divided Korea: One nation, two countries and three flags. *International Review for the Sociology of Sport, 43*(3), 289–311.

Miller, R. K. (2006). *The 2009 travel & tourism market research handbook*. Richard K. Miller & Associates.

Moilanen, T., & Rainisto, S. (2009). *How to brand nations, cities and destinations*. Palgrave Macmillan.

Mukarromah, B., Rihhadatul'aisy, J. A., & Pandin, M. G. R. (2022). *Axiological analysis on Netflix Series "Squid Game" as an effort to increase awareness of social issues among generation Z*.

Nye, J., & Kim, Y. (2013). Soft power and the Korean wave. In Y. Kim (Ed.), *The Korean Wave: Korean media go global* (pp. 31–42). Routledge.

Nye, J. S. (1990). Soft power. *Foreign Policy, 80*, 153–171.

O'connor, N., & Bolan, P. (2008). Creating a sustainable brand for Northern Ireland through film induced tourism. *Tourism, Culture & Communication, 8*(3), 147–158.

OECD Tourism Trends and Policies 2020. Recovered from https://www.Oecd-Ilibrary.Org/Sites/6e8b663c-En/Index.Html?Itemid=/Content/Component/6e8b663c-En

Oh, Y. (2018). *Pop city: Korean popular culture and the selling of place*. Cornell University Press.

Oktaviani, N., Ramadhani, R. A., & Chia, Y. P. K. (2022). South Korea tourism advertising semiotics Seoul X Bts version. *Budapest International Research and Critics Institute-Journal (Birci-Journal), 5*(3), 21472–21484.

Ooi, C. (2016). Soft power, tourism. In J. Jafari & H. Xiao (Eds.), *Encyclopedia of Tourism* (pp. 1–2). Springer.

Otmazgin, N. K. (2008). Contesting soft power: Japanese popular culture in east and Southeast Asia. *International Relations of the Asia-Pacific, 8*(1), 73–101.

Rawnsley, G. D. (2016). Reflections of a soft power agnostic. In X. Zhang, H. Wasserman & W. Mano (Eds), *China's media and soft power in Africa* (pp. 19–32). Palgrave Macmillan.

Roche, M. (1994). Mega-events and urban policy. Annals of Tourism Research, 21(1), 1–19. The end of liberal international order? *International Affairs, 94*(1), 7–23.

Roesch, S. (2009). *The experiences of film location tourists. Aspects of tourism*. Channel View Publications.

Seabrook, J. (2012). Factory girls: Cultural technology and the making of K-pop. *The New Yorker* (Vol. 88). Available at: www.Newyorker.com/Magazine/2012/10/08/Factory-Girls-2. Accessed December 20, 2018.

Shambaugh, D. (2015). China's soft-power push: The search for respect. *Foreign Affairs, 94*(4), 99–107.

Suh, Y. K., & Suh, Y. G. (2004). The effects of Korean fever in influencing the image of Korea as a prime destination, and the tourist decision making process. *The International Journal of Tourism Sciences, 28*(3), 47–64.

Suntikul, W. (2019, March 1). Bts and the global spread of Korean soft power. Diperoleh Dari The Diplomat: https://thediplomat.com/2019/03/Bts-And-Theglobal-Spread-Of-Korean-Soft-Power/

The Korea Foundation Annual Report. (2013). Recovered from: https://www.Kf.Or.Kr/Kfeng/Cm/Cntnts/Cntntsview.Do?Mi=2129&Cntntsid=1630

Trends, O. T. (2018). Policies. *Energy Policy, 36*(11), 4012–4021.

Trisni, S., Nasir, P. E., & Isnarti, R. (2019). South Korean government's role in public diplomacy: A case study of the Korean Wave boom. *Andalas Journal of International Studies (ajis), 8*(1), 31–42.

Unicef. (2021). "Bts and big hit renew commitment to "Love Myself" Campaign To Support Unicef In Ending Violence And Neglect As Well As Promoting Self-Esteem And Well-Being", Unicef.Org, Updated On: 05.03.2021, Accessed on: September 10, 2021. https://www.Unicef.Org/Lac/En/Press-Releases/Bts-And-Big-Hit-Renew-Commitment-To-Love-Myselfcampaign-To-Support-Unicef-In-Ending-Violence

Ward, S., & O'regan, T. (2009). The film producer as the long-stay business tourist: Rethinking film and tourism from a gold coast perspective. *Tourism Geographies, 11*(2), 214–232.

World Bank. (2020). Recovered from: https://www.Data.Worldbank.Org/Indicator/St.Int.Arvl?Locations=Kr

Xu, H., Wang, K., & Song, Y. M. (2020). Chinese outbound tourism and soft power. *Journal of Policy Research in Tourism, Leisure and Events, 12*(1), 34–49.

정민, 오준범, 신유란, & 류승희. (2018). 방탄소년단 (BTS) 의 경제적 효과-연평균 생산유발효과 약1 조원, 부가가치유발효과 약 1.4 조원.이슈리포트 _2018(15), 1–15.

Tourists' Perceptions of Service Quality: Using Text for Tourism Hospitality Industry Insights

António J. D. V. T. Melo, Eduardo Cordeiro Gonçalves, and António José Pinheiro

Abstract

More and more consumers describe their tourism experiences through online reviews. Online reviews create a wealth of textual data that is extremely important in supporting consumers' purchasing decisions and, as a result, in helping hotel managers understand customer preferences. The authors analyze 7248 online tourist reviews of 109 Portuguese rural accommodation establishments (RTA) of a leading international travel website. The results show that tourists' perceptions of overall service quality reflected in their online reviews depend on their perceptions of the accommodation and of the surroundings. These discoveries have implications for academics, as well as for RTA owners who want to understand precisely the factors that affect tourists' evaluations and perceptions of their services. Lastly, the authors conclude by stating some potential areas for future work.

Keywords

Online reviews • Text analysis • Tourism • Hospitality • Services • Qualitative content analysis

1 Introduction

Most consumer reviews and information searches take place online through search engines such as Google or on social media (like YouTube, Instagram, Facebook, or microblogs like Twitter), infomediary websites, or blogs (Chang et al., 2018; Doval-Fernández & Sánchez-Amboage, 2021; Duffett, 2022; Melo et al., 2022b; Sotiriadis & van Zyl, 2013; Wong et al., 2022). Therefore, this research seeks to find the factors that drive consumers online using content analysis coding.

The authors started with a brief discussion about rural tourism establishments, highlighting the relevance of extrinsic signals, and specific feedback from consumers of online services and the tourism sector.

Then, the authors list the hypotheses to detail the relationship between online reviews of an establishment and the key factors pointed out by consumers when they assess the service delivered. The authors describe the data, the sample, and the respective measures and, finally, present the results. Next, they present some conclusions and some limitations of the research. The findings help to fill the research gap related to online reviews (comments) about RTAs, and they also answer management requests about the key drivers of tourists' evaluations. The model proposed by the authors integrates online reviews at the RTA level, offering the potential for replication outside the rural tourism industry because the variables in the model can also be applied to other services.

2 Literature Review

2.1 Signaling Theory

Signaling theory helps to explain the two-part behavior (individuals and organizations) of strategic actions—communications or signals—taken by a sender to influence the perceptions and behaviors of receivers when they access different information (Jean et al., 2021; Mavlanova et al., 2012; Spence, 1974; Tóth et al., 2022). The importance of signaling theory lies in the fact that it reduces information asymmetry. Service purchase decisions, because services are dominated by intangible attributes, lack pre-purchase information, and consumers suffer some anxiety about the risk

A. J. D. V. T. Melo (✉) · E. C. Gonçalves · A. J. Pinheiro
University of Maia, Maia, Portugal
e-mail: ajdvtm@gmail.com

E. C. Gonçalves
e-mail: egoncalves@umaia.pt

A. J. Pinheiro
e-mail: ajpinheiro@umaia.pt

© The Author(s), under exclusive license to Springer Nature Switzerland AG 2024
J. Chica-Olmo et al. (eds.), *Sustainable Tourism, Culture and Heritage Promotion*,
Advances in Science, Technology & Innovation, https://doi.org/10.1007/978-3-031-49536-6_23

involved as they cannot try the product in advance (Kirmani & Rao, 2000; Melo et al., 2017, 2022a; Wells et al., 2011; Wirtz et al., 2013).

2.2 Online Review and Signaling

Currently, consumers and service providers regularly turn to numerous online reviews posted by a large sample of consumers who have already tried services to reduce the lack of information and learn about services, evaluate potential providers, and compare experiences of existing providers. The World Wide Web supports an up-to-date and enriched digital business environment in which potential consumers and business managers have the means to talk about services before consumers decide to buy them and allows managers to be aware of the perceived level of service (Melo, 2020). Online reviews enable multi-directional exchanges of information at exceptional circulation speeds (Cheung & Thadani, 2012). King et al. (2014) defined the main characteristics of online reviews as much greater volume, enormous dispersion, unidentified and deceit, persistence and observability, with community engagement. Online reviews include positive and/or negative signals from the customer–supplier relationship experience. Signals of good and bad experiences are conveyed from current buyers to the entire market, including new prospects, competitors, suppliers, and other stakeholders through online assessments (Tóth et al., 2022). The perceived value of an online review is defined as the reviews' helpfulness (Fang et al., 2016; Mudambi & Schuff, 2010). This definition suggests that this value is determined exclusively by their readers (Vargo & Lusch, 2008). Signaling theory shows how people, based on quality signals, make decisions, especially when quality is difficult to verify, as in the service sectors (Situmeang et al., 2020). These online reviews provide perceived usefulness in the form of signaling cues to perceive the evaluation of the quality of the products (Melo et al., 2017). Online reviews have become a significant source for regularly and consistently providing customers with product details information, opinions, goods and services experiences, and reviewers' identity cues convey information about the person. The received informational cues can create a favorable or unfavorable attitude toward products.

Online reviews accumulate information that includes both consumers' feedback and signaling cues for potential consumers, reflecting variations in service ratings from current consumers and revealing their perceptions of changes in service quality (Melo et al., 2017; Zuo et al., 2022). Consumers share their perceptions, their thoughts, and their reviews of goods and services, as well as the interactions they have had with service providers and the surrounding community (Li & Liang, 2022).

For service providers, reliable customer feedback offers a unique competitive advantage, and informational signaling opportunities, as a tool to improve positioning among potential customers (Lin & Kalwani, 2018; Liu & Park, 2015; Melo et al., 2022b). Online reviews are crucial sources to understand the real "voice of the customer" and are particularly useful for service providers and can help us more accurately detect consumer wants and preferences (Guo et al., 2017; Melo et al., 2022b).

Signals, visible cues used to communicate perceptions about product quality, are critical in services because they are fundamentally made up of attributes that are very difficult to assess before purchase because they are intangible, representing important purchase risk (Filieri et al., 2020). Consumers before purchasing service offerings suffer from anxiety about information asymmetry caused by the scarcity of pre-purchase information because they cannot try the product beforehand (Melo et al., 2017). The primary motivation for customers to seek third-party reviews online, and the reason their purchasing decisions are significantly influenced by third-party opinions, is to reduce their risks and information asymmetries about goods or services (Korfiatis et al., 2019).

2.3 Product Quality Perceptions Exposed in Online Reviews

Online reviews indicate information cues that entail the "wisdom of the crowd" (Andoy et al., 2022; Filieri, 2015). Therefore, in a closely related stream of existing literature, that of services, reviewing online reviews has recently been considered a key process for understanding overall customer satisfaction (Mittra & Khamkar, 2021; Shin & Perdue, 2022). Availability, its power to reach large numbers of people at exceptional circulation speed on a global scale and often allowing multi-directional exchanges of information, makes online reviews the most influential and powerful tool at our disposal today (Filieri et al., 2020; Tóth et al., 2022). Online reviews lead the reader, potential consumer, or business manager, to the perception of different attributes that the customer shared when expressing their positive and negative perceptions and emotions associated with the service experienced (Cheung & Thadani, 2012; Hennig-Thurau et al., 2004; Melo, 2020; Tóth et al., 2022).

From the existing literature, online reviews can be categorized into two types: the first focus on finding attributes in the content of these online reviews and in their stay experiences; the second is dedicated to the technical side of online assessments. Online reviews are unstructured user-generated content that reflects detailed customer experience and insights (Mittra & Khamkar, 2021; Situmeang et al., 2020; Zhao et al., 2019). Gao et al. (2018) stated online reviews are rich opinion resources to recognize competitors to gain competitiveness.

Customer satisfaction, a "subjective evaluation of a consumption experience based on some relationships between the customers' perceptions and objective attributes of the product," with hospitality service is a sum of satisfaction with the attributes that the consumers consider important (Oliver, 2010; Pizam & Ellis, 1999).

Satisfaction and perceived quality, generally strongly related, are foundations of businesses, for their ability to generate positive effects for the institutions, how to increase product quality, and profitability, help to gain competitive advantages, improve entrepreneurs' satisfaction, and improve the behavior of entrepreneurs in relation to the consumer market (Ahani et al., 2019; Mariani & Borghi, 2020; Melo, 2021; Melo et al., 2022b).

Tracking changes in consumer perceptions of service quality is crucial to maintaining a strong and healthy relationship with them and being able to analyze dynamic changes in their perceptions should be an important topic for managers (Melo et al., 2017; Zuo et al., 2022). Consumer perceptions have a direct impact on their decisions and, as such, on the performance of companies (Guo et al., 2017; Hendar et al., 2020; Mariani & Borghi, 2020; Melo et al., 2017, 2022b).

In the service environment, especially in tourism, online reviews tend to refer to some of the service attributes like comfort, feeling welcomed, gastronomy, sleep quality, environment, amenities, and attractive room or decoration (Barnesa & Jacobsen, 2014; Melo et al., 2017; Raksmey & Lai, 2019). Based on their research, the authors propose:

Hypothesis 1: In the comments provided in the online reviews, the overall quality of service perceptions of tourists in a specific RTA reflect their perceptions of the service attributes experienced in the lodgings.

Hypothesis 2: Online tourism reviews include comments that address tourists' perceptions of the service attributes experienced in relation to the surroundings of a specific RTA.

3 Methodology

In this investigation, the authors combined qualitative and quantitative analysis to expose factors that could describe the satisfactory and the dissatisfied experience of guests during their stay in RTAs in Portugal. The authors used content analysis to classify all the consumers' online reviews, using NVivo 12. Thus, they detected what kind of aspects of the tourists' satisfactory and dissatisfactory perceptions of the RTA in Portugal referred to the service provided, to help RTA owners better understand and improve their services. The authors analyzed the data with SPSS 27 quantitative data analysis software. Finally, for each RTA, the authors ran the model using SmartPLS 3 (Sarstedt et al., 2022).

3.1 Data

The authors convened and analyzed data about the online consumer reviews of Portuguese RTAs published on the Vrbo website, a leading specialized travel website for independently owned RTAs in Southern Europe. TravelBI by tourismo de Portugal (2019) identify 1,687 RTAs in 2019. From a convenience sample of 255 Portuguese RTAs, from all regions of the country, after data purification, the authors obtained 109 RTAs (Table 1). The authors filtered the data and obtained a sample of 109 RTAs, with a total of 7248 online reviews, posted in 2017–2019. As expected, based on the literature review, most of these evaluations were positive (95.95%).

The authors choose data from RTAs in Portugal, because, for the fifth time in the last six years, Portugal was again considered "Best Tourist Destination in Europe" at the World Travel Awards 2022.

Vrbo offered presence on the site at different levels (Table 2): exclusive, prestige, gold, and silver. As it represents less than 1%, the exclusive tier was excluded.

As for the number of messages per hosting, for less than 5 messages, we have 32% of RTAs, and with 5 or more messages we have 68% of RTAs (Table 3).

Table 1 Variables

Variable	Description	Measures	
Consumer reviews	Content of consumer online reviews	22 classes	
RTA types		No.	%
– Rural house rooms	Rural house rooms rental	58	53.31
– Whole rural house	Whole rural house rental	44	40.37
– Rural hotel	Rural hotel rooms rental	7	6.42

Table 2 Promotion by type of RTA

Promotion	Type of RTA						Total	
	Whole house rental		House rooms rental		Hotel rooms rental			
Prestige	12	25.00%	12	21.82%	0	0.00%	24	22.02%
Gold	25	52.08%	29	52.73%	5	83.33%	59	54.13%
Silver	9	18.75%	9	16.36%	0	0.00%	18	16.51%
None	2	4.17%	5	9.09%	1	16.67%	8	7.34%
	48		55		6		109	

Table 3 Average value of the variables by type of RTA

	RTA type		
	Whole rental	Rooms rental	
	House	House	Hotel
	Average	Average	Average
Price per night	31.98	34.26	31.82
Seniority in Vrbo (months)	72.44	88.91	76.67
Implication	11.19	12.18	15.17
Consumer satisfaction (%)	96.41	97.14	97.14

3.2 Measures

The authors followed a method proposed by Miles and Huberman (1984) and opened the process with a sample of 50 online customer reviews to identify recurring factors, with four coders relying on their knowledge gained from previous research (O'Connor, 2010). The coders evaluated and examined each review inspecting any disparities in their perceptions to reach an agreement.

The authors have established a dictionary of terms in eight languages, according to online reviews (Portuguese, English, Spanish, German, French, Dutch, Catalan, and Italian) which they used with NVivo 12 software. This process was content-driven, focusing on analyzing the information provided in each customer's online feedback. Subsequently, the codes were merged into 22 main classes that tourists mention when evaluating their rural experiences in online reviews with a qualitative content analysis (Table 5).

3.3 Results

The authors identified 22 main themes, using the qualitative content analysis that tourists raised when they evaluate their RTA experiences in online reviews. In the positive online evaluations of tourists, the five themes with the highest percentage of mention are (Table 4): the establishment (25.28%), the environment (21.96%), the feeling welcomed by the staff (13.48%), the decoration (8.08%) and gastronomy, speaks of breakfast, lunch, snacks, etc. (5.78%).

Table 4 Five themes most cited by satisfied and dissatisfied guests

Satisfied consumers (positive online reviews)	Dissatisfied consumers (negative online reviews)
Establishment	Establishment
Environment	Environment
Feeling welcomed	Attractive room
Decoration	Feeling welcomed
Gastronomy	Quality of sleep

In the negative online evaluations of tourists, the five themes with the highest percentage of mention are the establishment (21.78%), the environment (12.07%), attractive room—If they like the room, if it has a bathroom (12.07%), feeling welcomed by the team (9.71%) and tranquility, possibility to rest, quality of sleep (6.59%).

Initially, the authors used qualitative content analysis to identify the top 22 themes that tourists used when communicating evaluations of their experiences at RTAs through online reviews. As Melo et al. (2017) have done, these 22 classes can be further grouped hierarchically into three latent variables: accommodation perception (fourteen classes), surrounding perception (five classes), and overall perception of service quality (four classes) (Fig. 1). Lodging perceptions concern multimedia (Internet availability), space, food, accommodation, maintenance, price, equipment, decoration, cleanliness, parking, temperature, information, attractive room, and attention to special requests. Surrounding perceptions entail seasonality, environment (localization), access, distance, and activities. These perceptions on their

Table 5 Number of messages and percentages per classes

Classes	Number of satisfaction messages (4–5)	Number of dissatisfaction messages (1–3)
Host	926 (13.48%)	37 (9.71%)
Environment	1508 (21.96%)	46 (12.07%)
Food	397 (5.78%)	19 (4.99%)
Rest	93 (1.35%)	25 (6.56%)
Seasonality	191 (2.78%)	16 (4.20%)
Attractive room	287 (4.18%)	46 (12.07%)
Attention to special guests/special requests	312 (4.54%)	11 (2.89%)
Activities	320 (4.66%)	23 (6.04%)
Comfort	105 (1.53%)	3 (0.79%)
Cleanliness	57 (0.83%)	20 (5.25%)
Decoration	555 (8.08%)	2 (0.52%)
Equipment	33 (0.48%)	3 (0.79%)
Distance	0 (0.00%)	0 (0.00%)
Accommodation	1736 (25.28%)	83 (21.78%)
Temperature	46 (0.67%)	16 (4.20%)
Space	62 (1.08%)	7 (1.84%)
Multimedia	53 (0.77%)	1 (0.26%)
Information	30 (0.44%)	2 (0.52%)
Access	51 (0.74%)	4 (1.05%)
Maintenance	4 (0.06%)	4 (1.05%)
Price	74 (1.08%)	10 (2.62%)
Parking	27 (0.39%)	3 (0.79%)
Totals	6867	381

Note For each class, the class weight reflects the number of RTAs for which class is mentioned at least once in online reviews, relative to the total number of RTAs (109)

part influence the tourists' overall quality of service perceptions (host, rest, comfort, and satisfaction). Astonishingly, the tourists' lodging perceptions reveal a non-significant relationship with space, food, accommodation, maintenance, price, equipment, parking, attractive room, and attention to special request classes. However, surrounding perceptions do not show a significant relationship between activities and distance classes. Delving, tourists' lodging perceptions explained variability in five main classes: multimedia (0.145), decoration (0.272), cleanliness (0.367), temperature (0.451), and information (0.755). On the side of the tourists' surrounding perceptions, the variability is explained in three main classes: seasonality (0.677), environment (0.214), and access (0.164).

In line with the study of Melo et al. (2017), the authors' study reveals an overall quality of service perceptions construct, which reflects the perceptions of an overall assessment of the RTA and gives indications as to whether it meets the requirements of the guests.

Concerning the impact of lodging and surrounding perceptions on the overall quality of service perceptions, the authors find that both have significant and similar positive influences, respectively, 0.505 (4.974) and 0.512 (5.035), consistent with hypotheses 1 and 2. In the model proposed by this study, from the tourists' perspective, the variance in overall quality of service perceptions is explained in 72.5% by the lodging and surroundings perceptions.

Tourists' overall quality of service perceptions exhibited excellent composite reliability (0.935).

The model reveals the reliability of internal consistency of $\rho_c > 0.7$, convergent validity with average variance extracted > 0.5, absence of collinearity among its indicators, the relevance and significance of external loadings, and discriminant validity (by the results of the reflective construct evaluation exposed Table 6). Specifically, in this model, the square root of the average variance extracted from the construct is greater than its highest correlation with any other construct (Hair, 2021). Consequently, the reliability and validity of the construct measures are confirmed by the study and state the appropriateness of their inclusion in the path model (Hair et al., 2014). When the authors perform a formal analysis of the detection tolerance and

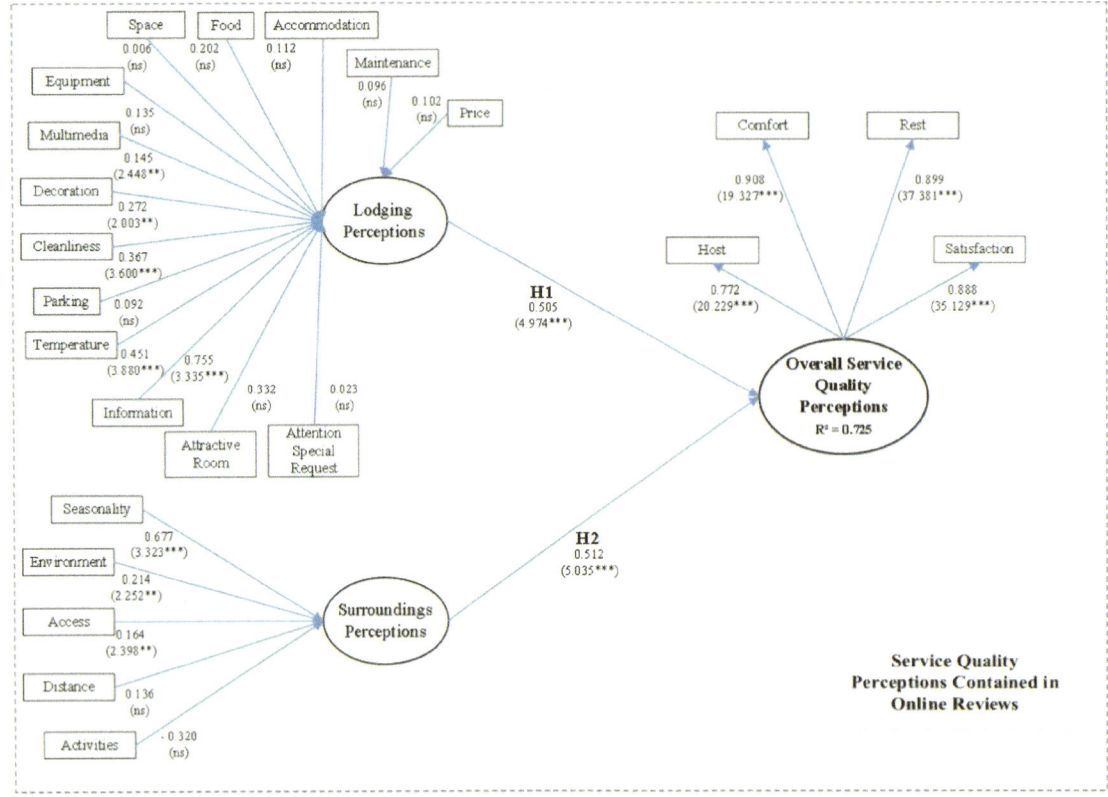

Note: *p > .1. **p > .05. ***p > .01. ns = non-significant.

Fig. 1 Model

Table 6 Reflective construct assessment

Latent variable	Indicators	Outer loadings	t value	Average variance extracted	Composite reliability	Indicator reliability	Cronbach's alpha	Discriminant validity
Overall quality of service perceptions	Host	0.900232	20,229	0.799239	0.934887	0.81041765	0.916256	Yes
	Satisfaction	0.893538	35,129			0.79841016		
	Comfort	0.883587	19,327			0.78072599		
	Rest	0.891533	37,381			0.79483109		

variance inflation factor, they found no multicollinearity among the formative construct indicators, so the model achieved predictive relevance.

4 Discussion

In this investigation, one of the two different analysis approaches used previously from the online reviews was followed: content analysis to look for different classes of information and relations between different kinds of information.

The authors of this research try to consider, for each key content class, the number of online reviews, since the sensitive perception of consumers is disparate for each class that

the online reviews refer to. It is necessary to identify the classes of content that reflect consumers' evaluations of services, so that business managers can devise strategies to improve their services by giving appropriate importance to each specific type of information they provide.

Theoretically, the authors' findings fill a research gap concerning the analysis of content classes in online reviews and the key service elements that consumers perceive and communicate in their online reviews. Further, the proposed model also specifies that the overall perception of service quality is formed by the conjunction of perceptions of the accommodation and the environment.

From a managerial perspective, this study leads service providers to deepen their understanding of what drives online consumer feedback. Illuminating the key aspects that guests use in their evaluation of the service experience, as RTA stays, is fundamental in delivering service excellence from the guest's point of view. The authors show that tourists' overall perceptions of service quality embody a sense of welcome, satisfaction, comfort, and tranquility. Tourists shape these integrated perceptions of service quality based on two head groups of perceptions: one related to the accommodation and the other related to the surroundings. With the content analysis of this study, 12 key factors that integrate online customer reviews were identified, as well as 11 factors with non-significant influences. Complementarily, the authors pointed out that tourists' overall perceptions of service quality embody a sense of welcome, satisfaction, comfort, and tranquility.

Based on the results of this study, lodging and surroundings perceptions contribute similarly to the overall quality of service perceptions, and the authors recommend that RTA managers evaluate both controllable key drivers related to their lodging and uncontrollable key drivers related to the surroundings.

The authors also show that in view of the different classes in online reviews, the perspective should be aggregated to better reflect how most people perceive, interpret, and aggregate these same perceptions. This perspective still allows the associated complexity to be reduced, in the same sense as consumers do when dealing with such an abundant amount of information. From the point of view of the entire community involved (managers, government agencies, industry associations, partners, intermediaries, tourists, and websites), the group of classes that reflect the perception of service quality should be the primary consideration in understanding, analyzing, and managing service quality.

With this understanding, RTAs deepen the understanding of what guests perceive about the service they provide, which can help them improve their services, increasing their perception of quality. Service providers should pay more attention to online consumer reviews, analyzing and monitoring them to better understand what drives consumer perceptions in their assessment of the quality of service provided. To consequently adjust their strategic service delivery decisions to the detected tourist needs and desires.

These key factors that result from guest perceptions and communication about the service encounters help us replicate the positive experiences and avoid the negative experiences allowing us to ensure that this does not happen again.

Tourism organizations, associations, and local and central governments, as well as websites, should periodically evaluate not only the quality of services provided by tourism establishments but also the relevant classes identified in this paper by building their own evaluation frameworks.

The authors believe that this type of research is very important for demonstrating the specific content of consumer messages, helping the representation and evaluation of these classes properly and independently in order to meet consumer demands.

Some of the limitations of this study may suggest some potential areas for future work. First, try to find the link between the consumer perception classes in service evaluation and the business performance measures. Second, include other content analyses, such as sentiment analysis, to better understand the key factors that drive consumers' evaluation of the service. Third, this research referred to RTAs in only one country, Portugal, and the authors only include information published on the Vrbo travel website. This is a specific website and only one country, so these results cannot be generalized to other sources of information and to other countries without careful consideration. Replicating this survey using multiple countries and online platforms could be interesting.

References

Ahani, A., et al. (2019). Revealing customers' satisfaction and preferences through online review analysis: The case of Canary Islands hotels. *Journal of Retailing and Consumer Services, 51*, 331–343. https://doi.org/10.1016/j.jretconser.2019.06.014

Andoy, A. D., et al. (2022). The effects of online reviews on purchase intention in the Shopee fashion industry. *Journal of Business and Management Studies, 4*(2). https://doi.org/10.32996/jefas.2022.4.2. 12

Barnesa, N. G., & Jacobsen, S. L. (2014). Missed eWOM opportunities: A cross-sector analysis of online monitoring behavior. *Journal of Marketing Communications, 20*(1–2), 147–158. https://doi.org/10.1080/13527266.2013.797788

Chang, J.-H., et al. (2018). Would you change your mind? An empirical study of social impact theory on Facebook. *Telematics and Informatics, 35*(1), 282–292. https://doi.org/10.1016/j.tele.2017.11. 009

Cheung, C. M. K., & Thadani, D. R. (2012). The impact of electronic word-of-mouth communication: A literature analysis and integrative model. *Decision Support Systems, 54*(1), 461–470. https://doi.org/10.1016/j.dss.2012.06.008

Doval-Fernández, T., & Sánchez-Amboage, E. (2021). *Instagram Como Herramienta De Marketing De Destinos: Funcionalidades y Utilidades.* https://doi.org/10.26807/rp.v25i111.1786

Duffett, R. (2022). Influence of YouTube commercial communication on organic eWOM, purchase intent and purchase associations among young consumers. *International Journal of Web Based Communities, 18*, 87–107. https://doi.org/10.1504/IJWBC.2022. 122394

Fang, B., et al. (2016). Analysis of the perceived value of online tourism reviews: Influence of readability and reviewer characteristics. *Tourism Management, 52*, 498–506. https://doi.org/10.1016/j. tourman.2015.07.018

Filieri, R. (2015). What makes online reviews helpful? A diagnosticity-adoption framework to explain informational and normative influences in e-WOM. *Journal of Business Research, 68*(6), 1261–1270. https://doi.org/10.1016/j.jbusres.2014.11.006

Filieri, R., et al. (2020). Extremely negative ratings and online consumer review helpfulness: The moderating role of product quality signals. *Journal of Travel Research*, 1–19. https://doi.org/10.1177/0047287520916785

Gao, S., et al. (2018). Identifying competitors through comparative relation mining of online reviews in the restaurant industry. *International Journal of Hospitality Management, 71*, 19–32. https://doi.org/10.1016/j.ijhm.2017.09.004

Guo, Y., et al. (2017). Mining meaning from online ratings and reviews: Tourist satisfaction analysis using latent dirichlet allocation. *Tourism Management, 59*, 467–483. https://doi.org/10.1016/j.tourman.2016.09.009

Hair, J. F. (2021). Reflections on SEM: An introspective, idiosyncratic journey to composite-based structural equation modeling. *SIGMIS Database, 52*(SI), 101–113. https://doi.org/10.1145/3505639.3505646

Hair, J. F. J., et al. (2014). *Multivariate data analysis (7th edition)* (7th edn.). Pearson Education.

Hendar, H., et al. (2020). Market intelligence on business performance: The mediating role of specialized marketing capabilities. *Journal of Intelligence Studies in Business, 10*(1), 42–58.

Hennig-Thurau, T., et al. (2004). Electronic word-of-mouth via consumer-opinion platforms. *Journal of Interactive Marketing, 18*(1), 38–52. https://doi.org/10.1002/dir.10073

Jean, R.-J.B., et al. (2021). E-platform use and exporting in the context of Alibaba: A signaling theory perspective. *Journal of International Business Studies, 52*(8), 1501–1528. https://doi.org/10.1057/s41267-020-00396-w

King, R. A., et al. (2014). What we know and don't know about online word-of-mouth: A review and synthesis of the literature. *Journal of Interactive Marketing, 28*(3), 167–183. https://doi.org/10.1016/j.intmar.2014.02.001

Kirmani, A., & Rao, A. (2000). No pain, no gain: A critical review of the literature on signaling unobservable product quality. *Journal of Marketing, 64*, 66–79. https://doi.org/10.1509/jmkg.64.2.66.18000

Korfiatis, N., et al. (2019). Measuring service quality from unstructured data: A topic modeling application on airline passengers' online reviews. *Expert Systems with Applications, 116*, 472–486. https://doi.org/10.1016/j.eswa.2018.09.037

Li, J., & Liang, X. (2022). Reviewers' identity cues in online product reviews and consumers' purchase intention. *Frontiers in Psychology, 12*.https://doi.org/10.3389/fpsyg.2021.784173

Lin, H.-C., & Kalwani, M. U. (2018). Culturally contingent electronic word-of-mouth signaling and screening: A comparative study of product reviews in the United States and Japan. *Journal of International Marketing, 26*(2), 80–102. https://doi.org/10.1509/jim.17.0016

Liu, Z., & Park, S. (2015). What makes a useful online review? Implication for travel product websites. *Tourism Management, 47*, 140–151. https://doi.org/10.1016/j.tourman.2014.09.020

Mariani, M. M., & Borghi, M. (2020). Online review helpfulness and firms' financial performance: An empirical study in a service industry. *International Journal of Electronic Commerce, 24*(4), 421–449. https://doi.org/10.1080/10864415.2020.1806464

Mavlanova, T., et al. (2012). Signaling theory and information asymmetry in online commerce. *Information & Management, 49*(5), 240–247. https://doi.org/10.1016/j.im.2012.05.004

Melo, A. J. D. V. T. (2020). *Ewom effects on business performance in rural tourism accommodation.* (Doctor of Philosophy—Business Economics Scientific research dissertation), University of Salamanca. Retrieved from https://gredos.usal.es/bitstream/handle/10366/143814/TDEX_MeloA.pdf?sequence=1&isAllowed=y

Melo, A. J. D. V. T. (2021). *Doctoral thesis—Ewon effects on business performance in rural tourism accommodation* (I. Press Ed.). Universidade da Maia.

Melo, A. J. D. V. T., et al. (2017). Service quality perceptions, online visibility, and business performance in rural lodging establishments. *Journal of Travel Research, 56*(2), 250–262. https://doi.org/10.1177/0047287516635822

Melo, A. J. D. V. T., et al. (2022a). Customer perceptions and strategies for rural tourism accommodation. *Pasos—Journal of Tourism and Cultural Heritage, 20*(Special issue 2022), 1171–1190. https://doi.org/10.25145/j.pasos.2022.20.079

Melo, A. J. D. V. T., et al. (2022b). Effects of market intelligence generation, online reviews, and management response on the business performance of rural accommodation establishments in France. *Cultural Sustainable Tourism. Advances in Science, Technology & Innovation Book Series (ASTI)* (pp. 105–119), Springer.

Miles, M. B., & Huberman, A. M. (1984). Drawing valid meaning from qualitative data: Toward a shared craft. *Educational Researcher, 13*(5), 20–30. https://doi.org/10.3102/0013189X013005020

Mittra, S., & Khamkar, P. R. (2021). To understand customer's perception towards hotel amenities, services quality and online reviews while booking hotels. *ANWESH: International Journal of Management & Information Technology, 6*(1), 38–42.

Mudambi, S. M., & Schuff, D. (2010). What makes a helpful online review a study of customer reviews on AMAZON.COM. *MIS Quarterly, 34*(1), 185–200. https://doi.org/10.2307/20721420

O'Connor, P. (2010). Managing a hotel's image on TripAdvisor. *Journal of Hospitality Marketing & Management, 19*, 754–772. https://doi.org/10.1080/19368623.2010.508007

Oliver, R. L. (2010). *Satisfaction: A behavioral perspective on the consumer* (2nd ed.). Routledge.

Pizam, A., & Ellis, T. (1999). Customer satisfaction and its measurement in hospitality enterprises. *International Journal of Contemporary Hospitality Management, 11*(7), 326–339. https://doi.org/10.1108/09596119910293231

Raksmey, S., & Lai, P.-C. (2019). *Cross-cultural posting online review behavior: Service attributes for hotels in Cambodia*

Sarstedt, M., et al. (2022). Progress in partial least squares structural equation modeling use in marketing research in the last decade. *Psychology & Marketing, 39*(5), 1035–1064. https://doi.org/10.1002/mar.21640

Shin, H., & Perdue, R. R. (2022). Hospitality and tourism service innovation: A bibliometric review and future research agenda. *International Journal of Hospitality Management, 102*, 103176. https://doi.org/10.1016/j.ijhm.2022.103176

Situmeang, F., et al. (2020). Looking beyond the stars: A description of text mining technique to extract latent dimensions from online product reviews. *International Journal of Market Research, 62*(2), 195–215. https://doi.org/10.1177/1470785319863619

Sotiriadis, M. D., & van Zyl, C. (2013). Electronic word-of-mouth and online reviews in tourism services: The use of twitter by tourists. *Electronic Commerce Research.* https://doi.org/10.1007/s10660-013-9108-1

Spence, A. (1974). *Market signaling: Informational transfer in hiring and related screening processes.* Harvard University Press.

Tóth, Z., et al. (2022). B2B eWOM on Alibaba: Signaling through online reviews in platform-based social exchange. *Industrial Marketing Management, 104*, 226–240. https://doi.org/10.1016/j.indmarman.2022.04.019

TravelBI by tourismo de Portugal, I. (2019). Turismo no Espaço Rural/Habitação. Instituto Nacional de Estatística. Retrieved 21.02.2023, 2023, from https://travelbi.turismodeportugal.pt/alojamento/turismo-no-espaco-ruralhabitacao

Vargo, S. L., & Lusch, R. F. (2008). Service-dominant logic: Continuing the evolution. *Journal of the Academy of Marketing Science, 36*(1), 1–10. https://doi.org/10.1007/s11747-007-0069-6

Wells, J. D., et al. (2011). What signal are you sending? How website quality influences perceptions of product quality and purchase intentions. *MIS Quarterly, 35*(2), 373–396. https://doi.org/10.2307/23044048

Wirtz, J., et al. (2013). *Essentials of services marketing, 2nd ed.* (P. E. S. A. P. Ltd Ed.). Pearson Education South Asia

Wong, S. F., et al. (2022). Sense of place: Narrating emotional experiences of Malaysian Borneo through Western Travel Blogs. *Tourism and Hospitality, 3*(3), 666–684. https://doi.org/10.3390/tourhosp3030041

Zhao, Y., et al. (2019). Predicting overall customer satisfaction: Big data evidence from hotel online textual reviews. *International Journal of Hospitality Management, 76*, 111–121. https://doi.org/10.1016/j.ijhm.2018.03.017

Zuo, W., et al. (2022). Changes in service quality of sharing accommodation: Evidence from airbnb. *Technology in Society, 71*, 102092. https://doi.org/10.1016/j.techsoc.2022.102092

Religious Tourism in Covid-19 Period: The Event of the Festival of Crosses, Barcelos (Portugal)

Hugo Martins

Abstract

The events have a very relevant role in the development of the territories. In the context of events, events at sacred sites stand out. The Festival of Crosses in Barcelos is known as the first Pilgrimage in the Minho region, in Barcelos (Portugal). This festival usually begins at the end of April, and ends on the 3rd May, which is a municipal holiday. Considering the increasing visitors that attracts and the quantity of highlights that it has in its programme, the Festival of Crosses has gained recognition, rivalling other popular festivities. This one stands out, naturally, for its religious component, but also for its ethnocultural aspects. The purpose of this investigation is to identify the impacts that the Covid-19 has affected in terms of the occupancy rate in the city and what changes its programming has undergone. Methodologically, we sought to analyse statistical data from between 2019 and 2022, characterising the Festival of Crosses with its various moments/events before and during the pandemic. The data was analysed and the respective conclusions were described. It was found that these festivals suffered many changes in their programme, mainly resorting to the digital aspect. In 2022, the demand values exceeded the pre-pandemic values.

Keywords

Religious tourism • Covid-19 • Festival of crosses • Barcelos • Portugal

H. Martins (✉)
Centre of Studies in Geography and Spatial Planning (CEGOT),
University of Maia, Maia, Portugal
e-mail: hugomartins@umaia.pt

1 Introduction

The tourism sector is very sensitive to problem situations that may affect tourist safety (Araujo et al., 2020), since, disasters and catastrophes create enormous challenges to the recovery of this industry. The media is full of reports of negative events that threaten the tourism industry. In fact, many destinations are affected by events such as natural disasters, cyclones, terrorist attacks, financial crises, forest fires, earthquakes… The tourism industry "has experienced many crises and disasters in the last decade, including political instability, economic recession, biosecurity threats" (Mair et al., 2016, p. 2), war and other negative events.

Over the last two decades, the tourism sector has suffered numerous crises namely the impacts of the terrorist attacks on the USA in 2001, the Indian Ocean tsunami, in 2005, and, "previous pandemics such as the 2009 swine flu crisis" (Higgins-Desbiolles, 2020, p. 613). According to Abbaspour et al. (2021, p. 141) "health pandemics, such as the case of Covid-19, are some of the various types of crises that the tourism industry faces, and pandemics are found to be different from many other disasters and crises that tourism has experienced".

The impact caused by Covid-19 worldwide is undeniable, with the tourism segment considered one of the most negatively impacted. The pandemic caused by Covid-19 is of a much greater magnitude than previous crises due to the global scale and widespread paralysis of travel, business and everyday activities (Martins, 2022a), significantly impacting economic, political and socio-cultural systems (Sigala, 2020).

Tourism is one of the economic sectors where it brings together several "elements that influence the territory and a wide range of activities" (Martins, 2022b, p. 39) not only of economic but also of social character. Thus, local entities and various stakeholders, consider that tourism can be a way for regional or even local development. Consequently, the material and immaterial religious heritage has been, to a certain extent, appropriated and even mercantilised by tourism, becoming part of the sector designated "religious tourism". Religious tourism "has worked as a promoter of social and cultural interaction, enhancer of social and personal development, as well as a stimulator of the territory in economic terms, of the preservation and enhancement of religious heritage" (Martins, 2022b, p. 39). Due to a high number of religious infrastructures and respective traditions, religious tourism has contributed to an effective development in the northern region of Portugal. This sector "has been, in recent years, the target of attention of several economic, political and scientific agents" (Silva, 2018, p. 1315).

As a result of cultural identity, there are hundreds of festivals and festivities that are part of popular Portuguese culture. Local identity is also reflected in the countless cultural manifestations, devotions and superstitions that proliferate all over the territory. More religious pilgrimages and festivals, others more folkloric, they include solemn processions, band parades, the arraial, the "cantares ao desafio", traditional songs, the "cabeçudos and Zés Pereiras", traditional games, horse racing, traditional sweets and snacks, fireworks and the arches of pilgrimage.

In Barcelos, the Festival of Crosses has its high point on 3 May, a municipal holiday. This festival is related to the old tradition of making carpets of flowers at festivals and pilgrimages in the municipality. Allied to all this tradition is the people's dedication and pride for their pilgrimage, and the concern to decorate and enrich in the best way the procession of the patron saint. Fortunately, this tradition has passed from generation to generation, without ever losing its traditional links.

Therefore, this work seeks to identify the impact that the Festival of Crosses and what consequences the pandemic caused, in Barcelos (Portugal). Regarding the method of this study, our research question was formulated, to which we aim to seek responses: "What impact did the SARS-CoV2 pandemic provoke in the Festival of Crosses in Barcelos (Portugal) and to what extent did it influence its celebration?". We have attempted to perform a statistical analysis of the data between 2019 (pre-pandemic situation) until 2022. Furthermore, we aimed to characterise the festivity with its various moments before the Covid-19 period to realise the adjustments that the pandemic affected in the programme of the Festival of Crosses.

2 Literature

Events may be grouped taking into account their theme. According to Raj et al. (2008), the main categories into which they may be classified are religious, cultural, musical, business, political/governmental, personal/private and sports events. Events may also be classified according to another set of criteria: (i) by purpose (institutional or promotional); (ii) by frequency (sporadic, periodic or opportunity); (iii) by target audience; and (iv) by level of participation (Pedro et al., 2009).

There is also the classification of events as to their scope; thus local, regional, national, international and mega events are found. Another type of classification is as to its relationship with the public and an event may be deemed open (when it has no defined public) or closed (with specific target audience) (Simões, 2012).

Alongside the definition of events, the typologies themselves also do not generate consensus between authors. Getz and Page (2019) classified events into three major groups: Hallmark Events, Special Events and Mega Events. Hallmark Events are catalogued as events of relevant traditional importance, attractiveness, quality and/or publicity, becoming the brand image of a destination and/or local community. Allen et al. (2022) state that Brand Events are so relevant and identifiable of a city, region or country that they end up being associated with the place (Carnival to Rio de Janeiro or the Tour de France to France). Special Events are defined as events that take place exclusively or outside the schedule set by a company or event sponsors. They may also fall into this category when these prove to be a different experience and outside the participants' routine (Getz & Page, 2019). The third major group, Mega Events are of great scale and importance, consequently generating high levels of tourism, media coverage and economic impact, both for the venue and the local community. According to Bowdin et al. (2006), these events are something so grand that they affect entire economies and are reflected in the global media (examples of the Olympic and Paralympic Games and the World Championships). This category generates great disagreement among the various authors since some argue that for an event to be considered "mega" it is necessary that it exceeds one million visitors. On the other hand, some argue that it should only be able to attract publicity on a global scale (Bowdin et al., 2006; Simões, 2012).

Events take place and have the most varied purposes. They are seen as a means of socialisation since respondents, who share common interests, can socialise and exchange ideas. They are moments of relaxation, leisure and relaxation and constitute extra-routine experiences and a form of entertainment, sports, music and culture among others.

Given that "tourists are increasingly looking for authentic products, religious tourism is clearly on the rise, as it combines the cultural and spiritual experience" (Martins, 2022b, p. 40). Religious tourism "can be understood as a confluence of opposite polarities, in which, on the one hand, there is the sacred aspect and, on the other, the profane aspect, in which visitors go there partly or solely for religious reasons" (Rinschede, 1992, p. 54).

However, events can be conditioned by factors of various origins. This is what happened with the propagation of the SARS-CoV-2 coronavirus that resulted in a pandemic. This pandemic affected from the first quarter of 2020 millions of people and numerous countries. The effects of the pandemic "will certainly have, in the long term, an immense impact on the economy, education, health, religion, culture, tourism, sports and social and psychological behaviours of people all over the world" (Mróz, 2021, p. 626). Europe's largest "Catholic shrines have encouraged pilgrims to deepen their ties to the shrine by participating in online services and prayers and to pursue spiritual and virtual pilgrimages" (Mróz, 2021, p. 626).

The pandemic certainly caused total devastation (Lew et al., 2020), and the coronavirus outbreak designated Covid-19, threw the world economy off balance, in claiming that the emergence of this pandemic (which paralysed international travel) exemplifies the vulnerability of the tourism industry to crises. The World Tourism Organization declared tourism as one of the sectors most affected by Covid-19 and Abbaspour et al. (2021) stated that this outbreak presented unprecedented challenges to tourism due to the global nature of its pandemic effect.

Indeed, Yang et al. (2020) demonstrated that outbreaks of infectious diseases (including Covid-19) cause no growth in tourism due to the industry's reliance on human mobility. Since March 2020, the pandemic caused by Covid-19 has been causing economic damage globally (Pan et al., 2021), and unlike recent epidemic outbreaks, e.g. SARS, Ebola and H1N1, the coronavirus remains the world's deadliest outbreak, accompanied by a global health crisis, financial crisis and economic slowdown, known as the Covid-19 recession ((Vassiliadis et al., 2021). The pandemic outbreak has forced many businesses to close, leading to unprecedented disruption of trade in most industry sectors (Donthu & Gustafsson, 2020). All events, whether sport, cultural or religious, were constrained: many were postponed and others cancelled. The example of the Festival of Crosses in Barcelos (Portugal) was no exception.

2.1 The Festival of Crosses: Historical Background

Barcelos is a Portuguese city in the sub-region of Cávado, belonging to the northern region and the district of Braga. It is the seat of the municipality of Barcelos, with a total area of 378.9 km^2, subdivided into 61 parishes. It is located in a place with archaeological remains since prehistoric times, but it was in the twelfth century that its history began, first when D. Afonso Henriques granted it a charter and made it a town.

At the end of the fourteenth century, a period of great development and dynamism began for Barcelos, which was revealed with the construction of the bridge, the wall (of which the Torre da Porta Nova is still standing), the Paço dos Duques and the Matriz Church. It is these monuments that today make up the historic centre of the city, which maintains a mediaeval atmosphere, punctuated by manor houses and historical houses such as the Solar dos Pinheiros or the Casa do Condestável. In terms of historical and cultural heritage, the Pottery Museum and the Barcelos Handicraft Centre provide a good insight into Minho artistic expression. Of all the pieces produced in the municipality, the colourful Barcelos Cockerel is the most representative, not forgetting the brass bands and figures depicting the region's habits and customs. The 18th-century churches of Our Lady of the Rosary and Good Jesus of the Cross (Fig. 1) also stand out.

The origins of the temple of the Good Jesus of the Cross date back to 1504. In that year, on Friday, December 20th, the shoemaker João Pires was going to hear the morning mass in the Chapel of Salvador and, while crossing the Barcelos Fairground, he came across a black cross drawn on the ground in sharp contrast with the colour of the earth that surrounded it. These crosses appeared a little everywhere. They had the peculiarity that, when dug up, no matter how hard they were dug, they did not disappear. The miracle of the Cross of the Barcelos Fairground was repeated on the 3rd May (day consecrated to the Invention of the Holy Cross) and the 14th September (day of the exaltation of the Holy Cross). The population has various versions explaining the event. In one of them, it all happened when an elderly person wanted to bake bread in the oven heated with wood that he had gone to collect on the banks of the river. The wood refused to burn and jumped out of the oven, not without catching some scorches that turned it somewhat blackened. After several attempts, the wood was made a cross by the hands of a neighbouring craftsman and taken in procession to the Chapel of Barcelos Fairground. According to others, the image of the Lord of the Cross from Barcelos was part of a set of three images that the impious from faraway lands threw into the sea in an act of religious treachery. The sea was not ungodly and respectfully took one of the images to the beach of the Lord of Matosinhos, another to the mouth of the river Cávado, where one of them was then left there to be the Lord of Fão, and another went up the river until it reached Barcelos. The one that was found was piously gathered in a chapel near the river. The popular connection to the cult of the Good Jesus of the Cross with the tradition of the three images that came to Matosinhos and to the mouth of Cávado must be understood as a result of the religious mentality of the fishermen nuclei of the North, as well as the mercantile activities by sea of the seventeenth and eighteenth centuries. This explains the Good Jesus of Fão, Matosinhos and Barcelos sanctuaries.

Before the chapel was built, the people erected a stone cross that soon aroused great devotion and many more offerings. These offerings served to start the construction of a wayside shrine, in 1505. The success of this religious faith led to the addition of the primitive Ermida and then in the eighteenth century to the construction of the current temple. The construction works started in 1705. In 1710, the church was finished. Located in Largo da Porta Nova is the octagonal-shaped Baroque building of considerable dimensions known as Church of the Lord of the Cross or the Good Lord of the Cross. In 1958, the Portuguese State classified it as a building of public interest. Its interior is made up of three altars in gilded woodcarvings. Devotion to the place has been growing. At the same time, each year the Festival of Crosses acquired more popular roots, greater expansion and, above all, extraordinary pomp. It is the epicentre of the Festival of Crosses.

2.2 The Programme of the Festival of Crosses

In the nineteenth century, the festivities had a markedly religious character. Hundreds of pilgrims from the rural parishes of Barcelos, from all over the country and neighbouring Galicia, sang and danced, some barefooted, with the "countess" on their head, on which they carried their lunch.

In the twentieth century, the religious component was mixed with remaining elements of a profane nature, visible in the merry-go-rounds, food and drink stalls, horse races, ethnographic parades, fireworks in the river Cávado and challenge singing in the city streets.

Nowadays, the "Festival of Crosses" is a point of pilgrimage for national and international visitors. Barcelos comes even more alive with the pilgrimage and its arches, the fair, the handicrafts, the procession, the carpets of natural flower petals and the folklore performances. The traditional battle of the flowers, the music bands and the "Cabeçudos and Zés Pereiras" reflect the pulse of the culture and traditions of Barcelos. The Royal Brotherhood of the Lord of the Cross is responsible for the meticulous execution of the Tapestries in the Temple of the Lord of the Cross. One of the ex-libris of this pilgrimage is the floral carpets. Every year, their execution mobilises many people, involving many hours of work and thousands of flower petals (RTP, 2022). The Battle of Flowers, the Arches of Pilgrimage, the Carpets of Natural Petals and the Procession of the Invention of the Holy Cross are the high points of the Festival of Crosses (Municipality of Barcelos, 2019b).

Before the pandemic period, the Festival of the Crosses was usually held over twelve days, when the town was filled with colour, joy and good humour with a programme full of concerts, drums, folklore and pyrotechnics, the high points being the Battle of Flowers, the concerts and the majestic Procession of the Invention of the Holy Cross. The success is a reflection of the work that the municipality has been doing in consolidating a perfect and balanced alliance between its religious and popular, sacred and profane roots, and already constitutes a benchmark in cultural, regional and national traditions (Martins, 2022a).

One of the attractions of the festivities is the Battle of Flowers which takes place on the first day of May. The city becomes small for so many people who want to participate in the "war" of petals that fly over Avenida da Liberdade. This Battle of Flowers involves twenty-four associations from the municipality of Barcelos and thousands of outsiders who, picking the flowers from the ground, also join in the fun.

Another attraction, created mainly to attract the younger public, is the "Bamos às Cruzes" party, where they can have fun and dance to the sound of music promoted by DJs.

The carpets of petals in the Church of the Lord of the Cross, the fireworks (Fireworks from the Air, Fireworks from the River, Fireworks from the Bridge and Closing Fireworks), the extensive amusement park, the fair, as well as the constant and varied entertainment throughout the city, the national and Galician folklore and the Zés P'reiras, bring thousands of people to Barcelos, contributing in the same way to the success of another edition of the festival.

The Procession of the Invention of the Holy Cross, on 3 May, a municipal holiday, celebrates the Miracle of the Crosses and marks the end of the great pilgrimage with an imposing parade of 89 crosses from the parishes of the municipality through the streets of the historic centre.

The Festival of Crosses coincided with being one of the first great pilgrimages to suffer from the consequences of the Covid-19 pandemic, as in March 2020, Portugal "closed" and the preparations for this pilgrimage, which takes place in May, also had to stop.

3 Methodology

For the present study, we sought to elaborate our research question to guide us in the research that we were going to carry out. Thus, we sought to answer the following question: "What impact did the SARS-CoV2 pandemic provoke in the Festival of Crosses in Barcelos (Portugal) and to what extent did it influence its celebration?". From this point onwards, we sought to delve into specialised literature, following the research steps suggested by Carmo and Ferreira (1998).

With regard to the literature, we have attempted, critically and carefully, to investigate the state of the art of the main research topic. We sought to realise the impact that the pandemic caused on events in general through the analysis of recent scientific articles. In addition, our focus was also centred on making a statistical analysis of the data between 2019 and 2022. It was also our intention to characterise the Festival of Crosses event in its various moments, before the pandemic to understand the variations provoked concerning the programme and the occupancy rate.

To response our research question, a set of points were identified: (a) to characterise the traditionally held Festival events (pre-pandemic period); (b) to characterise the ways in which the Festival was held in the period during the

pandemic until 2022; and (c) to identify what the tourist demand of the Barcelos municipality was in the pre-pandemic and through the pandemic periods (2019–2022).

4 Results

Regarding the pre-pandemic period, in terms of hotels and tourism in rural areas, the accommodation units throughout the municipality usually register an average occupancy rate of around 100% during the festivities, and there are also records of above average occupancy rates in neighbouring municipalities due to the event. This fact is also visible, right from the first moment, by the affluence of public in the sampling spaces of the municipality in the city centre, such as the Tourism Booth and the Medieval Tower, which together usually receive over ten thousand visitors between April 22nd and May 3rd, passing through the city about one million people from different countries. Before the pandemic period, the Festival of Crosses is held for around 11 days, starting at the end of April and culminating on 3 May.

In March 2020, a succession of events called into question the holding of the Festival of Crosses. At the very beginning of March 2020, several events were cancelled or postponed. The first major event Portugal was the Lisbon Tourism Fair (postponed initially to the end of May and then to the end of the year). In the following months, postponements and cancellations of events were announced, from sports to cultural, including all the summer music festivals. In the Algarve, 60% of hotel reservations for the coming months were cancelled, the sector's main association announced, with cancellations increasing in the following months throughout the country and tourism dropping to almost zero. Airline companies, as the example of the Portuguese Air Carrier announced the cancellation of more than a thousand flights between March and April, cancelling and suspending investments and placing employees on unpaid leave. Only in June does it slowly start some regular flights.

In mid-March, when the state of emergency (which includes compulsory confinement and restrictions on circulation on public roads) begins, schools and discos are closed, the maximum capacity of restaurants is reduced, people are restricted in shopping centres and public services, and the disembarkation of cruise passengers is prohibited. Air traffic itself, rail and river links between Portugal and Spain are suspended. Religious celebrations and other events involving a concentration of people are banned, as are movements of people outside their municipality of residence. Only from mid-May will the process of deconfinement begin in stages, with some measures including the opening of cafés and restaurants. However, several events such as religious and other celebrations are held without public, "behind closed doors" (example of 13 May in Fátima which was celebrated without the physical presence of pilgrims (DN/Lusa, 2020).

In 2020, in early April, the City Council decided to suspend all cultural, sports and recreational activities, as a measure to reduce contagion, prevent and combat the pandemic, in accordance with the decreed State of Emergency. The Festival of Crosses has thus been cancelled due to the Covid-19 pandemic. For the first time in 500 years, the Festival of Crosses was not held in its religious and popular aspect, due to the Covid-19 pandemic and the state of emergency in which the country found itself. In fact, there were years when its realisation was in question, being limited to religious ceremonies. It was especially during the period of World War I (1914–1918) and in the following years, when the health crisis (Spanish or pneumonic flu) and economic crisis affected the holding of the festivities. In the nineteenth century, one of the editions was postponed for a few days: in 1852, the visit to Barcelos on 6 and 7 May by D. Maria II and her husband D. Fernando II, as well as their children, Prince D. Pedro (future D. Pedro V) and Prince D. Luís (future D. Luís I) caused the Festival of Crosses of that year to be moved to the last of those days, so that the monarch, Perpetual Judge of the Royal Brotherhood of the Lord of the Cross, could attend the festivities. Another year when the Festival of Crosses was suspended/postponed was 1938, due to a tragedy in Viana do Castelo that killed more than 20 people from the south of Barcelos. Only the following week did the religious ceremonies presided by the Archbishop of Braga take place (Municipality of Barcelos, 2020).

However, alternatives were sought, so that the event would not go unnoticed. From April 30th until May 3rd, the City Hall marked the pilgrimage by posting several videos allusive to the event on its digital platforms (website and social networks). On these days, short videos of the Festival of Crosses of previous years were published (Fig. 2), the hoisting of the municipal flag, messages from the Mayor and the President of the Municipal Assembly and, on May 3, at 12:00, the solemn mass of the Festival of Crosses, live from the Lord of the Cross Church.

The aim was to highlight the cultural and religious traditions associated with the most important annual event in the city and municipality of Barcelos. The parish of Barcelos, in turn, asked the faithful to put crosses decorated with flowers in their homes on the municipal holiday to mark the day of the Invention of the Holy Cross (Redação, 2020). The videos focus on the Barcelos Market, the Pilgrimage Arches, the Carpets of Petals, the "Rusgas", the Folklore and Musical Bands, the Battle of Flowers, the Procession of the Crosses and the Fireworks.

In 2021, the Festival of Crosses will not be held again. However, Barcelos City Hall decided to prepare some initiatives for four days, so that people could celebrate the event in a symbolic way (Fig. 3). All the moments were broadcast essentially in streaming, so that everyone could follow through the digital platforms of the municipality, and other moments could be witnessed, mainly from the religious aspect.

Thus, on 30 April, street lighting was inaugurated and a pilgrimage arch was placed on Avenida da Liberdade. Until 8 May, the natural flower petal carpet could be appreciated in the Temple of the Good Lord Jesus of the Cross, made by artisans who have been making this construction for years.

Regarding music, concerts were scheduled and could be seen on the Facebook of the municipality, near the Temple

Fig. 2 Videos allusive to the Festival of Crosses 2020. *Source* Municipality of Barcelos (2020)

Fig. 3 Programme of the Festival of Crosses 2021. *Source* Municipality of Barcelos (2019a)

of the Good Lord Jesus of the Cross on May 1 and 2 in the morning.

On the 3rd, the concert took place in another space, the Largo da Porta Nova. On that same day, Zés P'reiras toured the city in the morning. In the afternoon, the solemn mass commemorating the Miracle of the Crosses was celebrated outside the Temple of the Good Lord Jesus of the Cross (so that it could hold more people).

In addition, throughout the week, people were able to relive the essence of the Festival of Crosses, through messages and videos, available on the municipal digital platforms—website and social networks—, which mark the most outstanding moments of the first great pilgrimage of Minho, similar to those of 2020: the raising of the municipality's flag by the Mayor, and messages from the Mayor and the President of the Municipal Assembly.

In 2022, the Festival of Crosses was again held in the characteristic moulds of the pre-pandemic period, but with a shorter period. It was held over a five day period with concerts with renowned singers to meet the public and attract more visitors. In addition to the music, there were firework displays during all the days.

Starting with the historical component, Barcelos City Council and the Royal Brotherhood of the Lord Jesus of the Cross decided to commemorate this return to the festivities in a special way by holding an exhibition in the Paços do Concelho (Town Hall) dedicated to part of the collection from the temple of the Lord of the Cross, the theme of which was "Gold and Silver in the Royal Brotherhood of the Lord Jesus of the Cross". In addition to this exhibition, there were also three other exhibitions in other places in the city: the exhibition "Camiños. O camiño e as artes", in the Municipal Art Gallery; the illustration exhibition "Raimundo Canta Barcelos"; and the exhibition "O mundo colorido de Mina Gallos", in the Medieval Tower.

The programme once again privileged religiosity, honoured tradition, promoted tourism, stimulated the economy and exalted entertainment and conviviality. Thus, on the first day, right after the mortar fire and the band of Zés P'reiras, at 11:00 a.m., the pilgrimage arches that decorate the festival area were inaugurated. In the afternoon, another inauguration marked the first day: the Tapestries of Petals of Natural Flowers opened to the public at the Temple of the Lord Jesus of the Cross. The highlight of the day took place at night with a concert on the Frente Ribeirinha stage. After that, there was the "Bamos às Cruzes" activity, a popular party in Jardim das Barrocas, with the performance of guest DJs. At midnight, in Largo da Porta Nova, there was the first pyrotechnic session of the Crosses.

On the second day, for most of the day, the television programme "Aqui Portugal" was broadcast from the city of Barcelos, giving greater prominence to the municipality's biggest festival. At night, there was another concert with a renowned artist and dancing with the activity "Bamos às Cruzes" with Dj and the Medieval Bridge fire session with the banks of the Cávado River illuminated by thousands of "living lights".

On the third day, there were several activities: the National Violin Competition "Pequenos Galinhos", the reception and concerts of the Philharmonic Bands and one of the highest points of the pilgrimage, the Battle of Flowers, in Avenida da Liberdade, whose theme was the "Battle for Peace". After the Battle of Flowers, the Lusogalaico Festival started, ending the day with the activity "Bamos às Cruzes" and another fireworks session, this time pyromusical, in the Porta Nova Square.

On the fourth day, there was the inauguration of the Pilgrim's Park and in the afternoon the conference "The Pilgrim Virgin on the Way to Santiago", at the Temple of Lord Good Jesus of the Cross, and the inauguration of the exhibition "Camiños. The Way and the Arts", in the Municipal Art Gallery. At the end of the day, there was entertainment again with another concert, "Bamos às Cruzes" and the fireworks show.

Fig. 4 Crosses made by the 89 parishes of the municipality. *Source* Notícias do Minho (2022)

On the fifth and last day, the 3rd, a municipal holiday, the Grand Procession of the Invention of the Holy Cross took place in the afternoon, in which the crosses of the 89 parishes of the municipality paraded. It was the highest point of the Festival (Fig. 4).

The city council immediately sought to create conditions in terms of accessibility for those who wanted to come to the festivities: it appealed to those arriving in Barcelos to park in the peripheral car parks and take advantage of the free and fast Barcelos BUS buses to get to the centre of the Festival. The Festival of Crosses 2022 closed with another musical concert, on the stage of Avenida da Liberdade and a Fogo Preso session, in the Largo da Porta Nova (Municipality of Barcelos, 2022).

In 2022, the Festival of Crosses brought back the huge crowds of people that flocked to Barcelos every day of the festivities, filling the hotels and restaurants. In the 2022 edition of the Festival of Crosses, more than 420 traders were present on the fairground, roughly divided into 240 spaces dedicated to the textile and footwear sectors, and more than 70 spaces dedicated to food, bread, confectionary and cakes. There were 45 traders in the fruit and vegetable sector, while in the fairground, nearly 60 horse-drawn carriages and other equipment entertained many of the visitors (Municipality of Barcelos, 2022).

In concrete terms, making a comparison in terms of days, the Festival of Crosses used to take place over several days. In the period before the pandemic, the number of days was around 12. However, after the interregnum of 2020 and 2021, in 2022, a wide variety of activities were sought but more concentrated, over five days (Table 1).

Table 1 Number of days that the Festival of Crosses takes place

Year	Number of days
2016	12
2017	15
2018	12
2019	11
2020	0
2021	0*
2022	5

*Although no event took place in physical presence, there were activities that were disseminated through social networks over 4 days
Source Self elaboration

To identify the demand for tourism in the periods before and during the pandemic, we sought to determine the effects of the coronavirus in terms of overnight stays, in national, regional and local terms, with a focus for the month of May, when the Festival of Crosses takes place.

Consequently, according to National Statistics Institute, regarding overnight stays in accommodation, it can be seen that in the year of 2019, before the pandemic period, Portugal proved to be a country extremely dependent on foreign visitors (non-residents), representing about 70% of the occupancy rate. The regions of Algarve (southern Portugal) and Lisbon (capital of Portugal) represent about 62% of total non-residents. The regions of Northern Portugal and Madeira represent 26% of the occupancy rate of non-residents (Table 2).

Regarding residents, representing only 30% of the occupancy rate in 2019, the values are more dispersed. Although

Table 2 Overnight stays in tourist accommodation establishments by geographical location in 2019 (%)

	Residents	Non-residents
Portugal	30	70
North	20	13
Centre	19	6
Lisbon Metropolitan Area	19	30
Alentejo	9	2
Algarve	24	32
Azores	5	3
Madeira	4	13
Total	100	100

Source National Institute Statistics—adapted (National Institute of Statistics, 2022)

the Algarve presents 24% of the occupancy rate, the regions of the North, Centre and Lisbon represent each about 19% (Table 2).

Given the problem caused by SARS-CoV-2, the figures changed significantly in 2020. It is crucial to note that the majority of the accommodation was closed. In Table 3, 47% of overnight stays are by non-residents and 53% are by residents. Compared to Table 2, in percentage terms, "residents contributed a lot to the hospitality sector, helping to offset the number of non-residents" (Martins, 2022b, p. 44): the change stood at − 63% in the total of residents and non-residents, as there was a significant reduction. The domestic market (residents) has compensated a certain part of this reduction.

In 2020, residents compensated the decrease in the non-resident market. The regions that registered the lowest occupancy rate by residents were Madeira (22%), the Lisbon region (37%) and the Algarve (48%). The remaining areas

recorded percentages above 50%, with the Alentejo region and the Centre area having the highest occupancy rate. In terms of variation between 2020 and 2019, the regions that recorded the largest decrease in occupancy rate were Lisbon (− 72%) and Algarve (− 62%) (Table 3).

The year 2021 registered a modest recovery. The variation in the occupancy rate between 2021 and 2020 was positive, with the internal market (residents) continuing to compensate for the lack of non-residents. The region with the highest variation rate was the Azores (123%). However, making an analysis between the data from 2021 and 2019 (period before the pandemic), it is possible to verify that there was no effective recovery, as the values remain negative (− 47%) (Table 3).

Analysing only the month of May 2020, the period when the Festival of Crosses takes place, the tourist accommodation sector had a variation in the whole Portuguese territory of − 651%, compared to the homologous period of 2019 (Table 4). The most penalised regions, above the Portuguese average, were the Algarve (− 745%) and the Azores (− 809%). In 2021, compared with the previous homologous period (2020), there was a positive recovery: 57% in the whole Portuguese territory. By regions, the one with the greatest recovery was the Alentejo (157%), the Azores (184%) and the Central region of Portugal (145%). The region with the weakest recovery was Madeira (11%) and Lisbon (25%). In 2022, it was a year of total recovery being the values of the overnight stays similar to the pre-pandemic period: in 2022 the variation was 364% compared to 2021. Comparing the variations between 2022 and 2019, the values, although negative (− 3%), are very close to the pre-pandemic values.

By regions, those that in 2021 had a strong positive variation compared to 2020 were Algarve (502%), Madeira (524%) and Lisbon (484%). Making an analysis between

Table 3 Overnight stays at tourist accommodation establishments by geographical location—change between 2019 and 2021 (%)

	2020			2021			
	Residents	Non-residents	Variation 2019 versus 2020 (%)	Residents	Non-residents	Variation 2020 versus 2021 (%)	Variation 2019 versus 2021 (%)
Portugal	53		− 63		50	45	− 47
North	63	37	− 60	58	42	41	− 43
Centre	78	22	− 53	75	25	32	− 38
Lisbon Metropolitan Area	37	63	− 72	35	65	47	− 59
Alentejo	81	19	− 38	80	20	25	− 22
Algarve	48	52	− 62	48	52	38	− 48
Azores	70	30	− 71	64	36	123	− 36
Madeira	22	78	− 67	25	75	80	− 41

Source National Institute Statistics—adapted (National Institute of Statistics, 2022)

2022 and 2019, it is possible to observe that only three regions exceeded in percentage terms the number of overnight stays of 2019, having positive values: Madeira (11%) North (5%) and Alentejo (3%). Algarve (− 11%), Centre (− 5%) and Lisbon (− 5%), on the other hand, show values below 2019. We conclude that, in 2022, the values of overnight stays are close to those of 2019, the year before the pandemic (Table 4).

In what concerns the territory under analysis, the municipality of Barcelos, inserted in the northern region, we seek firstly to make an analysis in annual terms and then focus on the period when the event takes place, the month of May.

According to Table 5, it is possible to verify that the overnight stays also registered a decrease: in 2019, there were 76,726 residents and 97,412 non-residents, ending the year with 176,157 overnight stays. In 2020, there was a negative variation in overnight stays by non-residents, − 387%: 97,412 in 2019 to 20,022 in 2020. Although the residents also registered a negative variation rate to the homologous period, − 96% (76,726 overnight stays in 2019 to 39,196 in 2020), we can affirm that the domestic market assured the occupancy rate making up for the lack of non-residents. In 2021, there was an increase in relation to 2020 of around 41% (total 103,483 overnight stays). Making a comparison between 2021 and 2019, the overnight stays are still negative, with a total negative variation of − 70%. The difference between residents and non-residents is still

very large: non-residents represent − 136% and residents − 23% in relation to the 2019 period (Table 4).

In a monthly overview, the average number of overnight stays in Barcelos during the month of May, the period when the Festival of Crosses takes place, generates a mean between 18 and 20% of the annual value. In the month of May 2020, as Portugal was completely "closed", there are no overnight stays recorded. In May 2021, there was a sharp decrease between the homologous period of 2019, − 74% of non-residents, − 508% of non-residents, totalling an average of − 197% (Table 6).

In May 2022, according to data provided by the National Statistics Institute, it is possible to observe that compared to the period of 2019, pre-pandemic, there was an increase of around 18%, both in terms of residents (19%) and non-residents (16%). We can affirm that the values of demand in 2022 exceeded the values of the period before the pandemic.

5 Conclusions

Religious tourism "has functioned as a promoter of social and cultural interaction, enhancer of social and personal development, as well as a stimulator of the territory in economic terms" (Martins, 2022b, p. 49). In this pandemic period, we are going through, religious festivities had to adapt and try to do what was possible within the conditions

Table 4 Changes in overnight stays at tourist accommodation in May between 2019 and 2022 (%)

	Variation 2020 versus 2019	Variation 2021 versus 2020	Variation 2022 versus 2021	Variation 2022 versus 2019
Portugal	− 651	57	364	− 3
North	− 611	91	291	5
Centre	− 659	145	196	− 5
Lisbon Metropolitan Area	− 672	25	484	− 5
Alentejo	− 439	157	116	3
Algarve	− 745	25	502	− 11
Azores	− 809	184	211	− 3
Madeira	− 522	11	524	11

Source National Institute Statistics—adapted (National Institute of Statistics, 2022)

Table 5 Overnight stays at tourist accommodation in Barcelos—Annual

Barcelos	2019	2020	2021	Variation (%) 2019 versus 2020	Variation (%) 2020 versus 2022	Variation (%) 2019 versus 2021
Residents	76,726	39,196	62,285	− 96	37	− 23
Non-residents	97,412	20,022	41,198	− 387	51	− 136
Total	176,157	61,238	103,483	− 188	41	− 70

Source National Institute Statistics—adapted (National Institute of Statistics, 2022)

Table 6 Overnight stays at Barcelos tourist accommodation in Barcelos—May

Barcelos	mai/19	mai/20	mai/21	mai/22	Variation (%) 2019 versus 2021	Variation (%) 2019 versus 2022
Residents	5908	–	3390	7285	− 74	19
Non-residents	8148	–	1341	9754	− 508	16
Total	14,056	–	4731	17,039	− 197	18

Source National Institute Statistics—adapted (National Institute of Statistics, 2022)

required as a result of the pandemic. Unlike some festivals and pilgrimages that took place in the month of May, the Festival of Crosses in 2020 did not take place, as two months before, in March 2020, it was a complete lock down in general, not only in Portugal but also in most countries of the world.

Taking into account the initial research question, it was possible to perceive that the Festival of the Crosses tried to adjust itself to the new context. While in 2019, before the pandemic, the city of Barcelos was visited by thousands of visitors, in 2020 and 2021 what can be accomplished was circumscribed, not to the physical territory of the festivities, but to the social networks. However, we consider that the Festival of Crosses is in fact a polarising factor in the tourist attraction of the region, showing why it is the first great pilgrimage in Portugal. In 2022, there is a return to possible normality, with fewer days of festivities (5 days), more concentrated, but with the capacity to attract more visitors with concerts, dances and entertainment, surpassing expectations in terms of numbers: in terms of tourist demand, it surpassed the values of the pre-pandemic period. Traditionally, it used to take place over 12 days. With the interregnum of two years without activities, in 2022, the organisation tried to carry out a varied set of activities, but more concentrated in time. The event lasted 5 days. Despite the fact that the number of days was reduced by more than half, the tourism values regarding overnight stays were higher than the pre-pandemic period. We therefore consider that this was the year of the upturn in the tourism sector with regard to events.

As a suggestion for future research, it would be interesting to obtain the perception of visitors to the Festival of Crosses in relation to the years before the pandemic, in order to understand whether they are satisfied with the changes provided by those responsible for the religious event.

References

Abbaspour, F., Soltani, S., & Tham, A. (2021). Medical tourism for COVID-19 post-crisis recovery? *Anatolia, 32*(1), 140–143. https://doi.org/10.1080/13032917.2020.1815067

Allen, J., Harris, R., Jago, L., Tatrai, A., Jonson, P., & D'Arcy, E. (2022). *Festival and special event management, essentials edition.* Wiley Online Library.

Araujo, E. J. S., Melchán, J. A. S., Bermejo, B. R., & Del Río, J. J. T. (2020). Comportamiento del sector turístico colombiano durante la pandemia, una luz al final del camino:¿ Lamentación o llamado a la acción? *Revista Ibérica de Sistemas e Tecnologias de Informação, E36,* 295–308.

Bowdin, G., O'Toole, W., Allen, J., Harris, R., & McDonnell, I. (2006). *Events management.* Routledge. https://doi.org/10.4324/9780080457154

Carmo, H., & Ferreira, M. M. (1998). *Metodologia da investigação: guia para auto-aprendizagem.* University of Alberta. https://books.google.pt/books?id=GdLlPQAACAAJ

da Silva, J. L. F. (2018). O Turismo Religioso no Noroeste de Portugal: As atividades económicas dos principais santuários na sua relação com o território envolvente. *Revista Turismo & Desenvolvimento, 3* (17/18), 1309–1324.

DN/Lusa. (2020). Cronologia de uma pandemia em português. *Diário de Notícias.* https://www.dn.pt/vida-e-futuro/cronologia-de-uma-pandemia-em-portugues-os-tres-meses-que-mudaram-o-pais-12259916.html

Donthu, N., & Gustafsson, A. (2020). Effects of COVID-19 on business and research. *Journal of Business Research, 117,* 284–289. https://doi.org/10.1016/j.jbusres.2020.06.008

Getz, D., & Page, S. J. (2019). *Event studies* (4th ed.). Routledge. https://doi.org/10.4324/9780429023002

Higgins-Desbiolles, F. (2020). Socialising tourism for social and ecological justice after COVID-19. *Tourism Geographies, 22*(3), 610–623. https://doi.org/10.1080/14616688.2020.1757748

Lew, A. A., Cheer, J. M., Haywood, M., Brouder, P., & Salazar, N. B. (2020). Visions of travel and tourism after the global COVID-19 transformation of 2020. *Tourism Geographies, 22*(3), 455–466. https://doi.org/10.1080/14616688.2020.1770326

Mair, J., Ritchie, B. W., & Walters, G. (2016). Towards a research agenda for post-disaster and post-crisis recovery strategies for tourist destinations: A narrative review. *Current Issues in Tourism, 19*(1), 1–26. https://doi.org/10.1080/13683500.2014.932758

Martins, H. (2022a). Os impactos económicos da Covid-19 em eventos. *Revista Turismo & Desenvolvimento, 38,* 265–280. https://doi.org/10.34624/rtd.v38i0.25863

Martins, H. (2022b). Religious tourism during the COVID-19 period: The Case of our lady of agony festival, Viana do Castelo, Portugal. In A. Mandić, R. A. Castanho, & U. Stankov (Eds.), (pp. 39–50). Springer International Publishing.

Mróz, F. (2021). The impact of COVID-19 on pilgrimages and religious tourism in Europe during the first six months of the Pandemic. *Journal of Religion and Health, 60*(2), 625–645. https://doi.org/10.1007/s10943-021-01201-0

Municipality of Barcelos. (2019a). *Festa das Cruzes 2021.* Online. https://www.cm-barcelos.pt/2021/04/festa-das-cruzes-2021/

Municipality of Barcelos. (2019b). *Festivals, fairs and pilgrimages.* https://www.cm-barcelos.pt/visitar/festas-feiras-e-romarias/festa-das-cruzes/

Municipality of Barcelos. (2020). *Festa das Cruzes 2020 em sua casa_Tapetes de Pétalas Naturais.* https://www.youtube.com/watch?v=uLSAJVu77Xw

Municipality of Barcelos. (2022). *Festa das Cruzes—um palco de animação alegria e vida*. https://www.cm-barcelos.pt/2022/04/festa-das-cruzes-um-palco-de-animacao-alegria-e-vida/

National Institute of Statistics. (2022). *Instituto Nacional de Estatística*. http://www.ine.pt

Notícias do Minho. (2022). *Festa das Cruzes em Barcelos*. https://www.noticiasdominho.pt/2022/02/festas-das-cruzes-esta-de-volta-barcelos.html

Pan, T., Shu, F., Kitterlin-Lynch, M., & Beckman, E. (2021). Perceptions of cruise travel during the COVID-19 pandemic: Market recovery strategies for cruise businesses in North America. *Tourism Management*, *85*, 104275. https://doi.org/10.1016/j.tourman.2020.104275

Pedro, F., Caetano, J., Christiani, K., & Rasquilha, L. (2009). *Gestão de eventos* (3.ª). Quimera.

Raj, R., Walters, P., & Rashid, T. (2008). *Events management: An integrated and practical approach*. SAGE Publications. https://books.google.pt/books?id=mDiAeUgIw6IC

Redação. (2020). Pela primeira vez em 500 anos não há Festa das Cruzes em Barcelos. *O Minho*. https://ominho.pt/pela-primeira-vez-em-500-anos-nao-ha-festa-das-cruzes-em-barcelos/

Rinschede, G. (1992). Forms of religious tourism. *Annals of Tourism Research*, *19*(1), 51–67.

RTP. (2022). *Dia Santo*. https://www.rtp.pt/programa/tv/p19236/e3

Sigala, M. (2020). Tourism and COVID-19: Impacts and implications for advancing and resetting industry and research. *Journal of Business Research*, *117*, 312–321. https://doi.org/10.1016/j.jbusres.2020.06.015

Simões, M. (2012). *Os eventos e a atractividade e competitividade turística das cidades: o caso de Lisboa* [Escola Superior de Hotelaria e Turismo do Estoril]. http://hdl.handle.net/10400.26/4458

Tourism of Portugal. (2013). *Igreja do Senhor Bom Jesus da Cruz—Barcelos*. https://www.visitportugal.com/pt-pt/content/igreja-do-senhor-bom-jesus-da-cruz-barcelos

Vassiliadis, C. A., Mombeuil, C., & Fotiadis, A. K. (2021). Identifying service product features associated with visitor satisfaction and revisit intention: A focus on sports events. *Journal of Destination Marketing & Management*, *19*, 100558. https://doi.org/10.1016/j.jdmm.2021.100558

Yang, Y., Zhang, H., & Chen, X. (2020). Coronavirus pandemic and tourism: Dynamic stochastic general equilibrium modeling of infectious disease outbreak. *Annals of Tourism Research*, *83*, 102913. https://doi.org/10.1016/j.annals.2020.102913

Event Tourism: Analysis of Residents' Perceptions of the São Bento Festivities, Santo Tirso (Portugal)

Rita Teles, Ana Raquel Barbosa, and Hugo Martins

Abstract

Popular events have proven to be a key factor for the development of economic and cultural strategies in the regions while contributing to the distinction and prosperity of destinations. In Portugal, festivals and pilgrimages are prominent festivities. Each city, town, and village are full of customs and traditions that combine the sacred and the profane. Therefore, tourists are attracted to visit the territory and simultaneously participate in the various activities, processions, concerts, marches, entertainment, and handicrafts. The *São Bento* Festivities are secular, and its program includes several activities that combine the religious and the profane. This research has as main goal to analyze the perceptions of residents in the municipality of Santo Tirso regarding its most iconic festivities, the São Bento Festivities. In order to ensure this achievement, data were collected through a questionnaire, conducted between May and June 2022, which had 211 respondents. A survey by interviewing representatives of the various municipal departments responsible for the event was also conducted. The results obtained show that, in general, residents' perceptions of the São Bento Festivities and its program are positive, meeting the expectations of the event's organizers.

Keywords

São Bento festivities • Event tourism • Residents' perceptions • Residents • Santo Tirso • *Religious tourism*

R. Teles · A. R. Barbosa
University of Maia, Maia, Portugal

H. Martins (✉)
Centre of Studies in Geography and Spatial Planning (CEGOT), University of Maia, Maia, Portugal
e-mail: hugomartins@umaia.pt

1 Introduction

More and more, we come across different types of events, whether small or large. Currently, the relevance of events is highlighted in social and economic terms since they contribute to the social harmony of a city, region, or country. Some examples of major revenue generating events in Portugal are (a) summer festivals, which every year unite a large number of people both nationally and internationally; (b) rally events, which attract mostly a young/adult male generation that are avid supporters of this sport; and finally, (c) fashion shows, designed for a predominantly female audience interested in new trends (Simões, 2012).

At the local level, there are many events mainly linked to religion that also work "as promoters of social and cultural interaction, enhancers of social and personal development, as well as dynamizers of the territory in economic terms, preservation, and enhancement of religious heritage" (Martins, 2022a, p. 266). These events, linked to the religious character, have been, in recent years, the target of attention of several economic, political, and scientific agents (Martins, 2022b).

Thus, this article arises from the need to understand how a municipality can be affected both positively and negatively by the various events that occur in it. Events can generate socio-economic improvements for regions and for the mechanisms involved in their management, whether public or private (Martins, 2022a; Smith, 2012). However, events also bring about a wide range of negative aspects, namely damage to the quality of life of locals, the environment and cultural heritage (Small, 2008). It becomes necessary that event tourism management be appropriate and responsible, to enable the regeneration of the places where events are held, thus driving the implementation of better infrastructures (transport, accommodation) and services (gastronomy, attractions), the stimulation of new economic activities, the increment of stronger tourism, and the consolidation of a better brand image of the city. Therefore, event tourism can emerge as a key factor to combat seasonality (Simões, 2012) since it instigates the displacement of people without it happening in a specific period.

In this work, we intend to study one of the most important events in the municipality of Santo Tirso—the *São Bento* Festivities. Our intention was to analyze the perceptions of residents about this event and its impacts to the municipality. To this end, a set of specific objectives was defined, namely: (i) to assess the interest, importance, and satisfaction of the event for residents, as well as (ii) to identify its impacts, both positive and negative. This paper aims to contribute to fill the shortage of studies related to this type of events, festivals, and pilgrimages.

The patron saint of Santo Tirso is *São Bento*, celebrated annually on the eleventh of July. The *São Bento* Festivities are a pagan pilgrimage of very deep roots, which attract to the city of Santo Tirso devotees from all over the country. As a rule, these festivities have a duration of 4/5 days. Therefore, depending on the year and the day of the festivities, the organization proposes different dates for the program of the festivities in order to try to add weekends to extend and promote more activities.

This chapter is divided into three section, apart from the Introduction and Conclusion. Section 2 presents a literature review on event tourism, focusing on events in Portugal and residents' perceptions. Section 3 presents the research methodology, characterizing the content analysis and the research techniques developed and used. Section 4 presents the results and discusses them.

2 Literature Review

2.1 Event Tourism

In generally, events fit into the strategic plans for the development of a tourist destination, proving to be a fundamental means for the promotion, dissemination, and attraction of visitors/tourists to a territory. The first authors to attribute value to events were Getz (1991) and Goldblatt (1990), defining them as something special, unique, and out of the ordinary, becoming a unique moment that occurs in a given place making it irreplaceable. An event is an exceptional occurrence that has as its purpose the satisfaction of the concrete needs of an entity or group (Goldblatt, 1990). For Getz (1991), an event is a social, cultural, or leisure experience that occurs outside the routine activities of an individual. Getz (1997) also states that events are temporary eventualities and may or may not be planned. When they are planned, as a rule, they have a preliminary schedule established and disseminated, they can have a defined periodicity or be held only once. However, each one takes place in its own environment,

since it is instigated by several agents, such as: date, duration, location, promoter, and participants. In this type of event, there is a constant interaction between stakeholders, guests, and participants. In an unplanned event, there is neither an organizing entity nor an established program: the participants' intentions are not clear.

According to Allen et al. (2022), the designation event is identifying characteristic rituals or celebrations that have been planned with the main purpose of achieving previously stipulated goals. Events can also have positive economic impacts, contributing to the development of local communities and companies and driving the creation of new businesses (Raj et al., 2009). In economic terms, events are perceived as stimulants for attracting tourists, increasing their length of stay and spending. Events are the ones that drive infrastructure development and stimulate the development of the place as a tourist destination, creating business and employment opportunities (Mogollón et al., 2017). From a negative perspective, price inflation, exploitation and opportunity costs, and changes in the quality of life of locals can be highlighted. Events can often have unintended consequences, which in turn can cause impacts on the environment: increased traffic, noise pollution, vandalism, disrespect for the local community, and increased volume of garbage, among others.

Events are great promoters of a tourist destination and can be a factor in attracting visitors (Simões, 2012). On the other hand, when an event gets a negative mark, the city where it took place becomes known as a place where a ruinous event occurred and its name is talked about in the media for less pleasant reasons. With the growing notoriety of events and consequently their impacts both on the territory and on communities, there are several research regarding the possible impacts caused by events, namely Bowdin et al. (2006) and Small (2008) (Table 1).

The socio-cultural impacts of events are somewhat difficult to quantify. They can encompass a shared experience between tourists and locals and may contribute to the strengthening of local pride, legitimization or broaden cultural horizons (Allen et al., 2022). Some authors also argue that events are a great way to publicize the particularities of a region. However, it is necessary to outline a strategy, so that events do not jeopardize the place where they are held, protecting the cultural heritage and the needs of the people (Getz, 1991).

In terms of political impacts, events favor the improvement of a tourist destination's image (Table 1). They attract visitors, generate economic benefits and consequently jobs; this factor is quite relevant making some governments become receptive in joining new events (Allen et al., 2022). The loss of control over the community when a mismanaged event is held is a concern since it can have considerable effects on the lives of the host communities.

Table 1 Summary of impacts caused by events

Core impacts	Positives	Negatives
Social e cultural	• Shared experience • Reviving traditions • Strengthen community pride • Increased community participation • Broadening cultural perspectives • Improved quality of life • Better intra-community communication • Preservation of cultural identity	• Community disinterest • Community manipulation • Negative community image • Aggressive behavior • Abuse of alcohol, drugs, and prostitution • Changes in values and customs • Exclusion of resident population • Crimes and vandalism; • Mercantilization of culture
Physical and environmental	• Propaganda for the environment • Establishment of better habits • Increased environmental awareness • Improved means of transportation and communication • Transformation and urban renewal	• Environmental damage • Pollution • Destruction of heritage • Traffic jams • Noises
Political	• National and international prestige • Image promotion • Promotion of investments • Social cohesion • Administrative development	• Risk of failure of the event • Negative publicity • Lack of accountability • Loss of community ownership and control
Economic and touristic	• Promotion of the destination • Increase in the number of tourists • Increase in the length of stay • Creation of new jobs • Increased profit • Increased revenue rate	• Community resistance to tourism • Loss of authenticity • Exploitation • Inflated prices • Opportunity costs

Source Bowdin et al. (2006) and Small (2008), adapted by the authors

Despite the many advantages provided by events, they also have disadvantages for host regions. The impacts generated in the community will be greater the more popular an event is. The main negative aspects are the congestion caused by the large affluence in the various existing infrastructures such as hotel establishments. Seasonality represents another negative aspect, since most mega-events (such as festivals) take place only in high season, in other words, in summer, which can frustrate the economic expectations of the inhabitants. Finally, pollution represents another adversity when referring to events, this happens due to the great load of visitors and the lack of concern in the preservation of natural resources and heritage existing in the tourist destination.

In Portugal, in strategic terms, events also have an importance that has been consolidating. The National Strategic Tourism Plan considers three main categories: mega-events, major promotional events, and local entertainment. The strategy defined was to hold one or two mega-events per decade, in order to ensure better projection and notoriety of a destination. In 2017, the Tourism Strategy 2027, when determining the strategic assets and the axes and lines of action to achieve the maximum exponent of tourism, artistic-cultural, sports, and business events appear as one of the strategic assets defined (Tourism of Portugal, 2017). The Tourism Strategy 2027 points out that events have a wide coverage in the Portuguese territory and that many of them

occur in areas where there is no significant demand. Events have an undeniable contribution to the international projection of the country, contributing, synchronously, to the economic stimulation of sparsely populated territories, extending the tourist practice to the whole country and fighting seasonality (Tourism of Portugal, 2017).

2.2 Residents' Perceptions of Events

A large number of authors investigating residents' perceptions of tourism impacts from events use Social Exchange Theory as their theoretical basis (Ap, 1992; Gursoy et al., 2002). This theory states that local residents form their perceptions based on expectations before events take place. Theoretically, these perceptions serve as a reference for future reassessments of impacts, according to the Prospect Theory's. Subsequent to events, residents feel compelled to re-evaluate the exchange values initially assigned. In this way, values below the reference points will be considered as losses, in other words, generating negative perceptions and disappointment. Values higher than initially assigned are seen as a gain, thus generating positive perceptions. This re-assessment of expectations helps residents not only to establish a reference point, but also to determine where they stand in relation to new events. Residents believe that events contribute to cultural progress and increased local purchasing power (Negruşa et al., 2016), validating traditions and contributing to the preservation of the natural environment, landscapes, and historic sites that would otherwise be neglected (Pereira et al., 2021). In addition to these aspects, communities can benefit from increased employment, improved taxes, and additional sources of revenue (Getz, 1997).

Events can lead to negative aspects such as traffic congestion, problems with the law, and increased crime (Kurland, 2019). Events can also damage the image of the host community due to lack of infrastructure, adequate support equipment, and poor organization/staff performance (Moisescu et al., 2019). Due to the benefits of events, negative impacts are universally ignored by communities when planning them, as residents glorify only the expected benefits. Despite the difficulties that may be faced, regarding the organization and planning of events, communities show a tendency to adopt a decision that combines two factors: technical reasoning and participation in the overall planning (Maheshwari et al., 2019).

In conclusion, the tourism literature states that there are three different opinions about an event. Some consider it to have advantages and disadvantages; the second group tends to focus only on the negative aspects, and finally, the last one only considers the positive impacts ignoring any negative aspect, such as pollution.

2.3 The Events in Santo Tirso: The Case of *São Bento* Festivities

Events serve to stimulate visitors to travel to a tourist destination because they capture attention, animate attractions and infrastructure, enhance the use of local resources and structures, and develop destinations as poles of tourist attraction. The greater the offer of tourism events, the greater the attraction of interest investments in a territory. In this way, it is expected that a tourism industry complementary to the one previously offered will emerge, generating economic revenue and creating spaces for leisure and work (Ribeiro et al., 2006). Events become a means of economic development as they amplify the tourism demand of a region, offer entertainment for the autochthones, and promote cultural and artistic development (Martins, 2022b). Tourists from foreign countries who attend events usually have greater economic power and are more willing to spend large amounts of money during their stay. This factor boosts the economy and subsequently constitutes a factor for the development of the region where the event is inserted.

The city of Santo Tirso receives events mostly of artistic-cultural and sporting nature, since the City Council established a protocol with the Handball Federation of Portugal which aims to hold "four major events" in the next four years in Santo Tirso (Santo Tirso TV, 2022). As events of extreme relevance to the county can be highlighted the *São Bento* Festivities (Fig. 1), the Rally of Santo Tirso, *Santo Tirso a Cores*, the Feast of Foam, the Exhibition of Education Training and the "Forum Educa" that take place annually. There is also the Nazarene Market on Easter weekend and the Gastronomic Weekends, an initiative of *Turismo Porto e Norte*. The urban markets and shows have a monthly basis, *Viva a Rua*, an initiative that came into effect this year and several other smaller events.

The *São Bento* Festivities are secular and have their origin in the commemoration of the city's Miracle Saint, *São Bento*. People would come to the Monastery of Santo Tirso

Fig. 1 Solemn mass|*São Bento* Festivities. *Source* Santo Tirso Municipality (2022)

Fig. 2 *São Bento* monastery.
Source Own Elaboration (July 10, 2022)

(Fig. 2) with the main purpose of celebrating the Saint by offering him eggs, carnations, olive oil, and wheat bread, and these offerings are associated with the popularity of Saint Benedict in curing warts of the skin, the commonly called "carnations" and help in difficult births.

Tradition has it that there used to be two weekly festivals held in honor of *São Bento*: one on March 21, when the saint died, and another on July 11, based on the idea of the translation of his body. While the first has fallen into oblivion, the second date lasts until the present day.

The *São Bento* Festivities combine the sacred and the profane, so the program is not exclusively religious. These festivities present a program with very varied proposals trying to appeal to the participation of the entire local population. Of the innumerable activities present in the program, the following stand out: the *Baile dos Carvalhais* (Fig. 3),

Fig. 3 *Baile dos Carvalhais*.
Source Santo Tirso Hall (2022)

Fig. 4 Calema concert. *Source* Santo Tirso Hall (2022)

some Ceremonies (Day of Elevation of Santo Tirso to the city, Solemn Mass and Sunday Masses, Presidential Speech), the Competition of the Window Shops, the Concerts (Fig. 4), the Parades through the city, the Diversions (Children/Adults), the Fireworks, the *Há Baile no Largo*, the Pilgrimage, and the *Praça Colorida*. The City Hall of Santo Tirso also takes advantage of the municipal holiday to celebrate its friendship with the twinned cities attracting foreign public to participate in the festivities of *São Bento*. The twinnings intend to approach and to foment the cooperation between people and citizens of different countries, developing joint projects that aim at the promotion of exchanges at educational, cultural, economic, and social level. Currently the municipality of Santo Tirso has twinning protocols with the cities of Gross-Umstadt (Germany), Clichy-La-Garene (France), Celanova (Spain), Cantagalo (S. Tomé and Príncipe), Macôn (France), Alcazar de S.Juan (Spain), Nova Friburgo (Brazil), and Saint-Péray (France).

3 Methodology

In the context of the research methods, we tried to answer the following research question: "What are the residents' perceptions of the *São Bento* Festivities?" Once the research question was defined, it was possible to look at the specialized literature in the area, therefore starting the second stage of the research phases suggested by Pizam (1994) and Tuckman (2012). Regarding the literature review, we sought to analyze the state of the art of the main theme of the research topic critically and carefully, event tourism and residents' perceptions of events. This review included the reading of several articles published in the main international scientific journals, books by authors of national and international reference, institutional websites, and publications of official entities.

In methodological terms, the option focused on a questionnaire survey applied to residents of Santo Tirso municipality based on Martins et al. (2022) and Faria et al. (2021), with the purpose of assessing residents' perceptions regarding *São Bento* Festivities. The sampling technique implemented was non-probabilistic by convenience. To achieve this purpose, an online questionnaire was conducted between May 4th and June 4th. Some questionnaires were also conducted face-to-face due to a low incidence of online questionnaires in the older age groups (these questionnaires were conducted in the city center of Santo Tirso, namely at *Fábrica de Santo Thyrso, Largo da Feira, Parque D. Maria II, Praça Conde São Bento, and Praça 25 de Abril*).

This questionnaire was divided into two parts: the characterization of the resident and the perception regarding the *São Bento* Festivities. In the characterization, we aimed to understand the age, gender, marital status, education, parish of residence, and profession of the residents. In the perception regarding the *São Bento* Festivities, the respondents evaluated the festivities considering several parameters on a Likert scale of five points (relevance, interest, importance for the municipality, number of days, and most successful year). They were also questioned about the means by which they seek information about the festivities, the importance of the activities that take place during the program of the festivities, and their degree of satisfaction. A total of 215 questionnaires were obtained, of which 211 were valid. Of the total questionnaires, 181 were filled out online, and 30 were conducted in person.

We also sought to understand the impact of situations that are directly and indirectly related to the event because, as mentioned earlier, the event may bring positive aspects, but also negative ones, as mentioned by Bowdin et al. (2006) and Small (2008). The interview survey was also used to compare residents' opinions and draw conclusions. Three representatives of the municipality's departments involved in the organization of the festival were interviewed face-to-face: the Cultural Programming Service, the Tourism Service, and the Event Support Service. The interviews took place on May 6, May 17, and June 2, respectively, months before the *São Bento* Festivities. These interviews were semi-structured and consisted of eighteen questions that seek to understand the different stages of the organization of the *São Bento* Festivities. The main themes addressed during these interviews were the importance of the *São Bento* Festivities, the perception of the organization regarding these festivities, the main obstacles and challenges experienced in planning the event, the justification of the locations of the program activities, the funding and tourism generated by the festivities, and the benefits to the population.

4 Analysis and Discussion of the Results

4.1 Socio-demographic Profile of the Respondents

In order to answer the research question, we analyzed the sample consisting of 211 valid questionnaires. The age of the respondents is divided by the most varied age groups. The 18–24 age group was the most expressive totaling 42.2% of the sample. The second age group was the 25–34 (16.1%). Respondents aged 45–54 had a response rate of 15.2%. The 35–44 age group had 12.3% of the sample. The 15–17, 55–64, and 65 or older age groups are less significant adding up to a total of 14.2% (Fig. 5).

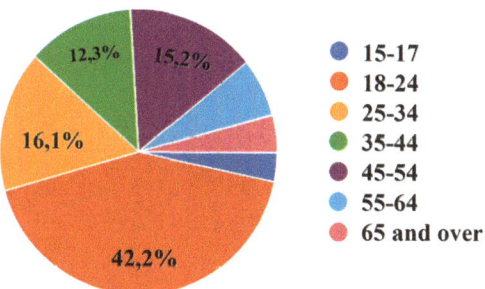

15-17
18-24
25-34
35-44
45-54
55-64
65 and over

Fig. 5 Age of the sample (%). *Source* Elaborated by the authors

Of the total number of respondents, 76.8% are female and 23.2% are male. The marital status of most respondents is single (59.2%), with about a quarter being married (27.5%). The sample also reveals that 5.7% of respondents live in a consensual union, 4.7% are divorced, and 2.8% are widowed (Table 2).

Out of the 211 respondents, 42.7% have completed secondary school and 30.3% have a degree beyond compulsory education. Of the respondents, 15.1% have only basic education, this is due to the fact that a large majority of the sample is aged 45 or older, and at the time they attended school the current mandatory level of education did not exist. Finally, 11.8% of respondents have master's degrees, doctorates, and post-graduate degrees (Table 2).

Of the sample, 37.4% are employees, 24.6% are students, and 11.4% are teachers. Of the respondents, 10% are senior managers or professionals, 6.2% are entrepreneurs or self-employed, and 2.8% are in the public sector. Finally, 5.2% of the respondents are retired and 2.4% are unemployed (Table 3). The Union of Parishes where the city of Santo Tirso is located, the Union of Parishes of Santo Tirso, Couto and Burgães, is the most significant, with 59.7% of respondents living in this locality. The remaining respondents with variations between 0.5 and 8.5 are distributed by the remaining parishes.

4.2 Residents' Perceptions

When asked about their evaluation of the *São Bento* Festivities, 38.4% of the participants make a positive assessment (97.2%), considering them very relevant, while 37.9% of the respondents evaluated them as extremely relevant. About one fifth of the sample (20.9%) was in the intermediate level. In contrast, only 2.8% rated the festivities as not very relevant (Table 4).

Regarding the importance that these festivities have for the municipality, the results are similar. Of the respondents, 45.5% considered it an extremely relevant event and 37.4% considered the festivities very relevant. In parallel, 16.6% thought it was a relevant festival for the municipality and 0.5% of the respondents rated the festivals as not very relevant (Table 4). It is important to point out that it was evident as to the evaluation and importance of the parties that none of the respondents considered them to be not at all relevant.

When classifying the festivals as to their interest, 44.1% of the respondents said that the festivals are very interesting. With a percentage of 28.4% and 25.2% came the ratings extremely interesting and interesting, respectively. In a minority, 2.4%, rated the festivals as not very interesting (Table 4).

Table 2 Socio-demographic profile of the participants (%)

Sex		Marital status		Qualifications	
Female	76.8	Single	59.2	High school	42.7
Male	23.2	Married	27.5	Graduation	30.3
		Consensual union	5.7	Elementary education	15.1
		Divorced	4.7	Master's, doctorate, and post-graduate	11.8

Source Elaborated by the authors

Table 3 Professional status (%)

Employment status/occupation	%
Employee	37.4
Student	24.6
Teacher	11.4
Senior manager or liberal profession	10.0
Entrepreneur or self-employed	6.2
Retired	5.2
Civil servant	2.8
Unemployed	2.4

Source Elaborated by the authors

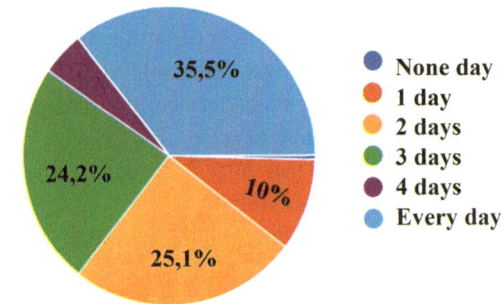

Fig. 6 Duration and participation in the event (%). *Source* Elaborated by the authors

These festivals last 4 or 5 days, depending on the year. Of the total respondents, 35.5% said that they participate in all the days of the festivities.

About half of the respondents set as number of days of participation two and three days, 25.1% and 24.2%, respectively. Only 10% of respondents usually visit the festival grounds on only one day. In a minority, those approached who visit the festivals for four days constitute 4.7 and 0.5% say they do not participate in any of the days. We consider, therefore, that the festivities are very appreciated by the residents (Fig. 6).

We tried to understand how the residents were informed about the festivities. Most respondents answered more than one option, totaling 388 answers. About 52% of respondents said that they do this search mainly from social networks such as Facebook (38.1%) and Instagram (13.9%).

It is evident, therefore, that those responsible for organizing the festivals should attend to this form of dissemination, as it is currently the best way to advertise. However, there is still a considerable number of respondents, about a quarter of the sample (25.5%), who obtain information through pamphlets, posters, and billboards. Although we

live in the digital era, there is still a significant group that seeks information on paper. Therefore, it is still important to disseminate information in this communication format. In addition, there is a good part of the sample that indicates being informed by friends and family (21.7%) (Fig. 7).

When asked about the most outstanding year of the *São Bento* Festivities, opinions disperse. Although it is an open answer, most identified the last two years in which the festivities took place, in the pre-pandemic period (2019—28.4% and 2018—18.5%) (Fig. 8).

This is probably due to the fact that these events were the last and those that respondents remember the most. Another outstanding year, according to the sample was the year 2016 (11.8%). It can be considered that these years were more voted, since there were also new dynamics, with emphasis on: the *Baile do Largo*, which attracted many young people, the projection of the final of the European Football Championship, and the hiring of artists with national and international notoriety. It should be noted that this question had some invalid answers because the answer was free.

In the context of the *São Bento* Festivities, there is in the program a set of activities that are held every year and range from entertainment, religious ceremonies to fireworks.

Table 4 Evaluation and importance of the *São Bento* festivities (%)

	1—nothing relevant	2	3	4	5—extremely relevant
Evaluation	0	2.8	20.9	38.4	37.9
Importance	0	0.5	16.6	37.4	45.5
Interest	0	2.4	25.1	44.1	28.4

Source Elaborated by the authors

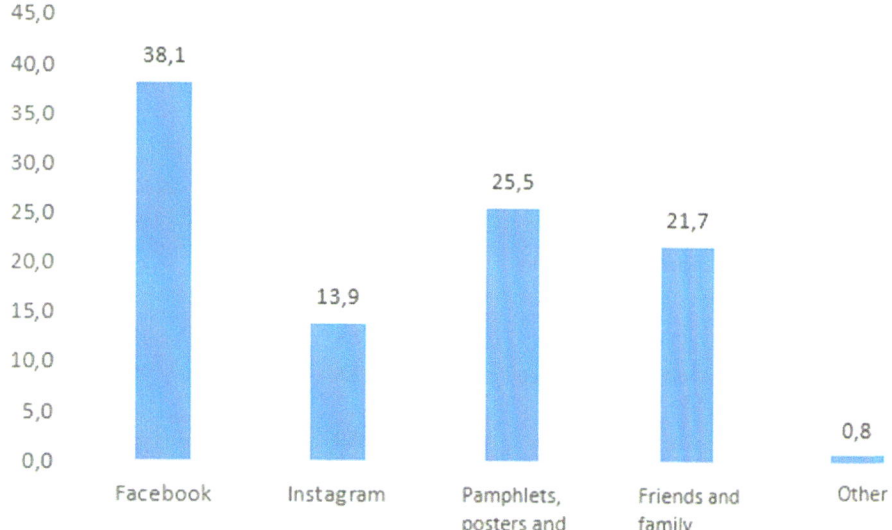

Fig. 7 Search for information about the festivities (%). *Source* Elaborated by the authors

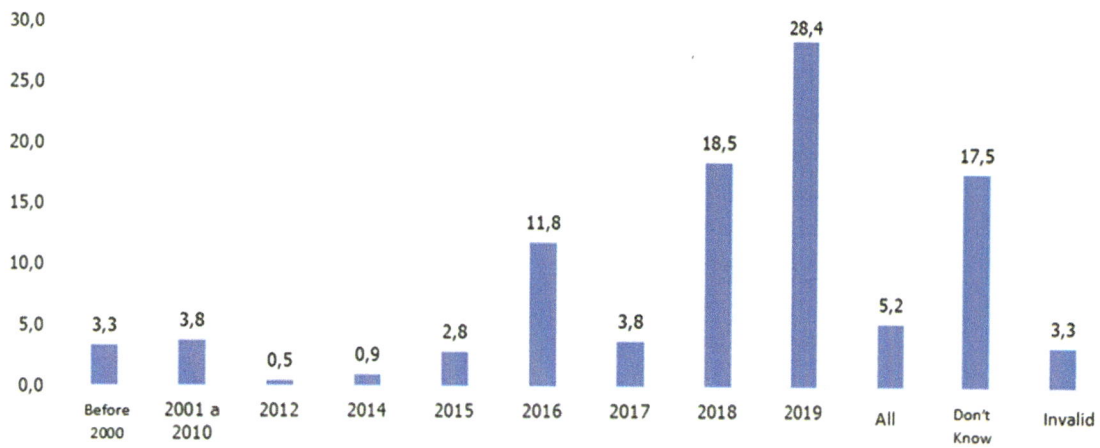

Fig. 8 Most successful year(s) for the event (%). *Source* Elaborated by the authors

Therefore, we considered it pertinent to understand, from the residents' point of view, which activities stand out the most, on a five-point Likert scale. Regarding the activities that were considered extremely important, it can be observed that most of the sample mentioned entertainments (55.5%), concerts (51.7%), and fireworks. Almost half of the sample also considered extremely important the activity *Há Baile no Largo* (48.3%) (Table 5). Another activity on the program, the *Praça Colorida*, with the flower carpets in *Conde São Bento* Square, most respondents considered the activity very important (34.1%). However, it should be noted that 33.6% of the sample also considered it extremely important. Of the remaining activities, most considered them to be in an intermediate level, namely the *Baile dos Carvalhais* (36.5%), the shop window competition (37.9%), and the ceremonies (Santo Tirso's elevation day, solemn mass, presidential speech), with 29.9% (Table 5).

In the level of unimportant activities, those most mentioned by the sample were the *Baile dos Carvalhais* (24.6%), the shop window competition (18.5%), and the ceremonies (16.1%) (the same as in the intermediate level). Of the activities that were considered not important, the most mentioned were the same: the *Baile dos Carvalhais* (12.8%), ceremonies (11.8%), and the shop window competition (10.4%) (Table 5). Based on this information, we consider that the sample places more emphasis on the more cultural and artistic part than on the religious aspect.

We also tried to understand the impact of situations that are directly and indirectly related to the event because, as mentioned before, the event can bring positive aspects, but also negative ones, as mentioned by Bowdin et al. (2006) and Small (2008). In terms of average, most of the sample was in the intermediate level (3—I agree), with 26.2%, followed by point 5—I strongly agree, with 25.6%. Of all

Table 5 Importance of the activities in the program of the *São Bento* festivities (%)

	1—not important	2	3	4	5—extremely important
Baile dos Carvalhais	12.8	24.6	36.5	16.6	9.5
Ceremonies (Santo Tirso's elevation day, solemn mass, presidential speech)	11.8	16.1	29.9	22.7	19.4
Shop window competition	10.4	18.5	37.9	22.7	10.4
Concerts	1.4	4.3	13.3	29.4	51.7
Entertainments (children/adults)	0.9	5.2	11.4	27.0	55.5
Fireworks	3.8	5.2	14.7	24.6	51.7
Há Baile no Largo	2.8	6.6	17.5	24.6	48.3
Praça Colorida (flower carpets at *Conde São Bento* Square)	2.8	6.6	22.7	34.1	33.6
Average	5.9	10.9	23.0	25.2	35.0

Source Elaborated by the authors

Table 6 General aspects of the event (%)

The event...	1—strongly disagree	2	3	4	5—strongly agree
Has been well advertised	9.0	17.1	38.9	19.0	16.0
Has been well organized	1.4	18.5	36.0	24.2	19.9
Is the most important in the municipality	1.9	7.1	20.4	20.4	50.2
Attracts many visitors and tourists	2.8	7.7	22.7	30.3	36.5
Generates employment	16.0	27.5	28.0	18.5	10.0
Contributes to increased pollution	12.3	20.9	32.2	17.1	17.5
Benefits local commerce	4.3	8.0	19.9	34.6	33.2
Helps local craftsmanship	2.8	10.0	25.1	30.8	31.3
Affects the lives of residents	12.8	24.2	28.4	18.0	16.6
Increases traffic problems	10.0	22.7	22.7	20.9	23.7
Helps preserve culture and identity	1.9	4.7	22.7	28.0	42.7
Increases insecurity and vandalism	34.5	29.0	17.5	10.0	9.0
Average	9.2	16.4	26.2	22.6	25.6

Source Elaborated by the authors

the aspects we aimed to survey, the only aspect that most of the sample highlighted was the fact that this event is the most important in the municipality (50.2%). Next, most of the sample considered that the event helps preserve the culture and identity of the region (42.7%). Furthermore, still within point 5 (strongly agree) of the Likert scale, we can highlight as positive the fact that the event attracts many visitors (36.5%), benefits the local commerce (33.2%) and local crafts (31.3%) (Table 6).

Within point 4 of the Likert scale (strongly agree), it was possible to observe that three aspects stood out, above thirty percent, namely the event benefits local commerce (34.6%), helps local handicraft (30.8%), and attracts many visitors and tourists (30.3%) (Table 6).

It was also possible to verify that most of the affirmations had more percentage in the intermediate level (3—agree). In

fact, most only agree that the event is well publicized (38.8%), followed by the good organization (36%). The other three statements that the sample considered to be in the middle range were the fact that the event contributes to an increase in the population (32.2%), affects the lives of residents during the period of the festivities (28.4%), and generates employment (28%) (Table 6). At level 2 (partially agree) of the Likert scale, it can be observed that the sample highlights two affirmations: 27.5% partially agree that the event generates employment and 20.9% partially agree that it contributes to an increase in pollution. In addition to these two, the negative aspects disclosure (18.5%) and organization (17.1%) stand out (Table 6). At level 1 (strongly disagree), it is possible to observe that the sample emphasized one affirmation above all: most respondents do not agree that the event increases insecurity and vandalism (34.5%)

Table 7 Degree of satisfaction (%)

Satisfaction with…	1—not satisfied	2	3	4	5—extremely satisfied
Responsible team	0.0	4.3	44.5	30.8	20.4
Activities schedule	0.5	5.7	33.2	38.4	22.3
Duration (number of days)	0.9	5.7	36.5	31.8	25.1
Publicity	3.8	11.4	41.2	29.4	14.2
Spaces where the event takes place	5.7	6.6	33.6	30.8	23.2
Parking conditions	24.2	32.7	28.0	6.1	9.0
Event (in general)	0.9	0.5	35.5	38.4	24.6
Average	5.1	9.5	36.1	29.4	19.8

Source Elaborated by the authors

(Table 6), being therefore a positive aspect for the event in particular and for the city in general. Table 6 also presents that two affirmations were spread over the five points of the scale in percentage terms and therefore did not meet with much consensus: the event affects the lives of residents and aggravates traffic problems.

Finally, we sought to analyze the degree of satisfaction of respondents with the event in general and with some aspects that directly affect it. In terms of average, using the five-point Likert scale, most of the sample fell into level 3 (satisfied), with 36.1%, followed by level 4 (very satisfied) with 29.4%. Only a percentage below twenty percent was dissatisfied: 5.1% not at all satisfied and 9.5% not very satisfied (Table 7).

Regarding level 4, the statements that stand out in terms of satisfaction were the schedule of activities and the event in general, both with 38.4% (Table 7). It was also possible to verify that most of the statements had a higher percentage in the intermediate level (3—agree). In fact, most were "only" satisfied with the team in charge (44.5%) and the way the event was publicized (41.2%). This level of satisfaction was also observed regarding the duration, 4–5 days, (36.5%), as well as the spaces where the activities of the event take place (33.6%) (Table 7). Regarding levels 1 (not at all satisfied) and 2 (not very satisfied), the one that stood out most negatively was the parking conditions with an overall percentage of 56.9% (Table 7).

4.3 Interviews with People in Charge of Municipal Departments

To complement the perceptions of the municipality's residents, we also tried to understand the perceptions of those who organize the event. In this case, we used the interview survey technique. It was possible to interview some elements responsible for departments of the Municipality of Santo Tirso that are directly involved in the preparation and implementation of the *São Bento* Festivities, namely an element responsible for the Cultural Programming Service, an element of the Tourism Service, and another element of the Event Support Service.

All interviewees consider that the *São Bento* Festivities are one of the biggest events that Santo Tirso promotes. The three interviewees referred that these festivities attract many visitors of various ages to the city. The program of the festivities is also an attraction of interest since the respondents can enjoy various activities, under the various aspects of the event, either cultural, artistic, or religious level. The Service of Cultural Programming and the Events Support Service also stated that these festivals are the ones that have a greater involvement of the entire population of the municipality, covering a greater range of the annual budget of the Municipality of Santo Tirso. Comparing with the perspective of the respondents, in Table 4, it can be established that they fully agree that the *São Bento* Festivities are one of the most important events of the Municipality of Santo Tirso.

Regarding the importance that the *São Bento* Festivities play in the municipality, the three interviewees gave a rating of 5, extremely important. Confronted with the opinion of residents (Table 4), about 83% consider the festivities are very and extremely relevant to the municipality.

The *São Bento* Festivities are the result of methodical work between the Festival Committee, the Municipality of Santo Tirso, and a vast number of labor, externally contracted and voluntary work. The representative of the Service of Cultural Programming referred that the majority of the population does not have the perception of the work involved, so that the parties take place, because they only see the parties "assembled." The Events Support Service also mentioned that a few years ago a team was created that is exclusively dedicated to this event, which makes the process of organizing the Festivities more agile. As each element knows its function, it is only necessary to make a few decisions and then proceed to implement them on the ground.

The activities of the program over the years have undergone several changes giving it an innovative character.

From the 80s on, the entertainment aspect was inserted, namely with cavaquinho concerts, entertainment, and the implementation of the Cultural Week. More recently activities such as *Baile dos Carvalhais* and *Há Baile no Largo* have been promoted. Taking into account the activities that currently take place during the Festivities of São Bento program, the interviewees consider that the concerts are the main focus of attracting visitors, followed by the fireworks sessions and *Há Baile no Largo*. Allusion was also made to the religious aspect and the restaurants, encompassing "ages from 8 to 80 years old." Comparing the information from the interviewees with the sample data, the answers of the interviewees are in line with what most of the respondents mentioned, except for entertainment, these being the most important.

Regarding the contacts made with street vendors and artists, it is concluded that they work differently. When it comes to street vendors, they must make an application to the City Hall in order to guarantee their space, called *terrado*. These merchants have to follow imposed rules and comply with the regulations stipulated by the City Council. As for the artists, there is a direct contact between the City Hall and the agencies/managers, and later an evaluation is made as to the values requested. Throughout the year, it is also made an evaluation of which would be the best artists to integrate the poster. As mentioned by the Cultural Programming Service, when hiring artists, we try to meet the expectations of residents and visitors. However, it is important not to exceed the stipulated budget because it is necessary to consider the cost/benefit ratio, because the greater the notoriety of the artist, the more expensive it becomes, but also the more public it brings.

When asked about the most challenging moment in the dynamics of the *São Bento* Festivities, the points raised by the three departments were the organization, setting up the spaces, and contact with artists. The Tourism Service also mentioned that it is necessary to contact the urban services in order to ensure garbage collection and street cleaning, so that everything is presentable for the next day. This was one of the aspects that most of the sample emphasized, as 32.2% agreed or strongly agreed, and there is a need for improvement in this part.

The interviewees have the perception of what needs to be done to publicize the event, they know the means of dissemination on which they should focus, namely social networks such as Facebook and Instagram and through posters and billboards. The municipality also has other forms of dissemination, such as the use of advertising broadcasted by the ATM network and through a partnership with *Comboios de Portugal*, where the *São Bento* Festivities are advertised on the trains, in the period preceding the event.

Regarding the time needed to prepare for the event, the response was unanimous among the three interviewees, stating that the minimum preparation time is approximately six months, which can be extended to eleven months in case of need to sign contracts with artists or logistics companies.

As for the most common problems during the preparation of the event, the most common were the meeting of deadlines, the lack of material, and several events that occur at the same time as the *São Bento* Festivities. It is also pointed out by the Tourism Services and Events Support Service that the weather is a constraining factor in the course of the event, since it is a variable that cannot be controlled and that can both benefit and hinder the activities of the event given that many of them are held outdoors.

The interviewees refer that, as a rule, there are not many unforeseen events during the course of the event. However, safety is highlighted by the Cultural Programming Service and the Tourism Service. It is perceived that security is a key factor for the realization of this event, using means such as the Municipal Police, PSP, Fire Department, and Civil Protection, and thereby conveying a safe environment for the inhabitants. This information corroborates the perception of most of the sample when they refer that the event does not increase insecurity and vandalism (63.5%). It was also mentioned, by the Event Support Service, the need to hire more human resources and the weather conditions, particularly the heat, during the assembly of equipment.

Considering the budget foreseen/spent for the *São Bento* Festivities, none of the interviewees mentioned concrete numbers. However, they mentioned that it is rare that the initial budget is exceeded. It is evident that they try to make sure that there is no budgetary slippage. Specifically, the source of income for the festivities is guaranteed taking into account four aspects: (a) a fee attributed by the City Council; (b) fundraising by the Festivities' Committee; (c) sponsorships (sponsors and patrons); and (d) the rental of the space and the sale of products (like beer).

The most expensive service is, without a doubt, the hiring of the artist for the concert, because the payment and the fees are increasingly high taking into account their notoriety. The acquisition of this service takes the largest slice of the budget, since it is the activity that attracts the largest number of people (Table 5). Fireworks were also mentioned as one of the main attractions. Most of the sample considered concerts and fireworks to be extremely important. Curiously, the entertainment is not mentioned, since the Festival Commission only makes profit with this activity, since the entertainment entrepreneurs pay to stay on the *terrado* (place where the entertainment is located), with no expenses on the part of the organization.

The City Council perceived that there is a need to not only meet the visitors, but also the inhabitants, seeking to maintain tradition, but creating conditions for the event to take place. It was mentioned that there is a special interest in returning the festivities to their origin, to the area surrounding the monastery, and that the spaces for activities, including entertainment, restaurants, and crafts were gradually changed. The City Council and the Festival Committee understand that the ideal would be to have a single space where all the activities would take place. This is not possible, so with the current arrangement of activities seeks to ensure an optimal circulation through the city, starting at the *Praça dos Carvalhais*, going to the *Praça 25 de Abril*, then the *Largo Coronel Baptista Coelho*, next to the *Parque D. Maria II*, ending in front of the monastery. From the perspective of the resident population (Table 7), most of the sample is satisfied, emphasizing that the current arrangement is not the best, but the possible.

When asked about the ability to attract visitors, the interviewees consider that the *São Bento* Festivities attract mostly visitors from neighboring municipalities and immigrants. This statement is further corroborated by the majority of the sample (Table 6) who fully agree. It is noteworthy that the more international the main artist, the more visitors are attracted to the event, an example of this is the group "The Stranglers." Unfortunately, this event cannot attract foreign tourists, apart from the twinned cities' partners (Germany, France, Spain, Cape Verde, S. Tomé, and Brazil, among others).

When asked about which year was the most successful of *São Bento* Festivities, the interviewees said that we should not look at the years individually, since each year has a different success. However, some eras were referenced: the 1980s with the implementation of several changes in the program of the festivities and from 2014 with the creation of *Há Baile no Largo* which attracted more young people. It should also be noted that from that year onward the festivities should be observed considering an upward direction. According to the data collected through the surveys, the majority of the sample refers to the years after 2014, with 2018 and 2019 standing out as the best years of the *São Bento* Festivities. This achievement is probably due to the fact that people have more present in their memory these last two editions of the festivities.

During the interview, the different departments were asked about the strategies adopted to get young people to visit the *São Bento* Festivities rather than the Sebastianas, which takes place at the same time in a neighboring municipality. Although all respondents consider that it is difficult to establish a comparison between the two festivals, since they are distinct, they mention that the Municipality of Santo Tirso created a set of initiatives in order to attract the younger age groups, namely the *Há Baile no Largo* and the hiring of young artists from Santo Tirso to perform at *the Baile dos Carvalhais*.

Finally, when asked if with the end of the *São Bento* Festivities, the city and the county would lose visitors and

tourists in the summer season, all respondents answered affirmatively, since these are the most important festivals of the city, which would cause a significant cultural and traditional loss. They also added that in the summer season people are more inclined to spend time outdoors, so the *São Bento* Festivities are an excellent event for socializing.

5 Conclusions

In Portugal, festivals and pilgrimages are prominent festivities, being events that promote social and cultural interaction (Martins, 2022a). Considering the increased competition between different events and destinations, it is important that organizing entities and destinations promoters know the motivations and behaviors of both residents and non-residents, to ensure adequate communication to the different segments, ensure that the needs and expectations of residents are met and promote and develop the territory where the events take place (Faria et al., 2021).

The *São Bento* Festivities are a festivity where the religious and profane aspects prevail, offering a diversified program that aims to attract various types of public, from the youngest to the oldest. Over the years, this event has been adapted to broaden and attract more people, betting on diversity, but trying to maintain its authenticity. The results obtained in this study show that, in general, residents' perceptions about the *São Bento* Festivities and their program are positive, meeting the expectations of the event's organizers.

The municipality and the event organizers keep in mind that events bring impacts both in the territory and to the communities and try to minimize the negative effects, such as traffic, and potentiate the positive effects identified by Bowdin et al. (2006) and Small (2008). Regarding the data collected from residents, it is observed that respondents visit the event on most of the days it takes place and repeat it annually. They learn about the event through family members, social networks, and advertising posters. Sometimes, the bet on concerts with nationally and internationally known artists attracts more public but makes the event more expensive. Amusements, on the other hand, which also work as a pole of attraction for visitors, are the most profitable, with no inherent expenses. Here too, the organizers are attentive and try to publicize the event in different media. However, there are aspects to be improved, such as the cleaning of the venues where the event takes place.

In terms of limitations, we believe that this research would be richer if we had interviewed a larger number of departments of the City Council, something that was not possible, but that would allow us to have a more comprehensive view of the dynamics for the realization of the event. As suggestions for future research, we consider it pertinent to assess the perceptions of fairgoers and entrepreneurs of the entertainment sector on the realization of this type of events.

References

Allen, J., Harris, R., Jago, L., Tantrai, A., Jonson, P., & D'Arcy, E. (2022). *Festival and special event management*. Wiley.

Ap, J. (1992). Residents' perceptions on tourism impacts. *Annals of Tourism Research, 19*(4), 665–690.

Bowdin, G., Allen, J., O'Toole, W., Harris, R., & McDonnell, I. (2006). *Events management* (3ª). Butterworth-Heinemann.

Faria, D., Vareiro, L., & Malheiro, A. (2021). Autenticidade e Motivações em Eventos Tradicionais: O caso da Festa das Cruzes. *Revista Turismo & Desenvolvimento, 36*(1), 409–425. https://doi.org/10.34624/rtd.v1i36.9671

Getz, D. (1991). *Festivals, special events, and tourism*. Van Nostrand Reinhold.

Getz, D. (1997). *Event management and event tourism*. Cognizant Communication Corporation.

Goldblatt, J. (1990). *The art and science of celebration*. Van Nostrand Reinhold.

Gursoy, D., Jurowski, C., & Uysal, M. (2002). Residents' attitudes: A structural modeling approach. *Annals of Tourism Research, 29*(1), 79–105.

Kurland, J. (2019). Arena-based events and crime: An analysis of hourly robbery data. *Applied Economics, 51*(36), 3947–3957. https://doi.org/10.1080/00036846.2019.1587590

Maheshwari, V., Giraldi, J., & Montanari, M. (2019). Investigating residents' attitudes of 2016 Olympic Games: Examining socio-cultural, economic and environmental dimensions. *Journal of Place Management and Development, 12*.https://doi.org/10.1108/JPMD-08-2018-0059

Martins, H. (2022a). Os impactos económicos da Covid-19 em eventos. *Revista Turismo & Desenvolvimento, 38*, 265–280. https://doi.org/10.34624/rtd.v38i0.25863

Martins, H. (2022b). Religious tourism during the COVID-19 period: The case of our lady of agony festival, Viana do Castelo, Portugal. In: A. Mandić, R. A. Castanho, U. Stankov. (Eds.), *Cultural sustainable tourism. Advances in science, technology and innovation*. Springer. https://doi.org/10.1007/978-3-031-10800-6_4

Martins, H., Carvalho, P., & Almeida, N. (2022). O turismo em Áreas Protegidas: uma análise ao perfil do turista no Parque Nacional da Peneda-Gerês (Portugal). *Cadernos de Geografia, 46*, 77–91. https://doi.org/10.14195/0871-1623_46_6

Mogollón, J. M. H., Fernández, J. A. F., & Cerro, A. M. C. (2017). Eventos baseados em reconstituições militares históricas como estratégia dinâmica para o turismo local: O caso da batalha de La Albuera (Espanha). *Revista Turismo & Desenvolvimento, 1*(27/28), 1071–1082. https://doi.org/10.34624/rtd.v1i27/28.9749

Moisescu, O., Gică, O., Coros, M., & Yallop, A. (2019). The UNTOLD story: Event tourism's negative impact on residents' community life and well-being. *Worldwide Hospitality and Tourism Themes, 11*, 492–505. https://doi.org/10.1108/WHATT-06-2019-0036

Negruşa, A., Toader, V., Rus, R., & Cosma, S. (2016). Study of perceptions on cultural events' sustainability. *Sustainability, 8*(12), 1269. MDPI AG. https://doi.org/10.3390/su8121269

Pereira, L., Jerónimo, C., Sempiterno, M., Lopes da Costa, R., Dias, Á., & António, N. (2021). Events and festivals contribution for local sustainability. *Sustainability, 13*(3), 1520. MDPI AG. https://doi.org/10.3390/su13031520

Pizam, A. (1994). Planning a tourism research investigation. In J. R. B. Ritchie & C. R. Goeldner (Eds.), *Travel, tourism and hospitality research: A handbook for managers and researchers* (2nd ed., pp. 91–104). Wiley.

Raj, R., Walters, P., & Rashid, T. (2009). *Events management: An integrated and practical approach*. Sage.

Ribeiro, J. C., Vareiro, L. C., Fabeiro, C. P., & de Blas, X. P. (2006). Importância da celebração de eventos culturais para o turismo do Minho-Lima: um estudo de caso. *Revista portuguesa de estudos regionais*, (11), 61–76. https://hdl.handle.net/1822/5143

Santo Tirso Hall. (2022 April 7). *Facebook*. https://www.facebook.com/CMSantoTirso/photos/?ref=page_internal

Santo Tirso TV. (2022, April 28). *Santo Tirso recebe Final 4 da Taça de Portugal de Andebol Feminino*. https://www.santo-tirso.tv/artigo/1/10622/santo-tirso-recebe-final-4-da-taca-de-portugal-de-andebol-feminino/

Simões, M. (2012). *Os eventos e a atractividade e competitividade turística das cidades: o caso de Lisboa* (Doctoral Thesis, Estoril Higher School of Hospitality and Tourism). http://hdl.handle.net/10400.26/4458

Small, K. (2008). Social dimensions of community festivals: An application of factor analysis in the development of the social impact perception (SIP) scale. In *Event management* (pp. 45–55). https://doi.org/10.3727/152599508783943219

Smith, A. (2012). *Events and urban regeneration: The strategic use of events to revitalise cities*. Routledge.

Tourism of Portugal. (2017). *Estratégia Turismo 2027—Liderar o turismo do futuro*. http://fortis.pt/files/2017/03/estrat%C3%A9gia-turismo-2027.pdf

Tuckman, B. (2012). *Manual de investigação em Educação* (4th ed.). Fundação Calouste Gulbenkian.

Pilgrimage Tourism for Enhancement of Heritage Conservation Management: Study of Potential, Possibilities in Kurnool, Anantapur Districts of Andhra Pradesh, India

Vasanta Sobha Turaga

Abstract

Religious places attract devotees in huge numbers to take up pilgrimages as prescribed by their respective faiths. Like Mecca or Vatican for Muslims and Christians, sacred places of large, mega scale to medium/small centres and temples spread across India are 'teertha-sthalas'/'punya-kshetras' are holy places for 'teerth yatras' (pilgrimages) by millions of Hindus. Pilgrims/devotees visit them all through the year normally, and much more in numbers on particular festivals/auspicious days. These kshetras/places provide well-developed visitor amenities, hospitality facilities provided by the government and private operators, contributing to regional/local economies. Historical and natural sites located in proximity of religious centres often stand neglected for lack of patronage and resources. Well-planned tourist circuits and loops, if designed to link popular pilgrimage destinations, can help support heritage conservation management. Kurnool and Anantapur districts of the southern Indian State of Andhra Pradesh are case examples, where popular large and medium-scale pilgrim centres are located alongside less-visited historical and natural sites, having high tourism potential. In this paper, the study of Kurnool, Anantapur region of Andhra Pradesh for the potential and possibilities of enhancing heritage conservation management and historic/natural sites tourism by connecting them to popular large/medium pilgrim centres of the region is presented with illustrations, examples along with the results of a brief survey conducted to demonstrate the application of hub and spoke model of tourism development.

Keywords

Pilgrimage tourism • Heritage conservation • Heritage management • Religious tourism • Cultural tourism • Andhra Pradesh

1 Introduction

Religious/spiritual sites and historical/scenic sites have their own patrons and set of tourists/visitors with specific interests. Surveys have recorded tremendous growth in the religious tourism sector in India in recent years, especially after the pandemic. With the growing demand, there is special focus from the government on the sector through enhanced investment and schemes to boost religious tourism further.

While social trips (non-religious visits to historical/natural places, recreation spaces, etc.) and business trips have also shown growth trends (NCAER, 2014; Shaikh, 2017), the condition of monuments and historical places remains in an unsatisfactory state of preservation, without the support and investment necessary for conservation and management.

The objective of this paper is to present a case for application of the 'hub-spoke' phenomenon, exploring possibilities of extending the growing religious (Hindu pilgrimage) tourism benefits to neglected monuments/historical (non-religious) places in proximity, in the geographically defined areas of Kurnool and Anantapur districts of Andhra Pradesh, a state in south India. To verify and demonstrate the possibility of application of the hub-spoke model for sharing tourism benefits to less-visited historical sites, two case-study sites are selected and surveyed, one each in Kurnool and Anantapur districts and the results are presented in this paper.

V. S. Turaga (✉)
Conservation Architect and Urban-Regional Planner, Vasaamaha Consultants, Hyderabad, India
e-mail: vasantasobhaturaga@gmail.com

V. S. Turaga
School of Planning and Architecture, Jawaharlal Nehru Architecture and Fine Arts University, Hyderabad, Telangana, India

© The Author(s), under exclusive license to Springer Nature Switzerland AG 2024
J. Chica-Olmo et al. (eds.), *Sustainable Tourism, Culture and Heritage Promotion*,
Advances in Science, Technology & Innovation, https://doi.org/10.1007/978-3-031-49536-6_26

The structure of this paper is as follows:

2 Tourism Trends in India—A Brief Overview

2.1 Religious/Spiritual Tourism Trends in India

Pilgrimages are periodically taken up by devotees of different faiths as prescribed by their respective religions, an age-old, popular concept existing in all parts of the world. Religious tourism in India is on a growth trajectory both in terms of the number of persons visiting and money spent on the visits, especially after the pandemic.

"Spiritual tourism is one of the biggest untapped markets for domestic travel: nearly 60% of domestic tourism in India is religion-based. According to travel marketplace Ixigo, there has been month-on-month, double-digit growth in hotel bookings on its site for a number of cities with religion. These include Puri (60% growth) which is famed for its Jagannatha Temple, Varanasi (48%), Tirupati in Andhra Pradesh (34%) and Shirdi in Maharashtra (19%)" (warc.-com, 2018).

As per the survey carried out by the National Sample Survey Organisation in 2014–15, "more than 4.8 million out of 58.4 million overnight trips i.e., 8.29% were pilgrimages or undertaken for religious purposes. The average expenditure on a religious trip was Rs. 2717/- per day ($33.97)— higher than the daily expenditure on a secular trip, Rs. 1068/- ($13.35)" (Shaikh, 2017).

Among the sectors that suffered big hits due to the Covid-19 pandemic, tourism and hospitality rank high, but eventually, it was religious tourism that helped to redeem the losses, as pilgrimages were the first kind of travel that people started to make as the restrictions of the pandemic were being gradually eased. Travel agencies reported that as travel picked up, religious destinations were the most sought after. And the period saw a transformation in the way pilgrimage tourists approached religious travel, as they were also done as much for a brief getaway as for *teerth yatras*.

The expenditure spent on religious travel has more than doubled in recent years, religious tourism offers strong opportunities (IndBiz, 2019). As a matter of policy, as governments in India have started initiatives to showcase India as a pluralistic, multi-cultural society as well as an ancient land of rich cultural and religious heritage, the potential of India to become a hub for religious tourism has also been strengthened.

2.2 Non-religious (Secular/Heritage/Social) Tourism Trends in India

Non-religious (Secular/Heritage/Social) tourism also has its own patrons and interest groups within domestic and international tourists. According to a survey conducted in 2009–10 (NCAER, 2014), within the international tourism sector, for foreigners, including Non-Resident Indians and Persons of Indian Origin, historical places (monuments, forts, palaces, museums), and natural sites (hill stations, mountains, beaches, scenic places) top the list of motivational factors for visiting India, followed by religious places, spiritual healing, medical tourism, etc. and social trips and visits to non-religious places have also recorded highest numbers (Shaikh, 2017).

The two visitor/tourist groups (religious and non-religious) comprise a considerable chunk of the tourism market. It is important that the potential and possibilities of enhancement of the existing tourism market are explored by integrating the religious and non-religious interest groups. It is a point to be noted that, India, being the second most populated country in the world, the number of persons is huge, when the percentages of population of survey findings are converted to absolute numbers. What is required is an appropriate strategy and planning that optimally utilizes tourism infrastructure, increases economic opportunities, helps preserve, promotes cultural heritage, and provides visitor experience satisfaction.

3 *Teerth Yatras*: Hindu Pilgrimages in India

In Sanskrit language, "*Tirtha* refers to any place, text or person that is holy. It particularly refers to pilgrimage sites and holy places in Hinduism. The process or journey/*yatra* associated with a *Tirtha* is called *Tirtha-yatra*" (Gangashetty, 2019). A *Tirtha* or *Punya-Kshetra* means a place of 'cross-over'; pilgrimage essentially is a travel undertaken by an individual to 'cross-over—from physical to subtle nature of self for spiritual evolution' (Gumma, 2021).

Pilgrimages are made to individual *Punya-Kshetras*, specific temples or holy cities such as Varanasi (Kashi) (Fig. 1b), Vaishno Devi, Prayagraj (Allahabad) in north India; Tirupati, Srirangam, Arunachalam and many others in south India, for example, or on circuits, based on themes and concepts prescribed in holy texts, with each place having its significance and connotation.

The most popular Hindu pilgrimage circuits include:

Char-Dham Yatra, pilgrimage to four temples in Badrinath, Dwaraka, Rameswaram and Puri located across the country and also a smaller circuit of temples of Kedarnath, Badrinath, Gangotri and Yamunotri located in Uttarakhand, in north India.

Dwadasa Jyotirlingas, 12 Shiva temples spread across India, including Sri Mallikarjuna Swamy temple in Srisailam located in Kurnool district of Andhra Pradesh, discussed in this paper (Fig. 1a).

Shaktipeethas, 18 and 52 temples goddess Shakti/Devi, spread across India and outside, including Sri Bhramarambika temple, also in Srisailam located in Kurnool district of Andhra Pradesh, discussed in this paper.

Pancharama Kshetras connecting five Shiva temples in Amaravati, Draksharamam, Bhimavaram, Palakollu and Samalkota are located in other districts of Andhra Pradesh. It is believed that five different deities installed five lingas (idols) and they are all scattered parts of a single '*lingam*'.

Panchabhoota Sthalas, five temples linked to five Elements, four of which are located in Tamil Nadu viz., Ekambareswara temple in Kanchipuram (Earth element), Jambukeswara temple in Tiruchirappalli (Water Element), Arunachaleswar temple in Tiruvannamalai (Fire Element) and Nataraja temple in Chidambaram (Space Element). The fifth Sri Kalahasteeswara Swamy temple (Vayu Lingam—Air element) is located in Chittoor district of Andhra Pradesh.

Divya-desas, 108 temples of Lord Vishnu, maximum number are located in south India, a few in north India and Nepal. Just two temples are in Andhra Pradesh including Thiru Singavel Kundram temple located in Ahobilam, an important pilgrimage centre in Kurnool district (Grandhi, 2015).

The above-mentioned pilgrimage circuits and temples are considered most important by Hindus across the country. Besides these exist innumerable sacred temples and places that are popular locally and regionally. Devotees visit temples all round the year and also particularly on auspicious dates determined by planetary positions, festivals and by their *sthala-puranas* (stories of historical significance, sacredness of the place).

For example, *Ekadashi*, the 11th day of the lunar month, twice a month, is associated with Lord Vishnu and considered the most auspicious day to keep a vrat/fasting and visit temples of Lord Vishnu. Devotees visit Shiva temples on Mondays, Lord Venkateswara Swamy temples on Saturdays. *Karteeka* month, lunar month around October/November, is associated with Lord Shiva, *Dussehra* festival is celebrated for nine days in September/October in *Ashwayuja masam* of lunar calendar worshiping goddess *Shakti/Devi/Durga* in different forms. Pilgrims take a holy dip during *pushakaralu*

(a)

(b)

Fig. 1 **a** Omkareswar Jyotirlinga temple on Narmada River in Madhya Pradesh State in Central India. **b** Varanasi (Kashi) ghats on River Ganges in Uttar Pradesh in North India

(a)   (b)

Fig. 2 a Preparations for holy dip in Ganga River/Triveni Sangam before Kumbh Mela in Prayagraj (Allahabad) in Uttar Pradesh State in north India. **b** Ganga Aarti (offering lamps to Ganga River), a daily evening ceremony performed on Ghats in Varanasi (Kashi) Ghats on River Ganges in Uttar Pradesh in north India, where thousands of visitors view from the riverside

in a particular river once in each year, rivers specified in a cycle of twelve years during the time planet Jupiter moves from one zodiac house to the next. The most famous of devotees gathering for taking the holy dip in River Ganga is the *Kumbhmela* held in Prayagraj city (Allahabad) in Uttar Pradesh State and other places, which sees massive attendance (Fig. 2a, b).

4 Kurnool and Anantapur Districts of Andhra Pradesh

4.1 Location and Introduction

Kurnool and Anantapur districts are located in the south-western part of south Indian State of Andhra Pradesh.

With a few helipads and minor airports, Kurnool and Anantapur districts are linked by air by international airports located in Hyderabad and Bengaluru, capital cities of the two neighbouring States of Telangana and Karnataka, and Tirupati in Andhra Pradesh. The highways connecting these three major cities form two loops from Kurnool city through the spread of Nallamala hills.

National Highway no. 44 cuts across the two districts, connecting two major metropolitan cities of Hyderabad and Bengaluru. Refer to Fig. 3a–c. Srisailam, a major Pilgrimage centre in India located in the northern part of Kurnool district in Nallamala hills and forests, is approached by a separate Srisailam Highway when travelling from Hyderabad city. Refer to Fig. 4a, b. The major cities, towns and two loops of highways are taken as reference in this paper for discussion of other places, proximity and approach roads.

(a)   (b)   (c)

Fig. 3 a Location of Andhra Pradesh State in India map. **b** Location of Kurnool and Anantapur districts in Andhra Pradesh and National Highway 44 connecting Hyderabad and Bengaluru, capital cities of neighbouring States of Telangana and Karnataka. **c** Map of Kurnool and Anantapur districts showing district boundaries, capital cities and major towns, rivers, railway lines and highways (Marking done on maps sourced from 'Maps of India' website—https://www.mapsofindia.com/)

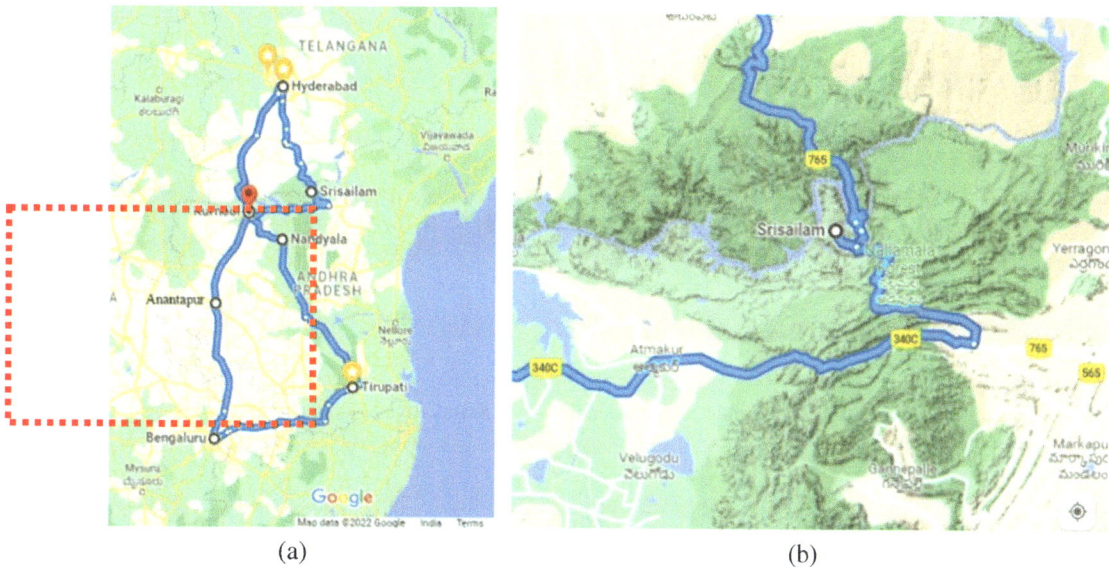

(a)   (b)

Fig. 4 a Major Highways connecting Hyderabad-Bengaluru via Kurnool and Anantapur, Hyderabad-Srisailam, Kurnool-Tirupati via Nandyala and Bengaluru-Tirupati. **b** Location of major Pilgrim centre Srisailam in Nallamala forests, approached by Srisailam Highway from Hyderabad—Map showing terrain of Nallamala Hills and Forests. *Maps Source* Routes marked on Google Maps

A few basic facts and figures of location, geography and demography of Kurnool and Anantapur districts are given in Appendix 1 and 2 of this paper.

4.2 Popular Religious Destinations in Kurnool and Anantapur Districts

4.2.1 Kurnool District

Srisailam

Sri Bhramaramba Mallikarjuna Swamy Devasthanam is a complex with two major pilgrim centres—Lord Shiva as deity *Mallikarjuna Swamy is* one of the *12 Jyotirlingas* and Goddess Shakti/Devi in the form of Sri Bhramarambika is one of the 18 *Shaktipeeths.*

Temples were built in second century AD and the present complex is spread across two acres. Srisailam is located in a picturesque setting of mountains, Krishna River and Nallamala forests. The number of visitors goes up to remarkably high numbers when the Srisailam dam gates are opened up. Pilgrim amenities are available for overnight stay and the place is well connected by road transport to travel crossing the forest and river.

While the temples have been built in second century AD, the present expansive complex is spread across two acres with a series of *Prakarams* (enclosing walls), *Mandapams* and Corridors with extravagant sculpture, leading up to the main shrines. The Eastern *Gopuram* (tower) standing at more than 70 ft height is the most imposing and the main shrine is an intricately carved, gold covered structure. The

other *gopurams*, the Maha Dwaram, the Tripuranthakam, and the Annapoorna *gopuram*, also stand on elaborately carved structures with iconography from Hindu mythological stories.

As the place is well developed as a pilgrim centre, a visit to Srisailam is most comprehensively accomplished if planned across two days, with one day dedicated to the religious part and the other to visit the scenic beauty around the temple complex, the magnificent course of the mighty River Krishna being the centrepiece, especially during monsoons when the river is in the spate and the dam presents the river in all its roaring glory. Even the drive to Srisailam is a picturesque journey, curving through the lush greenery of the Nallamala forest across the Eastern Ghats, with a tiger reserve and almost-guaranteed sightings of wildlife, tribal habitations along the way and small temples dotting the route (Fig. 5a, b).

Thousands of devotees visit the temples daily all round the year and the numbers exceed more than tenth of a million per day on festival auspicious days—Mondays, *Shivaratri* and *Karteeka* month of the lunar calendar falling during October–November months. As per official data (see Table 1), the number of visitors has shown high turnouts across the last few years, with the exception of 2020 when the Covid-19 pandemic forced a lockdown. While 2.9 million people visited Srisailam in the year 2018, there was a turnout of 2.79 million in 2019, 0.8 million in 2020 which grew to 1.2 million in 2021. More number of people throng this pilgrim destination in the months of February–March around the time of Maha Shivaratri, in the month of *Karthika* (October–November) and holidays.

Fig. 5 **a** *Sri Bhramaramba Mallikarjuna Swamy* temple complex-aerial view. **b** Krisha River, Nallamala hills and Srisailam dam. *Photos Source* Website of Srisailam devasthanam

Table 1 Number of Visitors to Srisailam—four years month-wise data

Months and years	I—2018	II—2019	III—2020	IV—2021
January	110,500	160,325	210,620	130,280
February	350,000	146,180	435,000	160,455
March	500,000	610,350	NIL	220,515
April	210,000	175,000	NIL	NIL
May	120,000	182,000	NIL	25,300
June	138,000	164,300	NIL	25,000
July	92,000	178,250	NIL	42,000
August	485,130	185,320	6000	63,000
September	316,000	214,200	8540	120,650
October	320,000	250,160	25,260	135,410
November	230,180	280,310	38,150	158,427
December	120,000	250,656	80,450	170,455
Total annual	**2,991,810**	**2,797,051**	**804,020**	**1,251,492**

Source District Tourism and Culture Officer, AP Tourism Authority, Govt. of AP

Ahobilam

Ahobilam, located 112 km southeast of Kurnool and 63 km from Nandyal town on the Eastern Ghats mountain range, is famous for the Lakshmi Narasimha Swamy temple built by the Vijayanagara rulers in the sixteenth century. It is believed that this is the place Lord *Vishnu* appeared in his reincarnation *avatar* as the man-lion *Narasimha Swamy* to save his young devotee *Prahalad.*

The Vedadri hills are the abode of the Nava Narasimhas, the nine different incarnations of Lord Narasimha, and the *Kshetra* attains unparalleled sacred significance due to the unique galaxy of deities present here. The Nava Narasimhas are *Jwala, Ahobila, Malola, Kroda, Karanja, Bhargava, Yognanda, Chatravata* and *Pavana Narasimha.* Ahobilam is all the more interesting as a religious site due to the location of the 'Nava'—nine shrines, spread across the hills with just two temples on the plains and the rest entailing a trek through thick forests, with challengingly steep climb at places. It is

considered auspicious to visit *nava* (nine) *Narasimha* temples in a single day, a total distance of 5 km, motorable up to a distance and a trek all the way up the hills. It is common to find not just the devout but also enthusiastic youngsters racing up the trekking paths. The paths are also picturesque with the greenery getting greener and impromptu waterfalls at many places during monsoon (Fig. 6a–c).

Mahanandi

Mahanandi is a village 21 km from Nandyal, where Mahanandiswara Swamy temple is located. Built more than 1000 years ago, the temple belongs to the Eastern Chalukyan period. Mahanandi is one of the nine Nandis, located on a route between Nandyal and Mahanandi. It is considered to be auspicious to complete a visit to all the *nava* (nine) Nandis one go. The Nava Nandi circuit is not as popular as the Nava Narasimha circuit at present, but has the potential to be promoted. The nine Nandis are Maha Nandi, Siva

Fig. 6 a *Diguva* (lower) *Ahobilam* temple on plain ground. **b** Trek up on rocky Vedadri hill through forest. *Photo Source* Blog: Rohini @ https://rohini.us/ahobilam-narasimha-swamy-temple/. **c** Trekking path to visit nine Narasimha temples (marking on Google Earth Image)

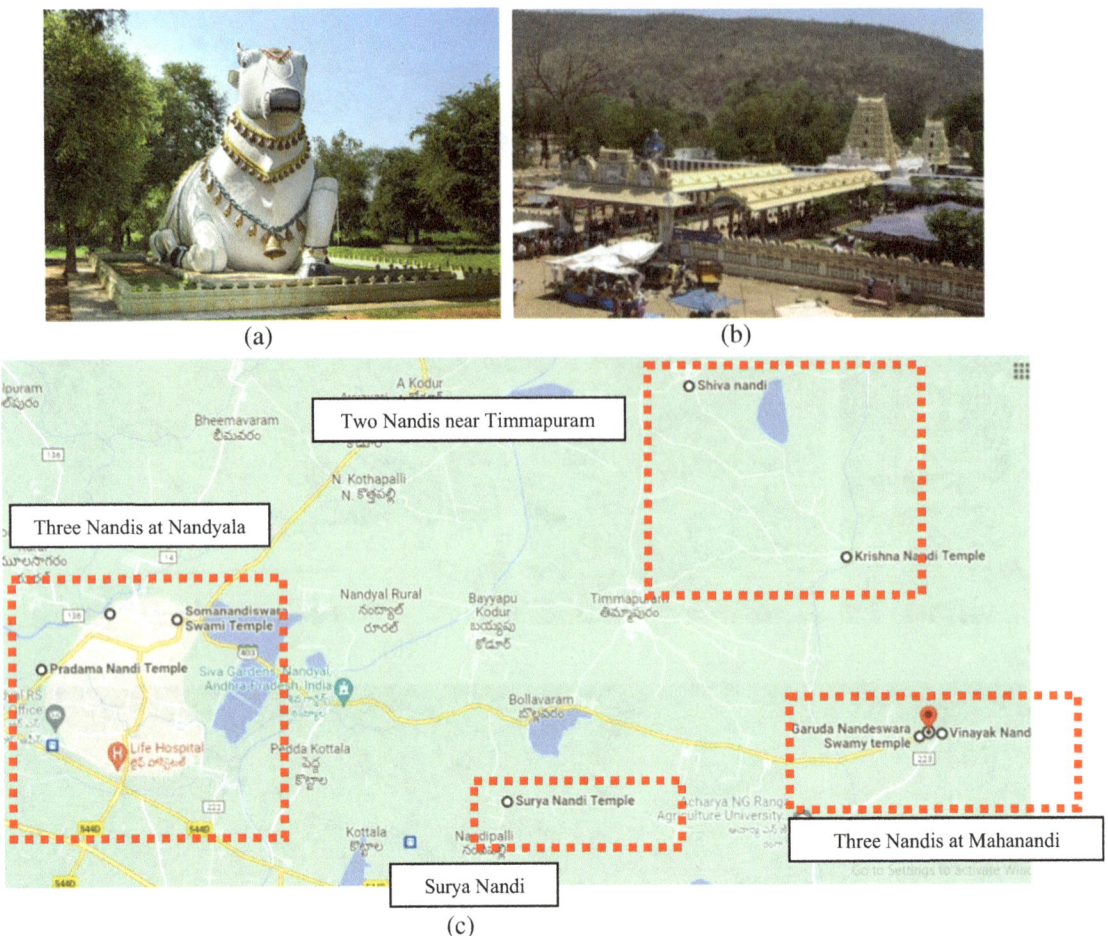

Fig. 7 **a** *Mahanandi*, giant size statue of *nandi*, the bull. *Photo Source* walkthroughindia.com. **b** View of Mahanandi temple. *Photo Source* hansindia.com. **c** Location of nine Nandis between Nandyala and Mahanandi marked on Google Maps

Nandi, Vinayaka Nandi, Soma Nandi, Prathama Nandi, Garuda Nandi, Surya Nandi, Krishna Nandi (also called Vishnu Nandi) and Naga Nandi.

Maha Nandi itself is a large complex, extensively developed with many amenities. Other temples such as Shiva Nandi and Vishnu Nandi are also marked by elaborate sculpture and architecture and have specific religious ceremonies on designated days, attracting local devotees in large numbers but lack the infrastructure for stay. While the temple collection itself has a lot of potential, absence of proper signage and information regarding history, legend, location and services, leaves them as isolated shrines getting only local devotees rather than as a worthwhile circuit for the interested (Fig. 7a–c).

Mantralayam

Mantralayam is a pilgrim village 90 km west of Kurnool. Located on the banks of Tungabhadra River, the village is famous for the *Brindavan* (last resting place) of seventeenth-

century Saint Sri Raghavendra Swamy. Raghavendra Swamy Mutt is a religious institution established following the lineage of *Madhva philosophy of Dvaita Vedanta*. The Mutt has a great number of followers, who visit Mantralayam frequently and on auspicious days (Fig. 8a, b).

Yaganti

Yaganti is a stunningly picturesque Shiva shrine with the temple complex built into towering sandstone formations. The temples with detailed carvings, tall *Gopurams* and long pillared verandahs, steep steps and holy water bodies with Hindu gargoyles are of immense beauty. The temple is famous for its natural caves and an attractive Nandi. Devotees take a holy dip in the *Pushkarini* (temple pond) in the fresh water flowing out of the hills.

The temples are believed to have existed from fifth to sixth century AD, and developed by Pallavas, Cholas, Chalukyas and significantly by the Vijayanagara rulers in the sixteenth–seventeenth century. With the principal deity

Fig. 8 **a** Raghavendra Swamy Mutt at Mantralayam. **b** Devotees at Mantralayam temple on special occasion *Rayara Mahostava*. *Photo Source* Facebook page 'Mantralayam'

Fig. 9 **a–c** Photos of Yaganti *Uma Maheswara Swamy* temple

Shiva in the form of Uma Maheswara Swamy, the complex also has a galaxy of other deities, including Lord Venkateswara Sri Subrahmanyeswara Swamy and Uma, the consort of Shiva (Fig. 9a–c).

A mere 90 km away from the town of Kurnool, Yaganti is not on any popular Shiva circuit and sees large crowds only on special occasions. Yaganti also lacks infrastructure beyond the rudimentary though it makes for a very interesting study in terms of both religious importance and tourist potential, with Banaganapalli, Kurnool, Orvakal and the scenic Owk Lake being at a very convenient driving distance.

Yaganti is one of the two case-study sites selected for conducting survey for this paper, explained in sections below.

4.2.2 Anantapur District

Lepakshi

Lepakshi village is located 100 km south of Anantapur, 120 km north of Bengaluru, on the west of National Highway and is driving distance from the town of Puttaparthi. Reachable by train and bus from Hindupur located 15 km away, Lepakshi is a popular religious and tourist destination. Lepakshi temple, a National Protected Monument, is in the process of being nominated to the Tentative List of UNESCO World Heritage Sites.

The iconic Nandi has been a prominent state symbol for Andhra Pradesh since the inception, especially for its handicrafts industry and development. Lepakshi the town of a tiny town with its only highlight the temple, representing Vijayanagara architecture and built around the sixteenth century. The main deity in the temple is Virabhadra and the monolithic Nandi stands guard over the temple, built with elaborate sculpture, murals and ceiling paintings. The temple is famous for its Hanging pillar, standing with a small gap from the floor, tested by spreading a paper or a thin cloth through and through from under the pillar (Fig. 10a–f).

Lepakshi is the second of the two case-study sites selected for conducting survey for this paper, explained in the below sections.

Gooty

The fort at Gooty is located 50 km north of Anantapur. Standing on a granite hill, it was established by the Vijayanagara rulers in the sixteenth century. Sir Thomas Munro (1761–1827), Governor of Madras Presidency during British rule, is buried here (Maddaiah, 2014). Gooty fort is a National Protected Monument (Fig. 11a).

Puttaparthi

Puttaparthi, a town located on the banks of Chitravathi River, 80 km from Anantapur, east of National Highway, is known for being the birthplace of spiritual guru Sri Sathya Sai Baba (1916–2011) and his *Ashram*/centre, Prashanti Nilayam. Many of his disciples reside in the town serving the *ashram* and its spiritual and social welfare activities. The town is also known for institutes of higher education and a super speciality charitable hospital, developed by the Puttaparthi religious trust. Many hotels, guest houses are available for visitors to stay at Puttaparthi (Fig. 11b).

Kadiri

Kadiri town is located 92 km southeast of Anantapur and 44 km from Puttaparthi. Khadri Lakshmi Narasimha Swamy temple, built by Vijayanagara rulers in the fifteenth–sixteenth century, is located here. The temple is a State Protected Monument of Andhra Pradesh (Fig. 11c).

Tadipatri

Tadipatri town is located 55 km northeast from Anantapur, east of National Highway, on the southern bank of Pennar River. Two temples, built by Vijayanagara rulers in the sixteenth century and considered to be among the "most ornate of this period", are both National Protected Monuments. These temples, though well-known, living temples with daily *pujas*/worship rituals on, and in spite of the high potential, do not find a place among the most religiously important pilgrim places nor on top tourist circuits (Fig. 12a–f).

The two temples, Chintala Venkataramanaswamy temple, dedicated to Lord Venkateswara Swamy, and Bugga Ramalingeswara Swamy temple, a Shiva temple are built of "local gray-green granite, intricately worked through-out" (Michell, 2015). The beautifully sculpted walls and pillars depicting mythological stories are a feast to eyes.

4.3 Existing Hospitality and Transport Facilities in Kurnool and Anantapur Districts

4.3.1 Transport Facilities

International airports at Hyderabad, Bengaluru and Tirupati, domestic airports/helipads at Puttaparthi, Anantapur, Kurnool, Kadapa are well connected by rail and roads internally from National, State highways.

4.3.2 Stay and Eat Facilities

Private hotels (3 star and below) and lodges are available in the cities of Kurnool, Anantapur, Srisailam and Nandyala. Major temples have attached guest houses, rooms and dormitory accommodation called Satrams with and without air conditioning facilities. Cottages, hotels and restaurants are also provided by the Tourism Department of Govt of Andhra Pradesh at popular tourist spots. Eat facilities, restaurants are available at most of the places, towns and villages and on highways, with more vegetarian places near the shrines (Fig. 13a, b).

4.4 Historical Monuments, Natural and Archaeological Sites in Kurnool, Anantapur Districts

4.4.1 Protected Monuments

In India, archaeological sites and monuments are 'Protected' by law by National and State governments. Of the temples and sites discussed above, Yaganti temples of Kurnool district, Lepakshi temple in Anantapur district are National Protected Monuments. State Protected Monuments include Diguva Ahobilam temple, Siva Nandi temple of Nava Nandis and Belum caves in Kurnool; and Kadiri temple in Anantapur (Fig. 14a; Table 2).

(a)

(b)　　　　(c)　　　　(d)

(e)　　　　(f)

Fig. 10 **a–d** Photos of *Veerabhadra Swamy* temple and *Nandi* the big bull at Lepakshi, Anantapur district. **e** Ceiling Paintings with natural dyes and **f** Famous Hanging pillar with a gap on the floor

(a)   (b)

(c)

Fig. 11 **a** Gooty fort on granite hill. *Photo Source* AP Tourism Website. **b** Puttaparthi Prashanti Nilayam. *Photo Source* prashanthinilayam.in.
c Four *Gopurams* (towers) of Khadri temple. *Photo Source* journeybeckons.blogspot.com

4.5 Ancient Rock Formations

The region around Kurnool city has ancient prehistoric igneous rock formations, mainly Silica and Quartz rocks. Tourist facilities and gardens are developed at Orvakal rock formations and Belum caves. Temples of Yaganti and Ahobilam are also built on rocky hills (Fig. 15a, b).

Fig. 12 **a**, **b** *Chintala Venkataramana* and *Bugga Ramalingeswara Swamy* temples, Tadipatri. **c–e** Intricate carvings, sculpture on temples' walls, gates. **f** *Gopuram* of the temple

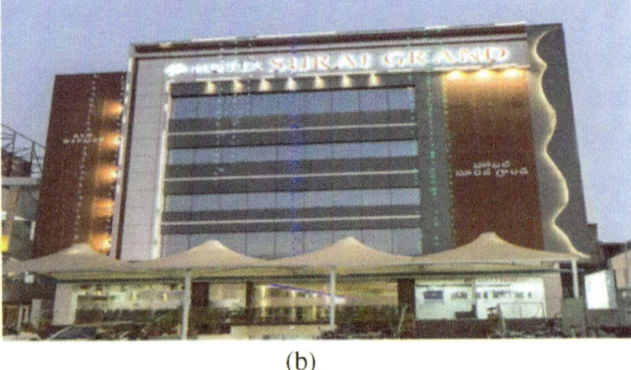

(a)   (b)

Fig. 13 **a** Accommodation at Mantralayam, Kurnool district. **b** Suraj Grand hotel, a 3-star hotel in Kurnool. *Photo Source* Justdial.com

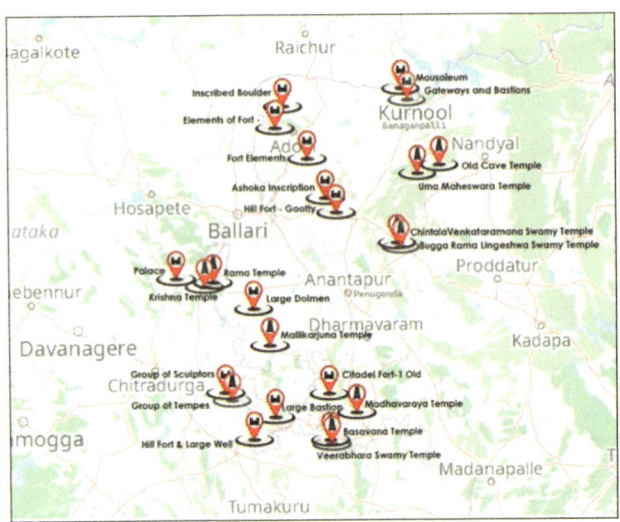

Fig. 14 Combined map of Kurnool and Anantapur districts showing location of National and State Protected Monuments Source: https://bhuvan-app1.nrsc.gov.in/culture_monuments/

Table 2 Numbers of National and state protected monuments in Kurnool and Anantapur districts and distribution of temples and historical, archaeological sites

Categories of sites	Kurnool	Anantapur
Total state protected sites	30	46
Temples	23	34
Historical/archaeological sites	7	12
Total national protected sites	10	23
Temples	2	11
Historical/archaeological sites	8	12

4.6 Natural Sites

Eastern Ghats: Hills, waterfalls, rivers, Nallamala forests, Srisailam dam and wildlife sanctuary are just a few natural sites to name in Kurnool, Anantapur districts.

4.7 Unprotected Heritage Sites

India with its long and rich history has historical sites spread across the country, which have no legal protection status or conservation done. These sites are of different building types, from different eras, owned by the government and many others under private ownership.

"A two-hour drive from Bundi…a few meters inside the town Nainwa, driving past the government offices and colonies, your eyes suddenly notice a fort. There is whole world residing within the fort", authors described 'the local cultural landscape' of Nainwa, a small town in the State of Rajasthan (Chakraborty & Chaudhery, 2013). This applies to most parts of the country, as well.

In the publication 'FORTS & PALACES IN INDIA: Encyclopaedia of 4000+ Forts and Palaces across India' author Pramod Mande listed 19 forts and palaces in Kurnool district and 64 in Anantapur district (Mande, 2022), most of which are not protected nor conserved.

According to the State of Built Heritage of India (SoBHI) study report, "INTACH has listed 66,803 built heritage sites spread over 432 towns. Considering there is not less than 8000 urban settlements in the country, of various historical and cultural dimensions, about 7568 towns and substantial rural hinterland still remain to be scanned for identification and listing" (INTACH, 2020).

4.8 Lesser-Known/Less Popular/Unprotected Historic Temples

Besides civic buildings and structures (non-religious) and sites of archaeological/natural/historical importance, there also exist thousands of temples, large and small, which are lesser known or less popular, unprotected and neglected too. Temples are either living, where *pujas* (prayer rituals) are conducted or discontinued and abandoned. For example, out of the nine ancient Nava Nandi temples, only two temples

Fig. 15 **a** Rocky hills of Ahobilam. **b** Belum Caves

are State Protected Monuments, viz., Maha Nandi temple (seventh century AD) and Siva Nandi temple (tenth century AD). Other seven temples, though living and visited, remain unprotected/unlisted and not preserved as per conservation norms. Same is the case of Nava Narasimha temples of Ahobilam, where only two (Eguva and Diguva Ahobilam) of the nine temples are under State Protection and conserved and other seven are left out.

According to the State of Built Heritage of India (SoBHI) study report, "the most prevalent built heritage typology in India is that of religious structures, the national average being 33%. In comparison to this Bihar, Odisha have the highest heritage typology of religious structures at 78 and 77%. In Andhra Pradesh, religious structures typology is at 35%" (INTACH, 2020). Old temples, abandoned or in-use, have to be listed for protection and preserved holistically for their historical, architectural and cultural value. These smaller temples when promoted either in isolation or included in the popular circuits, footfalls can be increased and contribute to tourism economics and possibly to conservation too.

4.9 Present Condition of Monuments, Historical Sites—Penukonda Fort and Banaganapalli Palace

The condition of many monuments and historical places in most parts of the country remains neglected, needing attention and patronage. According to the SoBHI study report on 'state of conservation' of built heritage in India, "at a national average, 22% of built heritage show signs of deterioration. About 5% of our built heritage is in imminent danger of disappearance" (INTACH, 2020).

Two historically and architecturally significant sites of Penukonda fort, Anantapur district and Banaganapalli Palace, Kurnool district are presented here for understanding the potential and state of affairs.

4.9.1 Penukonda Fort, Anantapur District

Penukonda town is located on the Hyderabad-Bengaluru highway, 79 km south of Anantapur. The village itself is a small habitation, dotted with some historical buildings, including an archaeological museum. Penukonda Fort, accessible through a winding road up the hillock, is a surprise package that suddenly emerges out of rocks. The entrance stone walls do not seem to get due attention but the fort on the banks of a glassy lake is a historic sprawl that facilitates stunning views of the area (Fig. 16a–c).

Penukonda was a "strategic citadel for Vijayanagara emperors from the fourteenth century onwards. The capital was sacked and burnt in 1565" (Michell, 2015). Penukonda fortification along with its walls, gate and watch towers, has heritage structures of Gagan Mahal, Hindu and Jain temples.

Part of Penugonda village lives within the outer-fortification walls amidst a few other historic structures. In spite of its Protected Monument status, under the care of the Archaeological Survey of India, the living fort is not preserved and maintained to satisfactory level, as seen in photos (Fig. 17a–d).

4.9.2 Banaganapalli Palace, Kurnool District

Banaganapalli was a Princely State (sub-regional division ruled by Nawabs) under the British rule before Indian Independence in 1947. Banaganapalli village is located 73 km southeast of Kurnool, 38 km southwest of Nandyal and the palace stands majestically on the outskirts 7 km from the village, on the road leading to Yaganti temple, just 4 km away on the north-western side (Fig. 18a–k).

The palace is an isolated structure standing on a small hillock but visible from a distance. Built on a symmetrical plan, the main rooms are on upper level to be climbed up by exterior staircases from three sides of the building. It's an exposed stone structure built on masonry arches and iron beams having ornamentation limited to little towers over the parapet and sunshades over windows. The palace has beautiful interiors, a central long hall with high ceiling,

Fig. 16 **a** Penukonda fort and settlement—Google Earth image. *Photo Source* Google Maps. **b** Fort interiors. *Photo Source* trawell.in. **c** Fort interiors. *Photo Source* inspirock.com

Fig. 17 **a** Historic Gagan-mahal. **b** Penugonda village settlement—modern houses. **c** Condition of maintenance. **d** Outer-fortification—present condition

clerestory over surrounding rooms and verandah on its front and sides. In the inside, walls are plastered and floor tiles imported from Europe are laid. Locally available grey-coloured slate stones slabs are used for verandah flooring, for which a near-by town Bethamcherla, is famous for. Banaganapalli Palace belongs to the same period and architectural style of other nineteenth-century palaces built by Rajas and Zamindars (nobles, feudal lordsm sub-regional heads under the British/Princely states), but yet stands out by its unique architecture built of stone masonry and simplicity.

Banaganapalli Palace is a private property, owned by the Nawab's family, unprotected by the state. Part of the palace collapsed was badly vandalized and mutilated by trespassers. It is only in the recent years that the Nawabs, appointed security, kept the doors locked and visitors are allowed in during the day for a meagre ticket of Rs. 10/- per head (about one-tenth of US dollar). The palace is still vacant. The palace became famous after a Telugu super-hit movie 'Arundhati' was shot. The Nawabs feel there is no financial return by investment in conservation and repair and the palace continues to stand in ruins.

5   Hub and Spoke Model for Tourism Benefits—Surveys at Two Case-Study Sites

In 2012, Yes Bank-FICCI prepared a report with a 10-point road map for development of religious tourism sector in India. One of the recommendations included "development of circuits using 'Hub and Spoke model'—to create nodes near religious centers, where there is already a basic

Fig. 18 Banaganapalli Palace: **a** First floor plan. **b** Inverted ceiling plan showing ceiling. **c** Plan showing partially collapsed roof condition as indicated in the legend. **d** Exterior view from the road. **e** Front side (east) bifurcated staircase leading to verandah rooms on first floor. **f** Rear side (west) view with small rooms in the basement and shaded first floor windows. **g** Central hall with high ceiling. **h** Central hall— eastern side—roof built on masonry arch and iron beams. Banaganapalli palace—present condition: **i** Vandalism—scribbling over the wall damaging orginal lime plaster. **j** Central hall—partially collapsed wall. **k** Arched verandah and locally available grey slate slab stones, for which near-by town Bethamcherla is famous for

infrastructure present and plan day trips from there. As the influx of crowds is already sizeable in famous spots, a hub and spoke model will ensure tourists spread to all the nearby attractions" (SIGA, 2012).

The Hub and Spoke concept, originally used in transportation planning, especially in airlines connectivity distribution, involves infrastructure development of a large-scale centralized hub connecting main nodes and distributing geographically to other points with lesser traffic points, like spokes, similar to a cycle wheel. This concept is adopted by many other fields and it found application in tourism sector too.

A transport/tourism hub—a node when made multifunctional by adding possible market attractions, hold the passengers/tourists—tend to extend their duration of stay and engage with local tourism (Lohmanna & Pearce, 2010). 'Due to the locations of the hubs and the gateways (destinations) in the transport network, nodes obtain an intense flow of travelers who pass through them to reach other nodes, giving the added advantage of capturing the passing traffic, with some travelers stopping and becoming tourists' (Akkaya, 2017).

In this paper, it is attempted to prove that it is geographically and physically possible to apply the 'Hub and Spoke model' to bring in tourist footfalls from near-by popular pilgrim centres to Penukonda and Banaganapalli and extend tourism benefits and potentially and possibly leading to better heritage conservation.

5.1 Application of Hub and Spoke Model—Surveys at Two Case-Study Sites

For the application of the 'Hub and Spoke' model, the key factor being the geographical and physical proximity, from among the popular pilgrim destinations in Anantapur and Kurnool, the sites located closer to the two case-study sites of Penukonda and Banaganapalli are marked in the maps below.

Map (Fig. 19a): Lepakshi, Puttaparthi, Kadiri around Penukonda in Anantapur district.

Map (Fig. 19b): Nandyala, Mahanandi, Yaganti, Belum Caves around Banaganapalli in Kurnool district.

All of these places fall within a radius of 50 km, and district capitals and major cities of Kurnool and Anantapur in less than 75 km.

Lepakshi near Penukonda and Yaganti near Banaganapalli are selected for carrying out a survey for application of the Hub and Spoke model, wherein Lepakshi and Yaganti, two pilgrim destinations with considerable footfalls, are taken as 'Hubs', for increasing tourism in Penukonda and

Banaganapalli, both taken as points of 'Spokes'. The survey methodology is explained below.

5.2 Surveys—Details and Methodology

Surveys are conducted at two sets of case-study sites that include one hub and one spoke point in each site, i.e. Lepakshi-Penukonda and Yaganti-Banaganapalli.

It is to be noted that the hubs selected here, Lepakshi and Yaganti, are regionally popular religious destinations with regular footfalls. The size of footfalls at these places is much lesser (data given below) than major teeth-yatra destinations such as Srisailam, discussed above. Amenities have been provided only by the authorities/agencies concerned and economic activities have emerged naturally. Lepakshi and Yaganti are not developed to be central hubs catering to infrastructure needs of other sites in proximity and therefore surveyed accordingly. However, considering the proximity and popularity, there is scope for development of central hubs at these places, circuiting and networking with other potential sites such as Penukonda and Banaganapalli.

5.2.1 Objectives
The objectives of conducting the surveys are as following:

- To study pilgrim/visitor numbers—routine and seasonal.
- To assess basic visitor amenities at/near/to the case-study sites, including transport network and accessibility.
- To understand pilgrim/visitor and trip characteristics such as age, gender, residence, travel modes, etc.
- To know pilgrim/visitors' needs, preferences and expectations for visiting places.

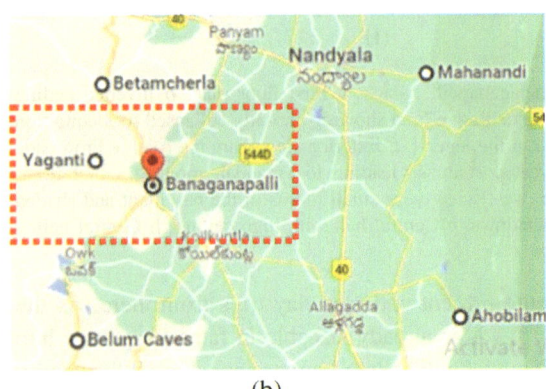

(a)   (b)

Fig. 19 **a** Location of Penukonda fort and nearby pilgrim centres with facilities. *Map Source* Google Maps. **b** Location of Banaganapalli Palace and nearby pilgrim centres with facilities. *Map Source* Google Maps

5.2.2 Methodology

The following methodology is adopted for data collection and surveys at the two case-study sites.

- Data Collection—from official and other sources—from departments of Government of Andhra Pradesh—AP Tourism Authority for Tourism/visitor data, AP State Road Transport Corporation for Public transport (public and private buses), Endowments Department for pilgrim statistics at temple sites, South Central Railways for trains data, Archaeological Survey of India and AP Department of Archaeology and Museums for Monuments/Heritage sites data.
- Amenities data—by interviews, site tours and observations, from websites and internet, Google Maps.
- Pilgrims/Visitors questionnaire survey: A questionnaire was prepared to gather data and seek opinion of visitors at the two hub points of Lepakshi and Yaganti.

5.2.3 Questions Asked

- Set 1: Basic details of name, age, gender, occupation, etc.
- Set 2: Trip Characteristics: Details of the present trip—place of stay, amount spent, mode of transport, number of accompanying persons, distance from lodge, purpose/motivation of the trip, etc.
- Set 3: Satisfaction of basic amenities available on site—parking, toilets, drinking water and eateries, guides and information, cleanliness, crowd management, barrier-free, etc.
- Set 4: Needs, preferences and expectations for visiting places—of distances, amenities, motivation, seasons, etc., in general and particularly to Penukonda and Banaganapalli.

The questionnaire responses are given in the survey findings section below.

5.3 Survey Findings

5.3.1 Data Collection—From Official and Other Sources—Findings

(a) **Location and Accessibility**:

The only approach ways to all four sites are by road, either directly by public road transport or by private vehicles; or by road from nearest railway stations and airports.

For visiting Lepakshi, one has to reach Hindupur railway station and bus terminus and drive further. Penukonda, however, is connected by both rail and road.

For reaching Yaganti, the nearest railway stations are Kurnool and Nandyal and then by road via Banaganapalle. Only two direct buses daily to Yaganti, one each in the morning and evening.

(b) **Number of Visitors**:

Yaganti—Daily pilgrims 200–250; Weekends, Holidays and Mondays—300 to 350 persons.

During festivals, Shivaratri (in February/March), more than 5000 devotees visit Yaganti and during Dusshera (in September/October), 2000 persons and above. About 0.15 million visit Yaganti through the year.

Lepakshi—Daily pilgrims 500 persons; Weekends, Holidays and Mondays—more than 1000 persons per day.

During festivals, big crowds gather there needing traffic and queue control. At Shivaratri (in February/March), more than 5000 devotees visit Lepakshi and during Dusshera (in September/October), number of visitors would be 2000 and above.

Penukonda—There is no entry fee to the fort; less than 50 persons come in a day on an average, 200 during weekends and holidays.

Banaganapalli—A small fee is collected informally for entering the building—no particular timings; less than 50 people come daily and 200–250 on weekends and holidays

(c) **Amenities**:

At Lepakshi, Yaganti temples, basic visitor amenities such as parking areas, toilets, drinking water, information signages, shops are available.

Overnight stay facilities and eating places of small and medium size are available at these places and also at Penukonda and Banaganapalli, and near-by Hindupur and on the highway. Higher class stay-hotels are only available in Puttaparthi, Nandyal, Anantpur, Kurnool and Bengaluru.

5.3.2 Findings from Interviews—Visitor Responses

Following are the findings from the questionnaire responses of Set 1 to Set 4 questions mentioned in Sect. 5.2.3:

(a) **Set 1 Questions—Visitor Characteristics—Sample Details**:
- Interviews were conducted at the two places during first weekend of February 2023. Visitors were selected randomly and included:

- Number of Pilgrims/Visitors interviewed: 31 persons in Lepakshi and 34 persons in Yaganti—A total of 65 persons.
- Age Groups: At Lepakshi and Yaganti—Maximum number i.e., about 45–50% of persons interviewed were in the Age group of 25–40 years, followed by 30–40% in the age group of 18–25 years, 15% of 40 years and above and 5% below 18 years age.
- Gender: Among the interviewed persons, at Lepakshi, 78 persons were male and 22 female. And at Yaganti, 60% male and 40% female.
- Occupations: At both places, a little less than 15% of the interviewees were employed (government and private), 40% self-employed and remaining (above 40%) were students and non-working members of families.
- These places got mostly Domestic visitors—from within Andhra Pradesh and States of Karnataka, Telangana and Tamil Nadu. At Lepakshi, 94% and at Yaganti two-thirds of the visitors interviewed were Telugu speaking (the language of Andhra Pradesh). Lepakshi is offered as a day-trip tour from Bengaluru.
- International travellers having interest in history, heritage and architecture visit for study purposes.

(b) **Set 2 Questions—Details of the Present Trip**:
- Purpose of the visit (see Fig. 20a): As expected at temples, for 60–80% of the interviewees, the purpose of the visit was Religious. Interestingly, it is found that one-third of visitors to Lepakshi and more than 20% of visitors to Yaganti said that they were visiting for recreation and study, giving hope that the tourism can be extended to historical places by hub and spoke model. Among the respondents were also persons belonging to non-Hindu faiths too.
- No. of accompanying persons (see Fig. 20b): While there were also solo and many couple visitors, three or four people generally travel together (35–55%) and group travel is popular too (25–30%).
- Travelling with (see Fig. 20c): People came with their families the most (35–55%), and then with friends (20–

25%), relatives (about 20%) and colleagues at Yaganti (15%).

- Mode of travel (Fig. 21a):

People who travelled by Public bus and private vehicle are almost the same at both places. At Lepakshi, 35% of the respondents said they came by bus and 45% by private vehicle. To Yaganti, 45% travelled by bus and 40% by private vehicle. About 20–25% took auto/cab (3, 4, 5 or 7 seaters) to reach their destinations.

Maximum of private vehicles were two-wheeler bikes, comprising 80–90% and remaining were four wheelers (cars and multi-seater jeeps/mini-buses). There are shops at Lepakshi, renting cars to travel to near-by places such as Penukonda and others.

Figure 21b shows the available parking facilities at Lepakshi and Fig. 21c shows a privately hired mini-bus used by visitors coming to Lepakshi as a group.

- Duration of the present trip (Fig. 22a): Almost 50% of the visitors have come on a day-trip and 25% on overnight trips, 20% on short trips of 3–5 days and less than 10% visitors on a more than 5 days trip.
- Place of stay and distance from the temple (Fig. 22b): Leaving the day-trip visitors, among the persons travelling overnight and more, about 25–40% stayed in hotels and others stayed at their relatives' places. At Lepakshi, 15 respondents said that they were staying with their relatives, which is two-thirds, if day-trip visitors are excluded. 80% of visitors stayed within 10 km distance from the temples and remaining stayed more than 10 km away.
- Amount spent on the present trip (Fig. 22c): Maximum no. of persons, 16 at Lepakshi (55%) and 26 (75%) at Yaganti spent an amount between Rs. 1000/- to Rs. 5000/- ($10 to $50), on this trip. Others (50% at Lepakshi and about 15% at Yaganti) spent less than Rs. 1000/- ($10) and about 5% spent more than Rs. 5000/- ($50) on this trip at both the places.

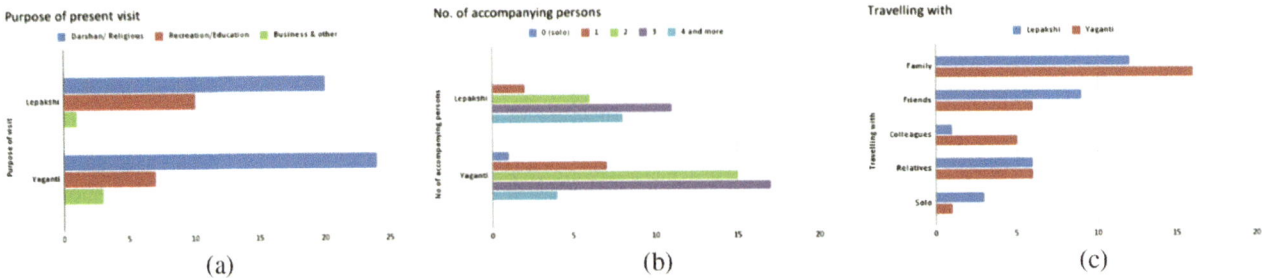

Fig. 20 Graphs showing responses of visitors at Lepakshi and Yaganti—Present trip. **a** Purpose of present trip **b** no. of accompanying persons **c** travelling with

Fig. 21 **a** Graph showing responses of visitors at Lepakshi and Yaganti—present trip—mode of travel. **b** Pic of parking lot at Lepakshi. **c** Mini-bus—privately hired—used for group tours

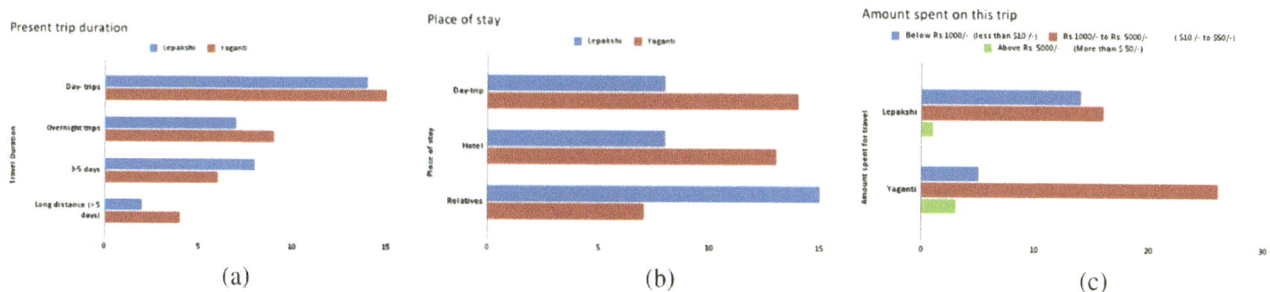

Fig. 22 Graphs showing responses of visitors at Lepakshi and Yaganti—present trip, **a** duration, **b** place of stay, **c** amount spent

- No. of visits to the same temple: 65–70% of visitors at Lepakshi and Yaganti have apparently visited these places earlier. Almost all of them visit once a year, 30–70% visit 2–3 times. About 10% of the visitors visit even more, 4 trips and more in a year to the same place.

(c) **Set 3 Questions—Assessment of Available Amenities**:
- Visitors expressed their satisfaction levels of basic amenities available on site at Lepakshi and Yaganti including, parking, toilets, drinking water, guides and information, cleanliness, crowd management, etc. Their responses ranged between very good, good, okay, bad and very bad, as seen the graphs below.
- While most of them have said good and okay for a few facilities, many persons expressed their dissatisfaction on parking facilities (35% at Lepakshi), toilets (45–60%), stay facilities (85% at Yaganti) and guides/information (35 and 70%)—see Fig. 23a, b.

(d) **Set 4 Questions—Travel Preferences and Expectations**:

On the questions asked about their preferences and expectations for visiting places, in general and particularly to Penukonda and Banaganapalli, visitors' responses are as follows:

- Preferred categories of sites (Fig. 24a): When choices of religious, historical and natural sites were given, 50% of respondents at Yaganti preferred natural sites and 50% at Lepakshi picked up Historical sites. 30–35% prefer religious sites and balance 20% is distributed between historical and natural sites, in swapped positions.
- Preferred months/seasons to travel (Fig. 24b): About 40% of the respondents prefer to travel between April to June, which is summer holidays time for schools and colleges in India; 15 to 35% between January to March and also July to September and less than 10% during October–December.
- Expectations at destination (Fig. 24c): Almost 90% of visitors at Lepakshi and 25% at Yaganti expressed their interest in visiting places with rich history and architecture. A 50% at Yaganti want to have basic physical amenities and 30% good guides and information at sites.
- Preferred travel duration (Fig. 24d): At Lepakshi, 55% of visitors and at Yaganti 35% of the respondents prefer travelling on short trips of 3–5 days; 25% prefer day-trips; and 20% like overnight trips. More than 30% at Yaganti want to go on longer duration trips (more than 5 days) and at Lepakshi, less than 5% preferred longer trips.
- Like to travel with (Fig. 24e): Maximum respondents preferred travelling with families and friends—50 and 40% at Lepakshi and 30 and 40% at Yaganti. About 15%

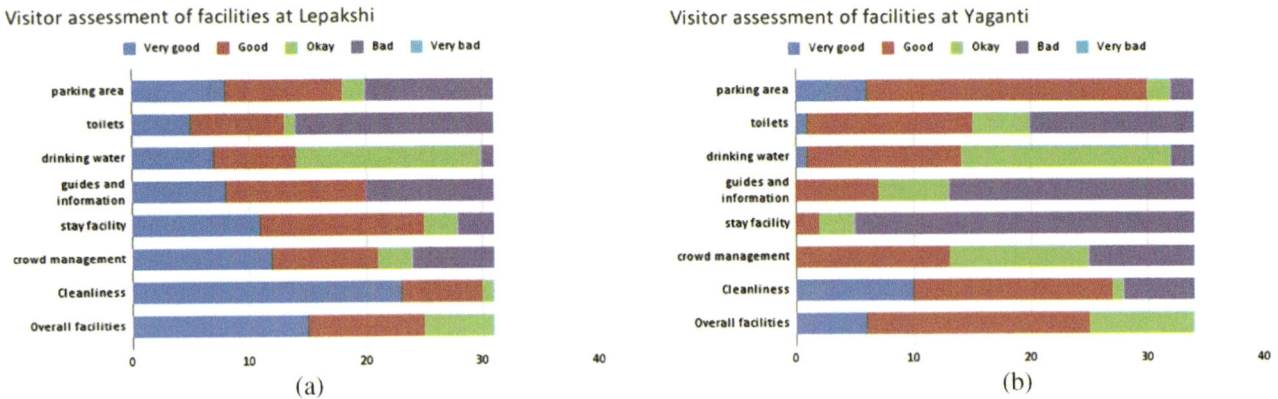

Fig. 23 **a** Graph showing responses of visitors at Lepakshi—Assessment of amenities. **b** Graph showing responses of visitors at Yaganti—Assessment of amenities

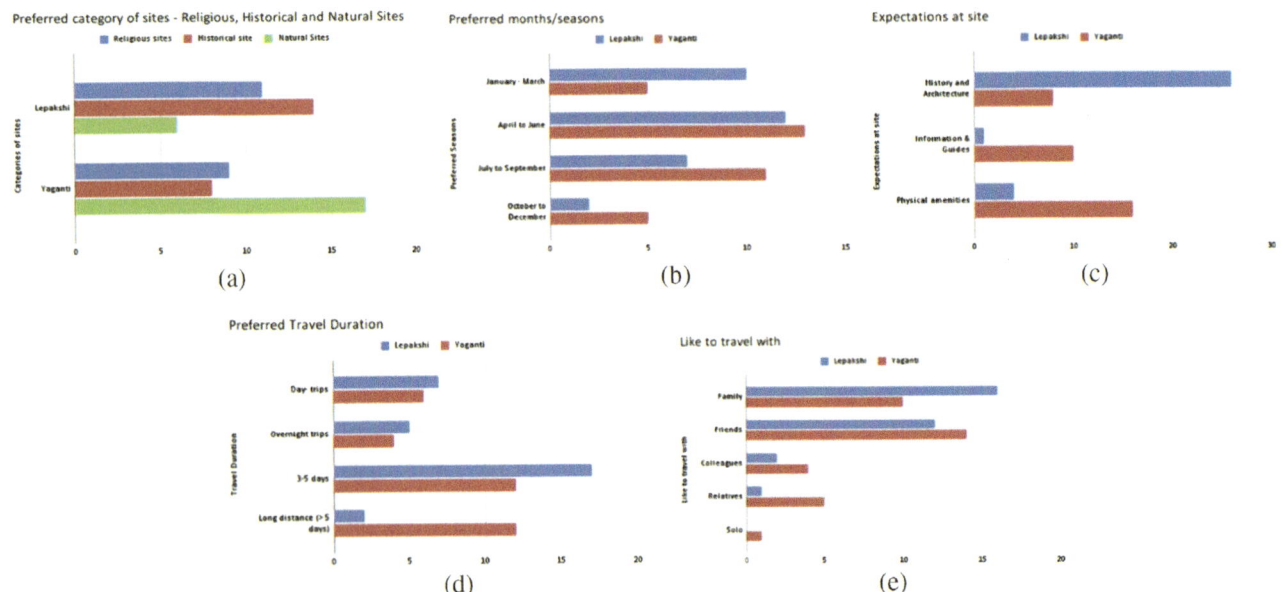

Fig. 24 Graphs showing responses of visitors at Lepakshi and Yaganti. **a** Preferred categories of sites. **b** Preferred months/seasons. **c** Expectations at destination. **d** Graph showing responses of visitors at Lepakshi—preferred travel duration. **e** Graph showing responses of visitors at Yaganti—like to travel with

of visitors at Yaganti prefer to go with their relatives and colleagues, which only 5% prefer at Lepakshi and solo trip is said to be preferred by only one person at Yaganti and none at Lepakshi.

And finally, visitors were asked at Lepakshi and Yaganti about visiting Penukonda and Banaganapalli. Two questions were asked—first, have you heard of Penukonda/Banaganapalli and second, would you like to visit these places. Below are their responses:

• Asked at Lepakshi about visiting Penukonda (see Fig. 25a): 95% of the visitors of heard about Penukonda and 90% said they would like to visit.

• Asked at Yaganti about visiting Banaganapalli (see Fig. 25b): Only 50% of the visitors heard about Banaganapalli but 90% said they would like to visit.

5.4 Findings—Summary

Lepakshi and Yaganti are two religious places having regular flow of devotees, though small when compared to big pilgrim centres across India, but regular and consistent all through the year, a little more during festivals and holidays. The visitors are majorly domestic and just a few international tourists whose interest lies in architecture and history.

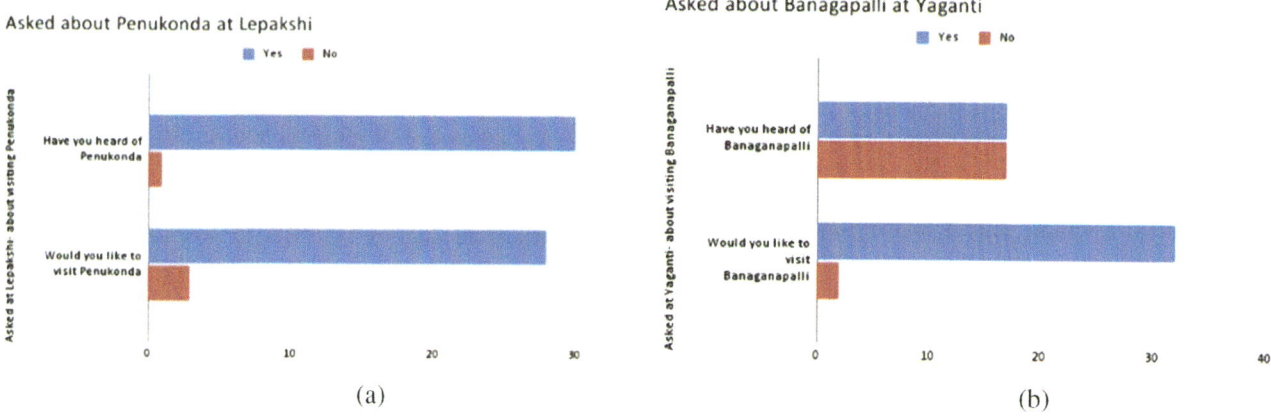

Fig. 25 Graphs showing responses of visitors **a** at Lepakshi—about visiting Penukonda **b** at Yaganti—about visiting Banaganapalli

These places are popular within the state and surrounding states and region—as the survey reiterated, 50% of visitors interviewed came on day-trips and 55–75% of them spent less than $50 on the trip.

People travelled in groups, with family, friends, relatives and colleagues. There are many young people visiting—employed and students, on bikes, on religious as well as recreation purposes, including non-Hindu visitors, from other religions and faiths.

Basic amenities are available at these two places, which the visitors assessed to be satisfied to an extent, but marked bad for condition of toilets, parking lots, stay facilities and guides/information.

When enquired on preferences and expectations, visitors answered positively on visiting historical, natural sites along with religious sites. Shown interest in visiting places rich in history and architecture and preferred places with basic amenities and good guides/information. A 90% of the visitors expressed their willingness to visit Penukonda and Banaganapalli, the two historical sites needing attention.

6 Extending Tourism Benefits to Heritage Conservation—Discussion

Tourism can capture the economic characteristics of the heritage and harness these for conservation by generating funding, educating the community and influencing policy. It is an essential part of many national and regional economies and can be an important factor in development, when managed successfully.

—ICOMOS INTERNATIONAL CULTURAL TOURISM CHARTER Managing Tourism at Places of Heritage Significance (1999)

Tourism is known to make active contribution to heritage conservation for reasons of bringing in funds, creation and maintenance of facilities, development of physical infrastructure and local economies, awareness and education and many more reasons. Not discounting the damages caused by excessive tourism, most of the historical sites in India are in need of dire attention and the benefits tourism is known to pull in are essential and useful.

Adaptive reuse of historic buildings—of re-purposing the old to suit the present needs—is also a popular mode of economically-feasible heritage conservation. For example, many Royal Palaces in India converted into Heritage hotels and restaurants, run successfully, serving the hospitality market. Heritage hotels in exotic locations in Rajasthan are well-known tourist sojourns. Nizam's Falaknuma Palace in Hyderabad is made into a 7-star hotel, Chettinad houses in southern State of Tamil Nadu are converted into homestays. When interviewed, a few of the *Zamindars* (descendents of erstwhile royal families/feudal lords) in Telangana and Andhra Pradesh were willing to explore the idea of opening up their palaces (in part or full) to public to raise revenue by heritage museums, hotels and tourism (Turaga, 2022).

There are also multiple examples in India and across the world of how tourism industry thrives with promotion and development of facilities at historic places, bringing in economic activities and revenue. Heritage and Tourism departments take up projects and programmes at monuments and public places. Historic Kondapalli Fort in Andhra Pradesh is one such recently completed Project.

A report on tourism in Jaisalmer, Rajasthan, India provided the statistics of doubling of Indian tourists and 15% rise in foreign tourists between 2009 and 2017, and gave details of the growing small businesses and local economics in the heritage fort city:

There are approximately 400 establishments operating within the Jaisalmer Fort, including household units, hotels and guest houses (37) and restaurants and food outlets (10-12). The resident population of the Fort is approximately 2500 persons. Approximately, 30% of the total establishments may be involved in tourism related retail, i.e. souvenir shops. The Fort residents are usually employed in business and public enterprises in the city. There are 187 hotels (127 according to the district website) ranging from 4-5 star establishments to Dharamsalas, of which 37 are located within the Fort. The fact that three times' the city's population arrives here each year indicates a significant input into the local economy. (Jaisalmer 2018)

Heritage Conservation and Tourism—Penukonda and Banaganapalli:

As is the case with the two sites in discussion, the condition of these two historical places, Penukonda fort, a National Protected Monument, and Banaganapalli Palace, a private property and unprotected heritage—both, in spite of having high historical and architectural values, show dire neglect and lack of preservation and maintenance. Among the many reasons for the state of preservation are, lack of promotion, visitor facilities, sufficient footfalls that fetch better revenue and would compel owners/government for better maintenance.

Tourism might not solve issues relating to heritage conservation and development, but tourism gives positive results aiding heritage conservation. Jaisalmer is a heritage fort, whose local economy has grown based on tourism activities. Heritage hotels offering elitist hospitality sell at exorbitant prices. The successful examples from across the world are to be followed for development of tourism and heritage conservation at Penukonda and Banaganapalli. Penukonda might not be as exotic as Jaisalmer, but it is as much a living fort with rich historical past and has the potential for involving the local residents in participating in tourism business by offering homestays, selling souvenirs, running cabs and sharing stories as guides. And Banaganapalli has the potential to be converted to a hotel or developed into cultural centre, convention place, museum and much more.

Development of facilities and visitor foot-falls are intertwined. Facilities pull in people and foot-falls compel development of facilities. As found by the surveys conducted at Lepakshi and Yaganti, people visiting religious places are willing to visit historical and natural sites, and expect basic amenities provided.

7 Conclusions

The 'hub and spoke' method of integrated tourism development offers tremendous possibilities for exploiting untapped potential of unexplored natural and historic sites, non-religious and lesser-known religious sites. When added with legal, management and revenue details, it is possible to extend benefits beyond historic preservation to development of local communities, businesses and economics. And the model is simple and replicable across the country.

Taking the case examples of religious and non-religious heritage sites within the specific geographical area of Kurnool and Anantapur districts of south Indian State of Andhra Pradesh, this paper explores the idea of extending the popularity and patronage of religious tourism to the lesser preserved heritage sites. For demonstrating the application of 'hub and spoke' model of tourism development, surveys have been conducted at two sets of case-study sites that include one hub and one spoke point in each site i.e., Lepakshi-Penukonda and Yaganti-Banaganapalli

A few positive points that have come out of the survey are:

- Historical sites are one of the preferred destination categories where about 25 to 50% of visitors picked up. At Lepakshi and Yaganti, places of religious-historical interest, 20–33% visitors stated their purpose of visit to be recreation/education.
- Almost 90% of the interviewees at Lepakshi said they desire to see places with rich history and architecture, and basic amenities (60% at Yaganti) and good guides/information (30% at Yaganti) is expected at these places.
- Of the visitors interviewed, more than 75% are younger than 40 years of age—young adults and students. About 40% of the private vehicles are two-wheeler bikes indicating visitors coming in from nearby places.
- More than 90% of the visitors at the two religious places of Lepakshi and Yaganti are willing to visit historical places of Penukonda and Banaganapalli.

These findings from the survey allow the scope to explore the application of 'hub and spoke' model, of pulling in visitors from popular religious places to neglected historical sites and extending tourism benefits. Though small in numbers, Lepakshi and Yaganti have regular flow of visitors. However, the amenities at all four places and connections to other places need to be systematically planned, developed and improved. Facilities at Lepakshi and Yaganti should be improved for being used as a small-scale regional tourist hub.

India, a country with an ancient history, has rich natural and cultural heritage as well, of which only a few thousands of sites are legally protected and preserved. There are innumerable unprotected heritage sites needing conservation. The hub and spoke model of tourism planning—linking religious and non-religious places, sharing facilities and development of circuits and networks—offers opportunities for extending tourism benefits to lesser visited historical places, possibly aiding heritage conservation.

Acknowledgements Figures: Photographs sourced from websites as acknowledged in captions and Google Earth images/maps.All other photos, images credit to Ms. Usha Revelli and the author.Data

Collection and Visitors' Surveys done in February 2023 by Author assisted by Teja Sai Chandra, Conservation Architect and Architecture students: Mukunda Priya, Vimala Sakshi, Abrar Hussain, Shiva Kumar and Venkat Nag.

Appendix 1

Kurnool District—Basic facts and figures[1]

"Kurnool District lies between the northern latitudes of 140° 54′ and 160°18′ and eastern longitudes of 760°58′ and 790° 34′. The altitude of the district varies from 100 ft above the mean sea level. This district is bounded on the north by Tungabhadra and Krishna rivers as well as Mahabubnagar district of Telangana State, on the south by Kadapa and Anantapur Districts on the west by the Bellary district of Karnataka State and on the east by Prakasam District. The district ranks 10 in population in Andhra Pradesh with 4,053,463 People accounting for 4.63% of the total Population of the state as per 2011 Population Census, while in area it occupies the 3rd place with 17,658 km^2, which account for 6.41% of the total area of the state.

Nallamalas and Erramalas are the two important mountain ranges in the district running in parallel from North to South. In 2016–17, the area covered by forest is 3.406 lakh Hects. which forms 19.29% to the total geographical area. The principal rivers flowing in the district are Tungabhadra (tributary Hundri) the Krishna and the Kunderu".

Appendix 2

Anantapur District—Basic Facts and Figures

"Anantapur district is located on 76°47′ and 78°26′ E of eastern longitudes and 13°41′ and 15°14′ N of northern latitudes. The district is bounded on the north by the Kurnool District, on the southeast by Chittor District, on the east by YSR District of Andhra Pradesh State and on the west and southwest by Karnataka state. Spread over 12,805 km^2, the district has population of 4,083,315 as per the 2011 census which accounts for 4.82% of the total population of the State with 12.16% decadal growth" (https://ananthapuramu.ap. gov.in/).

Anantapur District is part of Rayalaseema region of Andhra Pradesh. "Physiographic personality of the district altogether reflects in the form of basins, uplands and hilly terrain. The relief of the district is marked by an average height of 300–600 m in most part of the North, while it increases up to 1200 m towards the South. Six rivers flow within the district, viz., Penna, Chithravathi, Vedavathi, Papagni, Swarnamukhi and Thadakaleru" (District Census Handbook, 2014).

References

Akkaya, E. (2017). *Tourism theory: Concepts, models and systems.* https://doi.org/10.1079/9781780647159.0000

Anantapur District, AP Government Website. https://ananthapuramu. ap.gov.in/

Chakraborty, M., & Chaudhery, K. (2013). Indian cities in transition: Co-existence of multiple eras in the historical town of Nainwa. In Technical papers proceedings, 61st National Town and country planners congress. *Indian cities in transition* (pp. 32–37), Institute of Town Planners, India, New Delhi.

District Census Handbook, Anantapur. (2014). Census of India 2011—Andhra Pradesh—Series 29—Part XII A.

Gangashetty, R. (2019). *Thirtha Yatra—A guide to holy temples and Theertha Kshetras in India.* Notion Press.

Grandhi, L. (2015). *108 Vaishnava Divyakshetrala (Divyadesala) Yatra Darsini.* Victory Publishers.

Gumma, G. S. (2021). *Tirtha Yatra—pilgrimage: Hidden science of cross-over.* Author, Nandyala, Andhra Pradesh

ICOMOS. (1999). *International Cultural Tourism Charter Managing Tourism at Places of Heritage Significance*, International Council on Monuments and Sites, UNESCO, Paris.

IndBiz. (2019, June 24). *Religious tourism offer strong opportunities.* IndBiz|Economic Diplomacy Division. https://indbiz.gov.in/ religious-tourism-offer-strong-opportunities/

INTACH. (2020). *State of built heritage of India—Case of the unprotected.* Indian National Trust for Art and Cultural Heritage (INTACH), New Delhi.

Jaisalmer. (2018). *Adopt a heritage: Proposal for developing Jaisalmer Fort as am experiential heritage centre*, Jaisalmer—A unit of I Love Foundation, Rajasthan. https://www.ilovejaisalmer.com/ Adoptaheritage-JaisalmerFort.pdf

Kurnool District, AP Government Website. https://kurnool.ap.gov.in/ about-district/

Lohmann, G., & Pearce, D. G. (2010). Conceptualizing and operationalizing nodal tourism functions. *Journal of Transport Geography, 18*(2), 266–275. https://doi.org/10.1016/J.JTRANGEO.2009. 05.003

Maddaiah, K. (2014). *Sir Thomas Munro: Father of Rayalaseema: Governor of Madras Presidency.* Serials Publications.

Mande, P. M. (2022). *Forts & palaces in India: Encyclopedia of 4000+ Forts & palaces across India.* Aniket Enterprises.

Michell, G. (2015). *Southern India: A guide to monuments sites & museums.* Roli Books.

NCAER. (2014). *Motivational factors of visiting India—A study based on the international passenger survey, 2009–2010. A study commissioned by the Ministry of Tourism, Government of India.* National Council of Applied Economic Research. https://www. ncaer.org/image/userfiles/file/TSA/MotivationFactor-final-min.pdf

Shaikh, Z. (2017, September 5). Indians travel 4 times more for religious reasons than on business, indicates data from NSSO.

---

[1] Source: https://kurnool.ap.gov.in/about-district/.

The Indian Express. https://indianexpress.com/article/explained/indians-travel-4-times-more-for-religious-reasons-than-on-business-indicates-data-from-nsso-4828897

SIGA (Strategic Initiatives & Government Advisory) Team. (2012). *Diverse beliefs: Tourism of faith religious tourism gains ground*, Yes Bank—FICCI, New Delhi.

Spiritual tourism grows strongly in India. (2018, May 21). WARC. https://www.warc.com/newsandopinion/news/spiritual-tourism-grows-strongly-in-india/en-gb/40500

Turaga, V. S. (2022). A study of erstwhile Samsthan/Zamindari palaces of Telangana and Andhra Pradesh. *Nagarlok Quaterly Journal of Urban Affairs*, Indian Institute of Public Administration, New Delhi.

Correction to: Tourism as a Driver of Soft Power: The Case of South Korea

Jessica L. Quijano Herrera and Gema Pérez-Tapia

Correction to:
Chapter 22 in: J. Chica-Olmo et al. (eds.), *Sustainable Tourism, Culture and Heritage Promotion*,
Advances in Science, Technology & Innovation, https://doi.org/10.1007/978-3-031-49536-6_22

In the original version of the book, the following belated corrections have been incorporated: In chapter "Tourism As a Driver of Soft Power: The Case of South Korea", the affiliation "University of Málaga, Evangelical, Málaga, Spain; University of El Salvador, San Salvador, El Salvador" of the author "Jessica L. Quijano Herrera" has been changed to "University of Málaga, Málaga, Spain; Evangelical University of El Salvador, San Salvador, El Salvador". The correction to this book have been updated with the changes.

The updated version of this chapter can be found at https://doi.org/10.1007/978-3-031-49536-6_22

GPSR Compliance

The European Union's (EU) General Product Safety Regulation (GPSR) is a set of rules that requires consumer products to be safe and our obligations to ensure this.

If you have any concerns about our products, you can contact us on ProductSafety@springernature.com

In case Publisher is established outside the EU, the EU authorized representative is:

Springer Nature Customer Service Center GmbH
Europaplatz 3
69115 Heidelberg, Germany

The manufacturer's authorised representative in the EU is Springer
Nature Customer Service Centre GmbH, Europaplatz 3, 69115 Heidelberg,
Germany. If you have any concerns regarding our products, please
contact ProductSafety@springernature.com

Printed and bound by CPI Group (UK) Ltd, Croydon, CR0 4YY
19/05/2025
01875825-0001